Supersymmetry Demystified

Demystified Series

Supersymmetry Demystified

Patrick Labelle, Ph.D.

New York   Chicago   San Francisco   Lisbon   London
Madrid   Mexico City   Milan   New Delhi   San Juan
Seoul   Singapore   Sydney   Toronto

Copyright © 2010 by The McGraw-Hill Companies, Inc. All rights reserved. Printed in the United States of America. Except as permitted under the United States Copyright Act of 1976, no part of this publication may be reproduced or distributed in any form or by any means, or stored in a database or retrieval system, without the prior written permission of the publisher.

1 2 3 4 5 6 7 8 9 0   DOC/DOC   0 1 5 4 3 2 1 0 9

ISBN 978-0-07-163641-4
MHID 0-07-163641-2

Sponsoring Editor	**Copy Editor**
Judy Bass	James K. Madru
Editing Supervisor	**Proofreader**
Stephen M. Smith	Bhavna Gupta, Glyph International
Production Supervisor	**Indexer**
Pamela A. Pelton	WordCo Indexing Services, Inc.
Acquisitions Coordinator	**Art Director, Cover**
Michael Mulcahy	Jeff Weeks
Project Manager	**Composition**
Preeti Longia Sinha, Glyph International	Glyph International

Printed and bound by RR Donnelley.

McGraw-Hill books are available at special quantity discounts to use as premiums and sales promotions, or for use in corporate training programs. To contact a representative, please e-mail us at bulksales@mcgraw-hill.com.

Information contained in this work has been obtained by The McGraw-Hill Companies, Inc. ("McGraw-Hill") from sources believed to be reliable. However, neither McGraw-Hill nor its authors guarantee the accuracy or completeness of any information published herein, and neither McGraw-Hill nor its authors shall be responsible for any errors, omissions, or damages arising out of use of this information. This work is published with the understanding that McGraw-Hill and its authors are supplying information but are not attempting to render engineering or other professional services. If such services are required, the assistance of an appropriate professional should be sought.

CONTENTS

Contents

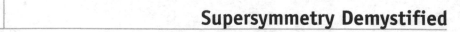

Contents

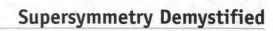

Contents

ACKNOWLEDGMENTS

My heartfelt thanks go to Judy Bass at McGraw-Hill and Preeti Longia Sinha at Glyph International for their guidance, patience, and dedication. I am indebted to Professor Kevin Cahill of the University of New Mexico for many insightful comments on an early draft. A special thanks goes to Professor Pierre Mathieu of Université Laval for introducing me to supersymmetry when I was an undergraduate and for his continuing support in my research career.

I am deeply grateful to my sisters, Micheline and Maureen, for being there for me throughout the years, through thick and thin. I also want to thank my three rescued cats, Sean, Blue, and Fanny, for keeping me company and keeping me entertained during countless hours of writing and proofreading. Adopt a rescued animal if you can!

I would like to dedicate this book to the memory of my mother, Pierrette, and the three siblings I have lost, Monique, Johnny, and Anne. They are no longer with us in body, but they are still very present in my heart.

ABOUT THE AUTHOR

Patrick Labelle has a Ph.D. in theoretical physics from Cornell University and held a postdoctoral position at McGill University. He has been teaching physics at the college level for 12 years at Bishop's University and Champlain Regional College in Sherbrooke, Quebec. Dr. Labelle spent summers doing research both at the Centre Européen de Recherches Nucléaires (CERN) in Geneva, home of the Large Hadron Collider, and at Fermilab, the world's second-largest particle accelerator near Chicago. He participated in the French translation of Brian Greene's *The Elegant Universe* (in the role of technical adviser).

CHAPTER 1

Introduction

1.1 What Is Supersymmetry, and Why Is It Exciting (the Short Version)?

The job of a particle theorist is pretty simple. It consists of going through the following four steps:

1. Build a theory that is mathematically self-consistent.
2. Calculate physical processes using the theory.
3. Call experimentalist friends to see if the results of the theory agree with experiments.
4. If they agree, celebrate and hope that the Nobel Committee will notice you. If they don't, go back to step 1.

That's pretty much all a particle theorist does with his or her life. However, if one randomly tries every theory that comes to one's mind, one soon realizes that most of them lead to results that have nothing to do with the real world. It also becomes clear that requiring certain specific properties and symmetries helps to weed out unphysical theories. One such requirement is that the theory must be Lorentz invariant.[*] Another symmetry that has proved unexpectedly powerful in the construction of the standard model of particle physics is gauge invariance.

However, the standard model is widely considered to be an incomplete theory, even at energies as low as the weak scale (about 100 GeV). One key reason is the so-called hierarchy problem that comes from the fact that the mass of the Higgs particle[†] receives large corrections from loop diagrams (see Section 1.3 for a more detailed discussion). These corrections, in principle, can be canceled by a fine-tuning of some parameters of the standard model, but this highly contrived solution seems very unnatural to most physicists (for this reason, the hierarchy problem is also often called a *naturalness* or *fine-tuning problem*). There is therefore a strong impetus to build new theories going beyond the standard model (while reproducing all of its successes, of course). For this, as always, it is useful to have some symmetry principle to guide us in designing those new theories. Supersymmetry is such a principle.

The basic idea is fairly simple to state. As the color $SU(3)$ group of the standard model reshuffles the color states of a given quark flavor among themselves and the weak $SU(2)$ group mixes fields appearing in weak doublets (e.g., the left-handed electron and neutrino states), supersymmetry is a symmetry that involves changing bosonic and fermionic fields into one another. Supersymmetry solves the hierarchy problem in a most ingenious manner: Fermion and boson loop corrections to particle masses cancel one another exactly! Actually, almost all the ultraviolet (high-momentum) divergences of conventional quantum field theory disappear in supersymmetric field theories (i.e., theories invariant under supersymmetry transformations), a fact that we will show explicitly with a few examples later in the book.

Unfortunately, exact invariance under supersymmetry implies that each existing particle should have a partner of the same mass and same quantum numbers but with a spin differing by one-half, which is obviously not observed in nature (there is no spin-zero or spin-one particle of the same mass as the electron, for example). However, there is no need to throw away the baby with the bath water. The same type of problem occurs with the standard model, which is based on symmetry under the direct group product $SU(3)_c \times SU(2)_L \times U(1)_Y$ (a brief review of the standard model will be provided in Chapter 15). We know that the world is *not* exactly invariant under this symmetry because this would imply that all gauge

---

[*] At least at "low energy" relative to the Planck mass, approximately 10^{19} GeV. At energies comparable to the Planck mass, a theory of quantum gravity is presumably necessary, and then all bets are off.

[†] The Higgs particle is special in two respects. It is the only fundamental particle of spin zero predicted by the standard model, and it is also the only particle in the standard model that has not been observed yet.

bosons would be massless, in contradiction with the observed properties of the $W^\pm$ and Z bosons. The solution in the standard model is provided by the spontaneous breaking of the symmetry via the Higgs mechanism. One might therefore hope to find some similar trick to spontaneously break supersymmetry. It turns out that the situation in supersymmetry is a bit more challenging than in the standard model, as we will discuss in Chapter 14.

In a different vein, something truly remarkable emerges from the simple idea of mixing bosons and fermions: Performing two successive supersymmetric transformations on a field gives back the same field but evaluated at a different spacetime coordinate than it was initially. Therefore, supersymmetry *is intimately linked to spacetime transformations*. This is unlike any of the internal symmetries of the standard model. In fact, if we impose invariance of a theory under *local* supersymmetric transformations, we find ourselves forced to introduce new fields that automatically reproduce Einstein's general relativity!* The resulting theory is called *supergravity*.

Supersymmetry has been showing up in some unexpected places. A notable example is superstring theory. When building a string theory that contains both bosonic and fermionic fields, supersymmetry automatically appears. The fact that the theory exhibits invariance under supersymmetry is the reason it is called *super*string theory.

Supersymmetry also has been used extensively in many modern developments in mathematical and theoretical physics, such as the Seiberg-Witten duality, the AdS-CFT correspondence, the theory of branes, and so on.

It is important to understand that, for now, supersymmetry is a purely mathematical concept. We do not know if it has a role to play in the description of nature, at least not yet. But it is very likely that we will find out in the very near future. Indeed, the recent opening of the Large Hadron Collider at the Centre Européen de Recherche Nucléaire (CERN) near Geneva offers the exciting possibility of finally testing whether supersymmetry plays a role in the fundamental laws of physics.

All this makes supersymmetry terribly exciting to learn. Unfortunately, it is quite difficult to teach oneself supersymmetry despite the vast literature on the subject. This brings us to the purpose of this book.

1.2 What This Book Is and What It Is Not

There are essentially three roadblocks to teaching oneself supersymmetry. The first is the highly efficient but at the same time horrendously confusing notation that permeates supersymmetry books and articles. It's hard not to get discouraged by the

---

* In roughly the same way that making a $U(1)$ global invariance local automatically leads to electromagnetism, for example.

dizzying profusion of indices that sometimes do not even seem be used consistently, as in χ_a, χ^a, $\chi^{\dot{a}}$, $\chi_{\dot{a}}$, $(\sigma^2)_{ab}$, $(\sigma^2)_{\dot{a}\dot{b}}$, $\bar{\sigma}^{\mu}_{a\dot{a}}$, and so on.

There is, of course, a very good reason for introducing all this notation: It helps immensely in constructing invariant quantities and making expressions as compact as possible. Unfortunately, this also makes it difficult for newcomers to the field because more effort must be put into making sense of the notation than into learning supersymmetry itself! It *is* possible to keep the notation very simple by not using those strange dotted indices and by not making any distinction between upper and lower indices, but the price to pay is that expressions are more lengthy, and some calculations become quite awkward. In itself, this is not such a big deal. But the goal of this book is not only to help you learn the rudiments of supersymmetry; it is also to provide you with the background required to "graduate" to more advanced references on the subject. And this necessitates familiarizing you with the notation of dotted and undotted spinors. For this reason, a compromise has been made. At first, the simpler and less compact notation will be used to ensure that the fancy indices do not get in the way of learning the basic concepts. After a few chapters of practice, however, the more advanced notation will be introduced with an emphasis on the rationale behind it.

A second difficulty in learning supersymmetry stems from the language used to describe the fermionic fields of the theory. Supersymmetry requires the use of Weyl or Majorana spinors, whereas most of us became familiar only with Dirac spinors in our introductory quantum field theory classes. The next two chapters therefore will be entirely devoted to introducing these types of spinors and explaining their relation to Dirac spinors.

A third difficulty is that there are actually two methods for constructing supersymmetric theories. As you might expect, one method, the so-called superspace or superfield formalism, is more powerful and compact but also much more abstract and difficult for a beginner than the second, down- and dirty-method. Alas, most references on supersymmetry use right away or very early on the superfield approach. In this book we will start with the less efficient but easier to learn approach and will only later introduce superfields and superspace.

Supersymmetry is a *huge* and formidably complex topic. In writing a (finite) book on the subject, one is faced with a gut-wrenching dilemma. Should one try to touch on as many aspects of the theory as possible, albeit in a necessarily superficial manner? Or should one restrict oneself to a very small (almost infinitesimal) subset of the theory with the goal of making the presentation, and especially the mathematical derivations, as thorough and detailed as possible?

Well, for the present book, the decision was not difficult because the goal of the Demystified series is to provide readers with self-teaching guides. This implies that the accent must be put on covering in depth the more basic concepts, even if it means paying the price of sacrificing several advanced topics. For example, we will not cover so-called extended supersymmetries (the meaning of which will become more clear later in this book), supergraphs, or applications in mathematical physics and superstring theory such as the Seiberg-Witten theory, BPS states, the AdS/CFT correspondence, and so on. Even the basic applications we will discuss cannot be given full justice in such a short book. For example, we will only be able to afford a cursory look at loop calculations in supersymmetry and to sample only some aspects of supersymmetry breaking and of the so-called minimal supersymmetric extension of the standard model (known affectionately as the *MSSM*).

The purpose of the present work is therefore not to make anyone an expert on supersymmetry or even on any particular application of supersymmetry. It is rather meant as a gentle introduction to the basic concepts, as a springboard toward more advanced references, whether they are on string theory, mathematical physics, or phenomenological applications. The goal is to give you a solid understanding of the basic concepts so that you will be able to read and understand more advanced references (some of which will be suggested in Section 1.4).

This brings us to the important question of prerequisites. It will be assumed that you have some basic knowledge of quantum field theory, at the level of doing calculations of simple scattering amplitudes. If you need your memory to be refreshed, the book by McMahon[30] in the Demystified series offers a gentle introduction to quantum field theory, whereas Maggiore's presentation[28] is only slightly more advanced but very pedagogical. In my humble opinion, the books by Srednicki[44] and Hatfield[24] are among the best references available on quantum field theory. The book by Peskin and Schroeder[37] is an invaluable complement. As always, the books by Weinberg[48] are filled with deep and unique insights but are at a more advanced level than the previous references.

Even though the essential elements of the standard model and the key formula for the calculations of amplitudes will be reviewed, they will be presented with the intention of refreshing your memory, not of actually teaching you the material. A very pedagogical introduction to the standard model is offered by Griffiths.[21] At a slightly more advanced level, the two volumes by Aitchison and Hey[2] are extremely well written and pedagogical. The many books by Greiner[19] and collaborators are filled with detailed calculations and clear explanations.

The intended audience for this book is therefore physicists with some basic background in quantum field theory and in particle physics who want to learn the nuts and bolts of supersymmetry but find themselves quickly stuck when reading on the subject because of the unfamiliar notation and the lack of computational details provided. We are pretty convinced that if you work your way through this book, you will have acquired the necessary basic tools to make other books or references on supersymmetry *much* easier to understand, and you will be able to make some progress in the study of whatever aspect of supersymmetry tickles your curiosity the most.

1.3 Effective Field Theories, Naturalness, and the Higgs Mass

Let's get back to the issue of the naturalness problem briefly mentioned in Section 1.1. The fact that the problem disappears completely in a supersymmetric theory is one of the most appealing aspect of supersymmetry (SUSY), so it deserves further elaboration.

In the last 30 years, our understanding of renormalization in the context of particle physics has undergone a radical shift (or a *paradigm shift* to use the oft-abused expression). To explain the difference between the "old" school of thought and the "modern" point of view, it's convenient to consider for regulator a "hard" cutoff Λ on the integration of momenta,

$$\int_0^\infty dk \rightarrow \int_0^\Lambda dk \tag{1.1}$$

Of course, such a regulator breaks almost all the symmetries usually encountered in particle physics and therefore is not practical, but for the purpose of building physical intuition about renormalization, nothing beats a good old momentum cutoff.

The old view of the cutoff was that it had no physical meaning, that it was a purely mathematical sleight of hand introduced to make the intermediate steps of the calculations mathematically well-defined and that, by consequence, it *had to* be taken to infinity at the end of the calculation. A renormalizable theory is, by definition, a theory in which any result for physical processes can be made cutoff-independent in the limit $\Lambda \rightarrow \infty$ by redefining, i.e., renormalizing, the finite number of parameters appearing in the lagrangian (the so-called bare parameters).

By contrast, the modern point of view is that all the quantum field theories we know (possibly with the exception of M theory) are to be treated as *effective field theories*, i.e., theories that are approximations of more fundamental quantum theories, and in that context, the cutoff must be viewed as a *physically meaningful parameter* whose value should *not* be taken to infinity. Instead, the finite value of the cutoff corresponds to the energy scale at which the effects of the "new physics" beyond the effective field theory become important. From that point of view, there is nothing wrong with nonrenormalizable theories. As long as the processes that are computed involve particles with a typical external (nonloop) momentum p much smaller than the scale of the new physics Λ_{NP}, the nonrenormalizable contributions appear in the form of an expansion of the form

$$\sum_n c_n \left(\frac{p}{\Lambda_{NP}} \right)^n$$

where the coefficients c_n are of order one. As long as the momentum p is much smaller than the scale of new physics, one can neglect all but the first few terms to a given precision. Of course, the desired accuracy also dictates the number of loops used in the calculation, so there are two expansions involved in any calculation.

Obviously, as p approaches Λ_{NP}, the whole effective field theory approach breaks down, and one needs to uncover the more fundamental theory (which might itself be an effective theory valid only up to some higher scale of new physics). In some sense, effective field theories are the "quantum equivalents" of Taylor expansions in calculus. The expression

$$1 - x + x^2/2 - x^3/6$$

is a good approximation to $\exp(-x)$ as long as x (the analogue of p/Λ_{NP}) is much smaller than 1. If we use $x = 2$, we are outside of the radius of convergence of the expansion, and we must replace the "effective" description $1 - x + x^2/2 \ldots$ by the more fundamental expression $\exp(-x)$.

However, things are not as obvious in the context of a quantum theory because even if the external momenta are small relative to the scale of new physics, the *loop* momenta have to be integrated all the way to that scale. So it might seem at first sight that no matter how small the external momenta are, since the loops will be sensitive to the new physics, the very concept of an approximate theory does not make sense in the context of quantum physics. However, the diagrams with external momenta much smaller than Λ_{NP} and internal momenta on the order of the scale of new physics are highly off-shell and, because of the Heisenberg uncertainty principle, are "seen" by the external particles as local interactions with coefficients suppressed by powers of the scale of new physics. A detailed and very pedagogical discussion of this point is presented in the paper by Peter Lepage,[26] who has pioneered the use of effective field theories in bound states and in lattice quantum chromodynamics (QCD). An excellent introduction to effective field theories is the paper by Cliff Burgess.[9]

Let us get back to the problem of naturalness. The key point to retain from the preceding discussion is that we must think of the cutoff as representing the scale of new physics, not as a purely mathematical artifact that must be sent to infinity at the end of the calculation.

Consider, then, an effective field theory that is to be regarded as an approximation of a more general theory, for energies much smaller than the scale of new physics. The question is whether it is natural to have dimensionful parameters in the effective theory, masses in particular, that are much smaller than Λ_{NP}. Of course, at tree level, there is never any naturalness problem; we may set the parameters of the effective theory to any value we wish. It's the loops that may cause some trouble (as always!). There is a naturalness problem if the dimensionful parameters of the effective field theory are driven by the loops up to the scale of new physics. In that

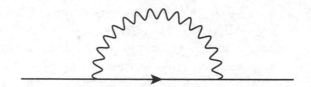

Figure 1.1   One-loop correction to a fermion propagator.

case, the only way to keep these parameters at their small values is to impose an extreme fine-tuning of the bare parameters to cancel the loop contributions.

To be more specific, consider the electron mass in quantum electrodynamics (QED), viewed as effective field theory of the electroweak model. The scale of new physics is therefore of order $\Lambda_{NP} \approx 100$ GeV, which is much larger than the electron mass, $m_e \approx 0.511$ GeV. The question is whether, if we introduce loop corrections, it is natural for the renormalized electron mass to remain so small. The one-loop correction to the electron mass, depicted in Figure 1.1, is roughly

$$\delta m_e \simeq \alpha \int \frac{d^4 k}{k^2} \frac{\slashed{k} + m_e}{k^2 - m_e^2} \simeq \frac{\alpha m_e}{4\pi} \ln\left(\frac{\Lambda_{NP}}{m_e}\right) \tag{1.2}$$

Naive power counting would lead us to expect a linear divergence, $\delta m_e \simeq \alpha \Lambda_{NP}$, but the linear term in k in the numerator actually vanishes on integration, leaving us with a much milder logarithmic divergence. Even if we set $\Lambda_{NP} \simeq 100$ GeV, we find that the correction to the bare mass is of order $\delta m_e / m_e \approx 19\%$ relative to its tree-level value. This is a small correction, and therefore, a small electron mass is natural in the context of QED as an effective field theory.

Let's look at the masses of gauge bosons. This time, the two one-loop diagrams of Figure 1.2 contribute if the gauge group is nonabelian, whereas only the first diagram exists if the gauge group is abelian. Schematically, the two diagrams give a correction to the gauge boson mass of the form

$$\delta m_g^2 = \alpha_g \int d^4 k \frac{k^2 + m}{(k^2 - m^2)^2} \tag{1.3}$$

Figure 1.2   One-loop corrections to a nonabelian gauge field propagator.

where m is the mass of whatever is circulating in the loop, and α_g is the usual $e_g^2/4\pi$, where e_g is the gauge coupling constant. Simple power counting predicts this time a quadratically divergent result, but again, this contribution actually vanishes on integration. If the gauge symmetry is unbroken, the correction to the gauge boson mass actually vanishes identically, $\delta m_g^2 = 0$. If the gauge symmetry is spontaneously broken, the correction is not zero, but the quadratic divergence is still absent, and the leading correction is only logarithmically divergent,

$$\delta m_g^2 \simeq \alpha_g m^2 \ln\left(\frac{\Lambda_{NP}}{m}\right) \tag{1.4}$$

Once more, the correction is small even if the scale of new physics is much larger than the physical mass of the gauge boson.

It is not fortuitous if the leading divergences in both our examples cancel out, leaving a milder logarithmic divergence. In both cases, a massless theory has more symmetry than the massive theory. The theory of a massless fermion has a chiral symmetry that ensures that the fermion will remain massless to all orders of perturbation theory. If we start with a massive fermion, the loop corrections no longer vanish, but they must go to zero in the massless limit. This is why the one-loop correction to the fermion mass could not be of the form $\delta m_e \simeq \Lambda_{NP}$ because this would not vanish as the limit m_e goes to zero. It is therefore chiral symmetry that "protected" us from the linear divergence.

In the gauge boson case, it is obviously the gauge symmetry that is restored when the gauge boson mass goes to zero. Therefore, all corrections to the gauge boson mass must vanish as the limit m_g goes to zero. This is why we did not generate a quadratic divergence, which would not have satisfied this criterion.

This definition of naturalness has been formalized by 't Hooft[45]:

> A theory is natural if, for all its parameters p which are small with respect to the fundamental scale Λ, the limit $p \to 0$ corresponds to an enhancement of the symmetry of the system.

Obviously, his Λ corresponds to our Λ_{NP}.

Let us now turn our attention to the mass of a scalar particle, e.g., the Higgs boson in the standard model. One one-loop correction arises from the fermion loop shown in Figure 1.3, and it corresponds to the integral

$$\delta m_h^2 \simeq \alpha^2 \int d^4k \, \frac{k^2 + m}{(k^2 - m^2)^2} \approx \alpha \Lambda_{NP}^2 \tag{1.5}$$

This time, taking the mass of the scalar to zero does not lead to any enhancement of symmetry, and we obtain a quadratically divergent correction. If we take the mass of the Higgs to be of order 10^2 GeV and the scale of new physics to be of

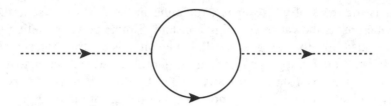

Figure 1.3   One-loop correction to the propagator of a scalar. The dotted lines represent
the scalar particle propagator, and the loop is a fermion loop.

order the usual grand unification scale of about 10^{15} GeV, we see that the one-loop
correction is roughly 24 orders of magnitude larger than the tree-level value of m_s^2
(taking $\alpha \approx 10^{-2}$). This means that the bare squared mass of the Higgs would have
to be fine-tuned to 24 digits to keep the renormalized mass equal to its tree-level
value! In other words, the mass of the Higgs is given by $m_H^2 = C_1 - C_2$, where
both C_1 and C_2 are of order 10^{28} GeV^2 and have the same first 24 digits!

This is a bit as if you were to throw a needle up in the air and watch it fall on
a marble floor exactly on its tip and *stay upright in that position*! You certainly
would be flabbergasted! It would not be *impossible* that it did land in exactly
the right way to have it be in equilibrium and then be undisturbed by any air
current after it landed, but it would seem so unlikely as to be preposterous. You
certainly would look for some explanation for why this happened. Maybe this part
of the floor is not marble, and the needle penetrated the surface? Maybe there is a
magnet under the surface, and the needle is magnetized in such a way as to stay
upright?

We are in a similar situation when pondering the mass of the Higgs in the standard
model. There has to be some way to obtain a reasonable mass of the Higgs without
invoking a mind-boggling accidental cancellation between two huge numbers!

So why not simply say that the mass of the Higgs could be huge, of the order of
10^{15} GeV? Because there are other constraints that this time put an *upper* bound
on the Higgs mass. For example, unitarity of the S-matrix leads to the constraint
$m_h < 780$ GeV (see Ref. 8 for more details).

This leaves us with only one possible way out: The scale of new physics must be
much smaller than 10^{15} GeV. As a criterion for absence of fine-tuning, it is natural
to demand that a one-loop correction must be smaller than the tree-level value.
Applying this to the Higgs mass in the standard model leads to $\Lambda_{NP} \le m_h/\sqrt{\alpha} \approx 1$
TeV. This makes the scale of new physics very low and accessible experimentally.

What could happen at this new scale? It could be revealed that the Higgs boson
is not an elementary particle but instead a bound state, as in technicolor theories.
Or maybe there are extra dimensions that can affect physics at the TeV scale, or
again, maybe particles are not pointlike but rather are excitations of strings.

But let us assume that the Higgs still would be present as an elementary point particle in some new, more fundamental, four-dimensional theory superseding the standard model in the TeV region. In that case, it would seem that the naturalness problem still would be present. The scalar masses now would be driven to the scale of new physics beyond this new theory, and we would be back to our starting point.

Unless this new theory would possess some symmetry that would protect even scalar masses from large corrections. And this is where supersymmetry makes it triumphant entrance! In theories with unbroken supersymmetry, quadratic divergences are altogether absent, so there is no naturalness problem! This makes supersymmetry a very attractive candidate for the new physics lurking around 1 TeV.

1.4 Further Reading

After you have finished this book, you will have the necessary background (and hopefully, the hunger) to consult more advanced references. Here are some suggestions.

We should first mention the ground-breaking papers that gave birth to supersymmetry. Supersymmetry was independently discovered in the west[16,33,39] and in the former Soviet Union.[17,46] This was followed by the seminal work of Wess and Zumino, who constructed explicit four-dimensional supersymmetric quantum field theories.[50] For more details, the first chapter of Ref. 47 is highly recommended.

If you are interested in the phenomenology of the MSSM, the best book to read after the present volume is the one by Aitchison.[1] It is presented at the same level as the present volume and complements it nicely.

The following books are slightly more advanced but should offer no difficulty after you have worked your way through *Supersymmetry Demystified*. Baer and Tata[5] offer a large number of cross sections and decay-rate calculations. The book by Binétruy[8] is quite exhaustive (with almost 400 references!) and discusses cosmologic implications as well as the MSSM in great depth. Weinberg's book[47] on supersymmetry is, as is all his work, a model of clarity and a great source of insights. It covers not only the MSSM but also supergravity, Seiberg-Witten, supergraphs, and several other topics.

Bailin and Love[6] focus on supersymmetry in the context of superstring theory. A book by Dine[12] also covers supersymmetry and its applications to superstring theory, but is more recent and has a wider scope (it covers Seiberg-Witten, cosmology, supergravity, and many other topics).

Although old and, unfortunately, hard to find, the book by Müller-Kirsten and Wiedemann[31] is a real gem because of the large number of derivations presented in great detail. A classic is the book by Wess and Bagger,[49] which is a good resource for learning about supergraphs and supergravity but is more advanced than the previous references.

In a different category, Ref. 15 contains a large number of the most important early research papers on supersymmetry and is an invaluable resource. In the same vein, Ref. 25 contains all the earlier Physics Reports on supersymmetry and remains a very useful reference.

In addition to books, a large number of excellent review papers or lecture notes on supersymmetry are available. For more details concerning the phenomenological aspects of the MSSM, some excellent resources are Refs. 10, 29, 32, 36, and 38. If you are interested in the Seiberg-Witten theory, Refs. 47 and 3 are good places to start. A couple of more formal reviews are Refs. 27 and 51. Reference 43 focuses on supersymmetry breaking. Reference 13 explains in great detail how to do calculations with Weyl spinors and provides a very large number of detailed calculations relevant to the MSSM.

This is far from being an exhaustive list (Ref. 8 alone contains almost 400 references!). In particular, very few of the original research papers are cited in the References, and for that we apologize to the large number of researchers who have discovered and refined supersymmetry over the years. You are invited to consult the bibliographies of the books and review papers mentioned here for a more complete list of references.

CHAPTER 2

A Crash Course on
Weyl Spinors

As mentioned in the Introduction, supersymmetric theories deal with Weyl or Majorana spinors, not the more familiar Dirac spinors. The first step to understand supersymmetry (SUSY) is therefore to become at ease with those two types of spinors. This chapter will focus on Weyl spinors and their connection with Dirac spinors. Along the way, we will work out many identities that will be used repeatedly throughout this book and will define some notation that will prove very handy in the building of supersymmetric theories.

This won't be glamorous stuff—only in Chapter 5 will we construct our first supersymmetric theory—but it is crucial to become proficient with the notation established in the first few chapters before tackling SUSY. Indeed, most physicists working in SUSY (whom we like to refer to as "superphysicists") will tell you that the brunt of the difficulty in learning SUSY is not understanding the theory itself but rather becoming at ease with Weyl and Majorana spinors and the daunting notation used to describe them. It is therefore imperative to first spend some time familiarizing ourselves with this notation; once this is achieved, calculations in SUSY will be a breeze (well, almost).

2.1 Brief Review of the Dirac Equation and of Some Matrix Properties

Let's start by reviewing a few basic relations and defining our convention for the Dirac matrices. The Dirac equation is

$$\gamma^\mu P_\mu \Psi = m\Psi \tag{2.1}$$

or, using Feynman's "slash" notation,

$$\slashed{P}\Psi = m\Psi$$

Here, P_μ represents the differential operator $i\partial_\mu$. Throughout this book we use the so-called natural units, in which $c = \hbar = 1$, unless stated otherwise.

It is also useful to recall the Dirac lagrangian:

$$\mathcal{L}_{\text{Dirac}} = \overline{\Psi}(\gamma^\mu P_\mu - m)\Psi \tag{2.2}$$

where the bar over the spinor denotes the Dirac adjoint, defined as $\overline{\Psi} \equiv \Psi^\dagger \gamma^0$. Equation (2.2) is actually not a lagrangian but a lagrangian *density* (because it must be integrated over spacetime to yield the action), but we will follow the incorrect but widespread convention of dropping the adjective *density* most of the time.

There are several different representations for the Dirac matrices γ^0 and $\vec{\gamma}$ used in the literature. We choose the representation

$$\gamma^0 = \begin{pmatrix} 0 & \mathbf{1} \\ \mathbf{1} & 0 \end{pmatrix} \qquad \vec{\gamma} = \begin{pmatrix} 0 & -\vec{\sigma} \\ \vec{\sigma} & 0 \end{pmatrix}$$

where, of course, $\mathbf{1}$ stands for the 2×2 identity matrix. These can be used to define

$$\gamma^\mu = (\gamma^0, \vec{\gamma})$$

The corresponding quantity with a covariant index is defined as

$$\gamma_\mu = \eta_{\mu\nu}\gamma^\mu = (\gamma^0, -\vec{\gamma})$$

where $\eta_{\mu\nu}$ is the flat spacetime metric for which we use the form

$$\eta_{\mu\nu} = \mathrm{diag}(1, -1, -1, -1)$$

This is sometimes referred to as the *mostly minus convention* as opposed to the *mostly plus convention* $(-1, 1, 1, 1)$. Usually, particle physicists adopt the mostly minus convention, whereas the mostly plus metric is preferred by general relativists. It is always important to check the convention used when consulting a new reference.

For the sake of completeness, let us also give the matrix γ_5, which will be needed shortly. We choose the following representation:

$$\gamma_5 = \begin{pmatrix} \mathbf{1} & 0 \\ 0 & -\mathbf{1} \end{pmatrix} \tag{2.3}$$

This is often called the *chiral* or *Weyl representation.*

Recall that the Pauli matrices are given by

$$\sigma^1 = \begin{pmatrix} 0 & 1 \\ 1 & 0 \end{pmatrix} \qquad \sigma^2 = \begin{pmatrix} 0 & -i \\ i & 0 \end{pmatrix} \qquad \sigma^3 = \begin{pmatrix} 1 & 0 \\ 0 & -1 \end{pmatrix}$$

We will make extensive use of the Pauli matrices, so it would be useful to recall some of their properties (which all can be checked using the explicit matrix representations). First, they are hermitian, i.e., $(\sigma^i)^\dagger = \sigma^i$. However, they do not behave so nicely under the separate operations of complex conjugation and transposition:

$$(\sigma^1)^* = \sigma^1 \qquad (\sigma^2)^* = -\sigma^2 \qquad (\sigma^3)^* = \sigma^3$$
$$(\sigma^1)^T = \sigma^1 \qquad (\sigma^2)^T = -\sigma^2 \qquad (\sigma^3)^T = \sigma^3 \tag{2.4}$$

Obviously,

$$\vec{\sigma}^* = \vec{\sigma}^T \tag{2.5}$$

The fact that σ^2 is singled out will play a key role later on.

The product of any two Pauli matrices is given by

$$\sigma^i \sigma^j = \mathbf{1}\delta^{ij} + i\epsilon^{ijk}\sigma^k \tag{2.6}$$

where δ^{ij} is the Kronecker delta, and ϵ^{ijk} is the totally antisymmetric Levi-Civita symbol. To be explicit,

$$\epsilon^{123} = \epsilon^{231} = \epsilon^{312} = 1 \qquad \text{and} \qquad \epsilon^{321} = \epsilon^{213} = \epsilon^{132} = -1$$

with all the other components being zero. The Levi-Civita symbol makes it possible to write the components of a cross-product between vectors in a compact manner. For example, the kth component of the vector $\vec{A} \times \vec{B}$ may be written in either of the following two forms:

$$\begin{aligned}
(\vec{A} \times \vec{B})^k &= \epsilon^{ijk} A^i B^j \\
&= \epsilon^{kij} A^i B^j
\end{aligned} \tag{2.7}$$

Note that we are using Einstein's summation convention throughout this entire book: Any repeated index is implicitly summed over. However, we will require only Lorentz indices to satisfy the rule that they must appear in the upper and lower positions to be contracted.

From Eq. (2.6) one can calculate the commutator of two Pauli matrices to be

$$[\sigma^i, \sigma^j] \equiv \sigma^i \sigma^j - \sigma^j \sigma^i = 2i\epsilon^{ijk}\sigma^k \tag{2.8}$$

The anticommutator is also easily computed:

$$\{\sigma^i, \sigma^j\} \equiv \sigma^i \sigma^j + \sigma^j \sigma^i = 2\delta^{ij}\mathbf{1} \tag{2.9}$$

From this anticommutator, we directly obtain

$$\sigma^i \sigma^j = -\sigma^j \sigma^i \qquad \text{for} \qquad i \neq j \tag{2.10}$$

$$(\sigma^i)^2 = \mathbf{1}$$

We noted earlier that σ^2 stands apart from the other two matrices when we consider complex conjugation or transposition. Let us focus on σ^2 again, for which we have

$$\sigma^2 \sigma^1 = -\sigma^1 \sigma^2 \qquad \sigma^2 \sigma^3 = -\sigma^3 \sigma^2 \qquad \sigma^2 \sigma^2 = \sigma^2 \sigma^2$$

the last one being, of course, trivial. What is interesting is that we can combine these results with Eq. (2.4) to get

$$\sigma^2 \vec{\sigma}^T = -\vec{\sigma}\sigma^2 \qquad (2.11)$$

$$\sigma^2 \vec{\sigma}^* = -\vec{\sigma}\sigma^2 \qquad (2.12)$$

Using the fact that $(\sigma^2)^2 = \mathbf{1}$, we get for free

$$\sigma^2 \vec{\sigma}\sigma^2 = -\vec{\sigma}^T = -\vec{\sigma}^*$$

$$\sigma^2 \vec{\sigma}^T \sigma^2 = \sigma^2 \vec{\sigma}^* \sigma^2 = -\vec{\sigma} \qquad (2.13)$$

where the second line is trivially obtained from the first line by sandwiching the latter between two σ^2 matrices. These identities will come in handy later on.

Consider now a vector $\vec{A}$ whose components are numbers, not matrices. We will soon encounter an expression of the form $\vec{A} \cdot \vec{\sigma}\, \sigma^j$. It will prove useful to rewrite this as

$$\begin{aligned}
\vec{A} \cdot \vec{\sigma}\, \sigma^j &= A^i \sigma^i \sigma^j \\
&= A^i \sigma^j \sigma^i + A^i [\sigma^i, \sigma^j] \\
&= A^i \sigma^j \sigma^i + 2i\epsilon^{ijk} A^i \sigma^k \\
&= A^i \sigma^j \sigma^i - 2i\epsilon^{jik} A^i \sigma^k \\
&= \sigma^j \vec{A} \cdot \vec{\sigma} - 2i(\vec{A} \times \vec{\sigma})^j \qquad (2.14)
\end{aligned}$$

2.2 Weyl versus Dirac Spinors

Let us now write the four-component Dirac spinor in terms of two-component spinors:

$$\Psi = \begin{pmatrix} \eta \\ \chi \end{pmatrix} \qquad (2.15)$$

In this book we will use capital Greek letters for four-component spinors and lowercase Greek letters for two-component spinors. There is no standard notation for the symbols used for the upper and lower two-component spinors; some authors switch the η and the χ or use different Greek letters altogether.

The two-component spinors appearing in Eq. (2.15) are called *Weyl spinors*. The reason it makes sense to decompose a Dirac spinor this way is that, as we will see in more detail in Section 2.4, η and χ transform independently under Lorentz transformations; i.e., they do not mix. The technical way to express this is to say that a Dirac spinor forms a *reducible* representation of the Lorentz group, whereas Weyl spinors form *irreducible* representations. This implies that from a purely mathematical point of view, Weyl fermions may be considered more "fundamental" than Dirac spinors. The reason why one has to put together two Weyl spinors to form a Dirac spinor will be discussed in Chapter 4.

Note that setting either η or χ equal to zero in a Dirac spinor yields eigenstates of γ_5:

$$\gamma_5 \begin{pmatrix} \eta \\ 0 \end{pmatrix} = + \begin{pmatrix} \eta \\ 0 \end{pmatrix}$$

$$\gamma_5 \begin{pmatrix} 0 \\ \chi \end{pmatrix} = - \begin{pmatrix} 0 \\ \chi \end{pmatrix}$$

The eigenvalue of γ_5 is called the *chirality* of the spinor. By a slight abuse of language, we will say that the upper two-component spinor η has a chirality of $+1$, whereas χ has a chirality of -1 (this is an abuse of language because it is actually the four-component spinor with two components set to zero that has a definite chirality, not the two-component spinors themselves).

A Weyl spinor with positive chirality is sometimes referred to as a *right-chiral spinor*, and a Weyl spinor with negative chirality is referred to as a *left-chiral spinor*. However, some references replace the adjectives *right-chiral* and *left-chiral* by *right-handed* and *left-handed*, which is quite unfortunate because, in general, knowing the chirality does *not* tell us the handedness of a particle! We will soon discuss how the handedness or, to be more technical, the *helicity* of a spinor is defined, but the key point to make here is that the notions of chirality and helicity of a spinor coincide only when a particle is massless. For a massive particle, both the left-chiral and the right-chiral states are, in general, linear combinations of left-handed and right-handed states. We will be careful about consistently using *left-chiral* and *right-chiral* to describe the eigenstates of the chirality operator. As we just mentioned, these two terms are synonyms of *left-handed* and *right-handed* only in the massless case.

Why has it become common to use *left-handed* and *right-handed* to describe the states of definite chirality, even for massive particles? We don't know for sure, but if we had to bet, our guess would be that people got so used to working with massless neutrinos in the standard model and talking about left-handed neutrinos and right-handed antineutrinos that they kept using this nomenclature even when they started considering massive neutrinos, and this habit has made its way into SUSY, where Weyl spinors are ubiquitous. However, we would not bet a lot of money.

But why not forget about chirality and just work with helicity? The answer is that, as already mentioned, *it is the chirality that tells us how a spinor behaves under Lorentz transformations*! When a spinor is massive, it is its chirality that specifies what representation of the Lorentz group it belongs to, not its helicity. Obviously, it is the behavior under Lorentz transformations that matters when constructing theories, so we will care much more about chirality than about helicity. Indeed, we will only discuss helicity in a few paragraphs in the entire book, whereas chirality will be omnipresent from the beginning to the end. We will discuss in detail the Lorentz transformations of the chiral states in Section 2.4.

We will therefore show explicit subscripts R and L on the spinors η and χ for the remainder of the chapter to indicate their *chirality*. Of course, if the spinors are massless, the labels also tell us the helicity. When we start constructing supersymmetric theories, we will be working exclusively with left-chiral spinors, so subscripts won't be necessary.

Be warned that when it comes to the different types of spinors, there is no standard notation. Most books or papers do not show explicitly the indices L and R, and there is no general consensus on what Greek letters to use for left- and right-chiral spinors. Some switch the χ and η symbols for left- and right-chiral spinors, whereas others use different Greek letters altogether. In addition, some equations and expressions will be different if other representations of the gamma matrices are used. On top of that, some authors place the right-chiral spinor in the lower half of the Dirac spinor rather than at the top. Always check the notation and representation used when consulting a new reference. *Caveat emptor*!

It is convenient to introduce the operators P_R and P_L, which, when applied to a full Dirac spinor, project out the right-chiral and left-chiral Weyl spinors. In other words,

$$P_R \Psi = P_R \begin{pmatrix} \eta_R \\ \chi_L \end{pmatrix} = \begin{pmatrix} \eta_R \\ 0 \end{pmatrix} \qquad P_L \Psi = P_L \begin{pmatrix} \eta_R \\ \chi_L \end{pmatrix} = \begin{pmatrix} 0 \\ \chi_L \end{pmatrix} \qquad (2.16)$$

which imply

$$\gamma_5 \left(P_R \Psi \right) = P_R \Psi \qquad \gamma_5 \left(P_L \Psi \right) = -P_L \Psi \qquad (2.17)$$

The explicit representations of these operators are obviously

$$P_R = \begin{pmatrix} \mathbf{1} & 0 \\ 0 & 0 \end{pmatrix} \qquad P_L = \begin{pmatrix} 0 & 0 \\ 0 & \mathbf{1} \end{pmatrix}$$

It is easy to verify that with the representation of γ_5 given in Eq. (2.3), we may write these projection operators as

$$P_R = \frac{\mathbf{1} + \gamma_5}{2} \qquad P_L = \frac{\mathbf{1} - \gamma_5}{2} \tag{2.18}$$

where now the $\mathbf{1}$ obviously represents the 4×4 identity matrix as opposed to the 2×2 identity matrix that it has represented so far. Most references are even less discerning than we are here and simply use the numeral 1 to represent the identity matrix in any dimension.

It is also easy to check that these two operators satisfy the following properties:

$$(P_L)^2 = P_L$$
$$(P_R)^2 = P_R$$
$$P_L P_R = P_R P_L = 0$$
$$\gamma_5 P_R = P_R \gamma_5 = P_R$$
$$\gamma_5 P_L = P_L \gamma_5 = -P_L$$

The first three identities are, of course, typical for projection operators.

If we substitute Eq. (2.15) into the Dirac equation, we obtain two coupled equations for the Weyl spinors:

$$(E \mathbf{1} - \vec{\sigma} \cdot \vec{p}) \eta_R = m \chi_L$$
$$(E \mathbf{1} + \vec{\sigma} \cdot \vec{p}) \chi_L = m \eta_R \tag{2.19}$$

Let us pause to introduce some new notation that will prove to be very useful later on. Let us define σ^μ to be the following set of four matrices:

$$\sigma^\mu \equiv (\mathbf{1}, \vec{\sigma}) \tag{2.20}$$

We also will need to introduce

$$\bar{\sigma}^\mu \equiv (\mathbf{1}, -\vec{\sigma}) \tag{2.21}$$

It's important to emphasize that the bar used here has nothing whatsoever to do with the one used to denote Dirac conjugation of a four-component spinor field, $\overline{\Psi} = \Psi^\dagger \gamma^0$. A different symbol over σ^μ—maybe a tilde—would have been more appropriate, but this is standard notation, and we are stuck with it.

Equation (2.13) imply the following identities:

$$\boxed{\begin{aligned} \sigma^2 (\sigma^\mu)^T \sigma^2 &= \bar{\sigma}^\mu \\ \sigma^2 (\bar{\sigma}^\mu)^T \sigma^2 &= \sigma^\mu \end{aligned}} \tag{2.22}$$

If we take the transpose of both equations (and use $(\sigma^2)^T = -\sigma^2$), we also get

$$\boxed{\begin{aligned} \sigma^2 \sigma^\mu \sigma^2 &= \bar{\sigma}^{\mu T} \\ \sigma^2 \bar{\sigma}^\mu \sigma^2 &= \sigma^{\mu T} \end{aligned}} \tag{2.23}$$

Since $(\sigma^i)^* = (\sigma^i)^T$, these equations are still valid with the transpose replaced by a complex conjugation.

The four identities in Eqs. (2.22) and (2.23) will be so useful to us that they have been included in Appendix A, which lists the most important identities used throughout this book.

An identity that is easy to prove (see Exercise 2.1) is

$$\boxed{\sigma^\mu \bar{\sigma}^\nu + \sigma^\nu \bar{\sigma}^\mu = 2\eta^{\mu\nu} \mathbf{1}} \tag{2.24}$$

Actually, doing the proof makes it obvious that the following also holds true:

$$\boxed{\bar{\sigma}^\mu \sigma^\nu + \bar{\sigma}^\nu \sigma^\mu = 2\eta^{\mu\nu} \mathbf{1}} \tag{2.25}$$

EXERCISE 2.1
Prove Eq. (2.24). *Hint*: Consider separately the cases $\mu = \nu = 0$, $\mu = 0$ and $\nu = i$, $\nu = 0$ and $\mu = i$ and finally $\mu = i$ and $\nu = j$ (where it is understood that a Latin index may take a value of 1, 2, or 3 only).

Using the definitions (2.20) and (2.21), the two coupled equations (2.19) may be written as

$$\begin{aligned} P_\mu \sigma^\mu \eta_R &= m \, \chi_L \\ P_\mu \bar{\sigma}^\mu \chi_L &= m \, \eta_R \end{aligned} \tag{2.26}$$

We also could have obtained Eq. (2.26) directly from the Dirac equation by noting that γ^μ may be written as

$$\gamma^\mu = \begin{pmatrix} 0 & \bar{\sigma}^\mu \\ \sigma^\mu & 0 \end{pmatrix} \qquad (2.27)$$

We also can write the Dirac lagrangian in terms of the Weyl spinors by substituting Eqs. (2.15) and (2.27) in (2.2):

$$\mathcal{L}_{\text{Dirac}} = \eta_R^\dagger \sigma^\mu i \partial_\mu \eta_R + \chi_L^\dagger \bar{\sigma}^\mu i \partial_\mu \chi_L - m \, \eta_R^\dagger \chi_L - m \, \chi_L^\dagger \eta_R \qquad (2.28)$$

Note one important point: Despite the fact that σ^μ and $\bar{\sigma}^\mu$ carry a Lorentz index, they are, of course, not themselves four-vectors (for the same reason that γ^μ is not a four-vector). However, we can form quantities with well-defined Lorentz properties by sandwiching them between Weyl spinors, as we will discuss in more detail below.

2.3 Helicity

Equation (2.26) shows clearly that the Dirac equation mixes η_R and χ_L. We cannot have one without the other. For example, if we set $\eta_R = 0$, the first equation yields $0 = m\chi_L$, which implies that χ_L must be zero as well, *unless the mass is zero*. If the mass is zero, the two equations for η_R and χ_L decouple to give

$$\frac{\vec{\sigma} \cdot \vec{p}}{|\vec{p}|} \eta_R = \eta_R$$

$$\frac{\vec{\sigma} \cdot \vec{p}}{|\vec{p}|} \chi_L = -\chi_L \qquad (2.29)$$

where we have set $E = |\vec{p}|$, which is valid for a massless particle (recall that in natural units, $c = 1$).

So, for a massless particle, the Dirac equation actually breaks up into two independent equations acting on two different two-component spinors. These two equations are known as the *Weyl equations*.

As we will discuss more later, this implies that when the mass is zero, we should think of η_R and χ_L as fields describing two different particles instead of two states of a single particle. In fact, one can even build theories in which only one of the two is present, whereas the second one is absent altogether (this is of course the case in the standard model which contains only left-handed massless neutrinos).

If we multiply those equations by $\hbar/2$ (showing explicitly $\hbar$ for the moment) and use the fact that $\hbar\vec{\sigma}/2$ is the spin operator $\vec{S}$, we get

$$\vec{S}\cdot\hat{p}\,\eta_R = \frac{\hbar}{2}\eta_R$$

$$\vec{S}\cdot\hat{p}\,\chi_L = -\frac{\hbar}{2}\chi_L$$

The operator $\vec{S}\cdot\hat{p}$ is called the *helicity operator*, and the corresponding eigenvalue is the helicity of the state. This eigenvalue corresponds to the component of the spin along the direction of the motion of the particle. We see that *in the massless limit, the Weyl spinors are eigenstates of the helicity operator*.

The fact that η_R has an helicity of $\hbar/2$ when $m=0$ tells us that its spin is aligned with its direction of motion. If spin were a classical angular momentum, this would correspond to a particle following the path of a right-handed helix, which is why we refer to the two upper components of a massless Dirac spinor as a *right-handed Weyl spinor*. Obviously, χ_L corresponds to a left-handed Weyl spinor.

There is a simple physical argument to explain why the two helicity states do not decouple for a massive state. To visualize, consider a particle with a classical angular momentum vector. Let's say that the particle moves to the right while tracing a right-handed helix around the positive x axis. If we boost ourselves (the observers) to a frame that is moving faster than the particle, we will now observe it moving to our left while tracing a left-handed helix. Therefore, boosting to a frame moving faster than the particle changes its helicity state from right-handed to left-handed, and vice versa. This has the implication that both states must be included to describe a massive spinor. The argument clearly fails when the particle is massless because it is impossible to boost to a frame moving faster than the particle. In that case, the two helicity states can never get mixed under Lorentz transformations, and they can be thought of as different particles altogether (as opposed to two states of a single particle in the massive case).

It's important to keep in mind that the chiral spinors χ_L and η_R only have well-defined helicities in the massless limit. Let's call ψ_+ and ψ_- the states of positive and negative helicities (i.e., the right-handed and left-handed states). Then, in general, we have

$$\chi_L = C_1\psi_+ + C_2\psi_-$$

$$\eta_R = C_2^*\psi_+ - C_1^*\psi_- \tag{2.30}$$

which can, of course, be inverted to give

$$\psi_+ = \frac{C_1^* \chi_L + C_2 \eta_R}{|C_1|^2 + |C_2|^2}$$

$$\psi_- = \frac{C_2^* \chi_L - C_1 \eta_R}{|C_1|^2 + |C_2|^2} \tag{2.31}$$

In other words, chirality and helicity offer two different bases on which the spinors may be expanded. The massless limit corresponds to C_1 being identically 0 and $C_2 = 1$.

Helicity, the projection of the spin on the direction of motion of the particle, is easier to picture physically than the abstract concept of chirality. But we care more about the chiral eigenstates χ_L and η_R because, as mentioned earlier, these are the states that have well-defined Lorentz transformation properties, as we will discuss next.

2.4 Lorentz Transformations and Invariants

Our next step is to look at the Lorentz transformations of spinors. Knowing how quantities transform under rotation and boosts is obviously crucial to building invariant lagrangians. The infinitesimal transformations of the right-chiral and left-chiral Weyl spinors are

$$\boxed{\begin{aligned} \eta_R \rightarrow \eta_R' &= \left(1 + \frac{i}{2}\vec{\epsilon} \cdot \vec{\sigma} - \frac{1}{2}\vec{\beta} \cdot \vec{\sigma}\right) \eta_R \\ \chi_L \rightarrow \chi_L' &= \left(1 + \frac{i}{2}\vec{\epsilon} \cdot \vec{\sigma} + \frac{1}{2}\vec{\beta} \cdot \vec{\sigma}\right) \chi_L \end{aligned}} \tag{2.32}$$

where $\vec{\epsilon}$ is the infinitesimal rotation vector (the direction of $\vec{\epsilon}$ gives the axis of rotation, and its magnitude gives the amount of rotation), and β is the infinitesimal boost parameter.

The transformations [2.32] show explicitly that, as we mentioned earlier, the left-chiral and right-chiral spinors transform independently under Lorentz transformations. They are therefore inequivalent irreducible representations of the Lorentz group and are the fundamental "building blocks" from which any other spinor representation can be constructed.

Because the Lorentz algebra is equivalent to the algebra $su(2) \times su(2)$, all representations can be labeled by two numbers that are multiples of one-half. The two fundamental representations then are described by the quantum numbers $(\frac{1}{2}, 0)$ and $(0, \frac{1}{2})$, which describe, respectively, the left-chiral and right-chiral spinors. We won't use this language in this book, but it is important to be aware of this notation because many SUSY references use it. By the way, even though the algebra acting on the Weyl spinors is $su(2) \times su(2)$, the *group* under which they transform is not $SU(2) \times SU(2)$ but instead $SL(2, C)$, the universal covering group of the Lorentz group $SO(3, 1)$. This might be useful to know if you want to impress your friends.

For the sake of comparison, consider the corresponding transformations for a four-vector, e.g., the four-momentum P^μ. The zeroth component $P^0 = E$ and the vector components $\vec{p}$ transform under rotations and under boosts in the following way:

$$E \rightarrow E' = E - \vec{\beta} \cdot \vec{p}$$
$$\vec{p} \rightarrow \vec{p}' = \vec{p} - \vec{\epsilon} \times \vec{p} - \vec{\beta} E$$

(2.33)

Now, let us ask: What Lorentz-invariant quantity can we build out of spinors? We can use the Lorentz transformations (2.32) to figure this out or simply use the Dirac lagrangian Eq. (2.2) as a guide. Let us first focus on the mass term. The Dirac lagrangian tells us that combination $\Psi^\dagger \gamma^0 \Psi$ is invariant. Writing this in terms of Weyl spinors, we find that

$$\overline{\Psi}\Psi = \Psi^\dagger \gamma^0 \Psi = \eta_R^\dagger \chi_L + \chi_L^\dagger \eta_R$$

is invariant under Lorentz transformations. Actually, it turns out that each term on the right-hand side is separately invariant.

EXERCISE 2.2
Show that $\eta_R^\dagger \chi_L$ is invariant under Lorentz transformations. Note that the Lorentz transformations of the spinors are given to first order in $\vec{\epsilon}$ and β so that all terms of higher order may be dropped.

Now consider the kinetic energy term of the Dirac lagrangian $\overline{\Psi}\gamma^\mu P_\mu \Psi$. Since this is invariant and P_μ is a four-vector, the quantity $\overline{\Psi}\gamma^\mu \Psi$ must transform as a four-vector (as a contravariant four-vector, to be precise). If we rewrite this expression in terms of Weyl spinors and use Eq. (2.27), we find that

$$\overline{\Psi}\gamma^\mu \Psi = \eta_R^\dagger \sigma^\mu \eta_R + \chi_L^\dagger \bar{\sigma}^\mu \chi_L$$

transforms as a four-vector under Lorentz transformations. Actually, it turns out that each term on the right separately transforms as a four-vector. As noted earlier, although σ^μ and $\bar{\sigma}^\mu$ carry a Lorentz index, they are not themselves four-vectors.

The matrices σ^μ must be sandwiched between an $\eta_R^\dagger$ and an η_R to yield a four-vector, and $\bar\sigma^\mu$ must be sandwiched between $\chi_L^\dagger$ and χ_L.

EXERCISE 2.3

Prove that $\eta_R^\dagger \sigma^\mu \eta_R$ transforms as a four-vector under a Lorentz transformation. *Hint*: Consider first $\eta_R^\dagger \sigma^0 \eta_R$ and show that it transforms like the energy E in Eq. (2.33). Then show that $\eta_R^\dagger \sigma^i \eta_R$ transforms like the momentum three-vector.

2.5 A First Notational Hurdle

Here we should pause to discuss a first (of many!) possible source of confusion owing to notation. This first hurdle concerns the use of the hermitian conjugate (or "dagger") symbol, $\dagger$. In linear algebra, taking the hermitian conjugate means taking the complex conjugate and transpose of a matrix or vector. However, in quantum field theory, the symbol $\dagger$ has in addition a *second* role: When applied to a quantum field, it consists of taking a complex conjugate of any complex number as well as replacing annihilation operators by creation operators (and vice versa). For example, writing a free complex scalar field $\phi(x)$ as

$$\phi(x) = \int \frac{d^3k}{2E(2\pi)^3}[a(k)e^{-ik\cdot x} + b^\dagger(k)e^{ik\cdot x}]$$

the hermitian conjugate of the field is given by

$$\phi^\dagger(x) = \int \frac{d^3k}{2E(2\pi)^3}[a^\dagger(k)e^{ik\cdot x} + b(k)e^{-ik\cdot x}] \tag{2.34}$$

Clearly, if the dagger is applied to a scalar field, it only represents taking the hermitian conjugate in the "quantum field operator sense" shown in Eq. (2.34).

Consider now quantum fields that are themselves components of column or row vectors, e.g., the spinor χ. Both components of χ—let's call them χ_1 and χ_2—are quantum fields, and it makes sense to talk about $\chi_1^\dagger$ and $\chi_2^\dagger$. This would be nonsensical for components which are simple complex numbers because we can't take the hermitian conjugate of a complex number. In the case of quantum spinor fields, taking the dagger of a component has a clear meaning: It stands for the hermitian conjugation in the quantum field operator sense (complex conjugation plus exchange of creation and annihilation operators).

The problem arises when we consider something like $\chi^\dagger$. What is meant by this is not completely clear, a priori. Does this represent the column vector with the hermitian conjugation in the quantum field operator sense applied to the components, namely,

$$\chi^\dagger = \begin{pmatrix} \chi_1^\dagger \\ \chi_2^\dagger \end{pmatrix} \tag{2.35}$$

or is the hermitian conjugate applied in both the operator and the linear algebra sense, as in

$$\chi^\dagger = \begin{pmatrix} \chi_1^\dagger, \chi_2^\dagger \end{pmatrix}$$

Most authors simply use $\chi^\dagger$ to represent both expressions, letting the context dictate which meaning is implied. With some practice, it is not difficult to get used to this double role of the dagger, especially when expressions are fairly short, but it is distracting to have to worry about making sense of an ambiguous notation when learning a new subject. For this reason, we will use the following convention in this book: When a dagger will be applied to a spinor field, it will represent the hermitian conjugation of each component of the spinor field in the quantum field operator sense *as well as* a transpose. Therefore, the quantity $\chi^\dagger$ will always represent the row vector with components $\chi_i^\dagger$, i.e.,

$$\chi^\dagger = \begin{pmatrix} \chi_1^\dagger, \chi_2^\dagger \end{pmatrix}$$

Note that this is the convention we used in the preceding section.

The question is, then, what do we do if we want to represent the column vector (2.35) unambiguously? We will simply have to "undo" the transpose included in the linear algebra dagger, so we will write

$$\boxed{\chi^{\dagger T} \equiv \begin{pmatrix} \chi_1^\dagger \\ \chi_2^\dagger \end{pmatrix}} \tag{2.36}$$

Of course, this notation would never be used if we were dealing with vectors made of complex numbers because we would simply write χ^* instead of $\chi^{\dagger T}$. Here, the quantum operator dagger is applied to each *component* of the spinor, which is why we cannot think of $\chi^{\dagger T}$ as χ^*.

This notation is admittedly a bit awkward, but it is a small price to pay for the gain in clarity we get by explicitly distinguishing the column and row spinors $\chi^{\dagger T}$ and $\chi^\dagger$.

Note that the possible confusion arises only when the dagger is applied to spinor quantum fields. When applied to matrices such as Dirac matrices or Pauli matrices, the dagger obviously stands for the usual hermitian conjugation of linear algebra, and when applied to a scalar field, it clearly represents the hermitian conjugation in the quantum field operator sense.

2.6 Building More Lorentz Invariants Out of Weyl Spinors

The simplest supersymmetric theory, which we will construct in Chapter 5, involves a single Weyl spinor (chosen by convention to be left-chiral) coupled to scalar fields. Later we will work with theories containing several left-chiral spinors. It is therefore important to figure out all the possible ways to construct Lorentz-invariant terms out of left-chiral spinors only. For the sake of completeness, we also will write down Lorentz-invariant expressions containing only right-chiral spinors or mixing left- and right-chiral spinors, although these won't be needed in the rest of this book.

Consider, then, a single left-chiral spinor χ_L. We already know one Lorentz-invariant expression that contains only χ_L: the kinetic term appearing in Eq. (2.28). On the other hand, the mass term that we obtained from the Dirac lagrangian mixes a left- and right-chiral spinor. So an interesting question arises: Is it possible to build a mass term out of a left-chiral spinor only? It turns out that it *is* possible, but not completely trivial.

We know that $\chi_L^\dagger \eta_R$ is invariant, so what we need is to somehow build out of the left-chiral field χ_L something that transforms like a right-chiral field η_R! If we then multiply $\chi_L^\dagger$ by this new quantity, we will have obtained a Lorentz-invariant mass term for the left-chiral field.

As a first guess to what this quantity could be, let's consider $\chi_L^{\dagger T}$. Recall from the preceding section that this is a column vector whose components are the dagger of the quantum fields [see Eq. (2.36)]. Keeping in mind that the hermitian conjugate is applied only to the quantum fields and that only a complex conjugation is applied to everything else, and using Eq. (2.32), we find that $\chi_L^{\dagger T}$ transforms as

$$\chi_L^{\dagger T} \to (\chi_L^{\dagger T})' = (\chi_L')^{\dagger T} = \left(1 + \frac{i}{2}\vec{\epsilon}\cdot\vec{\sigma} + \frac{1}{2}\vec{\beta}\cdot\vec{\sigma}\right)^* \chi_L^{\dagger T}$$

$$= \left(1 - \frac{i}{2}\vec{\epsilon}\cdot\vec{\sigma}^* + \frac{1}{2}\vec{\beta}\cdot\vec{\sigma}^*\right)\chi_L^{\dagger T}$$

Unfortunately, this does not transform like a right-chiral spinor [see first line of Eq. (2.32)]. There are two problems: The two signs are wrong, and the Pauli matrices are complexed conjugated. But it turns out that both problems can be solved in one shot if we recall Eq. (2.12):

$$\sigma^2 \vec{\sigma}^* = -\vec{\sigma}\sigma^2$$

which is exactly what the doctor ordered! To see why this helps, consider the Lorentz transformation of the quantity $i\sigma^2\chi_L$ (the inclusion of a factor of i is purely conventional but convenient when we relate particle and antiparticle states, as we will see in Chapter 4):

$$i\sigma^2\chi_L^{\dagger T} \rightarrow \left(i\sigma^2\chi_L^{\dagger T}\right)' = (i\sigma^2)^* \left(1 + \frac{i}{2}\vec{\epsilon}\cdot\vec{\sigma} + \frac{1}{2}\vec{\beta}\cdot\vec{\sigma}\right)^* \chi_L^{\dagger T}$$

$$= i\sigma^2\left(1 - \frac{i}{2}\vec{\epsilon}\cdot\vec{\sigma}^* + \frac{1}{2}\vec{\beta}\cdot\vec{\sigma}^*\right)\chi_L^{\dagger T}$$

$$= \left(1 + \frac{i}{2}\vec{\epsilon}\cdot\vec{\sigma} - \frac{1}{2}\vec{\beta}\cdot\vec{\sigma}\right)i\sigma^2\chi_L^{\dagger T}$$

which is the transformation of a right-chiral spinor! We have therefore built something that transforms like a right-chiral spinor out of a left-chiral spinor. This is a key result, so let us emphasize it:

$$\boxed{i\sigma^2\chi_L^{\dagger T} \text{ transforms like a right-chiral spinor}} \qquad (2.37)$$

The matrix $i\sigma^2$ will appear in so many of our calculations that it is useful to summarize some of the properties that will often be used:

$$(i\sigma^2)^T = -i\sigma^2 \qquad (i\sigma^2)^* = i\sigma^2 \qquad (i\sigma^2)^\dagger = -i\sigma^2 \qquad (i\sigma^2)^2 = -\mathbf{1}$$

Now we are ready to build Lorentz invariants that contain only left-chiral spinors. We consider only invariants with no Lorentz indices for now. Since the expression $\eta_R^\dagger \chi_L$ is invariant under Lorentz transformations and $i\sigma^2\chi_L^{\dagger T}$ transforms like η_R, we know that the following expression is also invariant:

$$\left(i\sigma^2\chi_L^{\dagger T}\right)^\dagger \chi_L = \chi_L^T (i\sigma^2)^\dagger \chi_L$$

$$= \boxed{\chi_L^T(-i\sigma^2)\chi_L} \qquad (2.38)$$

This is what we were seeking: a Lorentz invariant mass term containing only left-chiral spinors!

Since we know that $\chi_L^\dagger \eta_R$ is invariant, we obtain a second invariant containing only left-chiral spinors:

$$\boxed{\chi_L^\dagger i\sigma^2 \chi_L^{\dagger T}} \tag{2.39}$$

The two invariant combinations (2.38) and (2.39) play such an important role in the construction of supersymmetric theories that they are assigned special notations that we will discuss shortly. Before doing that, let us build other invariants made of right-chiral spinors alone or mixing left- and right-chiral spinors. First, we need to build something out of the right-chiral spinor η_R that will transform like a left-chiral spinor. It is easy to check that

$$\boxed{-i\sigma^2 \eta_R^{\dagger T} \text{ transforms like a left-chiral spinor}} \tag{2.40}$$

Again, the reason why we choose to include a factor of $-i$ will become clear when we discuss antiparticles in Chapter 4.

To build other invariants, we could proceed again by relying on the invariance of the terms appearing in the Dirac lagrangian. However it is also instructive to write down explicitly the Lorentz transformation properties of the various quantities we are dealing with.

Let's define the matrix A as

$$A \equiv \mathbf{1} + i\frac{\vec{\epsilon}\cdot\vec{\sigma}}{2} - \frac{\vec{\beta}\cdot\vec{\sigma}}{2}$$

With this definition, right-chiral spinors have for Lorentz transformation $\eta_R \to A\,\eta_R$.

Using the fact that the Pauli matrices are hermitian, we can easily find the hermitian conjugate of the matrix A:

$$A^\dagger = \mathbf{1} - i\frac{\vec{\epsilon}\cdot\vec{\sigma}}{2} - \frac{\vec{\beta}\cdot\vec{\sigma}}{2}$$

The inverse matrix is obtained simply by changing the sign of ϵ and η because these are infinitesimal quantities. Thus we find

$$A^{-1} = \mathbf{1} - i\frac{\vec{\epsilon}\cdot\vec{\sigma}}{2} + \frac{\vec{\beta}\cdot\vec{\sigma}}{2}$$

In turn, this implies

$$(A^{-1})^\dagger = (A^\dagger)^{-1} = \mathbf{1} + i\frac{\vec{\epsilon}\cdot\vec{\sigma}}{2} + \frac{\vec{\beta}\cdot\vec{\sigma}}{2}$$

where we stressed that the order of the dagger and inverse operations does not matter. This is the matrix that appears in the transformation of left-chiral spinors.

Thus the quantities we have at our disposal, together with their behavior under Lorentz transformations, are

$$\eta_R \;\to\; A\eta_R$$
$$i\sigma^2\chi_L^{\dagger T} \;\to\; A\big(i\sigma^2\chi_L^{\dagger T}\big)$$
$$\chi_L \;\to\; (A^\dagger)^{-1}\chi_L$$
$$-i\sigma^2\eta_R^{\dagger T} \;\to\; (A^\dagger)^{-1}\big(-i\sigma^2\eta_R^{\dagger T}\big) \tag{2.41}$$

It is then clear how to build invariants. If we call *type I* the expressions that transform with the matrix A and *type II* those that transform with the matrix $(A^\dagger)^{-1}$, then the possible invariants are either of the form

$$(\text{type I})^\dagger(\text{type II})$$

or of the form

$$(\text{type II})^\dagger(\text{type I})$$

This gives eight combinations (counting as different combinations related by hermitian conjugation), but it turns out that only six of those are independent, which may be taken to be

$$- \chi_L^T i\sigma^2\chi_L \qquad \chi_L^\dagger i\sigma^2\chi_L^{\dagger T}$$
$$\chi_L^\dagger \eta_R \qquad\qquad \eta_R^\dagger \chi_L$$
$$- \eta_R^\dagger i\sigma^2\eta_R^{\dagger T} \qquad \eta_R^T i\sigma^2\eta_R \tag{2.42}$$

The first two expressions involve only the left-chiral spinors that we obtained earlier. The next two expressions mix the two types of spinors and are the ones appearing in a Dirac mass term. The last two expressions contain only right-chiral spinors.

EXERCISE 2.4

Show that the two remaining combinations, involving both $i\sigma^2\chi_L^{\dagger T}$ and $-i\sigma^2\eta_R^{\dagger T}$, are equivalent to two of the expressions already listed.

2.7 Invariants Containing Lorentz Indices

Let us now consider terms that contain derivatives of spinor fields. We have seen, by looking at the Dirac equation written in terms of Weyl spinors [Eq. (2.26)], that (after using $P_\mu = i\partial_\mu$)

$$\sigma^\mu i\partial_\mu\eta_R \text{ transforms like a left-chiral spinor} \qquad (2.43)$$

whereas

$$\bar{\sigma}^\mu i\partial_\mu\chi_L \text{ transforms like a right-chiral spinor} \qquad (2.44)$$

Note how the matrix σ^μ "belongs" with right-chiral spinors and $\bar{\sigma}^\mu$ "belongs" with left-chiral spinors, in the sense that these are the combinations required to build quantities that have well-defined Lorentz transformation properties. Applying a σ^μ to a left-chiral spinor does not give anything with well-defined Lorentz properties, so it's not a useful thing to do. However, applying a $\bar{\sigma}^\mu$ to a left-chiral spinor and *then* applying a σ^μ does produce something interesting, as will be illustrated in Exercise 2.5.

To build Lorentz invariants containing derivatives, we simply have to use the expressions listed in Eq. (2.42) with the χ_L appearing in those equations replaced by Eq. (2.43) or the η_R replaced by Eq. (2.44).

As examples, consider building invariants containing only left-chiral spinors and only one derivative. These are obtained by taking the terms in Eq. (2.42), which contain both χ_L and η_R, and replacing η_R by Eq. (2.44). If we do this for the term $\chi_L^\dagger\eta_R$ of Eq. (2.42), we get

$$\chi_L^\dagger\bar{\sigma}^\mu i\partial_\mu\chi_L \qquad (2.45)$$

which we recognize as the kinetic term of the left-chiral spinor that appears in the Dirac lagrangian [Eq. (2.28)]. A second possibility is the term $\eta_R^\dagger\chi_L$, which, after replacing η_R by Eq. (2.44), gives

$$\left(\bar{\sigma}^\mu i\partial_\mu\chi_L\right)^\dagger\chi_L = -i\left(\partial_\mu\chi_L^\dagger\right)\bar{\sigma}^\mu\chi_L \qquad (2.46)$$

where we have used the fact that $(\bar{\sigma}^{\mu})^{\dagger} = \bar{\sigma}^{\mu}$ because the Pauli matrices are hermitian. As part of a lagrangian density, however, an integration by parts can be used to show that the term in Eq. (2.46) is equivalent to Eq. (2.45) after discarding a surface term.

We will frequently use integrations by parts on expressions appearing in lagrangian (lagrangian densities, to be exact). We will always assume that the surface terms vanish so that we may write

$$\int d^4x\, A\, \partial_{\mu} B = -\int d^4x\, (\partial_{\mu} A) B$$

where A and B are arbitrary expressions containing quantum fields. This then shows clearly that the right-hand side of Eq. (2.46) is equal to Eq. (2.45).

We also can construct invariants containing two derivatives of a left-chiral field by replacing both η_R in the last two terms of Eq. (2.42) by Eq. (2.44). We will never have to deal with these terms, however, because they are not renormalizable (an issue that we will address in Chapter 8). But there are plenty of other things of interest that we can build with spinors: Lorentz scalars, vectors, and tensors! Exercise 2.5 offers two examples.

EXERCISE 2.5
Build a contravariant vector (i.e., with one upper Lorentz index) and a second rank contravariant tensor using two left-chiral spinors and no derivatives.

In principle, this is all we need to start studying SUSY! Unfortunately, in practice, we need to introduce much more notation. However, as mentioned in the Introduction, the choice has been made to keep the notation as simple as possible until Chapter 11 in order to keep it from getting in the way of understanding the basic concepts of SUSY. Nevertheless, we will cheat a little and introduce a *tiny* bit of extra notation that will make many equations far less cumbersome. Before doing this, however, let's pause to mention a simple identity that will help us countless times.

2.8 A Useful Identity

One of the identities that will be used the most often is a simple modification of a relation familiar from elementary linear algebra.

Consider two plain column vectors v and w whose components are ordinary complex numbers ("ordinary" in the sense that they are commuting quantities) and a matrix A whose elements are also complex numbers. Then the following equality holds:

$$v^T A w = w^T A^T v \tag{2.47}$$

This can be checked easily by writing the indices explicitly. This identity follows from the fact that

$$v^T A w = (v^T A w)^T$$

which is true because the quantity $v^T A w$ is just a number, so it is trivially equal to its own transpose. Applying the the transpose to the parenthesis on the right-hand side, we recover Eq. (2.47).

Of course, if complex conjugates are taken on any of the quantities, they will appear on both sides of the equation. For example,

$$v^\dagger A^* w^* = w^\dagger A^\dagger v^*$$

We will not be dealing with vectors with complex numbers as components but with spinors, whose components are Grassmann quantities, which anticommute. In that case, we must introduce a minus sign when we switch the order of the spinors so that Eq. (2.47) is replaced by

$$\boxed{\alpha^T A \beta = -\beta^T A^T \alpha} \tag{2.48}$$

where α and β are arbitrary spinors. Of course, if a hermitian conjugate is applied to either spinor or to the matrix, we need to take this into consideration. For example,

$$\alpha^T A \beta^{\dagger T} = -\beta^\dagger A^T \alpha$$

We will use Eq. (2.48) or its equivalent form with hermitian conjugates applied to some of the terms repeatedly throughout this book.

2.9 Introducing a New Notation

We have achieved our goal for this chapter, which was to build Lorentz invariants out of Weyl spinors.

In principle, before moving on, we should discuss the shorthand notation of dotted and undotted indices, also called the *van der Waerden notation*, used by "superphysicists" when they do calculations in SUSY. This notation is very useful in that it makes it much easier to identify Lorentz invariant combinations of Weyl spinors, without the need to remember where to put the $i\sigma^2$ matrices or where to place hermitian conjugates. Unfortunately, it is also a real pain for beginners because each spinor now comes in four different versions.

Of course, we *could* avoid altogether this shorthand notation because it is not absolutely necessary to do SUSY calculations. But there would be two major drawbacks to this. One is that it would be more difficult for you to consult other publications on SUSY if you haven't been exposed to this notation, which is used widely. The second problem is that it *does* make calculations much simpler when we have to deal with complicated expressions (and expressions *do* get fairly messy in SUSY!). However, for the first several chapters, our calculations will be simple enough that using the shorthand notation would not provide a noticeable advantage and would make the topic harder to learn. On the other hand, starting in Chapter 11, where we begin working in superspace, calculations would become awkward if we would not take advantage of the shorthand notation of dotted and undotted indices.

So we have opted for a compromise. The van der Waerden notation is introduced in Chapter 3 because it is a logical extension of the material covered in Chapter 2, but it will not be used until Chapter 11 (aside from part of an exercise in Chapter 6). By the time we reach Chapter 11, the basic concepts and tools of SUSY hopefully will be sufficiently familiar that it should not be too much of a problem to add a new layer of notation. We would therefore invite the reader to skip Chapter 3 on a first reading and to get back to it before tackling superspace, in Chapter 11.

In the rest of this chapter we will only be introduced to the bare minimum of extra notation needed until Chapter 11.

In practice, SUSY theorists work almost exclusively with left-chiral spinors, even when considering supersymmetric extensions of the standard model (using some tricks that will be explained later), so we really only need to focus on the two invariants built out of a left-chiral spinor:

$$\chi_L^\dagger i\sigma^2 \chi_L^{\dagger T} \qquad \chi_L^T(-i\sigma^2)\chi_L \qquad (2.49)$$

From now on we will drop the subscript L; it will be implicit that χ always stands for a left-chiral spinor.

We *define* two new types of *spinor dot products* between left-chiral Weyl spinors as a shorthand notation for the two Lorentz invariants of Eq. (2.49):

$$\boxed{\begin{aligned} \chi \cdot \chi &\equiv \chi^T(-i\sigma^2)\chi \\ \bar\chi \cdot \bar\chi &\equiv \chi^\dagger i\sigma^2 \chi^{\dagger T} \end{aligned}} \qquad (2.50)$$

Let us write the two types of dot products explicitly in terms of the components of the spinors. It is conventional to write the components of a left-chiral Weyl spinor with lower indices, namely,

$$\chi \equiv \begin{pmatrix} \chi_1 \\ \chi_2 \end{pmatrix}$$

in terms of which the first invariant of Eq. (2.50) is

$$\chi \cdot \chi = \chi_2 \chi_1 - \chi_1 \chi_2$$

At first sight, this may seem to be identically zero, but recall that the χ_i are fermionic quantum fields, which *anticommute*. This is clear if we write the fields in terms of creation and annihilation operators. But what if we use the path integral approach, where quantum fields are not represented by operators? In that case, as you surely remember, the fermionic degrees of freedom must be represented by *Grassmann* variables, which also anticommute. In either case, we have

$$\chi_1 \chi_2 = -\chi_2 \chi_1, \qquad (\chi_1)^2 = (\chi_2)^2 = 0$$

Using the first of these two relations, we get three equivalent expressions for the dot product $\chi \cdot \chi$:

$$\begin{aligned} \chi \cdot \chi &= \chi_2 \chi_1 - \chi_1 \chi_2 \\ &= 2 \chi_2 \chi_1 \\ &= -2 \chi_1 \chi_2 \end{aligned} \tag{2.51}$$

Consider now the second invariant of Eq. (2.50). It is simply given by

$$\begin{aligned} \bar{\chi} \cdot \bar{\chi} &= \chi_1^\dagger \chi_2^\dagger - \chi_2^\dagger \chi_1^\dagger \\ &= 2 \chi_1^\dagger \chi_2^\dagger \\ &= -2 \chi_2^\dagger \chi_1^\dagger \end{aligned} \tag{2.52}$$

This is it. This is all we are going to introduce in terms of new notation in this section. In the next chapter, a different way to represent these dot products will be introduced, but for now, you should think of $\chi \cdot \chi$ and $\bar{\chi} \cdot \bar{\chi}$ purely as shorthand notations for the expressions appearing on the right of Eq. (2.50), nothing more.

To be honest, SUSY afficionados usually don't write any dot symbol ($\cdot$) between the spinors, but it makes expressions more clear to include it, so that's what we will do for the rest of this book. However, even the notation used here may cause some confusion, so let us issue a few warnings.

First, note that the bar appearing over the Weyl spinors has nothing whatsoever to do with the bar used to represent Dirac conjugation for a four-component spinor, $\overline{\Psi} = \Psi^{\dagger}\gamma^0$. We will discuss the bar notation for Weyl spinors in more depth later. For now, consider the bars in the second line of Eq. (2.50) to be simply a notation to distinguish the two types of dot products.

Second, those "spinor dot products" have nothing to do with the usual dot products between four-vectors or three-vectors familiar from introductory physics. It may be tempting to think of $\chi \cdot \chi$, say, as $\chi_1^2 + \chi_2^2$, but this, of course, would be incorrect (and equal to zero!). If we want to calculate these dot products explicitly in terms of components, we need to go back to their explicit definitions given in Eq. (2.50).

Third, it's very important to keep in mind that the difference between the two types of dot products is not only that one involves the hermitian conjugate of the spinors, whereas the other does not; there is also a sign difference between the two definitions, as is shown explicitly in Eqs. (2.51) and (2.52).

However, the two types of dot products obviously can be written in terms of one another using

$$\bar{\chi} \cdot \bar{\chi} = -\chi^{\dagger T} \cdot \chi^{\dagger T} \tag{2.53}$$

The right-hand side is admittedly rather ugly, which is why the bar notation of the left-hand side is preferred. The point of showing this relation is that the bar notation is redundant; we could write everything in terms of unbarred spinors, at the price of sometimes getting awkward expressions.

From Eq. (2.51), we see that the product of the two components of a spinor is related to the dot product of that spinor with itself in a very simple way:

$$\chi_1 \chi_2 = -\chi_2 \chi_1 = -\frac{1}{2}\chi \cdot \chi$$

whereas, of course, $\chi_1^2 = \chi_2^2 = 0$. These four relations can be summarized into a very useful identity:

$$\boxed{\chi_a \chi_b = -\frac{1}{2}(i\sigma^2)_{ab}\chi \cdot \chi} \tag{2.54}$$

Similarly, we have

$$\chi_1^\dagger \chi_2^\dagger = -\chi_2^\dagger \chi_1^\dagger = \frac{1}{2} \bar{\chi} \cdot \bar{\chi}$$

whereas $(\chi_1^\dagger)^2 = (\chi_2^\dagger)^2 = 0$, so the equivalent of Eq. (2.54) for the hermitian conjugated components is

$$\boxed{\chi_a^\dagger \chi_b^\dagger = \frac{1}{2}(i\sigma^2)_{ab} \bar{\chi} \cdot \bar{\chi}} \tag{2.55}$$

There is something important to repeat here: The hermitian conjugate in the last two equations is applied to the *components* of the spinor, not to the spinor itself. In that context, the dagger necessarily means the hermitian conjugate in the quantum field operator sense. If the components were simple complex numbers, it would make no sense to write, say, $\chi_1^\dagger$. We would simply write χ_1^*. Keep in mind that whenever we write components of a spinor with a dagger applied, as in $\chi_a^\dagger$, the quantum field operator sense is always implied.

There is another simple relation between the two types of products [much more elegant than Eq. (2.53)], but before demonstrating it, we first have to deal with a tricky issue. Consider taking the hermitian conjugate of $\chi \cdot \chi$:

$$(\chi \cdot \chi)^\dagger = (2\chi_2 \chi_1)^\dagger$$

We face a subtle problem. We know that the hermitian conjugate of the product of two matrices exchanges their order, but what do we do when we have Grassmann quantum fields (in other words, spinor quantum fields)? Do we include an extra minus sign owing to the exchange? The convention is to define the hermitian conjugate to exchange their order *without introducing a factor of minus one*. This is so important that we cannot resist the urge to put it in a box:

$$\boxed{\begin{array}{l} \text{The hermitian conjugate of the product of two spinor fields} \\ \text{exchanges their order without introducing a minus sign} \end{array}} \tag{2.56}$$

Again, an ambiguity arises if we use the path integral approach and write the fermion fields as Grassmann variables. In that case, instead of the hermitian conjugate of a product of fields, we would simply take the complex conjugate of that product. But, in order to be consistent with Eq. (2.56), we must define the complex conjugate of a product of Grassmann variables to change their order without introducing a minus sign! In other words, we define

$$(\alpha\beta)^* = \beta^*\alpha^* \qquad (2.57)$$

where α and β are some arbitrary Grassmann variables.

With this convention, we get

$$(\chi \cdot \chi)^\dagger = 2(\chi_2\chi_1)^\dagger = 2\chi_1^\dagger\chi_2^\dagger \qquad (2.58)$$

which is precisely $\bar\chi \cdot \bar\chi$! So we have the following two important identities:

$$\boxed{\begin{aligned} (\chi \cdot \chi)^\dagger &= \bar\chi \cdot \bar\chi \\ (\bar\chi \cdot \bar\chi)^\dagger &= \chi \cdot \chi \end{aligned}} \qquad (2.59)$$

where the second line is simply the hermitian conjugate of the first line.

So far we have only written invariants built out of a single left-chiral spinor. Of course, we also will need to consider invariants made of two different left-chiral spinors, let's say χ and λ. In this case, the two possible invariants mixing the two spinors are obviously

$$\chi \cdot \lambda \qquad \bar\chi \cdot \bar\lambda$$

At first sight it seems that we also should consider $\lambda \cdot \chi$ and $\bar\lambda \cdot \bar\chi$, but it turns out that the order of the spinors in spinor dot products does not matter, i.e.,

$$\boxed{\begin{aligned} \lambda \cdot \chi &= \chi \cdot \lambda \\ \bar\lambda \cdot \bar\chi &= \bar\chi \cdot \bar\lambda \end{aligned}} \qquad (2.60)$$

We have emphasized these identities by placing them in a box because they will be used repeatedly. They probably look wrong to you because the components of spinors anticommute, so shouldn't we add a minus sign when we switch their order? The answer is no because the dot products used here involve the matrix σ^2, which is antisymmetric. This introduces an extra minus sign that cancels the minus sign arising from the exchange of the spinor components. The following exercise will make this more explicit.

EXERCISE 2.6
Prove Eq. (2.60). *Hint*: Use the definitions (2.50) and the identity (2.48).

The generalization of Eq. (2.59) to the products between two different spinors is easily found to be

$$
\begin{aligned}
(\lambda \cdot \chi)^\dagger &= \bar\lambda \cdot \bar\chi \\
(\bar\lambda \cdot \bar\chi)^\dagger &= \lambda \cdot \chi
\end{aligned}
\tag{2.61}
$$

Consider now three left-chiral spinors χ, α, and β. The following identities are easy to prove:

$$
\begin{aligned}
\chi \cdot \alpha \, \chi \cdot \beta &= -\frac{1}{2} \chi \cdot \chi \, \alpha \cdot \beta \\
\bar\chi \cdot \bar\alpha \, \bar\chi \cdot \bar\beta &= -\frac{1}{2} \bar\chi \cdot \bar\chi \, \bar\alpha \cdot \bar\beta
\end{aligned}
\tag{2.62}
$$

We see that the net effect of these identities is to swap the spinors appearing in the dot products. Such identities are generically referred to as *Fierz identities*. These particular examples of Fierz identities will prove to be very useful later in this book.

EXERCISE 2.7
Prove Eq. (2.62). *Hint*: You may simply expand out completely both sides in terms of the components of the spinors and then move some of them around, or you may use Eqs. (2.54) and (2.55).

In the next chapter we will introduce the shorthand notation of dotted and undotted indices mentioned in the Introduction, but as also mentioned there, we will not make use of it until Chapter 11. So feel free to go directly to Chapter 4 and come back to Chapter 3 later.

2.10 Quiz

1. Write $\bar\chi \cdot \bar\lambda$ explicitly in terms of the components of the spinors.
2. Under what condition does a Weyl spinor have a definite helicity? Under what condition does it have a definite chirality?

3. Do left- and right-chiral spinors transform the same way or differently under rotations? What about their transformations under boosts?

4. Write down the expression for a Dirac mass in terms of Weyl spinors.

5. Consider the two components χ_1 and χ_2 of a Weyl spinor. What is $(\chi_1 \chi_2)^\dagger$ equal to ?

CHAPTER 3

New Notation for the Components of Weyl Spinors

As mentioned at the end of Chapter 2, you may safely skip this chapter on a first reading and come back to it before you tackle Chapter 11.

What we would like to do now is to introduce a notation for the components of the spinors that will make it obvious how to construct Lorentz invariants. Our notation will be inspired from the summation convention of relativity, where upper and lower indices can be contracted to form invariant quantities such as $A^\mu B_\mu$ (or, more generally, to form quantities that have definite transformation properties, such as tensors). We want our notation to be such that we simply contract the indices of the spinors to form invariants, without having to take hermitian conjugates or to insert $i\sigma^2$ matrices. This will force us to introduce more than one type of index, as we will soon find out.

Since in this section we will be showing the indices of the spinors explicitly in most equations, it would be very awkward to include the L or R labels as well. We will therefore drop these indices with the understanding that χ will always represent a left-chiral spinor, whereas η will consistently be used to represent a right-chiral spinor. When other spinors will be needed, their chirality will be defined when they are introduced.

As we have seen previously, we assign lower indices to the components of left-chiral spinors. There is no particular reason for assigning a lower index instead of an upper index and no particular reason to pick left-chiral spinors, but hey, we have to start somewhere! We will use indices taken from the first few letters of the Latin alphabet for the spinor components. So let us make the definition

$$\boxed{\text{Components of } \chi \equiv \chi_a}$$

What we mean by this is that in all the expressions we have written so far, the χ have to be considered as having lower indices, as is shown explicitly in Eqs. (2.51) and (2.52).

Next, we would like to introduce a notation for the components of right-chiral spinors. This is simple if we recall that a Lorentz invariant containing χ and the right-chiral field is $\eta^\dagger \chi$. In order to follow the usual convention of Lorentz invariants being built out of matched upper and lower indices, we *define* this invariant to be

$$\eta^\dagger \chi \equiv \eta^a \chi_a \tag{3.1}$$

You are probably wondering where the hermitian conjugate went! The point is that we *choose* the definition (3.1) to make the result as similar as possible to what we encounter in relativity, but the price to pay is that the quantities η^a are *not* the components of the spinor η themselves but *their hermitian conjugate*. It's important to keep this in mind, so let's emphasize it:

$$\eta^a \equiv \text{hermitian conjugate of the components of } \eta \tag{3.2}$$

The obvious question is then: What notation should we use to represent the components of the spinor η themselves? Well, we would like to keep an upper index, but it has to be different from a Latin letter, and we surely don't want to introduce a Greek letter, which could cause confusion with four-vector and tensor indices. The solution that has been chosen is to use an upper *dotted* Latin letter to represent the components of a right-chiral spinor, e.g., $\eta^{\dot{a}}$. This notation in terms of dotted and undotted indices is often referred to as the *van der Waerden notation.*[*]

Actually, this is not the end of the story. The notation $\eta^{\dot{a}}$ is fine as it is, as long as the indices are shown explicitly. But let's say that we want to write expressions without including explicit indices. It would be very confusing if, after dropping the indices, we were to write both η^a and $\eta^{\dot{a}}$ as η, so we introduce another layer of notation: We will include a bar symbol over spinors with dotted indices. Our final notation for the components of a right-chiral spinor is then

$$\boxed{\text{Components of } \eta \equiv \bar{\eta}^{\dot{a}}} \tag{3.3}$$

When the indices are shown, the bar notation is redundant, but it is conventional to include the bar whether or not the indices are explicitly written.

Again, what we mean by Eq. (3.3) is that in all the equations we wrote before this section, the η has to be thought of as having upper dotted indices. For example, Eq. (3.2) may be written as

$$\boxed{\eta^a \equiv (\bar{\eta}^{\dot{a}})^{\dagger}} \tag{3.4}$$

Of course, this also implies

$$\boxed{\bar{\eta}^{\dot{a}} = (\eta^a)^{\dagger}} \tag{3.5}$$

or, to be more explicit,

$$\boxed{\bar{\eta}^{\dot{1}} = (\eta^1)^{\dagger} \qquad \bar{\eta}^{\dot{2}} = (\eta^2)^{\dagger}}$$

We see that *the effect of taking the hermitian conjugate is to change a dotted index into an undotted index (and to add or remove a bar) without changing the position of the index.* Actually, we have only shown this for upper indices on right-chiral spinors, but as we will soon see, it will work for lower indices and for components of left-chiral spinors as well.

In case you are wondering, the bar in Eq. (3.3) is indeed the same bar that appears in the spinor dot product in Eq. (2.50), and again, this bar has nothing to do with Dirac conjugation. You may not like this notation at first (I sure didn't) because it may seem to create a potential source of confusion with Dirac conjugation, but

---

[*] Bartel Leendert van der Waerden was a Dutch mathematician who made important contributions to several fields of mathematics, most notably to group theory.

there is actually no problem because Dirac conjugation can only be applied to four-component spinors, not to Weyl spinors. In this book we will be working almost exclusively with Weyl spinors, which will always be denoted by lowercase Greek letters. The few four-component spinors we will need will always be denoted by an uppercase Ψ. So whenever a bar appears above a Weyl spinor, you will know that it is in the sense of indicating dotted indices, as in Eq. (3.5).

So far we have only introduced three types of indexed quantities: left-chiral spinors components with lower undotted indices χ_a and right-chiral spinors with upper dotted and upper undotted indices $\bar{\eta}^{\dot{a}}$ and η^a. Of our six Lorentz invariants listed in Eq. (2.42), the only one we can recover with our new notation is $\eta^\dagger \chi$, which corresponds to

$$\eta^\dagger \chi = (\bar{\eta}^{\dot{a}})^\dagger \chi_a = \eta^a \chi_a \tag{3.6}$$

after using Eq. (3.4).

Note that the combination $\bar{\eta}^{\dot{a}} \chi_a$ is *not* a Lorentz invariant [if it were, we would have $\eta^T \chi$ listed in Eq. (2.42)]. This is actually a nice feature that will turn out to be general: in order to obtain a Lorentz invariant, an upper index be matched with a lower index of the *same type*, i.e., both indices must be dotted or both must be undotted.

We clearly need to extend the notation in order to have expressions corresponding to all six invariants of Eq. (2.42). Consider the third invariant in Eq. (2.42):

$$\chi^\dagger \eta = (\chi_a)^\dagger \bar{\eta}^{\dot{a}} \tag{3.7}$$

Remember that what we would like is a notation where an upper index must be matched with a lower index *of the same type* to yield a Lorentz invariant, so it is natural to define (keeping the tradition that spinors with dotted components have a bar)

$$\chi^\dagger \eta \equiv \bar{\chi}_{\dot{a}} \bar{\eta}^{\dot{a}} \tag{3.8}$$

Comparing Eqs. (3.7) and (3.8), we see that we must define

$$\boxed{\bar{\chi}_{\dot{a}} \equiv (\chi_a)^\dagger} \tag{3.9}$$

This is a pleasing result because it has the same basic structure as Eq. (3.5): Taking the hermitian conjugate changes a dotted index, into an undotted index, and vice versa, without changing the position of the index.

To be explicit, Eq. (3.9) corresponds to

$$\bar{\chi}_{\dot{1}} = \chi_{1}^{\dagger} \qquad \bar{\chi}_{\dot{2}} = \chi_{2}^{\dagger} \qquad (3.10)$$

We now have defined left-chiral spinors with both dotted and undotted lower indices as well as right-chiral spinors with the two types of upper indices. What about left-chiral spinors with upper indices or right-chiral spinors with lower indices? To define those, we could continue with the other invariants in Eq. (2.42), but it is more instructive instead to go back to the Lorentz transformations given in Eq. (2.41).

Of course, we want to define quantities with the same type of indices to have the same Lorentz transformations, so we should define, e.g., the components $\bar{\chi}^{\dot{a}}$ to transform the same way as the $\bar{\eta}^{\dot{a}}$. Looking at Eq. (2.41), we see that it is the quantity $i\sigma^2 \chi^{\dagger T}$ that transforms the same way as a right-chiral spinor, i.e., with the matrix A. We therefore define

$$\bar{\chi}^{\dot{a}} \equiv \text{components of } i\sigma^2 \chi^{\dagger T}$$

In order for the indices to match, we must write

$$\bar{\chi}^{\dot{a}} = (i\sigma^2)^{\dot{a}\dot{b}} \chi_{\dot{b}}^{\dagger} \qquad (3.11)$$

or, using Eq. (3.9),

$$\bar{\chi}^{\dot{a}} = (i\sigma^2)^{\dot{a}\dot{b}} \bar{\chi}_{\dot{b}} \qquad (3.12)$$

where the indices of the $i\sigma^2$ were chosen in order to follow the rules that indices must be of the same type to be summed over. Note that in Eq. (3.11) we used a dotted $\dot{b}$ index on the Pauli matrix even if the spinor had an undotted index because of the presence of the hermitian conjugation, which we know changes an undotted index into a dotted one. However, some authors write that equation using $(i\sigma^2)^{\dot{a}b}$, so there is no universally accepted convention in this case.

Don't be alarmed by the freedom in the type of indices assigned to $i\sigma^2$; in contradistinction with spinor indices, the indices on σ^2 are mostly decorative. We will come back to this point shortly, but for now, just know that no matter what indices we are assigning to it, σ^2 always represents the usual Pauli matrix.

To be explicit, Eq. (3.12) gives

$$\bar{\chi}^{\dot{1}} = \bar{\chi}_{\dot{2}} \qquad \bar{\chi}^{\dot{2}} = -\bar{\chi}_{\dot{1}} \qquad (3.13)$$

Using Eqs. (3.10) and (3.13), we may express the components with superscripted dotted indices in terms of the components with subscripted undotted indices:

$$\bar{\chi}^{\dot{1}} = \chi_2^\dagger \qquad \bar{\chi}^{\dot{2}} = -\chi_1^\dagger$$

The result [Eq. (3.12)] shows that *the matrix $i\sigma^2$ may be used to raise dotted indices*. We have only shown this for left-chiral spinors, but we will soon prove that this is also valid for right-chiral spinors.

Now that we have defined the components of left-chiral spinors with upper dotted indices, the obvious next step is to define the components χ^a. From our earlier discussion, it seems natural to use the definition

$$\chi^a = (\bar{\chi}^{\dot{a}})^\dagger$$

Using Eq. (3.12), we find

$$\chi^a = \left[(i\sigma^2)^{\dot{a}\dot{b}} \, \bar{\chi}_{\dot{b}} \right]^\dagger$$
$$= [(i\sigma^2)^{\dot{a}\dot{b}}]^* (\bar{\chi}_{\dot{b}})^\dagger$$

where we used the fact that the $(i\sigma^2)^{\dot{a}\dot{b}}$ are just numbers, so applying a dagger on them amounts to a simple complex conjugation. Now we use Eq. (3.9) to get

$$\chi^a = [(i\sigma^2)^{\dot{a}\dot{b}}]^* (\chi_b^\dagger)^\dagger$$
$$= (i\sigma^2)^{ab} \chi_b \tag{3.14}$$

Since the components of $i\sigma^2$ are real, the complex conjugation did not do anything, except that we changed the dotted indices into undotted indices in order, once again, to ensure proper "index etiquette." As promised earlier, we will come back to the vexing question of the mutating indices of $i\sigma^2$ shortly.

Note that Eq. (3.14) implies that *the matrix $i\sigma^2$ may be used to raise undotted indices too*.

If the derivation of Eq. (3.14) does not satisfy you, we may repeat it in "matrix language" without explicitly showing the indices. We start with

$$\chi = (i\sigma^2 \, \bar{\chi})^{\dagger T}$$

where now we have to specify that the χ on the left has an upper index and the $\bar{\chi}$ has a lower index (we don't have to specify whether the indices are dotted or undotted because the presence or absence of a bar tells us that). We then get

$$\chi = (i\sigma^2 \, \bar{\chi})^{\dagger T}$$
$$= (\bar{\chi}^{\dagger}(i\sigma^2)^{\dagger})^T$$
$$= (\chi^T(-i\sigma^2))^T$$
$$= (-i\sigma^2)^T \chi$$
$$= i\sigma^2 \, \chi$$

It must be understood that the χ on the right has a lower index. This agrees with Eq. (3.14) if we reinstate the indices. The result [Eq. (3.14)] is important enough to deserve its own little box:

$$\boxed{\chi^a = (i\sigma^2)^{ab} \, \chi_b} \tag{3.15}$$

To be explicit,

$$\boxed{\chi^1 = \chi_2 \qquad \chi^2 = -\chi_1} \tag{3.16}$$

Now that we know how to find the components of left-chiral spinors with upper indices in terms of the components with lower indices using Eqs. (3.12) and (3.15), we may invert those two relations to express components with lower indices in terms of components with upper indices. All we need to use is the fact that the inverse of the matrix $i\sigma^2$ is $-i\sigma^2$, which is obvious if we recall that $(\sigma^2)^2 = 1$. Imposing the indices in the equations to match properly fixes uniquely the indices of $-i\sigma^2$, giving

$$\boxed{\bar{\chi}_{\dot{a}} = (-i\sigma^2)_{\dot{a}\dot{b}} \bar{\chi}^{\dot{b}}}$$

and

$$\boxed{\chi_a = (-i\sigma^2)_{ab} \chi^b}$$

Therefore, *the matrix* $-i\sigma^2$ *is used to lower both dotted and undotted indices.* We have only shown this for left-chiral spinors, but we will soon show that it is also valid for right-chiral spinors.

So far we have only defined right-chiral spinors with upper indices. If we impose that the η_a transform like the χ_a and look at Eq. (2.41), we see that the quantity that transforms like χ is $-i\sigma^2\eta^{\dagger T}$. Thus we define

$$\eta_a \equiv (-i\sigma^2)_{ab}(\eta^{\dot{b}})^{\dagger}$$
$$= (-i\sigma^2)_{ab}\eta^b$$

As promised, the matrix $-i\sigma^2$ is seen to also lower undotted indices on right-chiral spinors. Defining $\bar{\eta}_{\dot{a}} \equiv \eta_a^{\dagger}$ and following the same steps as in Eq. (3.14), we also can show that

$$\bar{\eta}_{\dot{a}} = (-i\sigma^2)_{\dot{a}\dot{b}}\bar{\eta}^{\dot{b}} \tag{3.17}$$

which shows that $-i\sigma^2$ may be used to lower dotted indices of right-chiral spinors as well.

Let us pause to comment on the "chameleon indices" of $i\sigma^2$ that change to whatever the equations require. This is something that may be quite confusing at first. Note that when we write spinors with different types of indices, e.g., η^a and $\eta^{\dot{a}}$, we really do mean different quantities. However, it is important to keep in mind that in all our equations, the components of the σ^2 matrix are always the usual components, no matter if the indices are dotted or undotted, no matter if they are upstairs or downstairs. For example, the following components are equal:

$$(i\sigma^2)^{12} = (i\sigma^2)^{\dot{1}\dot{2}} = \begin{pmatrix} 0 & 1 \\ -1 & 0 \end{pmatrix}_{12} = 1$$

whereas

$$(-i\sigma^2)_{12} = (-i\sigma^2)_{\dot{1}\dot{2}} = \begin{pmatrix} 0 & -1 \\ 1 & 0 \end{pmatrix}_{12} = -1$$

It may be confusing to have the matrices with upper and lower indices differing only by a minus sign. If we see the components $(i\sigma^2)^{ab}$ in an equation, what prevents us from rewriting this as $-(-i\sigma^2)_{ab}$? The key point is that we will never deal with σ^2 matrices in isolation; they will always be part of equations containing spinors and other quantities with well-defined Lorentz transformations, and consistency of the equations under those transformations will always clearly dictate what the correct indices are.

Let us summarize what we have accomplished so far. We have defined four types of components for both left- and right-chiral spinors. These are all related to one another following three simple rules:

1. Taking a hermitian conjugate switches dotted and undotted indices without changing their position (and of course adds or removes a bar since a bar always accompanies dotted indices).
2. The matrix $i\sigma^2$ may be used to raise indices (both dotted and undotted).
3. The matrix $-i\sigma^2$ may be used to lower indices (both dotted and undotted).

One last comment before moving on to the topic of Lorentz invariants: It is important to keep in mind that all four sets of components always may be written in terms of a single set and its hermitian conjugate. For example, all types of components of a left-chiral spinor may be written in terms of χ_1 and χ_2 and their hermitian conjugates $\chi_1^\dagger$ and $\chi_2^\dagger$, as we have seen explicitly in this section. So despite the fact that we have defined eight different components associated to a single left-chiral spinor, only four of them are independent.

3.1 Building Lorentz Invariants

Now let us use our new notation to build Lorentz invariants. As hinted before, the rule to obtain invariants turns out to simply be that an upper index must be matched with a lower index of the same type. We will show that this is the correct rule by recovering all the invariants of Eq. (2.42) this way and nothing else.

Following this rule, there are only eight combinations made out of one left-chiral and one right-chiral spinor that can be written:

$$\eta^a \chi_a \qquad \chi^a \eta_a$$
$$\bar{\eta}_{\dot{a}} \bar{\chi}^{\dot{a}} \qquad \bar{\chi}_{\dot{a}} \bar{\eta}^{\dot{a}}$$
$$\chi^a \chi_a \qquad \bar{\chi}_{\dot{a}} \bar{\chi}^{\dot{a}}$$
$$\eta^a \eta_a \qquad \bar{\eta}_{\dot{a}} \bar{\eta}^{\dot{a}} \qquad (3.18)$$

Of course, reversing the order of any of the spinors does not give a new invariant. For example, $\chi_a \eta^a$ is simply equal to $-\eta^a \chi_a$. In our list (3.18) we chose to list the quantities with undotted indices going diagonally down and dotted indices going diagonally up for a reason that will become clear later on. This does not mean that the quantities with the opposite order are not invariants, just that they are not independent.

Still, we seem to get eight invariants, whereas we had found only six in Chapter 2, so our eight expressions must not all be independent. This becomes evident if we recall that we may always write components with upper indices in terms of components with lower indices (without taking a hermitian conjugate) so that, for example, $\eta^a \chi_a$ and $\chi^a \eta_a$ must somehow be related. Indeed, we have

$$
\begin{aligned}
\eta^a \chi_a &= \left((i\sigma^2)^{ab} \eta_b \right) \chi_a \\
&= -(i\sigma^2)^{ab} \chi_a \eta_b \\
&= (i\sigma^2)^{ba} \chi_a \eta_b \\
&= \chi^b \eta_b
\end{aligned}
\tag{3.19}
$$

where the minus sign in the second line came from moving the η_b to the right of χ_a. In the third line, we have used the antisymmetry of $i\sigma^2$ to write $(i\sigma^2)^{ab} = -(i\sigma^2)^{ba}$, and in the last line, we have used the $i\sigma^2$ to raise the index of χ. To summarize, we have shown that

$$
\boxed{\eta^a \chi_a = \chi^a \eta_a}
\tag{3.20}
$$

EXERCISE 3.1

If the proof in Eq. (3.19) does not convince you, express explicitly $\eta^a \chi_a$ in terms of the components with lower indices only (χ_1, χ_2, η_1, and η_2). Do the same with $\chi^a \eta_a$, and show that the two expressions are equal.

Likewise, one can show that

$$
\boxed{\bar{\eta}_{\dot{a}} \bar{\chi}^{\dot{a}} = \bar{\chi}_{\dot{a}} \bar{\eta}^{\dot{a}}}
\tag{3.21}
$$

Therefore, we only have six independent invariants in the list in Eq. (3.18), which we will take to be

$$
\begin{array}{cc}
\eta^a \chi_a & \bar{\chi}_{\dot{a}} \bar{\eta}^{\dot{a}} \\[2mm]
\chi^a \chi_a & \bar{\chi}_{\dot{a}} \bar{\chi}^{\dot{a}} \\[2mm]
\eta^a \eta_a & \bar{\eta}_{\dot{a}} \bar{\eta}^{\dot{a}}
\end{array}
\tag{3.22}
$$

It is easy to show that these six invariants correspond indeed to the invariants of Eq. (2.42). The only tricky point is that we must remember that in all the formulas of Chapter 2 the left-chiral spinor χ is always written in terms of its components with lower undotted indices χ_a, whereas the right-chiral spinor η is always written in terms of its components with upper dotted indices $\eta^{\dot{a}}$. The claim then is that if we express all the spinors in Eq. (3.22) in terms of these components, we will recover the six expressions of Eq. (2.42).

For example, consider

$$\chi^a \chi_a \tag{3.23}$$

We need to express χ^a in terms of lower indices in order to compare it with Eq. (2.42). We know that the matrix $i\sigma^2$ raises indices, so we write

$$\chi^a \chi_a = \left[(i\sigma^2)^{ab} \chi_b) \right] \chi_a$$
$$= \chi_b \, (i\sigma^2)^{ab} \chi_a \tag{3.24}$$

where we have used the freedom to move the $(i\sigma^2)^{ab}$ around because they are simply numbers. Now, in order to compare with Eq. (2.42), we need to write Eq. (3.24) in matrix form, with the σ^2 matrix sandwiched between the two spinors. To do this, we have to switch the order of the indices of $i\sigma^2$, which brings in a minus sign because of the antisymmetry of $i\sigma^2$. Thus we get

$$\chi^a \chi_a = \chi_b \, (-i\sigma^2)^{ba} \chi_a$$
$$= \chi^T \, (-i\sigma^2) \, \chi$$

where in the last line it is understood that the indices on the spinors are all lower undotted indices. Sure enough, this is the first Lorentz invariant in Eq. (2.42). The other invariants can be obtained in a similar manner.

EXERCISE 3.2
Show that $\bar{\eta}_{\dot{a}} \, \bar{\eta}^{\dot{a}}$ corresponds to one of the invariants of Eq. (2.42).

EXERCISE 3.3
At some point we will need the relations equivalent to Eqs. (2.54) and (2.55) but written in terms of upper indices. Prove the two identities

$$\chi^a \chi^b = -\frac{1}{2} \, (i\sigma^2)^{ab} \chi \cdot \chi$$
$$\bar{\chi}^{\dot{a}} \bar{\chi}^{\dot{b}} = \frac{1}{2} \, (i\sigma^2)^{\dot{a}\dot{b}} \bar{\chi} \cdot \bar{\chi} \tag{3.25}$$

3.2 Index-Free Notation

In order to alleviate the notation, it is useful to be able to write expressions without explicit indices by writing, e.g., $\eta \, \chi$ instead of $\eta^a \chi_a$ or $\bar{\chi} \, \bar{\eta}$ instead of $\bar{\chi}_{\dot{a}} \, \bar{\eta}^{\dot{a}}$. This is what most SUSY references do, but in this book we will instead use the "spinor dot product notation" introduced in Chapter 2; i.e., we will write $\eta \cdot \chi$ and $\bar{\chi} \cdot \bar{\eta}$. This makes complex expressions easier to read by showing clearly which spinors are contracted together (e.g., it's easier to see which spinors are contracted together in $\chi \cdot \eta \lambda \cdot \chi$ than in $\chi \eta \lambda \chi$).

However, not showing explicit indices obviously leads to an ambiguity. We can distinguish spinors with dotted indices from those with undotted indices because the former have a bar over them, but once we drop indices, we can't tell if they are supposed to be in the upper or in the lower position. Consider, for example, $\eta \cdot \chi$. Does this represent $\eta^a \chi_a$ or $\eta_a \chi^a$? These two are *not* equivalent! They differ by a minus sign, as is easily shown:

$$\eta_a \chi^a = -\chi^a \eta_a$$
$$= -\eta^a \chi_a$$

where in the first step we have used the Grassmann nature of spin components and in the second step we made use of Eq. (3.20).

The same ambiguity arises if we drop the indices of any of the six invariants of Eq. (3.22). We therefore see that if we want to be able to not write explicit indices, we need a new rule to tell us when indices are in the upper position and when they are in the lower position.

The convention that has been chosen is that *a dot product between unbarred spinors corresponds to the undotted indices written diagonally downward*; i.e., $\eta \cdot \chi$ *always* stands for $\eta^a \chi_a$. This convention deserves to be emphasized:

The dot product between two unbarred spinors corresponds to the product of the components with the undotted indices going *diagonally downward*

What do we do if we want to write $\eta_a \chi^a$ as a dot product? We simply need to change the order first:

$$\eta_a \chi^a = -\chi^a \eta_a = -\chi \cdot \eta$$

As for barred spinors, the convention is the reverse: *A dot product between barred spinors corresponds to the dotted indices written diagonally upward*, i.e.,

$\bar{\eta} \cdot \bar{\chi}$ *always* stands for $\bar{\eta}_{\dot{a}} \bar{\chi}^{\dot{a}}$. Again, let us emphasize this important convention:

> The dot product between two barred spinors corresponds to the product of the components with the dotted indices going *diagonally upward*

It may seem confusing (and annoying) at first that the convention for the barred and unbarred spinors are reversed, but one way to see that it is natural is to remember that we have defined the original components of the left-chiral spinor χ as being χ_a and the original components of the right-chiral η as being $\bar{\eta}^{\dot{a}}$. Therefore, in both conventions stated above, the original component of the spinors appears on the right of the spinor dot product. In that sense, the two conventions are consistent with each other.

It is easy to check that the dot products we just introduced in terms of components with indices are in fact exactly equal to the dot products introduced in Eq. (2.50). For example, consider $\eta \cdot \chi$. Using the notation of upper and lower indices, we have

$$\eta \cdot \chi = \eta^a \chi_a = \eta^1 \chi_1 + \eta^2 \chi_2$$

On the other hand, in the notation of Chapter 2,

$$\eta \cdot \chi = \eta^T (-i\sigma^2) \chi = \eta_2 \chi_1 - \eta_1 \chi_2$$

Using Eq. (3.16), we see that these two expressions are indeed equal.

Now recall the identities (3.21) and (3.20). Translated into dot-product notation, these become

$$\eta \cdot \chi = \chi \cdot \eta$$
$$\bar{\eta} \cdot \bar{\chi} = \bar{\chi} \cdot \bar{\eta} \qquad (3.26)$$

which we already mentioned in Eq. (2.60) and which you proved in Exercise 2.6.

The fact that the order does not matter in a dot product is familiar from relativity or even from three-dimensional vector algebra, where $\vec{A} \cdot \vec{B} = \vec{B} \cdot \vec{A}$. But keep in mind that here the identities (3.26) are more subtle. There is one minus sign introduced by the exchange of the two spinor components, which are Grassmann quantities, and a second minus sign owing to the switch from upper to lower indices on both spinors (or, if you prefer, owing to the antisymmetry of the matrix $i\sigma^2$). The end result is that we can switch the order of the spinors in those dot products without introducing a minus sign, a result that will come in handy many times.

Using dot products, the six Lorentz invariants built out of a left-chiral and a right-chiral spinor of Eq. (3.22) therefore are written as

$$\eta^a \chi_a = \eta \cdot \chi$$

$$\bar{\chi}_{\dot{a}} \bar{\eta}^{\dot{a}} = \bar{\chi} \cdot \bar{\eta}$$

$$\chi^a \chi_a = \chi \cdot \chi$$

$$\bar{\chi}_{\dot{a}} \bar{\chi}^{\dot{a}} = \bar{\chi} \cdot \bar{\chi}$$

$$\eta^a \eta_a = \eta \cdot \eta$$

$$\bar{\eta}_{\dot{a}} \bar{\eta}^{\dot{a}} = \bar{\eta} \cdot \bar{\eta}$$

We now understand the choice of the order of indices in Eq. (3.22). They all have been written with undotted indices going diagonally downward and dotted indices going diagonally upward so that when the time would come to write them as dot products, we would not have to switch the order of anything and introduce minus signs.

It's important to repeat that we *could* write absolutely *everything* in this entire book in terms of the components with lower undotted indices and their hermitian conjugates only. Most equations would be more cumbersome because there usually would be a bunch of σ^2 matrices around and hermitian conjugate symbols all over the place, but this is not the main motivation for the introduction of the notation. The key point in using the van der Waerden notation is that it makes it easy to see how expressions transform under Lorentz transformations, which is of major importance in building lagrangians.

EXERCISE 3.4
Write down $\chi \cdot \chi \chi \cdot \eta$ in terms of components with lower undotted indices.

3.3 Invariants Built Out of Two Left-Chiral Spinors

In SUSY, it is conventional to work exclusively in terms of left-chiral spinors, so let's restrict our discussion to this type of spinor from now on. If we have only one such spinor, say, χ, we may write only two invariants: $\chi \cdot \chi$ and $\bar{\chi} \cdot \bar{\chi}$.

If we add a second left-chiral spinor, let's call it λ, then we now can write six Lorentz invariants:

$$\lambda^a \chi_a = \lambda \cdot \chi$$
$$\bar{\lambda}_{\dot{a}} \bar{\chi}^{\dot{a}} = \bar{\lambda} \cdot \bar{\chi}$$
$$\chi^a \chi_a = \chi \cdot \chi$$
$$\bar{\chi}_{\dot{a}} \bar{\chi}^{\dot{a}} = \bar{\chi} \cdot \bar{\chi}$$
$$\lambda^a \lambda_a = \lambda \cdot \lambda$$
$$\bar{\lambda}_{\dot{a}} \bar{\lambda}^{\dot{a}} = \bar{\lambda} \cdot \bar{\lambda} \tag{3.27}$$

Again, the identities (3.26) apply:

$$\lambda \cdot \chi = \chi \cdot \lambda$$
$$\bar{\lambda} \cdot \bar{\chi} = \bar{\chi} \cdot \bar{\lambda}$$

3.4 The ϵ Notation for $\pm i\sigma^2$

Writing the matrices $\pm i\sigma^2$ all the time makes the calculations very awkward, especially when indices are explicitly shown. To avoid this, we will therefore introduce one last layer of notation that is fairly common in SUSY and that consists of using the symbol ϵ to denote the matrices $\pm i\sigma^2$, following

$$(i\sigma^2)^{ab} \equiv \epsilon^{ab}$$
$$(i\sigma^2)^{\dot{a}\dot{b}} \equiv \epsilon^{\dot{a}\dot{b}}$$
$$(-i\sigma^2)_{ab} \equiv \epsilon_{ab}$$
$$(-i\sigma^2)_{\dot{a}\dot{b}} \equiv \epsilon_{\dot{a}\dot{b}}$$

We see that $\epsilon^{ab} = \epsilon^{\dot{a}\dot{b}}$ and $\epsilon_{ab} = \epsilon_{\dot{a}\dot{b}}$. It greatly simplifies the notation to use these definitions, but one must be careful about issues of signs. To be explicit, the notation implies

$$\epsilon^{12} = \epsilon^{\dot{1}\dot{2}} = 1$$
$$\epsilon^{21} = \epsilon^{\dot{2}\dot{1}} = -1$$

whereas the expressions with lower indices have the opposite signs:

$$\epsilon_{12} = \epsilon_{\dot{1}\dot{2}} = -1$$

$$\epsilon_{21} = \epsilon_{\dot{2}\dot{1}} = 1$$

Although the ϵ with upper indices can be seen as two-dimensional versions of the familiar Levi-Civita symbol ϵ^{ijk}, it is important to keep in mind that ϵ with lower indices has the opposite sign.

Now it is a simple matter to raise or lower indices on spinors using this new notation. We have, for example,

$$\chi^a = \epsilon^{ab} \chi_b$$

$$\bar{\chi}_{\dot{a}} = \epsilon_{\dot{a}\dot{b}} \bar{\chi}^{\dot{b}} \tag{3.28}$$

and, of course, similar expressions for the right-chiral spinor.

We see that the ϵ matrices act as a metric on the spinor indices. However, it is important to keep in mind that in contrast with the metrics of special or general relativity, the ϵ are antisymmetric under the exchange of their indices:

$$\epsilon^{ab} = -\epsilon^{ba}$$

$$\epsilon^{\dot{a}\dot{b}} = -\epsilon^{\dot{b}\dot{a}}$$

$$\epsilon_{ab} = -\epsilon_{ba}$$

$$\epsilon_{\dot{a}\dot{b}} = -\epsilon_{\dot{b}\dot{a}}$$

In particular, *note that the order of the indices in Eq. (3.28) must be respected*: The *second* index of the ϵ must match the index of the spinor with which it is contracted; otherwise, a minus sign must be present. For example,

$$\epsilon_{ab}\psi^a = -\epsilon_{ba}\psi^a$$

$$= -\psi_b$$

In our new notation, Eqs. (2.54), (2.55), and (3.25) take the simpler form

$$\chi_a\,\chi_b = \frac{1}{2}\,\epsilon_{ab}\chi\cdot\chi$$

$$\bar{\chi}_{\dot{a}}\,\bar{\chi}_{\dot{b}} = -\frac{1}{2}\,\epsilon_{\dot{a}\dot{b}}\bar{\chi}\cdot\bar{\chi}$$

$$\chi^a\,\chi^b = -\frac{1}{2}\,\epsilon^{ab}\chi\cdot\chi \tag{3.29}$$

$$\bar{\chi}^{\dot{a}}\,\bar{\chi}^{\dot{b}} = \frac{1}{2}\,\epsilon^{\dot{a}\dot{b}}\bar{\chi}\cdot\bar{\chi}$$

Obviously, the order in the indices of ϵ in Eq. (3.29) also must be respected.

Because of this need to pay very close attention to the order of the indices, the use of the ϵ is quite a bit more tricky than the use of the metric tensor $g_{\mu\nu}$ in general relativity. Still, it is worthwhile to use the ϵ notation for the computational simplicity it offers.

Since $-i\sigma^2$ is the inverse of $i\sigma^2$, we also have that

$$\epsilon^{ab}\,\epsilon_{bc} = \epsilon_{cb}\,\epsilon^{ba} = \delta_c^a$$

$$\epsilon^{\dot{a}\dot{b}}\,\epsilon_{\dot{b}\dot{c}} = \epsilon_{\dot{c}\dot{b}}\,\epsilon^{\dot{b}\dot{a}} = \delta_{\dot{c}}^{\dot{a}} \tag{3.30}$$

where the two δ are the usual Kronecker deltas. We may use the antisymmetry of ϵ to generate several new identities, e.g.,

$$\epsilon^{ab}\,\epsilon_{bc} = -\epsilon^{ba}\,\epsilon_{bc} = -\epsilon^{ab}\,\epsilon_{cb} = \epsilon^{ba}\,\epsilon_{cb} = \delta_c^a$$

Note again that the order of the indices is crucial: We get a positive Kronecker delta when the index that is common to the two symbols appears in *different positions* in the two ϵ (first and second or second and first positions).

Let us repeat the proof that $\eta \cdot \chi = \chi \cdot \eta$ [see Eq. (3.19)] to illustrate the power of our new notation:

$$
\begin{aligned}
\eta \cdot \chi &= \eta^a \chi_a \\
&= \epsilon^{ab} \eta_b \chi_a \\
&= -\epsilon^{ab} \chi_a \eta_b \\
&= \epsilon^{ba} \chi_a \eta_b \\
&= \chi^b \eta_b \\
&= \chi \cdot \eta
\end{aligned}
$$

where the first minus sign came from the exchange of the two spinor components, and the second minus sign arose from the exchange of the ϵ indices.

3.5 Notation for the Indices of σ^μ and $\bar{\sigma}^\mu$

Now let's find out what kind of indices we should assign to those two matrices. Recall the results in Eqs. (2.43) and (2.44), which we repeat here for convenience:

$$\sigma^\mu i\partial_\mu \eta \text{ transforms like a left-chiral spinor} \tag{3.31}$$

whereas

$$\bar{\sigma}^\mu i\partial_\mu \chi \text{ transforms like a right-chiral spinor} \tag{3.32}$$

Let's start with Eq. (3.32). Since we assigned a lower undotted index to χ and we assigned an upper dotted index to quantities transforming like a right-chiral spinor, we see that the matrix $\bar{\sigma}_\mu$ converts a lower undotted index into an upper dotted index, so it must carry two mixed upper indices as follows:

$$\boxed{\text{Indices of } \bar{\sigma}^\mu = (\bar{\sigma}^\mu)^{\dot{a}b}} \tag{3.33}$$

Note that the order is important; the dotted index comes first. As for σ^μ, Eq. (3.31) tells us that it converts an upper dotted index into a lower undotted index, so

$$\boxed{\text{Indices of } \sigma^\mu = (\sigma^\mu)_{a\dot{b}}} \tag{3.34}$$

We have seen that the matrices ϵ may be used to raise or lower indices on the spinors. It is natural to wonder how they may be used to move the indices of σ^μ and $\bar{\sigma}^\mu$. The key relations are provided in Eq. (2.23). Consider the first of these identities, which we'll repeat here:

$$\sigma^2 \sigma^\mu \sigma^2 = \bar{\sigma}^{\mu T} \tag{3.35}$$

According to Eqs. (3.33) and (3.34), σ^μ has two lower indices, whereas $\bar{\sigma}^\mu$ has two upper indices. We therefore need matrices to raise the two indices of σ^μ; i.e., we need two matrices $i\sigma^2$. We simply multiply both sides of Eq. (3.35) by i^2 to get

$$i\sigma^2 \sigma^\mu i\sigma^2 = -\bar{\sigma}^{\mu T} \tag{3.36}$$

Let's take advantage of the ϵ notation. Assigning lower indices ab to σ^μ, as in Eq. (3.34), we must write the left-hand side of Eq. (3.36) as

$$i\sigma^2 \sigma^\mu i\sigma^2 = \epsilon^{ca} (\sigma^\mu)_{ab} \epsilon^{\dot{b}\dot{d}} \tag{3.37}$$

Note that we do not have any choice on the positions of the indices; since Eq. (3.36) involves the product of three matrices, we *must* have the second index of the first ϵ be the same as the first index of the matrix σ^μ and the second index of the σ^μ be equal to the first index of the second ϵ. And since Eq. (3.37) carries upper indices $c\dot{d}$ (in that order), we must assign the same indices to the right-hand side of Eq. (3.36). We then obtain that, with all indices explicitly shown, Eq. (3.36) corresponds to

$$\epsilon^{ca} (\sigma^\mu)_{ab} \epsilon^{\dot{b}\dot{d}} = -(\bar{\sigma}^{\mu T})^{c\dot{d}}$$

Since a transpose switches the order of indices, we finally get

$$\epsilon^{ca} (\sigma^\mu)_{ab} \epsilon^{\dot{b}\dot{d}} = -(\bar{\sigma}^\mu)^{\dot{d}c}$$

Pay close attention to the order of the indices in this expression. This can be written in a more elegant form by using $\epsilon^{\dot{b}\dot{d}} = -\epsilon^{\dot{d}\dot{b}}$ to give us

$$\boxed{\epsilon^{ca} \epsilon^{\dot{d}\dot{b}} (\sigma^\mu)_{ab} = (\bar{\sigma}^\mu)^{\dot{d}c}} \tag{3.38}$$

This is what we were looking for: an expression showing us how the ϵ matrices can be used to raise the two indices of σ^μ.

Note two important points about this identity. First, the indices of σ^μ that are raised appear in the *second* position of both ϵ. Second, note how the order of the dotted and undotted indices changes in going from σ^μ to $\bar{\sigma}^\mu$, obeying the definitions in Eqs. (3.33) and (3.34).

The identity (3.38) will be very useful shortly. Of course, we can invert it [or, if you prefer, start from the second relation of Eq. (2.23)] to show that

$$\epsilon_{cb}\,\epsilon_{\dot{d}\dot{a}}\,(\bar{\sigma}^\mu)^{\dot{a}b} = (\sigma^\mu)_{c\dot{d}} \qquad (3.39)$$

Let us turn our attention to the ever-important issue of building quantities with definite Lorentz transformation properties (invariants or tensors). The index assignations (3.33) and (3.34) make it easy to write down such quantities. For example, what contravariant four-vectors can we write down if we have two left-chiral Weyl spinors χ and λ at our disposal? Simply by matching indices, we get the following combinations:

$$\bar{\chi}_{\dot{a}}\,(\bar{\sigma}^\mu)^{\dot{a}b}\,\lambda_b \qquad \bar{\lambda}_{\dot{a}}\,(\bar{\sigma}^\mu)^{\dot{a}b}\,\chi_b \qquad \chi^a\,(\sigma^\mu)_{a\dot{b}}\,\bar{\lambda}^{\dot{b}} \qquad \lambda^a\,(\sigma^\mu)_{a\dot{b}}\,\bar{\chi}^{\dot{b}} \qquad (3.40)$$

However, only two of those four expressions are independent, as we will demonstrate below. Before doing that, let's introduce an index-free notation for these expressions. Dropping the indices, we get, for example,

$$\bar{\chi}_{\dot{a}}\,(\bar{\sigma}^\mu)^{\dot{a}b}\,\lambda_b = \bar{\chi}\,\bar{\sigma}^\mu\,\lambda \qquad (3.41)$$

As always when we drop indices, some ambiguity might arise. If we were to see the right-hand side appearing in some formula, how would we know if we are to assign upper or lower dotted indices to $\bar{\chi}$? The answer is that *we have to remember that $\bar{\sigma}^\mu$ carries two upper indices*. If we keep this in mind, then it is clear that the right-hand side of Eq. (3.41) corresponds to the expression on the left-hand side. Similarly, we must remember that σ^μ *carries two lower indices*. If we remember this, no ambiguity is possible.

We now can rewrite the four terms of Eq. (3.40) unambiguously as

$$\bar{\chi}\,\bar{\sigma}^\mu\,\lambda \qquad \bar{\lambda}\,\bar{\sigma}^\mu\,\chi \qquad \chi\,\sigma^\mu\,\bar{\lambda} \qquad \lambda\,\sigma^\mu\,\bar{\chi} \qquad (3.42)$$

Now let's show that these expressions are not all independent. Let's start with the first one and use ϵ matrices to express the spinors in terms of components with upper indices:

$$\bar{\chi}\,\bar{\sigma}^\mu\,\lambda = \bar{\chi}_{\dot{a}}\,(\bar{\sigma}^\mu)^{\dot{a}b}\,\lambda_b$$
$$= \epsilon_{\dot{a}\dot{c}}\,\bar{\chi}^{\dot{c}}\,(\bar{\sigma}^\mu)^{\dot{a}b}\,\epsilon_{bd}\,\lambda^d$$

Now we contract the two ϵ with the $\bar{\sigma}^\mu$ using Eq. (3.39) and see where it leads us:

$$\epsilon_{\dot{a}\dot{c}}\,\bar{\chi}^{\dot{c}}\,(\bar{\sigma}^\mu)^{\dot{a}b}\,\epsilon_{bd}\,\lambda^d = \epsilon_{\dot{c}\dot{a}}\,\epsilon_{db}\,(\bar{\sigma}^\mu)^{\dot{a}b}\,\bar{\chi}^{\dot{c}}\,\lambda^d$$
$$= (\sigma^\mu)_{d\dot{c}}\,\bar{\chi}^{\dot{c}}\,\lambda^d \qquad \text{[using Eq. (3.39)]}$$
$$= -\lambda^d\,(\sigma^\mu)_{d\dot{c}}\,\bar{\chi}^{\dot{c}}$$
$$= -\lambda\,\sigma^\mu\,\bar{\chi}$$

where in the first step we have switched the indices of both ϵ, which produces a factor of $(-1)^2 = 1$.

To summarize, we thus have shown that

$$\boxed{\bar{\chi}\,\bar{\sigma}^\mu\,\lambda = -\lambda\,\sigma^\mu\,\bar{\chi}} \qquad (3.43)$$

By simply renaming $\lambda \leftrightarrow \chi$, this, of course, also implies

$$\boxed{\chi\,\sigma^\mu\,\bar{\lambda} = -\bar{\lambda}\,\bar{\sigma}^\mu\,\chi} \qquad (3.44)$$

This confirms the claim made earlier that only two of the four expressions in Eq. (3.42) are independent.

Let's now look at what happens if we take the hermitian conjugate of one of these expressions, let's say $\bar{\chi}\,\bar{\sigma}^\mu\,\lambda$. Keeping in mind the rule that when we take the hermitian conjugate we switch the order of the components of the spinors without introducing a minus sign, we get

$$(\bar{\chi}\,\bar{\sigma}^\mu\,\lambda)^\dagger = \lambda^\dagger\,(\bar{\sigma}^\mu)^\dagger\,(\bar{\chi})^\dagger$$
$$= \lambda^\dagger\,\bar{\sigma}^\mu\,(\bar{\chi})^\dagger \qquad (3.45)$$

using the fact that Pauli matrices are hermitian. But since taking the hermitian conjugate on the components of a spinor simply changed undotted indices into dotted indices, the right-hand side of Eq. (3.45) is simply $\bar{\lambda}\bar{\sigma}^\mu\chi$. We therefore have established that

$$\boxed{(\bar{\chi}\,\bar{\sigma}^\mu\lambda)^\dagger = \bar{\lambda}\,\bar{\sigma}^\mu\chi} \qquad (3.46)$$

Similarly, it is easy to show that

$$\boxed{(\chi\,\sigma^\mu\bar{\lambda})^\dagger = \lambda\,\sigma^\mu\bar{\chi}} \qquad (3.47)$$

We'll close this chapter with the derivation of (yet another!) identity (by the way, all these identities we are proving *will* be useful at some point, I promise!). This one starts with

$$(\lambda \, \sigma^\mu \bar{\chi})(\lambda \, \sigma^\nu \bar{\chi}) \tag{3.48}$$

The goal is to rewrite this as something proportional to $(\lambda \cdot \lambda)(\bar{\chi} \cdot \bar{\chi})$. At first sight, this might seem impossible. How can we get the spinors of the same type dotted together when they appear in different matrix products in Eq. (3.48)? What could bring about this kind of magic? The answer lies in the identities in Eq. (3.29).

To see how this works, we first write Eq. (3.48) in terms of components and move the components of the same type next to one another (recall that we pick up a minus sign when we move a spinor component through another one and that the components of the σ^μ matrices are plain numbers):

$$(\lambda \, \sigma^\mu \bar{\chi})(\lambda \, \sigma^\nu \bar{\chi}) = \lambda^a \, (\sigma^\mu)_{a\dot{b}} \, \bar{\chi}^{\dot{b}} \quad \lambda^c \, (\sigma^\nu)_{c\dot{d}} \, \bar{\chi}^{\dot{d}}$$

$$= -(\sigma^\mu)_{a\dot{b}} \, (\sigma^\nu)_{c\dot{d}} \lambda^a \, \lambda^c \, \bar{\chi}^{\dot{b}} \, \bar{\chi}^{\dot{d}}$$

Now, using Eq. (3.29), we may write this as

$$-(\sigma^\mu)_{a\dot{b}}(\sigma^\nu)_{c\dot{d}} \lambda^a \, \lambda^c \, \bar{\chi}^{\dot{b}} \, \bar{\chi}^{\dot{d}} = \frac{1}{4} \, (\sigma^\mu)_{a\dot{b}}(\sigma^\nu)_{c\dot{d}} \, \epsilon^{ac} \, \epsilon^{\dot{b}\dot{d}} \lambda \cdot \lambda \bar{\chi} \cdot \bar{\chi} \tag{3.49}$$

Switching the order of the indices in the two ϵ matrices and using Eq. (3.38), we get

$$\frac{1}{4} \, (\sigma^\mu)_{a\dot{b}}(\sigma^\nu)_{c\dot{d}} \, \epsilon^{ac} \, \epsilon^{\dot{b}\dot{d}} \lambda \cdot \lambda \bar{\chi} \cdot \bar{\chi} = \frac{1}{4} \, \epsilon^{ca} \, \epsilon^{\dot{d}\dot{b}} \, (\sigma^\mu)_{a\dot{b}}(\sigma^\nu)_{c\dot{d}} \lambda \cdot \lambda \bar{\chi} \cdot \bar{\chi}$$

$$= \frac{1}{4} \, (\bar{\sigma}^\mu)^{\dot{d}c} \, (\sigma^\nu)_{c\dot{d}} \lambda \cdot \lambda \bar{\chi} \cdot \bar{\chi}$$

By definition, we have

$$(\bar{\sigma}^\mu)^{\dot{d}c} \, (\sigma^\nu)_{c\dot{d}} = \text{Tr}(\bar{\sigma}^\mu \, \sigma^\nu)$$

This trace turns out to be equal to $2\eta^{\mu\nu}$ (see Exercise 3.5). Our final result therefore is

$$\boxed{(\lambda \, \sigma^\mu \bar{\chi})(\lambda \, \sigma^\nu \bar{\chi}) = \frac{1}{2} \, \eta^{\mu\nu} \, \lambda \cdot \lambda \bar{\chi} \cdot \bar{\chi}} \tag{3.50}$$

an identity which will be used several times.

It's clear that the same steps also lead to

$$(\bar{\chi}\,\bar{\sigma}^{\mu}\lambda)(\bar{\chi}\,\bar{\sigma}^{\nu}\lambda) = \frac{1}{2}\,\eta^{\mu\nu}\,\lambda \cdot \lambda\,\bar{\chi} \cdot \bar{\chi}$$

(3.51)

EXERCISE 3.5
Show that $\mathrm{Tr}(\bar{\sigma}^{\mu}\,\sigma^{\nu}) = 2\eta^{\mu\nu}$. *Hint*: Use Eq. (2.24).

3.6 Quiz

1. Write the invariant $\bar{\eta}^{\dot{a}}\,\bar{\chi}_{\dot{a}}$ in an index-free notation (two answers are acceptable).
2. What are $\epsilon_{\dot{1}\dot{1}}$ and $\epsilon_{\dot{1}\dot{2}}$ equal to?
3. What can the expression $\epsilon^{ac}\,\epsilon^{\dot{d}b}\,(\sigma^{\mu})_{a\dot{b}}$ be simplified to?
4. What is $\epsilon_{\dot{a}\dot{b}}\,\epsilon^{\dot{a}\dot{c}}$ equal to?
5. What is the relationship between components with dotted indices and components with undotted indices in the same position?
6. What matrix may be used to lower both dotted and undotted indices?
7. If we contract all the indices in $\epsilon_{ab}\epsilon^{cb}\chi^{a}$, what do we get?

CHAPTER 4

The Physics of Weyl, Majorana, and Dirac Spinors

In Chapter 2 we got acquainted with Weyl spinors and their behavior under Lorentz transformations. However, as noted before, supersymmetric theories are often written in terms of Majorana spinors instead. It is therefore important to understand the connection among Weyl, Majorana, and Dirac spinors both at the mathematical level and at the physical level. In addition, we will see how we can go back and forth between Weyl and Majorana spinors. The first step is to understand the operation of charge conjugation applied to spinors and the connection with antiparticle states.

4.1 Charge Conjugation and Antiparticles for Dirac Spinors

Charge conjugation is the operation that turns a field describing a certain particle into a field describing the corresponding antiparticle. To study the effect of charge conjugation, let us consider a Dirac spinor because this probably is more familiar to most readers. In this chapter and for the remainder of this book we will drop the L and R labels on the Weyl spinors to alleviate the notation. *It will be understood that η will always represent a right-chiral spinor and χ a left-chiral spinor.*

Consider, then, a Dirac spinor corresponding to a certain particle p (which could be an electron, a neutrino, a quark, etc.):

$$\Psi_p = \begin{pmatrix} \eta_p \\ \chi_p \end{pmatrix} \tag{4.1}$$

The antiparticle spinor is obtained by applying the charge-conjugation operator (which we will write down explicitly in an instant) to the particle state, so we may write

$$\Psi_p^c \equiv \Psi_{\bar{p}} \equiv \begin{pmatrix} \eta_{\bar{p}} \\ \chi_{\bar{p}} \end{pmatrix}$$

where the c denotes charge conjugation. Here, a bar over the name of a particle simply will represent the corresponding antiparticle (as in $\bar{\nu}$ for the antineutrino). It is unrelated to the bar used in Chapters 2 and 3.

Note that in a Dirac spinor, η_p and $\eta_{\bar{p}}$ are two *different* right-chiral spinors. The same comment applies to χ_p and $\chi_{\bar{p}}$; they are different left-chiral spinors. But clearly, those four spinors are related. To establish the connection, we need to work out explicitly the charge conjugation.

The operation of charge conjugation is defined as

$$\Psi_p^c \equiv C\overline{\Psi}_p^T \tag{4.2}$$

where the bar denotes here the usual Dirac adjoint, and the charge-conjugation matrix C may be taken to have the following representation:

$$C = -i\gamma^2\gamma^0 = \begin{pmatrix} i\sigma^2 & 0 \\ 0 & -i\sigma^2 \end{pmatrix}$$

With a little work, we may write Eq. (4.2) in a more convenient form:

$$\Psi_p^c = C\overline{\Psi}_p^T = C(\Psi_p^\dagger \gamma^0)^T = C(\gamma^0)^T \Psi_p^{\dagger T} = C\gamma^0 \Psi_p^{\dagger T} \equiv C_0 \Psi_p^{\dagger T}$$

where we have used $(\gamma^0)^T = \gamma^0$ and have introduced

$$C_0 = C\gamma^0 = -i\gamma^2 \gamma^0 \gamma^0 = -i\gamma^2 = \begin{pmatrix} 0 & i\sigma^2 \\ -i\sigma^2 & 0 \end{pmatrix}$$

Here we have used the fact that $\gamma^0 \gamma^0 = 1$. Note that $C_0 C_0 = 1$ as well.

Written explicitly, the charge-conjugated state is

$$\Psi_p^c = \begin{pmatrix} i\sigma^2 \chi_p^{\dagger T} \\ -i\sigma^2 \eta_p^{\dagger T} \end{pmatrix}$$

Using the fact that the charge-conjugated state is the antiparticle state, defined in Eq. (4.1), we finally find

$$\eta_{\bar{p}} = i\sigma^2 \chi_p^{\dagger T}$$

$$\chi_{\bar{p}} = -i\sigma^2 \eta_p^{\dagger T}$$

(4.3)

This is an interesting result: These expressions are exactly what we obtained in Eqs. (2.40) and (2.37)! What Eq. (4.3) tells us is that the *right-chiral antiparticle spinor* is the right-chiral quantity built out of the *particle left-chiral spinor*! Likewise, the left-chiral antiparticle state is given by the left-chiral expression built out of the right-chiral particle spinor. This is, by the way, the reason we introduced factors of i and $-i$ in Eqs. (2.40) and (2.37): it was in order to make the correspondence with charge conjugation exact.

It is easy to check that applying charge conjugation twice to a state gives back the initial state:

$$(\Psi_p^c)^c = C_0 (\Psi_p^c)^{\dagger T} = C_0 (C_0 \Psi_p^{\dagger T})^{\dagger T} = C_0 C_0^* \Psi_p = C_0^2 \Psi_p = \Psi_p$$

where we have used that the operation $\dagger T$ applied to a matrix (as opposed to a quantum field) is simply a complex conjugation. Applying the charge conjugation operation to both sides of Eq. (4.2), we find

$$\Psi_p = C \left(\overline{\Psi_p^c} \right)^T$$

(4.4)

which leads to

$$
\begin{aligned}
\eta_p &= i\sigma^2 \chi_{\bar{p}}^{\dagger T} \\
\chi_p &= -i\sigma^2 \eta_{\bar{p}}^{\dagger T}
\end{aligned}
\tag{4.5}
$$

Of course, this also can be obtained directly by inverting Eq. (4.3).

As expected, the particle left-chiral spinor can be written as the left-chiral quantity built out of the antiparticle right-chiral spinor. Likewise, the particle right-chiral spinor can be written as the right-chiral expression built out of the antiparticle left-chiral spinor.

Equations (4.3) and (4.5) teach us an important lesson: We are always free to express the left- and right-chiral states of an antiparticle in terms of, respectively, the right- and left-chiral states of the particle field (and vice versa). For example, we may rewrite a Dirac spinor *in terms of left-chiral fields only*, at the condition of introducing the left-chiral *antiparticle* state:

$$
\Psi = \begin{pmatrix} \eta_p \\ \chi_p \end{pmatrix} = \begin{pmatrix} i\sigma^2 \chi_{\bar{p}}^{\dagger T} \\ \chi_p \end{pmatrix}
\tag{4.6}
$$

This point will be crucial when we build the supersymmetric extension of the standard model because we will express all fermions in terms of left-chiral states.

It must be kept in mind here that even though only left-chiral fields appear, the upper two components still transform as a right-chiral spinor, so the four-component spinor always has the structure:

$$
\Psi = \begin{pmatrix} \text{quantity that transforms like a right-chiral spinor} \\ \text{quantity that transforms like a left-chiral spinor} \end{pmatrix}
$$

Of course, we could as well have written the Dirac spinor in terms of right-chiral fields only:

$$
\Psi = \begin{pmatrix} \eta_p \\ \chi_p \end{pmatrix} = \begin{pmatrix} \eta_p \\ -i\sigma^2 \eta_{\bar{p}}^{\dagger T} \end{pmatrix}
$$

4.2 CPT Invariance

A short but crucial point needs to be mentioned here. Realistic relativistic quantum field theories must be *CPT*-invariant (some authors write *TCP* or *PCT*); i.e., they must be invariant under the combined operations of charge conjugation *C*, parity *P*,

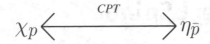

Figure 4.1   Two states related by *CPT*.

and time reversal T. A consequence of this requirement is that lagrangian densities must be "real" in the sense that they must satisfy

$$\mathcal{L}^\dagger = \mathcal{L} \tag{4.7}$$

This shows right away why we cannot have a theory containing a left-chiral spinor χ without including as well its hermitian conjugate $\chi^\dagger$ because a lagrangian containing only one of the two obviously could not respect Eq. (4.7).

Now, from Eq. (4.3) we see that $\chi^\dagger$ can be written in terms of the right-chiral antiparticle spinor $\eta_{\bar{p}}$. This means that the states χ_p and $\eta_{\bar{p}}$ always appear together in a *CPT* invariant field theory, which we illustrate in Figure 4.1.

In physical terms, this implies that if a theory contains a left-chiral spinor, it also necessarily contains the corresponding antiparticle, which will be right-chiral. When promoted to the status of quantum field operator, χ_p creates the right-chiral antiparticle and annihilates the left-chiral particle.

For similar reasons, the pair of fields shown in Figure 4.2 also must always appear together in a *CPT*-invariant quantum field theory.

In other words, if there is a right-chiral particle state, there also must be a corresponding left-chiral antiparticle state.

Actually, the two figures are totally equivalent because it is completely arbitrary what we call the particle and what we call the antiparticle. So we are free to switch the labels p and $\bar{p}$ in the first figure, which then becomes the second figure. What we have learned is that if a particle of a certain chirality is present in a theory, there will necessarily be a corresponding antiparticle of opposite chirality. If the particle is massless, *chirality* and *helicity* can be used interchangeably.

Note that *CPT* invariance does *not* constrain us to necessarily have all four states χ_p, η_p, $\chi_{\bar{p}}$, and $\eta_{\bar{p}}$ present. It is possible to build a completely consistent theory out of χ_p and $\eta_{\bar{p}}$ alone or out of $\chi_{\bar{p}}$ and η_p alone. We will discuss this key observation more fully in the next few sections.

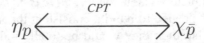

Figure 4.2   Two other states related by *CPT*.

4.3 A Massless Weyl Spinor

Let us start with the simplest fermion theory that we could possibly imagine: the theory of a noninteracting massless two-component Weyl spinor. We will take it to be left-chiral. Then our lagrangian contains only the kinetic term given by

$$\mathcal{L}_{\text{kin}} = \chi^\dagger \, \bar{\sigma}^\mu \, i \partial_\mu \chi \tag{4.8}$$

EXERCISE 4.1
Show that the lagrangian (4.8) is real in the sense of Eq. (4.7).

From what we learned in the preceding section, we know that one would observe in the lab a left-chiral particle and its corresponding right-chiral antiparticle (again, we could exchange the *particle* and *antiparticle* labels, but we choose to call the left-chiral state the *particle*). It is instructive to make a count of the degrees of freedom involved. A two-component spinor contains two complex variables, so there are four "off-shell" degrees of freedom (i.e., four degrees of freedom before the equations of motions are imposed). Physical massless spinors must obey the Weyl equations Eq. (2.29), which provide two constraints, leaving two degrees of freedom and therefore two types of observable states.

Since we know that the particles observed would be the left-chiral particle and the corresponding right-chiral antiparticle, we should be able to write the lagrangian in terms of χ_p and $\eta_{\bar{p}}$ instead of χ_p and $\chi_p^\dagger$. Indeed, this can be done using the second relation of Eq. (4.5):

$$\chi_p = -i\sigma^2 \, \eta_{\bar{p}}^{\dagger T}$$

so that

$$\chi_p^\dagger = \eta_{\bar{p}}^T \, (-i\sigma^2)^\dagger = \eta_{\bar{p}}^T \, (i\sigma^2)$$

The lagrangian then takes the form

$$\mathcal{L}_{\text{kin}} = \eta_{\bar{p}}^T \, i\sigma^2 \, \bar{\sigma}^\mu \, i \partial_\mu \chi_p$$

Note that this is a completely viable theory; *there is no need to include the states* η_p *and* $\chi_{\bar{p}}$. They are not needed to build a consistent theory. Of course, we could choose to add them in, but then they would be totally independent and therefore would correspond to a new particle in its own right.

4.4 Adding a Mass: General Considerations

Let's start again with a left-chiral Weyl spinor χ_p, but now we will make it massive.

Before proceeding, it's important to point out that some references restrict the term *Weyl spinors* to *massless* spinors, in which case Weyl spinors are always eigenstates of the helicity operator. Others define Weyl spinors as being the chirality eigenstates, which are irreducible representations of the Lorentz group. With the first definition, it is helicity that is used as the defining property of Weyl spinors, whereas in the second convention, it is behavior under Lorentz transformations that is judged as being more fundamental. For the proponents of the first definition, *massive Weyl fermions* is an oxymoron, whereas the other camp has no problem with that. Of course, in the massless case, the two schools of thought live in perfect harmony; it is just when discussing massive particles that things get ugly.

In this book we will follow the second convention; i.e., we will use the term *Weyl spinors* to denote the eigenstates of the chirality operator that have the Lorentz transformations given in Eq. (2.32). Therefore, with our convention, it does make sense to talk about massive Weyl spinors. The down side is that our massive Weyl spinors do not have a well-defined helicity, but this is a small price to pay because helicity will play very little role in this book.

Let's get back to the physics of massive fermions. Things are much more interesting than in the massless case. As we argued in Section 2.3, it is possible to use a Lorentz transformation to change the helicity of a massive spinor. This implies that we need both helicities to describe a general massive *particle* state. From Eqs. (2.30) and (2.31) and the discussion preceding them, we must conclude that if we need both helicity states to describe a massive particle, we need both chiralities as well. Therefore, *for a massive fermion, both chiralities of the particle must be present in the theory*.

We can summarize the situation by saying that if we start with a left-chiral particle state χ_p, Lorentz transformations (LT) necessitate the introduction of a right-chiral spinor η_p, as illustrated schematically in Figure 4.3. This figure is not to

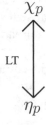

Figure 4.3 Two states of a massive spin 1/2 fermion required by Lorentz transformations.

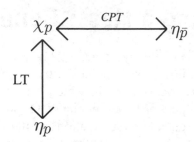

Figure 4.4 States associated to a massive left-chiral spinor through Lorentz and *CPT* transformations.

be interpreted as saying that a Lorentz transformation changes a left-chiral state into a right-chiral state—it is the helicity that is flipped by a Lorentz boost—but as meaning that Lorentz transformations require the existence of both chiral states for a massive fermion.

Now comes the interesting part. As we saw in the preceding section, *CPT* invariance forces us to introduce the charge-conjugate state $\eta_{\bar{p}}$, which is right-chiral. On the other hand, as we have just seen, for a massive particle, Lorentz invariance also forces us to introduce a right-chiral state η_p. The situation is summarized in Figure 4.4.

The questions that come immediately to mind are: Can we take those two right-chiral states forced on us to be the same? Can we take them to be different physical states? In other words, do we have a choice?

The answer to all these questions is yes! We do have a choice, and this leads to defining two different types of four-component spinors.

4.5 Adding a Mass: Dirac Spinors

First, let's consider the case where we choose the two right-chiral states to represent *different* physical states, $\eta_p \neq \eta_{\bar{p}}$. In that case, there are four physical (on-shell) distinct states: the left- and right-chiral states of the particle χ_p and η_p as well as the left- and right-chiral states of the antiparticle $\chi_{\bar{p}}$ and $\eta_{\bar{p}}$. This is illustrated in Figure 4.5.

We can combine the left- and right-chiral states of the particle into a four-component spinor that is the familiar Dirac spinor that we all know and love and which we encountered in the first chapter:

$$\Psi_D = \begin{pmatrix} \eta_p \\ \chi_p \end{pmatrix}$$

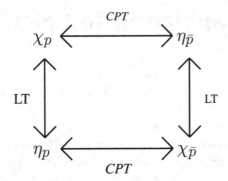

Figure 4.5   The four states of a massive Dirac fermion.

Since we assume the antiparticle states $\chi_{\bar{p}}$ and $\eta_{\bar{p}}$ to be distinct from χ_p and η_p, we have

$$\Psi_D^c \neq \Psi_D$$

Let us write the mass term for such a Dirac spinor. From the Dirac lagrangian Eq. (2.2), we know that the mass is

$$-m\overline{\Psi}_D\Psi_D = -m\Psi^\dagger\gamma_0\Psi$$
$$= -m\left(\eta_p^\dagger\chi_p + \chi_p^\dagger\eta_p\right) \tag{4.9}$$

This shows clearly that both χ_p and η_p are necessary in order to write down a Dirac mass term.

As noted earlier, we also could write the Dirac spinor in terms of left-chiral spinors only, using Eq. (4.6). The mass term then is

$$-m\overline{\Psi}_D\Psi_D = -m\left(\chi_{\bar{p}}^T\left(-i\sigma^2\right)\chi_p + \chi_p^\dagger i\sigma^2 \chi_{\bar{p}}^{\dagger T}\right) \tag{4.10}$$

which, using the spinor dot products defined in Eq. (2.50), can be written in the elegant form

$$-m\overline{\Psi}_D\Psi_D = -m(\chi_{\bar{p}} \cdot \chi_p + \bar{\chi}_p \cdot \bar{\chi}_{\bar{p}}) \tag{4.11}$$

EXERCISE 4.2
Show that the mass term in Eq. (4.11) is real in the sense of Eq. (4.7).

4.6 The QED Lagrangian in Terms of Weyl Spinors

Later in the book we will be constructing theories directly in terms of Weyl spinors, and it will be very useful to be able to make the connection with the familiar lagrangians of quantum electrodynamics (QED), quantum chromodynamics (QCD), and so on, which are usually all expressed in terms of four-component Dirac spinors. So let's see what the QED lagrangian looks like when expressed in terms of Weyl spinors.

Let's first consider the Dirac lagrangian for an electron (so the subscript p will be replaced by e), which we can obtain directly from Eq. (2.28):

$$\mathcal{L}_{\text{Dirac}} = \overline{\Psi}_D \big(\gamma^\mu i\partial_\mu - m\big)\Psi_D$$
$$= \chi_e^\dagger \, \bar{\sigma}^\mu \, i\partial_\mu \chi_e + \eta_e^\dagger \, \sigma^\mu \, i\partial_\mu \eta_e - m\big(\eta_e^\dagger \, \chi_e + \chi_e^\dagger \, \eta_e\big)$$

Again, it is useful to write the Dirac spinor in terms of left-chiral spinors only, using Eq. (4.6). This is so because we will later write all the interactions of the standard model and of the minimal supersymmetric standard model in terms of left-chiral spinors. We already took care of the mass term in Eq. (4.11), so let's concentrate on the kinetic energy:

$$i\overline{\Psi}_D \gamma^\mu \partial_\mu \Psi_D = i\Psi_D^\dagger \gamma^0 \gamma^\mu \partial_\mu \Psi_D$$

$$= i\big(\chi_{\bar{e}}^T (-i\sigma^2) \; \chi_e^\dagger \big) \begin{pmatrix} 0 & \mathbf{1} \\ \mathbf{1} & 0 \end{pmatrix} \begin{pmatrix} 0 & \bar{\sigma}^\mu \partial_\mu \\ \sigma^\mu \partial_\mu & 0 \end{pmatrix} \begin{pmatrix} i\sigma^2 \; \chi_{\bar{e}}^{\dagger T} \\ \chi_e \end{pmatrix}$$

$$= i\chi_{\bar{e}}^T \sigma^2 \sigma^\mu \sigma^2 \partial_\mu \chi_{\bar{e}}^{\dagger T} + i\chi_e^\dagger \bar{\sigma}^\mu \partial_\mu \chi_e \tag{4.12}$$

which may be written more elegantly as (see Exercise 4.3)

$$i\overline{\Psi}_D \gamma^\mu \partial_\mu \Psi_D = i\chi_{\bar{e}}^\dagger \bar{\sigma}^\mu \partial_\mu \chi_{\bar{e}} + i\chi_e^\dagger \bar{\sigma}^\mu \partial_\mu \chi_e \tag{4.13}$$

So the Dirac lagrangian is finally, using Eq. (4.11),

$$\mathcal{L}_{\text{Dirac}} = i\chi_{\bar{e}}^\dagger \bar{\sigma}^\mu \partial_\mu \chi_{\bar{e}} + i\chi_e^\dagger \bar{\sigma}^\mu \partial_\mu \chi_e - m(\chi_{\bar{e}} \cdot \chi_e + \bar{\chi}_e \cdot \bar{\chi}_{\bar{e}}) \tag{4.14}$$

In Chapter 13 we will build the supersymmetric generalization of this lagrangian.

EXERCISE 4.3
Derive Eq. (4.13) from Eq. (4.12). *Hint*: You will need Eqs. (A.36), and (A.5) and an integration by parts.

Let's now include the coupling of the fermion with the electromagnetic field. This is obtained from the Dirac lagrangian by replacing the partial derivative ∂_μ by the gauge covariant derivative D_μ defined by

$$D_\mu \equiv \partial_\mu + iq A_\mu$$

where A_μ is the gauge (photon) field, and q is the electric charge of the particle. For the electron, $q = -e$ (we use the convention that e is positive), so

$$D_\mu = \partial_\mu - ie A_\mu$$

In addition, we must add the kinetic energy of the photon, i.e.,

$$-\frac{1}{4} F_{\mu\nu} F^{\mu\nu}$$

The two new terms coupling the Weyl spinors to the gauge field can be obtained from the two terms of Eq. (4.12) by replacing ∂_μ by $-ie A_\mu$:

$$\text{Terms containing } A_\mu = e\, \chi_{\bar{e}}^T\, \sigma^2\, \sigma^\mu A_\mu\, \sigma^2\, \chi_{\bar{e}}^{\dagger T} + e\, \chi_e^\dagger\, \bar{\sigma}^\mu A_\mu\, \chi_e$$

$$= -e\, \chi_{\bar{e}}^\dagger\, \bar{\sigma}^\mu A_\mu\, \chi_{\bar{e}} + e\, \chi_e^\dagger\, \bar{\sigma}^\mu A_\mu\, \chi_e \qquad (4.15)$$

where in the second step we have used Eqs. (A.36) and (A.5) in the way described in Exercise 4.3.

Note that it would have been *incorrect* to replace ∂_μ by $-ie A_\mu$ in Eq. (4.14) (see Exercise 4.4).

EXERCISE 4.4
Explain why we would have obtained the wrong coupling of the gauge field with the spinors had we replaced ∂_μ by $-ie A_\mu$ in Eq. (4.14) instead of doing it in Eq. (4.12).

Adding Eq. (4.15) to the Dirac lagrangian Eq. (4.14), we finally obtain the QED lagrangian expressed in terms of left-chiral Weyl spinors:

$$\mathcal{L}_{\text{QED}} = i\chi_{\bar{e}}^\dagger\, \bar{\sigma}^\mu\, \partial_\mu \chi_{\bar{e}} + i\chi_e^\dagger\, \bar{\sigma}^\mu\, \partial_\mu \chi_e - e\, A_\mu\, \chi_{\bar{e}}^\dagger\, \bar{\sigma}^\mu\, \chi_{\bar{e}} + e\, A_\mu\, \chi_e^\dagger\, \bar{\sigma}^\mu\, \chi_e$$

$$-m(\chi_{\bar{e}} \cdot \chi_e + \bar{\chi}_e \cdot \bar{\chi}_{\bar{e}}) - \frac{1}{4} F_{\mu\nu} F^{\mu\nu} \qquad (4.16)$$

There is something quite interesting going on here. In the derivation of the kinetic term of the positron, we have used Eq. (A.36) and an integration by parts. Each operation introduces a minus sign, so the final kinetic energy of the positron has the same sign as the kinetic energy of the electron (which did not require any manipulation). On the other hand, in the case of the coupling of the positron to the electromagnetic field, we used Eq. (A.36) but, of course, did not perform any integration by parts (there are no derivatives!), so we ended up with the particle and antiparticle spinors having couplings to the electromagnetic field of opposite signs. This, of course, was to be expected because these two states have opposite charges! What is interesting is to see how we started from the Dirac lagrangian written in terms of a Dirac spinor, in which there is only one coupling to the electromagnetic field, and ended up with an expression containing two couplings to A_μ with opposite signs once we expressed the theory in terms of particle and antiparticle Weyl spinors.

We can write Eq. (4.16) even more elegantly by introducing two gauge covariant derivatives acting on the left-chiral particle and antiparticle spinors. Actually, it is customary to use a common symbol D_μ for both derivatives, with the understanding that when it is acting on the electron Weyl spinor χ_e, it is defined as

$$D_\mu \chi_e \equiv (\partial_\mu - ieA_\mu)\chi_e \tag{4.17}$$

whereas when it is acting on the positron Weyl spinor $\chi_{\bar{e}}$, it carries the opposite charge:

$$D_\mu \chi_{\bar{e}} \equiv (\partial_\mu + ieA_\mu)\chi_{\bar{e}} \tag{4.18}$$

With this convention, we may write Eq. (4.16) as

$$\boxed{\mathcal{L}_{\text{QED}} = i\chi_{\bar{e}}^\dagger \bar{\sigma}^\mu D_\mu \chi_{\bar{e}} + i\chi_e^\dagger \bar{\sigma}^\mu D_\mu \chi_e - m(\chi_{\bar{e}} \cdot \chi_e + \bar{\chi}_e \cdot \bar{\chi}_{\bar{e}}) - \frac{1}{4}F_{\mu\nu}F^{\mu\nu}}$$

$$\tag{4.19}$$

In Chapter 13 we will build the supersymmetric generalization of this lagrangian.

4.7 Adding a Mass: Majorana Spinors

Now consider taking the two right-chiral states η_p and $\eta_{\bar{p}}$ to be one and the same, as illustrated in Figure 4.6.

Identifying η_p and $\eta_{\bar{p}}$ only makes sense if *the particle is its own antiparticle*. Therefore, this quantum field cannot carry quantum numbers that can distinguish a

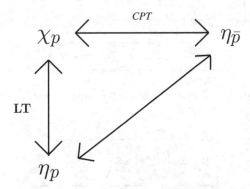

Figure 4.6 The two states of a massive Majorana fermion (η_p is defined to be equal to $\eta_{\bar{p}}$).

particle from an antiparticle (e.g., it must have no electric charge or lepton number). In this case, there are *only two degrees of freedom*, which we can choose to be χ_p and η_p with

$$\eta_p = \eta_{\bar{p}}$$
$$= i\sigma^2 \chi_p^{\dagger T}$$

where we have used Eq. (4.3).

It's easy to check that with this choice, the antiparticle left-chiral state $\chi_{\bar{p}}$ is indeed the same as the particle left-chiral state χ_p. Using Eq. (4.5) with the particle and antiparticle labels switched, we find

$$\chi_{\bar{p}} \equiv -i\sigma^2 \eta_p^{\dagger T}$$
$$= -i\sigma^2 \left(i\sigma^2 \chi_p^{\dagger T}\right)^{\dagger T}$$
$$= -i\sigma^2 (i\sigma^2)^* \chi_p$$
$$= -i\sigma^2 (i\sigma^2) \chi_p$$
$$= \chi_p$$

Instead of expressing everything in terms of χ_p and η_p, we may, of course, choose to express everything in terms of the left-chiral field χ_p and its hermitian conjugate $\chi_p^{\dagger}$. Note that in terms of degrees of freedom, there is no difference with a massless Weyl spinor. Here, we have all four quantities χ_p, η_p, $\eta_{\bar{p}}$, and $\chi_{\bar{p}}$, but they all can be expressed in terms of χ_p and $\chi_p^\dagger$ using

$$\chi_{\bar{p}} = \chi_p \qquad \eta_{\bar{p}} = \eta_p = i\sigma^2 \chi_p^{\dagger T}$$

We now can assemble χ_p and η_p into a four-component spinor that is called a *Majorana spinor* Ψ_M:

$$\Psi_M = \begin{pmatrix} \eta_p \\ \chi_p \end{pmatrix} = \begin{pmatrix} i\sigma^2 \chi_p^{\dagger T} \\ \chi_p \end{pmatrix} \tag{4.20}$$

By construction, a Majorana spinor is unchanged by charge conjugation:

$$\Psi_M^c = \Psi_M$$

which simply expresses the fact that $\chi_{\bar{p}} = \chi_p$ and $\eta_{\bar{p}} = \eta_p$. In fact, we could have used this as our starting point (and many authors do) in defining a Majorana spinor; i.e., we could have started from a Dirac spinor and imposed the antiparticle states to be the same as the particle states.

Let us write down the lagrangian for a free Majorana particle. We insert Eq. (4.20) into the Dirac lagrangian, but we also include an overall factor of 1/2 for reasons that will become clear later:

$$\mathcal{L}_M = \frac{1}{2}\overline{\Psi}_M \left(\gamma^\mu i \partial_\mu - m \right) \Psi_M \tag{4.21}$$

The theory of a Majorana particle is usually left in this form, but it is instructive to expand it in terms of the two-component spinors χ_p and $\chi_p^\dagger$:

$$\mathcal{L}_M = \frac{1}{2}\chi_p^\dagger \bar{\sigma}^\mu \, i\partial_\mu \chi_p + \frac{1}{2}\chi_p^T(-i\sigma^2)\sigma^\mu (i\sigma^2) \, i\partial_\mu \chi_p^{\dagger T}$$

$$-\frac{m}{2}\left[\chi_p^T(-i\sigma^2)\chi_p + \chi_p^\dagger i\sigma^2 \, \chi_p^{\dagger T} \right]$$

$$= \frac{1}{2}\chi_p^\dagger \bar{\sigma}^\mu \, i\partial_\mu \chi_p + \frac{1}{2}\chi_p^T \bar{\sigma}^{\mu T} \, i\partial_\mu \chi_p^{\dagger T} - \frac{m}{2}(\chi_p \cdot \chi_p + \bar{\chi}_p \cdot \bar{\chi}_p) \tag{4.22}$$

where we have again used Eq. (A.5). This expression can be simplified further using an integration by parts on the second term to write it as

$$\frac{1}{2}\chi_p^T \bar{\sigma}^{\mu T} \, i\partial_\mu \chi_p^{\dagger T} = -\frac{1}{2}\left(i\partial_\mu \chi_p^T \right) \bar{\sigma}^{\mu T} \chi_p^{\dagger T} \tag{4.23}$$

This is actually exactly equal to the first term of Eq. (4.22), as can be shown using Eq. (A.36):

$$-\frac{1}{2}\left(i\partial_\mu \chi_p^T\right)\bar{\sigma}^{\mu T}\chi_p^{\dagger T} = \frac{1}{2}\chi_p^\dagger \bar{\sigma}^\mu i\partial_\mu \chi_p$$

The Majorana lagrangian (4.22) therefore can be simplified to

$$\boxed{\mathcal{L}_M = \chi_p^\dagger \bar{\sigma}^\mu i\partial_\mu \chi_p - \frac{m}{2}\left(\chi_p \cdot \chi_p + \bar{\chi}_p \cdot \bar{\chi}_p\right)} \qquad (4.24)$$

Note the difference between the Majorana lagrangian and the Dirac lagrangian of Eq. (4.14), ignoring the difference of normalization of the mass terms for now. The Dirac lagrangian contains the kinetic terms of two distinct states, χ_p and $\chi_{\bar{p}}$, and it connects them through the mass term. By contrast, the Majorana lagrangian contains a single kinetic term and uses only χ_p and its hermitian conjugate to make up the mass term.

We see why we had included a factor of $1/2$ in front of the kinetic term in Eq. (4.21): This was in order to obtain the canonical normalization after combining the two kinetic terms of Eq. (4.22) into a single term. As for the factor of $1/2$ in front of the mass term, it is chosen so that the parameter m appearing in the lagrangian is indeed the mass of the particle, as we will demonstrate in Section 4.9.

4.8 Dirac Spinors and Parity

You might wonder why we consider Dirac spinors at all, given that Weyl spinors (or, equivalently, Majorana spinors) are more fundamental and simpler. Why bother with Dirac fields at all? The answer has to do with parity. Parity transforms χ_p and η_p into each other. Obviously, a theory containing a single Weyl spinor (or, equivalently, a single Majorana spinor) cannot be parity-invariant. In order to build a parity-invariant theory, such as QED or QCD, one needs *both* chiral states of the particle, and it is obviously convenient to group these states into a four-component spinor, which turns out to be nothing other than a Dirac spinor. Of course, the *CPT*-conjugate states are also needed to complete the theory.

Consider the Dirac mass term written in the form (4.9) (*before* expressing η_p in terms of $\chi_{\bar{p}}^\dagger$):

$$-m\left(\eta_p^\dagger \chi_p + \chi_p^\dagger \eta_p\right)$$

This is clearly parity-invariant because it is invariant under the exchange of η and χ. On the other hand, a Majorana mass term obviously violates parity (there is no η_p field to switch with the χ_p!).

4.9 Definition of the Masses of Scalars and Spinor Fields

If you are used to working with Dirac spinors, the presence of the factor of 1/2 in the mass terms of the Majorana and Weyl spinors may be somewhat unsettling. Before discussing this issue, let us first talk about scalar fields. The lagrangian of a free real scalar field A is

$$\mathcal{L}_A = \frac{1}{2}\, \partial^\mu A\, \partial_\mu A - \frac{1}{2}\, m^2\, A^2 \tag{4.25}$$

How do we know that the parameter m is indeed the mass of the particle? By working out the equation of motion of the field and comparing it with the Klein-Gordon equation. For a scalar field ϕ, the Euler-Lagrange equation is

$$\partial_\mu \left[\frac{\partial \mathcal{L}}{\partial(\partial_\mu \phi)} \right] - \frac{\partial \mathcal{L}}{\partial \phi} = 0 \tag{4.26}$$

Applying this to Eq. (4.25), we simply get

$$\partial_\mu \partial^\mu A + m^2 A = 0$$

which is the Klein-Gordon equation for a particle of mass m. The operator $\partial_\mu \partial^\mu$ is often represented by the symbol $\Box$ and is called the *d'Alembertian operator*.

Our result shows that the factor of 1/2 was necessary in Eq. (4.25) for m to be the mass of the particle. Now consider instead a complex scalar field ϕ whose lagrangian is given by

$$\mathcal{L}_\phi = \partial^\mu \phi \partial_\mu \phi^\dagger - m^2 \phi \phi^\dagger$$

Note that there are no factors of 1/2 now! We want to check that the m appearing there is indeed the mass of the particle. This time, in applying the Euler-Lagrange equation, we treat ϕ and $\phi^\dagger$ as independent fields. Applying Eq. (4.26), we get

$$\partial^\mu \partial_\mu \phi^\dagger + m^2 \phi^\dagger = 0$$

which is again the Klein-Gordon for the field $\phi^\dagger$ with mass m. On the other hand, the Euler-Lagrange equation for $\phi^\dagger$ yields again the Klein-Gordon equation, but this time for the field ϕ, still with mass m. It is clear why the factor of $1/2$ is required in the real scalar field lagrangian and not in the case of the complex scalar field: In the first case, the real scalar field appears twice in each term of the lagrangian.

Let us now turn our attention to spinor fields. Roughly speaking, a Majorana spinor is to a Dirac field what a real scalar field is to a complex scalar field. Indeed, a Majorana spinor obeys what is essentially a reality condition for spinors, $\Psi_M^c = \Psi_M$. By analogy with scalar fields, we would expect the free Majorana lagrangian to contain a factor of $1/2$ relative to the Dirac lagrangian. This is, of course, a purely hand-waving argument, but we can demonstrate that it is indeed correct by working out the Euler-Lagrange equations, which we write again as applied to the spinor components χ_a:

$$\partial_\mu \left(\frac{\partial \mathcal{L}}{\partial (\partial_\mu \chi_a)} \right) - \frac{\partial \mathcal{L}}{\partial \chi_a} = 0 \tag{4.27}$$

We now apply this to the lagrangian Eq. (4.24) [we will drop the label p on the spinors because there are no antiparticle spinors in Eq. (4.24) and therefore no possibility of confusion]. The only thing we must be careful about is that derivatives with respect to spinor components are Grassmann quantities, so they anticommute with spinors. For example,

$$\frac{\partial}{\partial \chi_a} (\chi \cdot \chi) = \frac{\partial}{\partial \chi_a} [\chi_b (-i\sigma^2)^{bc} \chi_c]$$

$$= -\left(\frac{\partial \chi_b}{\partial \chi_a} \right) (i\sigma^2)^{bc} \chi_c + \chi_b (i\sigma^2)^{bc} \frac{\partial \chi_c}{\partial \chi_a}$$

where we picked up a minus sign in the second term by passing the derivative through the spinor component χ_b. Now we use

$$\frac{\partial \chi_b}{\partial \chi_a} = \delta_b^a$$

where the δ is just the usual Kronecker delta; i.e., $\delta_1^1 = \delta_2^2 = 1$, whereas $\delta_1^2 = \delta_2^1 = 0$. We then have

$$\frac{\partial}{\partial \chi_a} \chi \cdot \chi = \delta_b^a (-i\sigma^2)^{bc} \chi_c + \chi_b (i\sigma^2)^{bc} \delta_c^a$$

$$= (-i\sigma^2)^{ac} \chi_c + \chi_b (i\sigma^2)^{ba}$$

$$= (-i\sigma^2)^{ac} \chi_c + \chi_b (-i\sigma^2)^{ab}$$

$$= -2(i\sigma^2)^{ac} \chi_c \qquad (4.28)$$

Of course,

$$\frac{\partial}{\partial \chi_a} \bar{\chi} \cdot \bar{\chi} = 0$$

Using Eq. (4.28), we find

$$\frac{\partial \mathcal{L}}{\partial \chi_a} = i\, m\, (\sigma^2)^{ab} \chi_b \qquad (4.29)$$

This takes care of the second term of Eq. (4.27). Now consider the first term:

$$\partial_\mu \left(\frac{\partial}{\partial(\partial_\mu \chi_a)} \left(\chi^\dagger \bar{\sigma}^\nu i \partial_\nu \chi \right) \right) = -i \left(\partial_\mu \chi_b^\dagger \right) (\bar{\sigma}^\mu)^{ba}$$

where the minus sign came from passing the derivative with respect to $\partial_\mu \chi$ to the right of $\chi^\dagger$, and we did not show explicit spinor indices to simplify the notation. This may be written as

$$-i \left(\partial_\mu \chi_b^\dagger \right) (\bar{\sigma}^\mu)^{ba} = -i\, (\bar{\sigma}^{\mu T})^{ab}\, \partial_\mu \chi_b^{\dagger T} \qquad (4.30)$$

Substituting Eqs. (4.29) and (4.30) into Eq. (4.27), the Euler-Lagrange equation for χ is finally

$$-i\bar{\sigma}^{\mu T} \partial_\mu \chi^{\dagger T} - i\, m\, \sigma^2 \chi = 0$$

Using the fact that $\sigma^2 \sigma^2 = \mathbf{1}$ and multiplying by i, we may write this as

$$\bar{\sigma}^{\mu T} \sigma^2\, \sigma^2\, \partial_\mu \chi^{\dagger T} + m \sigma^2 \chi = 0$$

If we multiply this from the left by yet another σ^2 and use Eq. (A.8), we finally get the Euler-Lagrange equation for χ:

$$\sigma^\mu \sigma^2 \, \partial_\mu \chi^{\dagger T} + m\chi = 0 \tag{4.31}$$

On the other hand, a similar calculation shows that the Euler-Lagrange equation for $\chi^\dagger$ is

$$\sigma^2 \bar{\sigma}^\mu \, \partial_\mu \chi - m \, \chi^{\dagger T} = 0 \tag{4.32}$$

If we now isolate $\chi^{\dagger T}$ from Eq. (4.32) and substitute it into Eq. (4.31), we find

$$\sigma^\mu \bar{\sigma}^\nu \, \partial_\mu \partial_\nu \chi + m^2 \chi = 0 \tag{4.33}$$

where we again used $\sigma^2 \sigma^2 = \mathbf{1}$. With the help of Eq. (A.9), we may write the first term as

$$\sigma^\mu \bar{\sigma}^\nu \, \partial_\mu \partial_\nu \chi = 2\partial^\mu \partial_\mu \chi - \sigma^\nu \bar{\sigma}^\mu \, \partial_\nu \partial_\mu \chi$$

But, after renaming the indices $\mu \leftrightarrow \nu$ in the last term, we get

$$\sigma^\mu \bar{\sigma}^\nu \, \partial_\mu \partial_\nu \chi = 2\partial^\mu \partial_\mu \chi - \sigma^\mu \bar{\sigma}^\nu \partial_\nu \partial_\mu \chi$$

which implies

$$\sigma^\mu \bar{\sigma}^\nu \partial_\mu \partial_\nu \chi = \partial^\mu \partial_\mu \chi \tag{4.34}$$

Using this in Eq. (4.33), we finally get that χ obeys

$$\partial^\mu \partial_\mu \chi + m^2 \chi = 0$$

the Klein-Gordon equation for a particle of mass m. One can easily show that $\chi^\dagger$ obeys the same equation. This completes the proof that with the factor of one-half included in Eq. (4.24), the parameter m is indeed the mass of the fermion.

4.10 Adding a Mass: Weyl Spinor

It is sometimes stated in the literature that it is not possible to write a mass term for a theory containing only a left-chiral particle (together with its right-chiral antiparticle), that it is necessary to introduce a right-chiral particle state. This is incorrect; we can easily write a mass term for a single left-chiral field. Actually, we have already done it in Eq. (4.24)! So the lagrangian for a massive left-chiral Weyl fermion is simply

$$\mathcal{L}_W = \chi_p^\dagger \bar{\sigma}^\mu i \partial_\mu \chi_p - \frac{m}{2}(\chi_p \cdot \chi_p + \bar{\chi}_p \cdot \bar{\chi}_p) \qquad (4.35)$$

We see that Eq. (4.35) *is exactly the same as the Majorana lagrangian expressed in two-component notation given in Eq. (4.24)*! There is no physical difference at all between the theory of a massive Weyl fermion and the theory of a Majorana particle. The only difference is a technical one: The Majorana theory is usually written in terms of the four-component Majorana spinor, as in Eq. (4.21), whereas the massive Weyl theory is written in terms of the two-component spinor, as in Eq. (4.35). In other words, we could say that the Majorana theory is simply the four-component version of a massive Weyl fermion (note that the degrees of freedom match; both theories contain two observables states).

It's because of this equivalence that sometimes people refer to Eq. (4.35) as the lagrangian of a Majorana spinor, not of a massive Weyl spinor, keeping the term *Weyl spinor* for *massless* spinors. But, as mentioned earlier, in this book we choose to define Weyl spinors through their Lorentz transformations (2.32), which hold regardless of the mass of the particles. We will reserve the expression *Majorana spinor* for the *four-component* spinor satisfying the condition of Eq. (4.7).

Let us close this section with an important remark. In the last two sections we chose to express everything in terms of the left-chiral state χ and its hermitian conjugate $\chi^\dagger$. It is, of course, possible to build Weyl spinors and Majorana spinors out of η and $\eta^\dagger$ instead, but it is conventional in SUSY to express all fermions in terms of left-chiral states (a choice most likely influenced by the habit of working with the standard model, in which there are no right-chiral neutrino states). This is why we focused on left-chiral spinors and did not include the corresponding expressions and identities for right-chiral spinors (which, however, can be worked out easily with the information that has been provided).

4.11 Relation Between Weyl Spinors and Majorana Spinors

Because of the equivalence between the Majorana and Weyl descriptions, we can always write *any* expression containing a Weyl spinor as one containing a Majorana spinor and vice versa. Indeed, all supersymmetry can be couched in either Weyl or Majorana language, and many references use only one of the two. Which one is preferred is obviously a matter of taste, but it could be argued that Weyl spinors are more fundamental because they are the simplest building blocks that can be used to construct supersymmetric theories. The main reason that Majorana spinors nevertheless are used widely is probably because calculations in that language resemble more those made with the familiar Dirac spinors, especially when it comes to working out scattering amplitudes. In this book the emphasis has been put on Weyl spinors, but the main results also will be expressed in the Majorana language, and some calculations of scattering amplitudes will be performed using the Majorana language.

In this section we will establish some basic relations between Weyl and Majorana spinors. Let's start with lagrangian terms without derivatives. We have already seen in Section 4.7 that

$$\overline{\Psi}_M \Psi_M = \chi \cdot \chi + \bar{\chi} \cdot \bar{\chi} \tag{4.36}$$

(we may drop the label p without causing any ambiguity because there is no distinction between the antiparticle and particle states for a Majorana fermion). Keep in mind that the bar over a Majorana spinor denotes the usual Dirac adjoint, $\overline{\Psi}_M = \Psi_M^\dagger \gamma^0$.

Obviously, we need a second relation if we want to be able to isolate $\chi \cdot \chi$ and $\bar{\chi} \cdot \bar{\chi}$ in terms of the Majorana spinor. What we are looking for is provided by sandwiching γ^5 between two Majorana spinors. It is easy to check that

$$\overline{\Psi}_M \gamma^5 \Psi_M = -\chi \cdot \chi + \bar{\chi} \cdot \bar{\chi} \tag{4.37}$$

Inverting Eqs. (4.36) and (4.37), we obtain

$$\chi \cdot \chi = \overline{\Psi}_M P_L \Psi_M$$
$$\bar{\chi} \cdot \bar{\chi} = \overline{\Psi}_M P_R \Psi_M \tag{4.38}$$

where the projection operators are defined in Eq. (2.18). Actually, it is straightforward to generalize these relations to products involving distinct spinors. Consider a second left-chiral Weyl spinor λ. Let's represent the Majorana spinor built out of λ as Λ_M. Then the generalizations of Eq. (4.38) are

$$\chi \cdot \lambda = \overline{\Psi}_M P_L \Lambda_M$$
$$\bar{\chi} \cdot \bar{\lambda} = \overline{\Psi}_M P_R \Lambda_M$$

Recall that the order in spinor dot products does not matter; i.e., $\chi \cdot \lambda = \lambda \cdot \chi$ and $\bar{\chi} \cdot \bar{\lambda} = \bar{\lambda} \cdot \bar{\chi}$. This implies that the order in the corresponding Majorana expressions does not matter either, i.e.,

$$\overline{\Psi}_M P_L \Lambda_M = \overline{\Lambda}_M P_L \Psi_M$$
$$\overline{\Psi}_M P_R \Lambda_M = \overline{\Lambda}_M P_R \Psi_M$$

which can be checked easily.

The next obvious thing to try is to sandwich γ^μ and $\gamma_5 \gamma^\mu$ between Majorana spinors. One can show that

$$\overline{\Psi}_M \gamma^\mu \Lambda_M = \chi^\dagger \bar{\sigma}^\mu \lambda - \lambda^\dagger \bar{\sigma}^\mu \chi$$
$$\overline{\Psi}_M \gamma^5 \gamma^\mu \Lambda_M = \chi^\dagger \bar{\sigma}^\mu \lambda + \lambda^\dagger \bar{\sigma}^\mu \chi$$

which imply

$$\chi^\dagger \bar{\sigma}^\mu \lambda = \overline{\Psi}_M P_R \gamma^\mu \Lambda_M$$
$$\lambda^\dagger \bar{\sigma}^\mu \chi = -\overline{\Psi}_M P_L \gamma^\mu \Lambda_M$$

Note that no σ^μ matrices appear here. These would appear if we were expressing all the spinors in terms of right-chiral states.

Of course, all these relations also can be verified explicitly by using Eq. (4.20) for a Majorana spinor in terms of Weyl spinors.

4.12 Quiz

1. What is the *CPT* conjugate state of a right-chiral particle?
2. What is the difference between a Dirac mass term and a Majorana mass term (after they are both expressed in terms of Weyl spinors)?
3. What is the difference between the theory of a Majorana spinor and the theory of a Weyl spinor?
4. Is it possible to write a Dirac spinor in terms of left-chiral fields only? What about in terms of right-chiral fields only?
5. Write down the mass term of a left-chiral Weyl spinor. Prove that it is real.
6. What symmetry, is a symmetry of a Dirac spinor but not a symmetry of a Majorana spinor (in the sense that a Dirac spinor has a well-defined transformation under this symmetry, whereas a Majorana spinor doesn't)?

CHAPTER 5

Building the Simplest Supersymmetric Lagrangian

We are finally ready to construct our first supersymmetric theory! This is where the fun really begins. It took a while to get here, but it was necessary to become familiar with Weyl and Majorana spinors and the notation used to describe them before we could attack supersymmetry (SUSY). This chapter will be short and sweet. The lagrangian considered truly will be the simplest one we could possibly imagine: There will be no mass terms and no interactions. Still, it will be sufficient to learn the essential concepts and to see SUSY in action (no pun intended).

5.1 Dimensional Analysis

Before building our first supersymmetric lagrangian, it will be useful to review how one works out the dimensions of the various quantities we will need. In natural units, both the speed of light and Planck's constant are set equal to 1. This implies that dimensions of time and length are equal, i.e., $[T] = [L]$. It also implies that mass and energy have the same dimensions because $E = mc^2$ becomes $E = m$ in natural units. In addition, since Planck's constant had dimensions of energy multiplied by time, setting it equal to 1 tells us that $[T] = [L] = [E]^{-1}$. We now can express the units of everything as powers of energy. Instead of writing $[E]^n$ for some exponent n, it is simpler to give the exponent. We therefore have the following dimensions for the fundamental scales of time, length, and mass:

$$[L] = [T] = -1 \qquad [M] = 1$$

Note that this implies that the dimension of a derivative ∂_μ is 1.

The action $S = -i(\int d^4x \mathcal{L})$ is a dimensionless quantity which implies that lagrangian densities (which we often simply call *lagrangians*) must have dimension 4. Using this, we can work out the dimensions of the fields we will deal with. For example, the kinetic energy of a scalar field is given by

$$\partial_\mu \phi \, \partial^\mu \phi \tag{5.1}$$

which tells us that a scalar quantum field has dimension 1 (because $[\partial_\mu] = 1$). The quantum electrodynamic (QED) lagrangian density is given by

$$\mathcal{L}_{\text{QED}} = \bar{\Psi}(P_\mu \gamma^\mu - m)\Psi - \frac{1}{4} F_{\mu\nu} F^{\mu\nu}$$

which tells us that a spinor quantum field has dimension $3/2$. We worked out the dimension of a Dirac field but of course Weyl and Majorana quantum fields have the same dimension. We also see that the energy-momentum tensor $F_{\mu\nu}$ has dimension 2. Since it is defined as

$$F_{\mu\nu} = \partial_\mu A_\nu - \partial_\nu A_\mu$$

we also find that the gauge field A_μ has dimension 1. The result is the same for nonabelian gauge fields such as the gluon fields.

Let us summarize our results:

$$[\partial_\mu] = [\phi] = [A^\mu] = 1$$

$$[\Psi_D] = [\Psi_M] = [\chi] = [\eta] = \frac{3}{2} \tag{5.2}$$

5.2 The Transformation of the Fields

The simplest supersymmetric model we may consider is composed of two free massless fields, with one bosonic and the other one fermionic. We therefore consider a Weyl spinor and a complex scalar field. By *free*, we mean that there is no interaction between the two fields. We could take the Weyl spinor to be either right- or left-chiral, but the convention is to use a left-chiral field. Our starting point therefore is the lagrangian

$$\mathcal{L} = \partial_\mu \phi \, \partial^\mu \phi^\dagger + \chi^\dagger i \bar\sigma^\mu \, \partial_\mu \chi \tag{5.3}$$

Now we are finally ready to introduce the SUSY transformations of the fields! At this point, we could simply throw at you the transformations of the fields, and we could check that they leave the lagrangian invariant (up to total derivatives). However, the transformations are quite complicated, and it's not obvious why anyone would have chosen these particular expressions in the first place. Since the goal of this book is to demystify SUSY, we will devote a fair part of this chapter to understanding the logic behind the choice of the supersymmetric transformations.

Consider first the transformation of the scalar field. We consider an infinitesimal transformation proportional to a small parameter ζ. Again, there is no standard notation for this parameter, so always check first the convention used when looking at a new reference. It might seem that ϵ would be a better choice of notation, but as we saw in Section 3.4, this letter is also used in most SUSY references to represent the matrices $\pm i\sigma^2$.

In the following calculations, always keep in mind that ζ is infinitesimal, so we will drop terms of order ζ^2 or higher in all our equations. Another important point is that we take ζ to be spacetime-independent; i.e.,

$$\partial_\mu \zeta = 0$$

a relation we will use repeatedly. We are therefore considering *global* SUSY transformations, as opposed to local transformations, for which ζ would be spacetime-dependent (another expression commonly used in the literature for global SUSY is *rigid supersymmetry*). Making ζ spacetime-dependent—in other words, *gauging SUSY*—forces one to introduce a gauge field that turns out to have the properties of a graviton (the force carrier of gravity), the same way that gauging the global $U(1)$ symmetry of the Dirac lagrangian leads to electromagnetism. For this reason, theories with local SUSY invariance are called *supergravity theories* (SUGRA for short). However, supergravity necessitates familiarity with general relativity, in particular the tetrad formalism, and is beyond the scope of this book.

We postulate that the variation of the scalar field is proportional to the Weyl spinor χ (this is, after all, the whole idea of SUSY: Bosonic fields, are transformed into fermionic fields, and vice versa). Let us then write, tentatively,

$$\phi \rightarrow \phi' = \phi + \delta\phi \tag{5.4}$$

with

$$\delta\phi \simeq \zeta\chi$$

We used the symbol $\simeq$ because this relation is still qualitative. We need a bit more work to make this precise.

When we write down transformations of fields, we must ensure two things: the two sides of the equation must have the same dimension, and they must have the same Lorentz transformations. Let's consider the second requirement first. Since ϕ is a scalar field, we must somehow build a Lorentz invariant out of ζ and χ. Since χ is a Weyl spinor, it is clear that we have no choice other than taking ζ to be a Weyl spinor too! This is a key result, so let us emphasize it:

> The parameter of a global SUSY transformation is a (constant) Weyl spinor

What does this mean? It means that ζ is a two-component quantity, namely,

$$\zeta = \begin{pmatrix} \zeta_1 \\ \zeta_2 \end{pmatrix}$$

that takes different values in different frames. We must now decide whether we should take ζ to be a right-chiral or a left-chiral spinor. We do have the choice, but again, the convention is to take it to be left-chiral. This tells us that if we do a Lorentz transformation, ζ transform as χ_L in Eq. (2.32). Note that ζ is not a quantum field, even though it is a spinor, so there is no particle associated to it.

Going back to the transformation of ϕ, we therefore must write down a Lorentz invariant made out of the two left-chiral spinors ζ and χ. But we already know how to do this; we simply take the spinor dot product $\zeta \cdot \chi$! Recall that the order does not matter, so we could as well have used $\chi \cdot \zeta$. We can finally write down the precise expression for the transformation of ϕ:

$$\boxed{\delta\phi = \zeta \cdot \chi}$$

(5.5)

Now let us turn to the question of the dimensions. Since ϕ has dimension 1 and χ has dimension 3/2, we obtain that

> The infinitesimal SUSY parameter ζ has dimension $-\dfrac{1}{2}$

Let's now consider the transformation of χ. We want it to be linear in ζ and ϕ or $\phi^\dagger$ (it would be extremely difficult to build a consistent theory with fields transforming nonlinearly). We will use $\phi^\dagger$ for a reason that will become clear in an instant. So we take

$$\delta\chi \simeq C \zeta \phi^\dagger$$

(5.6)

where C is an arbitrary constant that we will have to determine. The reason we did not include any such constant in the transformation of ϕ is that it could have been reabsorbed into the definition of ζ. However, once the normalization of ζ is fixed by the transformation of ϕ, we have no freedom left to get rid of the constant in the transformation of χ, and we will indeed see that C is not equal to 1.

Before going any further, we need to check whether the right-hand side of Eq. (5.6) has the same Lorentz transformation properties as the field χ (since the variation $\delta\chi$ must, of course, transform the same way as χ itself). In this respect, everything is fine: Since ϕ is a scalar, the right-hand side of Eq. (5.6) transforms like a left-chiral spinor because ζ is left-chiral.

But there is a problem: The left side has dimension 3/2, whereas the right side has dimension $1 - 1/2 = 1/2$. So we need to increase the dimension of the right side by one. We do not want to put in another factor of ϕ because we want the transformation to be linear in the fields. The only other quantity with dimension 1 at our disposal is ∂_μ, so we try

$$\delta\chi \simeq C \zeta \partial_\mu\phi^\dagger$$

Now the dimensions are right, but the Lorentz properties do not match! We need to contract the μ index with something, and it has to be something dimensionless. The only possible terms fitting the bill at our disposal are σ^μ and $\bar\sigma^\mu$. Looking at Eq. (2.44), we see that when applied to a left-chiral spinor, it is the matrices $\bar\sigma^\mu$ that have well-defined Lorentz transformations. We therefore try

$$\delta\chi \simeq C \bar\sigma^\mu \zeta \partial_\mu\phi^\dagger$$

but unfortunately, this still doesn't work! To see the problem and how to fix it, let's move $\partial_\mu \phi^\dagger$ in front (which we can do because it's not a matrix),

$$\delta\chi \simeq C\,(\partial_\mu \phi^\dagger)\,\bar{\sigma}^\mu\,\zeta \tag{5.7}$$

Recall that the quantity $\bar{\sigma}^\mu \partial_\mu \chi_L$ transforms as a right-chiral spinor [see Eq. (2.44)]. The fact that the derivative is acting on a scalar field instead of the spinor does not, obviously, affect the behavior of the expression under a Lorentz transformation, so $(\partial_\mu \phi^\dagger)\bar{\sigma}^\mu \zeta$ also transforms as a right-chiral spinor!

Thus the right-hand side of Eq. (5.7) transforms as a right-chiral spinor, whereas what we need is something that transforms as a left-chiral spinor. From Chapter 2, however, we know how to do make something out of a right-chiral spinor that will transform as a left-chiral spinor! The rule is to take the hermitian conjugate and transpose of the right-chiral spinor and then multiply it by $-i\sigma^2$ [see Eq. (2.40)].

Applying this to the first line of Eq. (5.7), we finally take the transformation of χ to be:

$$\delta\chi = -i\sigma^2 \big(C\,(\partial_\mu \phi^\dagger)\,\bar{\sigma}^\mu\,\zeta\big)^{\dagger T}$$

This has the right dimension and also transforms as a left-chiral spinor.

Recall that the operation $\dagger$ applies a hermitian conjugate both in the quantum field operator sense and in the matrix sense (a transpose followed by complex conjugation). Obviously, the transpose has no effect on the scalar field, so

$$(\phi^\dagger)^{\dagger T} = (\phi^\dagger)^\dagger = \phi$$

The only reason we chose to use $\phi^\dagger$ in our initial guess for the transformation of the spinor was so that the final expression would contain ϕ. Of course, this is purely a matter of taste; there would have been nothing wrong with having $\phi^\dagger$ in the final result instead!

On the other hand, the effect of $\dagger T$ on anything that is not a quantum field, *including the spinor* ζ, is to simply apply a complex conjugation. We therefore get

$$\delta\chi = -i\sigma^2\,C^*(\partial_\mu \phi)\,\bar{\sigma}^{\mu*}\,\zeta^*$$
$$= -C^*(\partial_\mu \phi)\,i\sigma^2\,\bar{\sigma}^{\mu*}\,\zeta^*$$

where

$$\zeta^* = \begin{pmatrix} \zeta_1^* \\ \zeta_2^* \end{pmatrix}$$

Using the fact that $\bar{\sigma}^{\mu*} = \bar{\sigma}^{\mu T}$ and $(\sigma^2)^2 = \mathbf{1}$, we may write

$$-C^* (\partial_\mu \phi) i\sigma^2 \bar{\sigma}^{\mu*} \zeta^* = -C^* (\partial_\mu \phi) i\sigma^2 \bar{\sigma}^{\mu T} \sigma^2 \sigma^2 \zeta^*$$

which, after using Eq. (A.8), gives our final result for the transformation of the spinor:

$$\boxed{\delta\chi = -C^* (\partial_\mu \phi) \sigma^\mu i\sigma^2 \zeta^*} \tag{5.8}$$

5.3 Transformation of the Lagrangian

We haven't even really started yet! All we have done is to write down transformations of the fields that have the correct dimensions and correct behavior under the Lorentz group. What we want to do now is to see if there is a choice of constant C such that the preceding transformations will leave the action invariant. Recall that the action is the integral of the lagrangian density:

$$S = \int d^4x \mathcal{L}$$

The action is invariant if $\delta S = 0$, which implies

$$\int d^4x \delta\mathcal{L} = 0$$

This does not imply that $\delta\mathcal{L}$ itself must be zero. If we assume that the fields vanish sufficiently rapidly at spatial infinity, the action will be invariant if either the variation of the lagrangian density is zero *or* if it is a total derivative of the fields:

$$\delta\mathcal{L} = \partial_\mu \mathcal{F}(\phi, \chi, \dots)$$

where $\mathcal{F}$ is some function of the fields, and the dots are included because later on we will have other fields in our lagrangians.

Now let us build our first supersymmetric theory! Again, the goal is to see if we can find a value of the constant C such that the variation of the lagrangian density will be zero after dropping any total derivative. The variation of the lagrangian

density Eq. (5.3) is

$$\delta\mathcal{L} = \partial_\mu(\delta\phi)\,\partial^\mu\phi^\dagger + \partial_\mu\phi\,\partial^\mu\delta(\phi^\dagger) + \delta(\chi^\dagger)\,i\bar\sigma^\mu\,\partial_\mu\chi + \chi^\dagger\,i\bar\sigma^\mu\,\partial_\mu\delta\chi$$

$$= \partial_\mu(\delta\phi)\,\partial^\mu\phi^\dagger + \partial_\mu\phi\,\partial^\mu(\delta\phi)^\dagger + (\delta\chi)^\dagger\,i\bar\sigma^\mu\,\partial_\mu\chi + \chi^\dagger\,i\bar\sigma^\mu\,\partial_\mu\delta\chi \quad (5.9)$$

where we have used $\delta(\phi^\dagger) = (\delta\phi)^\dagger$ and $\delta(\chi)^\dagger = (\delta\chi)^\dagger$. To see that this is true, let's take the dagger of Eq. (5.4):

$$\phi^\dagger \to \phi^\dagger + (\delta\phi)^\dagger$$

On the other hand, by definition, we have

$$\phi^\dagger \to \phi^\dagger + \delta(\phi^\dagger)$$

so, clearly, $\delta(\phi^\dagger) = (\delta\phi)^\dagger$, and we may simply write $\delta\phi^\dagger$ without any ambiguity.

So the variation of $\phi^\dagger$ and $\chi^\dagger$ are simply the hermitian conjugates of Eqs. (5.5) and (5.8). One finds (see Exercise 5.1):

$$(\delta\phi)^\dagger = \bar\chi \cdot \bar\zeta = \chi^\dagger\,(i\sigma^2)\,\zeta^* \quad (5.10)$$

and

$$(\delta\chi)^\dagger = C\,(\partial_\mu\phi^\dagger)\,\zeta^T\,i\sigma^2\,\sigma^\mu \quad (5.11)$$

EXERCISE 5.1
Prove Eqs. (5.10) and (5.11).

Using Eqs. (5.5), (5.8), (5.10), and (5.11) into Eq. (5.9), we get

$$\delta\mathcal{L} = \underbrace{(\partial^\mu\chi^\dagger)i\sigma^2\,\zeta^*\partial_\mu\phi}_{\text{I}} \underbrace{-(\partial^\mu\phi^\dagger)\,\zeta^T\,(i\sigma^2)\,\partial_\mu\chi}_{\text{II}}$$

$$+ \underbrace{C(\partial_\nu\phi^\dagger)\,\zeta^T\,(i\sigma^2)\,\sigma^\nu\,i\bar\sigma^\mu\,\partial_\mu\chi}_{\text{III}} \underbrace{-C^*\chi^\dagger\,i\bar\sigma^\mu\,\sigma^\nu\,(\partial_\mu\partial_\nu\phi)(i\sigma^2\,\zeta^*)}_{\text{IV}} \quad (5.12)$$

Note that in all the expressions we will write down, derivatives always will be acting only on the field that is immediately to their right. To make this explicit, we will always group together a derivative and the field it is acting on inside parentheses. For example, instead of writing $\partial_\mu\phi\chi$, which could cause confusion, we will write $(\partial_\mu\phi)\chi$.

Keep in mind that the scalar field (or the scalar field with any number of derivatives acting on it) can be moved freely through spinors and matrices. Also recall that two lagrangian terms differing by a total derivative are to be considered equivalent. These points are important to remember if you do some calculations and obtain a result that seems at first sight to be different from the one given in this book or in any other SUSY reference. In addition, several spinor identities we have seen in previous chapters, the most important of which are listed in Appendix A, may be used to rearrange expressions. It takes some practice to be able to tell quickly whether or not two apparently different expressions are actually equivalent.

Coming back to Eq. (5.12), we observe that the first and last term contain ζ^*, whereas the second and third term contain ζ. Thus each pair of terms must cancel separately (up to total derivatives).

Consider the first and last terms. At first sight, these two terms look very different: The last term contains $\bar{\sigma}^\mu$ and σ^ν, whereas these matrices do not appear at all in the first term. However, we have already shown how the two are related! Indeed, in Eq. (4.34) we proved that

$$\sigma^\mu \bar{\sigma}^\nu \partial_\mu \partial_\nu \chi = \partial^\mu \partial_\mu \chi$$

The proof makes it evident that the identity also holds if the derivatives are acting on a scalar field, so we have

$$\sigma^\mu \bar{\sigma}^\nu \partial_\mu \partial_\nu \phi = \partial_\mu \partial^\mu \phi$$
$$= \Box \phi \tag{5.13}$$

In Section 4.9 we proved this using the identity (A.9). It also can be proven by directly contracting the Lorentz indices, as you are invited to do in Exercise 5.2.

EXERCISE 5.2
Prove Eq. (5.13) by expanding explicitly the Lorentz indices (i.e., by writing $\sigma^\nu \partial_\nu = \sigma^0 \partial_t + \sigma^i \partial_i$ and so on).

Therefore, the sum of the first and last terms of Eq. (5.12) is

$$\text{I} + \text{IV} = (\partial^\mu \chi^\dagger) i \sigma^2 \zeta^* \partial_\mu \phi - i C^* \chi^\dagger \Box \phi i \sigma^2 \zeta^*$$

Of course, we can move $\Box \phi$ around because it is not a matrix quantity and write

$$\text{I} + \text{IV} = (\partial^\mu \chi^\dagger) i \sigma^2 \zeta^* \partial_\mu \phi - i C^* \chi^\dagger i \sigma^2 \zeta^* \Box \phi$$

We cannot make those two terms cancel because the derivatives act on different fields. But we have a trick to move derivatives around: integrations by parts! If we integrate by parts the derivative acting on $\chi^\dagger$ and use the fact that $\partial_\mu \zeta^* = 0$, we get

$$I + IV = -\chi^\dagger i\sigma^2 \zeta^* \Box\phi - iC^* \chi^\dagger i\sigma^2 \zeta^* \Box\phi \tag{5.14}$$

where, as usual, we have dropped the surface term. We see that this vanishes if we take $C = -i$.

Now consider the sum of the second and third terms of Eq. (5.12):

$$II + III = -(\partial^\mu \phi^\dagger) \zeta^T i\sigma^2 \partial_\mu \chi + C(\partial_\nu \phi^\dagger) \zeta^T i\sigma^2 \sigma^\nu i\bar\sigma^\mu \partial_\mu \chi$$

which, after integrating the derivatives acting on the spinor field by parts and using Eq. (5.13), may be written as

$$II + III = (\Box\phi^\dagger) \zeta^T i\sigma^2 \chi - iC (\Box\phi^\dagger) \zeta^T i\sigma^2 \chi \tag{5.15}$$

which again vanishes if we take $C = -i$!

We have therefore achieved our goal: We have introduced a SUSY transformation that leaves our action invariant! We have shown that the lagrangian

$$\mathcal{L}_{\text{free}} = \partial_\mu \phi \, \partial^\mu \phi^\dagger + \chi^\dagger i\bar\sigma^\mu \partial_\mu \chi \tag{5.16}$$

is invariant under the SUSY transformations

$$\delta\phi = \zeta \cdot \chi$$
$$\delta\phi^\dagger = \bar\zeta \cdot \bar\chi$$
$$\delta\chi = -i(\partial_\mu \phi) \sigma^\mu i\sigma^2 \zeta^*$$
$$\delta\chi^\dagger = -i(\partial_\mu \phi^\dagger)\zeta^T i\sigma^2 \sigma^\mu \tag{5.17}$$

Building our first supersymmetric theory was not so hard after all! Unfortunately, there is one subtlety that will require us to modify these results significantly in two ways: we will need to include a new field in the lagrangian (with its own SUSY transformation) and modify the transformation of the spinor. To understand why anyone would want to do that, we will first need to review the concept of symmetry charges and their algebra.

5.4 Quiz

1. What is *rigid supersymmetry*?
2. What is the dimension of the infinitesimal SUSY parameter ζ?
3. We want the transformation of the spinor field to be linear in the scalar field ϕ and in the SUSY parameter ζ. Using dimensional analysis alone, what form must the variation of the spinor field $\delta\chi$ take (the answer does not have to have the correct Lorentz transformation property)?
4. What is SUGRA?
5. What would be wrong with the transformation $\delta\chi = C(\partial_\mu\phi)\,\bar{\sigma}^\mu\zeta$?

CHAPTER 6

The Supersymmetric Charges and Their Algebra

When a lagrangian is observed to possess some continuous symmetry, the natural thing to do is to find the *charges* generating the symmetry, also called the *generators* of the symmetry. Actually, as we will see, what really matters is the *algebra* obeyed by the charges, i.e., the commutation or anticommutation rules, not the charges themselves (as an aside, when the algebra involves anticommutators, it is often referred to as a *graded Lie algebra*). We will start by reviewing some elementary results concerning charges and their algebra before tackling the supersymmetry (SUSY) algebra in Section 6.5. We will be led to modify our supersymmetric theory of the preceding chapter substantially in order for the SUSY algebra to be well-defined for off-shell fields (see Sections 6.6 to 6.8).

6.1 Charges: General Discussion

To explain what charges are, let's go back to the field transformations $\delta\phi$ and $\delta\chi$, which leave the action invariant. We have written these variations explicitly in terms of the fields and of the infinitesimal spinor ζ. Now, if we think in terms of canonical quantization, ϕ and χ are operators, and the calculation of any physical process always boils down to the evaluation of expectation values of some combination of the fields.

To simplify the discussion, consider for now a scalar field operator ϕ (unrelated to the scalar field ϕ that appeared in our supersymmetric lagrangian). If a certain transformation is a symmetry of the theory, it must leave unchanged the expectation value

$$\langle a|\phi|b\rangle \tag{6.1}$$

We will be mostly interested in spacetime transformations and, later, the implementation of the supersymmetric transformations as transformations in superspace. Let us then focus for now on spacetime transformations. Transformations that leave a classic scalar field unchanged should leave the expectation value (6.1) unaffected as well. To be specific, we should have

$$\langle a'|\phi(x')|b'\rangle = \langle a|\phi(x)|b\rangle \tag{6.2}$$

Note that there are two differences between the two expressions: The field operator is evaluated at different spacetime points, and it is sandwiched between different states. We now introduce a new field operator ϕ' that we define by

$$\langle a'|\phi(x')|b'\rangle \equiv \langle a|\phi'(x')|b\rangle \tag{6.3}$$

In other words, the effects of the change of states is contained in the definition of the new field operator. If we substitute the right-hand side of Eq. (6.3) back into Eq. (6.2), we get

$$\langle a|\phi'(x')|b\rangle = \langle a|\phi(x)|b\rangle$$

Since the states are arbitrary, we must conclude that

$$\phi'(x') = \phi(x)$$

This is not surprising, this is exactly how a classic scalar field behaves under a Lorentz transformation!

Let's write x' as $x - \delta x$ (the reason for the choice of sign of δx will become clear shortly). This gives us

$$\phi'(x - \delta x) = \phi(x)$$

or

$$\boxed{\phi'(x) = \phi(x + \delta x)} \tag{6.4}$$

We see that our choice of a negative sign of the variation in x' was so that we would get a positive sign on the right-hand side of Eq. (6.4).

We will use this in a moment, but first, let's talk about what happens if the field is not a scalar. Let's consider instead a field operator $\mathcal{F}$ that transforms under some representation of the Lorentz group (a vector, spinor, etc.). Then, Eq. (6.2) will be replaced by

$$\langle a'|\mathcal{F}(x')|b'\rangle = S(\Lambda)\,\langle a|\mathcal{F}(x)|b\rangle \tag{6.5}$$

where $S(\Lambda)$ is the Lorentz transformation matrix corresponding to the representation of the field operator. Correspondingly, Eq. (6.4) is replaced by

$$\mathcal{F}'(x) = S(\Lambda)\,\mathcal{F}(x + \delta x) \tag{6.6}$$

We will need to apply this result to the case of a spinor field later in this chapter.

Now we will work out a different expression for the left-hand side of Eq. (6.2), an expression that will allow us to establish a connection with the charges.

Physically meaningful transformations preserve the norm of states and therefore can be written in the form of unitary operators (or antiunitary operators in the case of time reversal), namely, operators satisfying $U^\dagger = U^{-1}$. We will take the transformation of the states to be

$$|a\rangle \to U^\dagger|a\rangle \qquad |b\rangle \to U^\dagger|b\rangle \tag{6.7}$$

so the left-hand side of Eq. (6.2) takes the form

$$\langle a'|\phi(x')|b'\rangle = \langle a|U\phi(x')U^\dagger|b\rangle$$

It is, of course, arbitrary whether we define the states to transform with the matrix U or with the matrix $U^\dagger$, and you will encounter both conventions in the literature. Now, using Eq. (6.3), we can relate this expression to the primed field ϕ':

$$\langle a|\phi'(x')|b\rangle = \langle a|U\phi(x')U^\dagger|b\rangle$$

which implies

$$\phi'(x') = U\phi(x')U^\dagger$$

or

$$\phi'(x) = U\phi(x)U^\dagger \tag{6.8}$$

Now we are ready to introduce the charges generating the transformations. Assume that the transformations we are interested in are parametrized by a set of n infinitesimal parameters $\epsilon_1, \epsilon_2, \ldots, \epsilon_n$, in which case there are necessarily n charges $Q_1, Q_2, \ldots, Q_n$. The charges are defined through

$$U \equiv \exp(\pm i\epsilon_i Q_i) = \exp(\pm i\epsilon \cdot Q) \tag{6.9}$$

where we introduced a shorthand dot notation for the sum $\epsilon_i Q_i$. We allowed for two possible signs in the argument of the exponential because we will encounter both cases in our calculations.

It is important to keep in mind that here the ϵ could either be ordinary (commuting) constants or Grassmann quantities, such as the Weyl spinor ζ that we introduced to parametrize supersymmetric transformations. The corresponding charges are then, respectively, bosonic or fermionic.

Substituting Eq. (6.9) in Eq. (6.8) and expanding to first order in ϵ, we get

$$\begin{aligned}
\phi'(x) &\approx (1 \pm i\epsilon \cdot Q)\,\phi(x)\,(1 \mp i\epsilon \cdot Q) \\
&= \phi(x) \pm i\epsilon \cdot Q\,\phi(x) \mp i\phi(x)\epsilon \cdot Q \\
&= \phi(x) \pm i[\epsilon \cdot Q, \phi(x)]
\end{aligned} \tag{6.10}$$

Let's define

$$\phi'(x) = \phi(x) + \delta\phi(x) \tag{6.11}$$

By comparing Eqs. (6.10) and (6.11), we obtain a very useful result:

$$\boxed{\pm[\epsilon \cdot Q, \phi] = -i\,\delta\phi} \tag{6.12}$$

where the sign on the left-hand side is the same sign as the argument of the exponential in the operator U.

In the case of spacetime symmetries, Eqs. (6.4) and (6.11) imply the obvious relation

$$\delta\phi = \phi(x + \delta x) - \phi(x) \tag{6.13}$$

For a general field $\mathcal{F}$, we have instead

$$\delta\mathcal{F}(x) = S(\Lambda)\,\mathcal{F}(x + \delta x) - \mathcal{F}(x) \tag{6.14}$$

Equations (6.12) and (6.4) [and its generalization, Eq. (6.6)] are the key results of this section. They show us how to relate the transformation of a field to a commutator with a charge and to the transformation of the coordinates, respectively.

As a very simple example, consider a lagrangian density that is a function of a single real scalar field $\mathcal{L}(\phi)$. The action $S = i\int d^4x\,\mathcal{L}$ is obviously invariant under a translation of the field

$$\phi(x^\mu) \to \phi'(x^\mu) = \phi(x^\mu + a^\mu) \tag{6.15}$$

where a^μ is a constant-displacement four-vector. Therefore, the variation of the field $\delta\phi$ under this symmetry transformation is

$$\delta\phi(x^\mu) = \phi(x^\mu + a^\mu) - \phi(x^\mu)$$
$$\approx a^\mu\partial_\mu\phi \tag{6.16}$$

This is an obvious invariance because the lagrangian density is integrated over all spacetime, so we can always do a shift of the integration variable $x^\mu \to x^\mu + a^\mu$, which implies that

$$\int d^4x\,\mathcal{L}(\phi) = \int d^4x\,\mathcal{L}(\phi'). \tag{6.17}$$

Since there are four parameters, the four a_μ, we introduce four charges that we suggestively call P^μ, and write the unitary operator as

$$U \equiv \exp(ia^\mu P_\mu) \tag{6.18}$$

Substituting Eqs. (6.16) and (6.18) into Eq. (6.12), we get

$$a^\mu[P_\mu, \phi] = -ia^\mu\partial_\mu\phi \tag{6.19}$$

Since this must be true for an arbitrary infinitesimal four-vector a_μ, we conclude that

$$\boxed{[P_\mu, \phi] = -i\partial_\mu\phi} \tag{6.20}$$

a relation that will soon prove useful.

As an aside, the charges of a symmetry sometimes are also called the *generators* of the transformation. For example, the P_μ are often called the *generators of translation* in spacetime.

As a second simple example, consider a theory containing a single complex scalar field and which is described by a lagrangian density that is invariant under a $U(1)$ transformation, $\phi \rightarrow e^{i\alpha}\phi$, with α being a constant parameter. This is an example of an *internal symmetry*, i.e., a symmetry that is not related to a change of coordinates but to an equivalence between different fields at the same spacetime point. In this case, the variation of the field is $\delta\phi = i\alpha\phi$. Writing $U = \exp(i\alpha Q)$, Eq. (6.12) simply gives

$$[Q, \phi] = \phi \tag{6.21}$$

A less trivial example is the object of Exercise 6.1.

EXERCISE 6.1

Consider the Lorentz transformation

$$x^\mu \rightarrow \Lambda^\mu{}_\nu x^\nu$$

For an infinitesimal transformation, we may write

$$\Lambda^\mu{}_\nu x^\nu \approx x^\mu + \omega^\mu{}_\nu x^\nu$$
$$= x^\mu + \omega^{\mu\nu} x_\nu \tag{6.22}$$

where $\omega^{\mu\nu}$ is antisymmetric, $\omega^{\mu\nu} = -\omega^{\nu\mu}$. Because of this antisymmetry, it is important to keep track of which index is the first one and which is the second one when we write $\omega^\mu{}_\nu$, which is why the lower index is shifted to the right.

We write the unitary operator implementing Lorentz transformation on a scalar field as

$$U = e^{-\frac{i}{2}\omega^{\mu\nu} M_{\mu\nu}}$$

where the charges $M^{\mu\nu}$ are also antisymmetric. The factor of $1/2$ is included to allow the sum to be over all values of the indices without double counting (since $\omega^{12} M_{12} = \omega^{21} M_{21}$, for example). Note the sign in the exponent! It is natural to take the Lorentz transformation to have the opposite sign to the translation transformation by analogy with classic physics (see Section 2.5 of Ref. 20 for more details).

The transformation of the field is given by

$$\delta\phi = \phi(x^\mu + \omega^{\mu\nu} x_\nu) - \phi(x^\mu)$$

First, prove that we may write

$$\delta\phi = \frac{1}{2}\omega^{\mu\nu}(x_\nu\partial_\mu - x_\mu\partial_\nu)\phi \qquad (6.23)$$

Then prove that the commutation relations between the charges $M^{\mu\nu}$ and a scalar field are

$$[M_{\mu\nu}, \phi] = i(x_\nu\partial_\mu - x_\mu\partial_\nu)\phi$$

The combination of the Lorentz transformations introduced in Exercise 6.1 and the translations generated by the P_μ form the *Poincaré group*. You will work out the Poincaré algebra, the set of commutation relations obeyed by the Lorentz and translation generators, in Exercise 6.3.

There is unfortunately no standard in the literature for the signs used in the exponents of the operators U, i.e., for the choice between U and $U^\dagger$ in the transformations of the states (6.7) or for the choice between active and passive transformations in the change of coordinates (which affects the sign of δx). A difference of convention in any of these aspects affects the signs of some of the equations containing charges. Because of this, it is typical to see equations differing in signs from one book to the next. I have chosen a set of conventions that maximizes agreement with most introductory references on SUSY, but it's impossible to agree with all of them because they don't even agree with one another!

Note that Eqs. (6.20) and (6.21) are formal expressions in the sense that they tell us what the commutations of the charges with the scalar field are without giving us explicit expressions for those charges. It is possible to obtain explicit representations of the charges, as we will discuss in the next section, but what is even more useful is to work out the algebra of the charges, a topic to which we now turn our attention.

6.2 Explicit Representations of the Charges and the Charge Algebra

Given that we already have figured out the SUSY transformations of the fields, it might seem like there is no point in finding the charges. However, the charges associated with a symmetry provide a powerful tool to learn a great deal about theories having this symmetry without even writing down any lagrangian! Actually, we don't even need to work out the explicit charges; all we really need is the *algebra* of the charges, i.e., the commutation rules (or anticommutation rules) they obey.

As an example of how the knowledge of mere commutation rules may be so useful, think about the harmonic oscillator in quantum mechanics. Given only the commutation rules of the raising and lowering operators, it is possible to find the allowed energies of the system without even working out any explicit wavefunction! Or consider the spin operators $S_\pm$ and S_z. Using only the algebra between those operators, it is possible to show that particles must have either integer or half-integer spins and that for a given spin s there are $2s + 1$ states. In a similar vein, the algebra of the SUSY charges will reveal many general properties of supersymmetric theories, such as the types of particles that must be grouped together in order to form SUSY-invariant theories without us having to write down any lagrangian.

Now that we have motivated the usefulness of knowing the algebra of the charges generating a symmetry, let us turn to the question of how to actually determine it.

One approach is to build explicit representations for the charges and then to compute by brute force their algebra. It is difficult to use Eq. (6.12) to guess the correct expressions for the charges because of the commutator on the left-hand side. Fortunately, a systematic way to obtain the charges is available. It consists of building a set of currents J_μ^i associated with the symmetry from which the corresponding charges are found by integrating over space (not spacetime) the zeroth component of the currents:

$$Q^i = \int d^3x \, J_0^i (\vec{x}, t)$$

It is important to note that in this approach, the charges (and the currents) are themselves *quantum field operators*. The algebra of the charges then can found using the equal time commutation relations of the fields present in the theory.

We will review the use of currents in detail in Section 7.7. For now, let us simply summarize the strategy underlying this approach. The starting point is a set of field transformations leaving the action invariant (i.e., leaving the lagrangian density invariant up to total derivatives). From that, one can work out the algebra of the charges through the following steps:

Field transformations $\Rightarrow$ currents $\Rightarrow$ charges as quantum fields $\Rightarrow$ algebra

(6.24)

Of course, as a double-check that the charges are correct, one then can use Eq. (6.12) to recover the transformations of the fields.

In the case of spacetime symmetries, there is a *second* explicit representation of the charges available: as *differential operators* acting on the fields. In this chapter we will use carets over the differential operator representation of the charges to distinguish them from their representation as quantum field operators. Let us now see how these differential operators can be found. The differential operator representation of the charges is defined via

$$\phi(x') \equiv \exp(\pm i\epsilon_i \hat{Q}_i)\, \phi(x)$$

$$\approx \phi(x) \pm i\epsilon_i \hat{Q}_i\, \phi(x) \tag{6.25}$$

where now we think of the field ϕ as an ordinary function of spacetime. If we know explicitly the transformation of the coordinates, we simply write x' in terms of x and Taylor expand $\phi(x')$ to first order in the change of coordinates, which allows us to read off the charges $\hat{Q}$ directly. Once we have differential operators representations for the charges, it is a simple matter to compute their algebra.

As an example, consider again spacetime translations, Eq. (6.15). Using $\hat{P}_\mu$ to represent the differential operator representation of the charges associated with this symmetry, we write

$$\hat{U} = \exp(-ia^\mu \hat{P}_\mu) \tag{6.26}$$

Note the sign difference with respect to Eq. (6.18). There is much freedom in the choices of signs in most formula we deal with in this chapter, which makes it quite frustrating to compare equations from different references that typically do not use the same conventions. The absence of standard notation means that the signs of the commutation relations and the signs in the definitions of the charges will vary from one source to another. This must be kept in mind when consulting the literature.

With our convention, we find

$$\phi(x) + a^\mu \partial_\mu \phi(x) = \phi(x) - ia^\mu \hat{P}_\mu \phi(x) \tag{6.27}$$

which leads to

$$\boxed{\hat{P}_\mu = i\partial_\mu} \tag{6.28}$$

Note the difference between Eqs. (6.28) and (6.20). It goes beyond the difference of sign on the right-hand side! In Eq. (6.20), the charge P^μ is a *quantum field operator*, not a differential operator.

EXERCISE 6.2
Define the action of the differential operators $\hat{M}_{\mu\nu}$ on a scalar field through

$$\phi(x') \equiv \exp\left(\frac{i}{2}\omega^{\mu\nu} \hat{M}_{\mu\nu}\right)\phi \tag{6.29}$$

with the change of coordinates given again by Eq. (6.22). Show that

$$\hat{M}_{\mu\nu} = i(x_\mu \partial_\nu - x_\nu \partial_\mu) \tag{6.30}$$

Be warned that most books do not use a different notation for the charges represented by quantum field operators and the charges represented by differential operators, which can lead to some confusion. Even worse, the same symbol is often used to represent the *eigenvalues* of the charges. For example, consider again P^μ. Depending on the context, this may represent a differential operator, a quantum field operator, or the actual four-momentum of a state! Always make sure that you know which of these meaning is implied when looking at an equation that contains a charge. In this chapter we use a caret, as in $\hat{P}_\mu$, for the differential operator representation and a simple capital letter, as in P^μ, for the quantum field operator. Later, when we will need to represent the eigenvalues of the momentum operator, we will use a lowercase letter, as in p_μ.

Once we know the representation of the charges as differential operators, it is a trivial matter to work out the algebra they obey (see Exercise 6.3 for the example of the Poincaré algebra).

To summarize, the strategy is to start from transformations of the coordinates that leave the lagrangian density invariant (up to total derivatives) and from there to go through the following steps:

Transformation of coordinates $\Rightarrow$ charges as differential operators $\Rightarrow$ algebra   (6.31)

At first, it may seem as if this approach is irrelevant to SUSY, a symmetry relating boson and fermion fields. However, as already mentioned in Chapter 1, and as we will show explicitly shortly, there is a deep connection between SUSY and Lorentz transformations, a connection that can be formalized by defining an extension of spacetime called *superspace*, which contains Grassmann coordinates in addition to the usual spatial and time coordinates. One then can write the SUSY charges as differential operators acting in that extended space. This approach will be pursued in Chapter 10.

If the algebra of the charges is known, one actually can reverse the steps illustrated in Eq. (6.31). One may use the algebra to work out the transformation of the coordinates. This is how we will use Eq. (6.31) in Chapter 10, where we will start from the algebra of the SUSY charges to work out the SUSY transformations of the superspace coordinates.

We have so far discussed two ways to obtain the algebra of the charges by building explicit representations for them. It turns out that it is also possible to obtain the algebra without ever writing down any explicit representation of the charges. In other words, one may start from the transformations of the fields leaving the action invariant and obtain the algebra directly:

$$\text{Field transformations} \Rightarrow \text{algebra} \qquad (6.32)$$

There is also a shortcut in the case of spacetime symmetries: One can start from the transformation of the coordinates and obtain the algebra of the charges directly:

$$\text{Coordinate transformations} \Rightarrow \text{algebra} \qquad (6.33)$$

In the case of SUSY, the only information we have so far is the field transformations that leave the action invariant, so the only two routes available to us to find the algebra are Eqs. (6.24) and (6.32) (as noted earlier, the representation of SUSY transformations as changes of coordinates will have to wait until Chapter 10, where we introduce superspace). In the next section we will work out the algebra using the shortcut of Eq. (6.32).

EXERCISE 6.3
Using the representations of the operators P^μ and $M^{\mu\nu}$ as differential operators [as given in Eqs. (6.28) and (6.30)], show that they obey the so-called Poincaré algebra, i.e.,

$$[\hat{P}_\mu, \hat{P}_\nu] = 0$$
$$[\hat{M}_{\mu\nu}, \hat{P}_\lambda] = i(\eta_{\nu\lambda}\hat{P}_\mu - \eta_{\lambda\mu}\hat{P}_\nu)$$
$$[\hat{M}_{\mu\nu}, \hat{M}_{\rho\sigma}] = i(\eta_{\nu\rho}\hat{M}_{\mu\sigma} - \eta_{\nu\sigma}\hat{M}_{\mu\rho} - \eta_{\mu\rho}\hat{M}_{\nu\sigma} + \eta_{\mu\sigma}\hat{M}_{\nu\rho}) \qquad (6.34)$$

Warning: Again, several different conventions are used in the literature, which leads to commutation relations differing by an overall sign.

6.3 Finding the Algebra Without the Explicit Charges

We will now show how the knowledge of the field transformations can be used to obtained the algebra of the charges directly, following Eq. (6.32). In this section, all the charges have to be thought of as *quantum field operators*, not differential operators.

To find the algebra of the charges generating a symmetry, we clearly need to consider *two* transformations in order to obtain an expression with a product of two charges. Let us then consider two successive transformations applied to a field ϕ: a first transformation with parameters α_i, followed by a second transformation with parameter β_j. We will use a dot-product notation for the combinations $\alpha_i Q_i$ and $\beta_j Q_j$.

We have

$$U_\beta\, U_\alpha\, \phi\, U_\alpha^\dagger\, U_\beta^\dagger = \exp(i\beta \cdot Q)\, \exp(i\alpha \cdot Q)\, \phi\, \exp(-i\alpha \cdot Q)\, \exp(-i\beta \cdot Q)$$

$$\approx \phi + i[\beta \cdot Q, \phi] + i[\alpha \cdot Q, \phi]$$

$$- [\beta \cdot Q, [\alpha \cdot Q, \phi]] + \cdots \tag{6.35}$$

where we neglected all terms of order α^2 or β^2 but kept the term bilinear in α and β.

By definition, Eq. (6.35) is simply a variation with parameter α followed by a variation with parameter β so that we can, of course, also write

$$U_\beta\, U_\alpha\, \phi\, U_\alpha^\dagger\, U_\beta^\dagger = \delta_\beta\, \delta_\alpha\, \phi \tag{6.36}$$

We now consider the commutator of two transformations, more precisely

$$\delta_\beta\, \delta_\alpha\, \phi - \delta_\alpha\, \delta_\beta\, \phi \tag{6.37}$$

This is equal to

$$U_\beta\, U_\alpha\, \phi\, U_\alpha^\dagger\, U_\beta^\dagger - U_\alpha\, U_\beta\, \phi\, U_\beta^\dagger\, U_\alpha^\dagger \approx -[\beta \cdot Q, [\alpha \cdot Q, \phi]] + [\alpha \cdot Q, [\beta \cdot Q, \phi]] \tag{6.38}$$

Setting Eq. (6.38) equal to Eq. (6.37), we get

$$\delta_\beta\, \delta_\alpha\, \phi - \delta_\alpha\, \delta_\beta\, \phi = -[\beta \cdot Q, [\alpha \cdot Q, \phi]] + [\alpha \cdot Q, [\beta \cdot Q, \phi]] \tag{6.39}$$

If we expand this out, being careful about the order of all the quantities involved, we find that half the terms cancel out, leaving

$$\delta_\beta \delta_\alpha \phi - \delta_\alpha \delta_\beta \phi = \alpha \cdot Q \ \beta \cdot Q \ \phi - \beta \cdot Q \ \alpha \cdot Q \ \phi - \phi \ \alpha \cdot Q \ \beta \cdot Q$$
$$+ \phi \ \beta \cdot Q \ \alpha \cdot Q$$
$$= (\alpha \cdot Q \ \beta \cdot Q - \beta \cdot Q \ \alpha \cdot Q) \phi$$
$$- \phi (\alpha \cdot Q \ \beta \cdot Q - \beta \cdot Q \ \alpha \cdot Q)$$
$$= [\alpha \cdot Q, \ \beta \cdot Q] \phi - \phi [\alpha \cdot Q, \ \beta \cdot Q]$$
$$= \big[[\alpha \cdot Q, \ \beta \cdot Q], \ \phi \big] \qquad (6.40)$$

The same result, of course, can be obtained using the Jacobi identity, which may be written in the form

$$\big[A, [B, C] \big] - \big[B, [A, C] \big] = \big[[A, B], C \big] \qquad (6.41)$$

Applying this to Eq. (6.39) gives right away the last line of Eq. (6.40).

The next step is to factor out the infinitesimal parameters, but the details depend on whether the charges are bosonic, which is the case for all continuous symmetries encountered in introductory quantum field theory classes, or fermionic, as is the case in SUSY. When the charges are bosonic, Eq. (6.40) leads to the *commutation* relations between the charges, whereas for fermionic charges, the equation yields *anticommutation* rules, as we will see explicitly soon.

6.4 Example

Before tackling the more complex SUSY case, it might be useful to pause and do a couple of examples with more familiar symmetries.

We will illustrate the use of Eq. (6.40) in the case of the algebra between the charges P^μ. First, consider the left side of the equation. If we apply a second translation to $\delta_a \phi(x_\mu)$ with an infinitesimal displacement b_μ this time, we get

$$\delta_b \delta_a \phi(x_\mu) = \delta_b \big(\phi(x_\mu + a_\mu) - \phi(x_\mu) \big)$$
$$= \delta_b \phi(x_\mu + a_\mu) - \delta_b \phi(x_\mu)$$
$$= \phi(x_\mu + a_\mu + b_\mu) - \phi(x_\mu + a_\mu) - \phi(x_\mu + b_\mu) + \phi(x_\mu) \qquad (6.42)$$

This result is symmetric under the exchange $a_\mu \leftrightarrow b_\mu$, which is not surprising because the order of the two translations does not matter. This implies that

$$\delta_b \, \delta_a \, \phi(x_\mu) = \delta_a \, \delta_b \, \phi(x_\mu) \tag{6.43}$$

so the left-hand side of Eq. (6.40) is zero.

Now let us work on the right-hand side of Eq. (6.40). We take the first transformation to be

$$U_a = \exp(i a^\mu P_\mu) \tag{6.44}$$

and the second one to be

$$U_b = \exp(i b^\nu P_\nu) \tag{6.45}$$

In Eq. (6.40) we therefore replace $\alpha \cdot Q$ by $a^\mu P_\mu$ and $\beta \cdot Q$ by $b^\nu P_\nu$, so Eq. (6.40) is finally given by

$$\left[[a^\mu P_\mu, b^\nu P_\nu], \phi \right] = a^\mu \, b^\nu \left[[P_\mu, P_\nu], \phi \right] = 0 \tag{6.46}$$

which implies

$$\left[[P_\mu, P_\nu], \phi \right] = 0 \tag{6.47}$$

Now we want to obtain from this the commutator $[P_\mu, P_\nu]$. Obviously, $[P_\mu, P_\nu] = 0$ is a possible solution, but is it the most general one?

This is one difficulty in using Eq. (6.40): The final step involves extracting from an expression of the form $[[Q_a, Q_b], \phi]$ the commutator between the charges (or the anticommutator for fermionic charges). This is not always trivial. One must make sure that one writes the most general commutator satisfying Eq. (6.40), and there is, at first sight, always more than one possible answer. However, symmetry arguments are usually sufficient to pin down the solution.

For example, one could argue that taking the commutator $[P_\mu, P_\nu]$ to be a constant still would satisfy Eq. (6.47). But we need to have something with two Lorentz indices, so we could try

$$[P_\mu, P_\nu] = c\eta_{\mu\nu} \tag{6.48}$$

This obviously satisfies Eq. (6.47), but there is a problem: The left-hand side of Eq. (6.48) is antisymmetric under the exchange of the two Lorentz indices, whereas the right-hand side is symmetric. This forces us to set $c = 0$, giving

$$[P_\mu, P_\nu] = 0 \qquad (6.49)$$

which is indeed the correct answer.

The key point of this example is that we obtained the commutation relations of the charges P_μ without ever writing them down explicitly! As a less trivial example, you are invited to prove again the second equation of Eq. (6.34), i.e.,

$$[M_{\mu\nu}, P_\lambda] = i(\eta_{\nu\lambda} P_\mu - \eta_{\lambda\mu} P_\nu). \qquad (6.50)$$

EXERCISE 6.4
Prove the commutation relations (6.50) using Eqs. (6.40) and (6.20).

6.5 The SUSY Algebra

We will now use Eq. (6.40) to find the algebra of the SUSY charges!

But first let us ask: how many charges do we have? Well, how many infinitesimal parameters are involved in SUSY tranformations? There is the two-component spinor ζ and its complex conjugate ζ^* for a total of four parameters. We therefore need four charges Q_1, Q_2, $Q_1^\dagger$, and $Q_2^\dagger$, which we obviously (see below) can group into a Weyl spinor, that we will simply call Q, and its hermitian conjugate $Q^\dagger$. These SUSY charges are also often referred to as *supercharges* (what a surprise). Until now, we have used Q to represent charges in general, but for the rest of this book it will be used to specifically represent the SUSY charges.

The argument of the exponential in the unitary operator U must be Lorentz-invariant, so we must combine Q and $Q^\dagger$ with ζ and ζ^* in a Lorentz-invariant way (which is why it was obvious that the supercharges had to be combined into a Weyl spinor). We choose to make Q a left-chiral spinor. We know from Chapter 2 that the two possible invariant combinations are then

$$Q \cdot \zeta = Q(-i\sigma^2)\zeta$$

and

$$\bar{Q} \cdot \bar{\zeta} = Q^\dagger i\sigma^2 \zeta^*$$

(recall that ζ is not a quantum field operator, so we may write $\zeta^{\dagger T}$ simply as ζ^*). Written in terms of the supercharges, the unitary operator U generating SUSY transformations therefore is given by

$$U_\zeta = \exp(i\, Q \cdot \zeta + i\, \bar{Q} \cdot \bar{\zeta}) \tag{6.51}$$

If we apply Eq. (6.12) to the SUSY transformations of ϕ and χ given in Eq. (5.17), we get

$$[\zeta \cdot Q + \bar{\zeta} \cdot \bar{Q}, \phi] = -i\zeta \cdot \chi$$

$$[\zeta \cdot Q + \bar{\zeta} \cdot \bar{Q}, \chi] = -i(\partial_\mu \phi)\, \sigma^\mu \sigma^2 \zeta^*$$

which implies

$$[\zeta \cdot Q, \phi] = -i\zeta \cdot \chi$$

$$[\bar{\zeta} \cdot \bar{Q}, \chi] = -i(\partial_\mu \phi)\, \sigma^\mu \sigma^2 \zeta^*$$

$$[\bar{\zeta} \cdot \bar{Q}, \phi] = [\zeta \cdot Q, \chi] = 0 \tag{6.52}$$

We now consider two successive SUSY transformations of the fields in order to use Eq. (6.40). We will use β as the infinitesimal parameter of the second transformation. Therefore, the first transformation is generated by the operator (6.51), and the second transformation is generated by

$$U_\beta = \exp(i\, Q \cdot \beta + i\, \bar{Q} \cdot \bar{\beta})$$

Let's apply Eq. (6.40) to the scalar field ϕ of our supersymmetric lagrangian (we will consider the spinor field χ later); we get

$$\delta_\beta \delta_\zeta \phi - \delta_\zeta \delta_\beta \phi = \left[[Q \cdot \zeta + \bar{Q} \cdot \bar{\zeta}, \; Q \cdot \beta + \bar{Q} \cdot \bar{\beta}], \; \phi \right]$$

$$= \left[[Q \cdot \zeta, \; Q \cdot \beta], \phi \right] + \left[[Q \cdot \zeta, \; \bar{Q} \cdot \bar{\beta}], \phi \right]$$

$$+ \left[[\bar{Q} \cdot \bar{\zeta}, \; Q \cdot \beta], \; \phi \right] + \left[[\bar{Q} \cdot \bar{\zeta}, \; \bar{Q} \cdot \bar{\beta}], \phi \right] \tag{6.53}$$

Right-Hand Side of Eq. (6.53)

The right-hand side of Eq. (6.53) can be brought into a much more useful form by writing all the commutators of the charges with the infinitesimal parameters factored out (it will become clear shortly why this is a useful thing to do). Let's consider the first commutator, which we rewrite explicitly in terms of the components:

$$[Q \cdot \zeta, \ Q \cdot \beta] = \left[Q \left(-i\sigma^2\right) \zeta, \ Q(-i\sigma^2) \beta \right]$$

$$= -\left[Q_a \left(\sigma^2\right)^{ab} \zeta_b, \ Q_c \left(\sigma^2\right)^{cd} \beta_d \right] \tag{6.54}$$

where it is understood that the whole expression is acting on the field ϕ.

The indices on the σ^2 have been chosen so that the usual summation convention of relativity is respected (if you have read Chapter 3, this should be familiar; if not, don't read too much into it, it's just a convention and these σ^2 matrices will drop out of our final result anyway).

Now we have to be careful. The ζ_i and β_i are Grassmann numbers, but the Q_i are Grassmann *operators*! Thus we can move the β_i and ζ_i through each other and through the Q_i on the condition of adding a minus sign, but we *cannot* switch the order of the Q_i (for the same reason that we cannot switch the order of $\hat{X}$ and $\hat{P}$ in quantum mechanics). As for the components of the σ^2, these are simple numbers, so we can move them anywhere we wish with no fear. We therefore can write

$$-\left[Q_a \left(\sigma^2\right)^{ab} \zeta_b, \ Q_c \left(\sigma^2\right)^{cd} \beta_d \right] = -\left(\sigma^2\right)^{ab} \left(\sigma^2\right)^{cd} \left(Q_a \zeta_b Q_c \beta_d - Q_c \beta_d Q_a \zeta_b \right)$$

$$= \left(\sigma^2\right)^{ab} \left(\sigma^2\right)^{cd} \zeta_b \beta_d \left(Q_a Q_c + Q_c Q_a \right)$$

$$= \left(\sigma^2\right)^{ab} \left(\sigma^2\right)^{cd} \zeta_b \beta_d \{ Q_a, Q_c \} \tag{6.55}$$

Note that we started with a *commutator* of $Q \cdot \zeta$ and $Q \cdot \beta$ but ended up with an *anticommutator* of the charges. This is obviously due to the fermionic (Grassmann) nature of the infinitesimal parameters.

Equation (6.55) is a useful form because it is written *explicitly* in terms of the anticommutator of the charges, which will make it easier for us to isolate them in the next section.

We still have three commutators to work out in Eq. (6.53). Consider the second one:

$$[Q \cdot \zeta, \ \bar{Q} \cdot \bar{\beta}] = \left[Q_a \left(-i\sigma^2\right)^{ab} \zeta_b, \ Q_c^\dagger \left(i\sigma^2\right)^{cd} \beta_d^* \right]$$

$$= \left[Q_a \left(\sigma^2\right)^{ab} \zeta_b, \ Q_c^\dagger \left(\sigma^2\right)^{cd} \beta_d^* \right]$$

If you have read Chapter 3, you know that, strictly speaking, we should be using dotted indices on the second σ^2, but again, this will not matter for the present calculation. Now we can simply repeat the steps that led to Eq. (6.55) to get

$$[Q \cdot \zeta, \ \bar{Q} \cdot \bar{\beta}] = -\left(\sigma^2\right)^{ab} \left(\sigma^2\right)^{cd} \zeta_b \beta_d^* \{ Q_a, Q_c^\dagger \} \tag{6.56}$$

The remaining two commutators of Eq. (6.53) are now just a repeat:

$$[\bar{Q} \cdot \bar{\xi}, Q \cdot \beta] = -(\sigma^2)^{ab} (\sigma^2)^{cd} \zeta_b^* \beta_d \{Q_a^\dagger, Q_c\}$$

$$[\bar{Q} \cdot \bar{\xi}, \bar{Q} \cdot \bar{\beta}] = (\sigma^2)^{ab} (\sigma^2)^{cd} \zeta_b^* \beta_d^* \{Q_a^\dagger, Q_c^\dagger\} \tag{6.57}$$

Putting it all together, we see that Eq. (6.53) may be written as

$$\delta_\beta \delta_\zeta \phi - \delta_\zeta \delta_\beta \phi = [\mathcal{O}, \phi] \tag{6.58}$$

with the operator $\mathcal{O}$ being given by the sum of Eqs. (6.55) and (6.56), and the two expressions of Eq. (6.57)

$$\mathcal{O} = (\sigma^2)^{ab} (\sigma^2)^{cd} \left(\zeta_b \beta_d \{Q_a, Q_c\} - \zeta_b \beta_d^* \{Q_a, Q_c^\dagger\} \right.$$
$$\left. - \zeta_b^* \beta_d \{Q_a^\dagger, Q_c\} + \zeta_b^* \beta_d^* \{Q_a^\dagger, Q_c^\dagger\} \right) \tag{6.59}$$

Left-Hand Side of Eq. (6.53)

Using the SUSY transformations (5.17), we find

$$\delta_\beta \delta_\zeta \phi = \delta_\beta (-\zeta^T i \sigma^2 \chi)$$
$$= -\zeta^T i \sigma^2 \, \delta_\beta \chi$$
$$= \zeta^T i \sigma^2 i \sigma^\mu i \sigma^2 \beta^* \, \partial_\mu \phi$$
$$= -i \zeta^T \sigma^2 \sigma^\mu \sigma^2 \beta^* \, \partial_\mu \phi$$
$$= -i \zeta^T \bar{\sigma}^{\mu T} \beta^* \, \partial_\mu \phi \qquad \text{using Eq. (A.5)}$$
$$= i \beta^\dagger \bar{\sigma}^\mu \zeta \, \partial_\mu \phi \qquad \text{using Eq. (A.36)} \tag{6.60}$$

where we have used the fact that $\partial_\mu \phi$ is not a matrix, so we are free to move it around.

The last line in Eq. (6.60) is clearly the most condensed and most elegant way to write the result, but the form shown two lines above (with the σ^2 still explicit) will be more useful to compute the SUSY algebra because Eq. (6.59) contains σ^2 matrices too.

By switching β and ζ, we obviously have

$$\delta_\zeta \delta_\beta \phi = -i \beta^T \sigma^2 \sigma^\mu \sigma^2 \zeta^* \, \partial_\mu \phi$$
$$= i \zeta^\dagger \bar{\sigma}^\mu \beta \, \partial_\mu \phi$$

Therefore, the left-hand side of Eq. (6.58) is equal to

$$\delta_\beta \delta_\zeta \phi - \delta_\zeta \delta_\beta \phi = -i(\zeta^T \sigma^2 \sigma^\mu \sigma^2 \beta^* - \beta^T \sigma^2 \sigma^\mu \sigma^2 \zeta^*)\, \partial_\mu \phi$$
$$= -i(\zeta^\dagger \bar\sigma^\mu \beta - \beta^\dagger \bar\sigma^\mu \zeta)\partial_\mu \phi \qquad (6.61)$$

Again, the first form will be the one we will use to obtain the SUSY algebra, but the second one is the most elegant (and will prove useful later on).

What we now have to do is to set this equal to the right-hand side of Eq. (6.58). In order to be able to extract explicit expressions for the anticommutators of the SUSY charges, we need to write Eq. (6.61) in the form of the commutator of something with the scalar field. The trick is simply to use Eq. (6.20), namely, $\partial_\mu \phi = i[P_\mu, \phi]$, which allows us to write the first line of Eq. (6.61) as

$$\delta_\beta \delta_\zeta \phi - \delta_\zeta \delta_\beta \phi = \left(\zeta^T \sigma^2 \sigma^\mu \sigma^2 \beta^* - \beta^T \sigma^2 \sigma^\mu \sigma^2 \zeta^*\right)\left[P_\mu, \phi\right]$$
$$= \left[\left(\zeta^T \sigma^2 \sigma^\mu \sigma^2 \beta^* - \beta^T \sigma^2 \sigma^\mu \sigma^2 \zeta^*\right)P_\mu,\ \phi\right] \qquad (6.62)$$

where the last step is possible because ζ, β and the sigma matrices commute with ϕ.

Finally, comparing Eq. (6.62) with Eq. (6.58), we get

$$[\mathcal{O}, \phi] = \left[(\zeta^T \sigma^2 \sigma^\mu \sigma^2 \beta^* - \beta^T \sigma^2 \sigma^\mu \sigma^2 \zeta^*)P_\mu,\ \phi\right] \qquad (6.63)$$

with $\mathcal{O}$ given by Eq. (6.59). It is tempting to conclude that this implies

$$\mathcal{O} = \left(\zeta^T \sigma^2 \sigma^\mu \sigma^2 \beta^* - \beta^T \sigma^2 \sigma^\mu \sigma^2 \zeta^*\right)P_\mu \qquad (6.64)$$

but we have only proved Eq. (6.64) when the two sides are applied to the scalar field ϕ. Before concluding that it is a valid identity in our theory, we must prove that it is also satisfied when applied to the spinor field χ. It turns out, as we will demonstrate in the next section, that Eq. (6.64) is *not* valid when applied to the spinor field. This will lead us to modify our lagrangian to ensure that Eq. (6.64) *does* apply to all fields. Since the end result is that Eq. (6.64) will be satisfied, let's go ahead and use it to work out the SUSY algebra.

The right-hand side of Eq. (6.64) is not yet in a suitable form for comparison with Eq. (6.59) because we still need to factor out the infinitesimal parameters and write explicit indices on everything. To facilitate the comparison, we will assign the same indices to ζ and β as they have in Eq. (6.59), i.e., b for ζ and d for β. We therefore write

$$\zeta^T \sigma^2 \sigma^\mu \sigma^2 \beta^* = \zeta_b (\sigma^2)^{ba} (\sigma^\mu)_{ac} (\sigma^2)^{cd} \beta_d^* \qquad (6.65)$$

and

$$\beta^T \sigma^2 \sigma^\mu \sigma^2 \zeta^* = \beta_d (\sigma^2)^{dc} (\sigma^\mu)_{ca} (\sigma^2)^{ab} \zeta_b^* \qquad (6.66)$$

Note that in Eq. (6.59) the indices of the two σ^2 matrices are, respectively, ab and cd. We would like to have the same indices in our two σ^2 matrices in Eqs. (6.65) and (6.66) because then we will be able to cancel them out in Eq. (6.64). This is easy to achieve because σ^2 is antisymmetric, so $(\sigma^2)^{ba} = -(\sigma^2)^{ab}$ and $(\sigma^2)^{dc} = -(\sigma^2)^{cd}$. In addition, we will move the ζ_b^* to the left of the β_d in Eq. (6.66) because this is the order in which these appear in Eq. (6.59). Using the fact that the components of the σ matrices are simply numbers that can be moved around freely, we can rewrite Eq. (6.65) as

$$\zeta^T \sigma^2 \sigma^\mu \sigma^2 \beta^* = -\zeta_b \beta_d^* (\sigma^2)^{ab} (\sigma^2)^{cd} (\sigma^\mu)_{ac} \qquad (6.67)$$

and Eq. (6.66) as

$$\beta^T \sigma^2 \sigma^\mu \sigma^2 \zeta^* = \zeta_b^* \beta_d (\sigma^2)^{ab} (\sigma^2)^{cd} (\sigma^\mu)_{ca}$$

Using these last two results in Eq. (6.64), we obtain

$$\mathcal{O} = -(\sigma^2)^{ab} (\sigma^2)^{cd} \left(\zeta_b \beta_d^* (\sigma^\mu)_{ac} + \zeta_b^* \beta_d (\sigma^\mu)_{ca} \right) P_\mu \qquad (6.68)$$

Using Eq. (6.59) for $\mathcal{O}$ and canceling out the components $(\sigma^2)^{ab} (\sigma^2)^{cd}$ appearing on both sides, we finally get

$$\zeta_b \beta_d \{Q_a, Q_c\} - \zeta_b \beta_d^* \{Q_a, Q_c^\dagger\} - \zeta_b^* \beta_d \{Q_a^\dagger, Q_c\} + \zeta_b^* \beta_d^* \{Q_a^\dagger, Q_c^\dagger\}$$

$$= -\zeta_b \beta_d^* (\sigma^\mu)_{ac} P_\mu - \zeta_b^* \beta_d (\sigma^\mu)_{ca} P_\mu \qquad (6.69)$$

Since the spinors ζ, β and their complex conjugates are arbitrary, Eq. (6.69) actually contains four independent equations. The terms in $\zeta_b \beta_d$ and in $\zeta_b^* \beta_d^*$ lead to

$$\boxed{\{Q_a, Q_b\} = 0} \qquad (6.70)$$

and

$$\{Q_a^\dagger, Q_b^\dagger\} = 0 \qquad (6.71)$$

If you have read Chapter 3, you know that using the dotted and bar notation, this also can be written as

$$\{\bar{Q}_{\dot{a}}, \bar{Q}_{\dot{b}}\} = 0$$

Now, if we consider the terms in $\zeta_b \beta_d^*$ in Eq. (6.69), we find

$$\{Q_a, Q_c^\dagger\} = (\sigma^\mu)_{ac} P_\mu$$

which is usually written with the index on $Q^\dagger$ labeled b:

$$\{Q_a, Q_b^\dagger\} = (\sigma^\mu)_{ab} P_\mu \qquad (6.72)$$

Again, using the bar and dotted notation of Chapter 3, this may be written as

$$\{Q_a, \bar{Q}_{\dot{b}}\} = (\sigma^\mu)_{a\dot{b}} P_\mu \qquad (6.73)$$

It is important to point out that the normalization of the supercharges is arbitrary, so we are free to rescale Eqs. (6.72) and (6.73) by a constant. Another way to put it is that we could have multiplied the supercharges in the exponential of U_ζ by an arbitrary factor. A normalization that is used often in the SUSY literature corresponds to rescaling our supercharges by $Q \to Q/\sqrt{2}$, which then transforms Eqs. (6.72) and (6.73) into

$$\{Q_a, Q_b^\dagger\} = 2(\sigma^\mu)_{ab} P_\mu$$
$$\{Q_a, \bar{Q}_{\dot{b}}\} = 2(\sigma^\mu)_{a\dot{b}} P_\mu \qquad (6.74)$$

We will use the normalization Eq. (6.72).

Let's finish with the terms in $\zeta_b^* \beta_d$ in Eq. (6.69). These yield the relation

$$\{Q_a^\dagger, Q_c\} = (\sigma^\mu)_{ca} P_\mu \qquad (6.75)$$

It is easy to see, however, that this is equivalent to Eq. (6.72). By the definition of an anticommutator, we may exchange $Q_a^\dagger$ and Q_c (without picking up any minus sign) and write Eq. (6.75) as

$$\{Q_c, Q_a^\dagger\} = (\sigma^\mu)_{ca} P_\mu$$

which, after renaming the indices, is exactly the same as Eq. (6.72).

The commutation relations (6.70), (6.71), and (6.72) are what we were seeking: the algebra of the SUSY charges. The fact that the anticommutators (6.72) contain the four-momentum operator is a crucial aspect of supersymmetry, and we will discuss this in more detail shortly.

To complete the algebra, we need to work out the commutators of the SUSY charges with the Poincaré generators P^μ and $M^{\mu\nu}$. It is easy to see that the supercharges commute with the momentum operator because the SUSY transformations do not generate any function of x on which the derivatives of P could act, so

$$\boxed{\begin{aligned}[Q_a, P_\mu] &= 0 \\ [Q_a^\dagger, P_\mu] &= 0\end{aligned}} \tag{6.76}$$

Things are a bit more interesting when it comes to the commutator with the Lorentz generators $M_{\mu\nu}$. To work this out, we first need to write down the Lorentz transformation of Weyl spinors. The Lorentz transformation of a spinor field has two contributions: The argument transforms in the same way as for a scalar field [see Eq. (6.23)], but in addition, the spinor gets multiplied by the matrices shown in Eq. (2.32).

Our first step is to write down a covariant expression for the transformation of the spinors given in Eq. (2.32) in terms of the parameters $\omega^{\mu\nu}$. For a left-chiral Weyl spinor, the answer is that Eq. (2.32) may be written as

$$\chi' = \exp\left(\frac{i}{2}\omega^{\mu\nu}\sigma_{\mu\nu}\right)\chi \tag{6.77}$$

where the matrices $\sigma_{\mu\nu}$ are defined as

$$\boxed{\sigma_{\mu\nu} \equiv \frac{i}{4}(\sigma_\mu\bar\sigma_\nu - \sigma_\nu\bar\sigma_\mu)} \tag{6.78}$$

Note that $\sigma_{\mu\nu}$ is antisymmetric under the exchange of the indices μ and ν. The exponential can be shown to reproduce the transformation of χ given in Eq. (2.32), with an appropriate choice of the coefficients $\omega^{\mu\nu}$.

As for a right-chiral spinor η, it transforms as in Eq. (6.77) but with $\sigma_{\mu\nu}$ replaced by

$$\bar{\sigma}_{\mu\nu} \equiv \frac{i}{4}(\bar{\sigma}_\mu \sigma_\nu - \bar{\sigma}_\nu \sigma_\mu) \qquad (6.79)$$

The complete Lorentz transformation of a left-chiral spinor therefore is

$$\chi \to \chi'(x) = \exp\left(\frac{i}{2}\omega^{\mu\nu}\sigma_{\mu\nu}\right)\chi(x^\mu + \omega^{\mu\nu}x_\nu)$$

$$\approx \chi + \frac{i}{2}\omega^{\mu\nu}\sigma_{\mu\nu}\chi + \frac{1}{2}\omega^{\mu\nu}(x_\nu \partial_\mu \chi - x_\mu \partial_\nu \chi)$$

$$= \chi + \frac{1}{2}\omega^{\mu\nu}(i\sigma_{\mu\nu}\chi + x_\nu \partial_\mu \chi - x_\mu \partial_\nu \chi) \qquad (6.80)$$

so the variation of the spinor is given by

$$\delta_\omega \chi = \frac{1}{2}\omega^{\mu\nu}(i\sigma_{\mu\nu}\chi + x_\nu \partial_\mu \chi - x_\mu \partial_\nu \chi) \qquad (6.81)$$

To work out the commutation relation of the supercharges Q_a with the Lorentz generators $M_{\mu\nu}$, we start again from Eq. (6.40), which now takes the form

$$-\frac{1}{2}\big[[\zeta \cdot Q, \omega^{\mu\nu}M_{\mu\nu}], \phi\big] = \delta_\omega \delta_\zeta \phi - \delta_\zeta \delta_\omega \phi \qquad (6.82)$$

where the factor of $-1/2$ is the one that accompanies the generators $M_{\mu\nu}$ (see Exercise 6.1). It is then a simple matter to show that

$$[Q_a, M_{\mu\nu}] = (\sigma_{\mu\nu})_a{}^b Q_b \qquad (6.83)$$

the proof of which is the subject of Exercise 6.5. The indices of $\sigma_{\mu\nu}$ are chosen this way in order to satisfy the usual summation convention. The matrices $\sigma_{\mu\nu}$ are antisymmetric, which is why we need to use spaces to indicate clearly that the upper index is the second index.

The only tricky aspect of the derivation is that we need at some point the commutator of the supercharges with the scalar field. This is easy to find by going back to Eq. (6.12), which, using the SUSY transformation of the scalar field $\delta\phi = \zeta \cdot \chi$, becomes

$$[\zeta \cdot Q, \phi] = -i\zeta \cdot \chi \tag{6.84}$$

This immediately gives us

$$[Q_a, \phi] = -i\chi_a$$

or, suppressing the spinor indices,

$$[Q, \phi] = -i\chi \tag{6.85}$$

Using this result, it is a simple matter to prove Eq. (6.83) starting from Eq. (6.82). This calculation is the object of the first part of Exercise 6.5.

The calculation of the commutation relation with the charges $Q_a^\dagger$ and $Q^{a\dagger}$ is much simpler using the van der Waerden notation of Chapter 3. We will simply quote the result and leave the derivation to the second part of Exercise 6.5. If you haven't covered Chapter 3 yet, you may skip this exercise for now and come back to it later. The final results are

$$\boxed{\begin{aligned} \left[\bar{Q}_{\dot{a}}, M_{\mu\nu}\right] &= -\bar{Q}_{\dot{b}}\,(\bar{\sigma}_{\mu\nu})^{\dot{b}}_{\ \dot{a}} \\ \left[\bar{Q}^{\dot{a}}, M_{\mu\nu}\right] &= (\bar{\sigma}_{\mu\nu})^{\dot{a}}_{\ \dot{b}}\,\bar{Q}^{\dot{b}} \end{aligned}} \tag{6.86}$$

We have achieved our goal of obtaining the complete algebra of the supercharges. However, as mentioned earlier, we have cheated in the step from Eq. (6.63) to Eq. (6.64) because we have not shown that Eq. (6.64) was valid when applied to the spinor field as well as to the scalar field. We will remedy this lapse in the next section.

EXERCISE 6.5

a. Prove the commutation relation (6.83).
b. Prove Eq. (6.86) to be done after you have gone through Chapter 3.

Hint: In order to get an expression for the supercharges $\bar{Q}$, which are related to the hermitian conjugates of the supercharges Q, we use Eq. (6.40), but with the transformations on the left side applied to $\phi^\dagger$ instead of ϕ. So the starting point

will be

$$-\frac{1}{2}\big[[\bar{Q} \cdot \xi, \omega^{\mu\nu} M_{\mu\nu}], \phi^{\dagger}\big] = \delta_{\omega}\delta_{\zeta}\phi^{\dagger} - \delta_{\zeta}\delta_{\omega}\phi^{\dagger} \qquad (6.87)$$

It will then be straightforward to prove the first commutator of Eq. (6.86). You can then use ϵ symbols to obtain the second commutator from the first. You will need to show that

$$\epsilon^{\dot{c}\dot{b}}\epsilon_{\dot{d}\dot{a}}(\bar{\sigma}^{\mu\nu})^{\dot{a}}{}_{\dot{b}} = (\bar{\sigma}^{\mu\nu})^{\dot{c}}{}_{\dot{d}} \qquad (6.88)$$

6.6 Nonclosure of the Algebra for the Spinor Field

As mentioned earlier, the SUSY algebra we have just obtained is, strictly speaking, valid only when applied to the scalar field. To prove that the algebra is valid in general, we have to check if we would have obtained the same result starting from the transformation of the spinor field. We will now see that it will *not* work without modifying the theory in a nontrivial way.

Consider the variation of the spinor field. Using Eq. (5.17), we get

$$\delta_{\beta}\delta_{\zeta}\chi = \delta_{\beta}(-i\sigma^{\mu}i\sigma^{2}\zeta^{*}\,\partial_{\mu}\phi)$$

$$= (\sigma^{\mu}\sigma^{2}\zeta^{*})\ \partial_{\mu}(\delta_{\beta}\phi)$$

$$= (\sigma^{\mu}\sigma^{2}\zeta^{*})\ \beta \cdot \partial_{\mu}\chi$$

where, obviously, $\beta \cdot \partial_{\mu}\chi$ stands for the usual spinor dot product between β and $\partial_{\mu}\chi$:

$$\beta \cdot \partial_{\mu}\chi = \beta^{T}(-i\sigma^{2})\partial_{\mu}\chi$$

Note that what we have is really two identities, one for each component of χ. To make this more clear, we can write

$$\delta_{\beta}\delta_{\zeta}\chi_{a} = (\sigma^{\mu}\sigma^{2}\zeta^{*})_{a}\ \beta \cdot \partial_{\mu}\chi \qquad (6.89)$$

To simplify the notation, let's define

$$\alpha \equiv \sigma^\mu \sigma^2 \zeta^* \tag{6.90}$$

so that we may write Eq. (6.89) as

$$\delta_\beta \delta_\zeta \chi_a = \alpha_a \, \beta \cdot \partial_\mu \chi \tag{6.91}$$

Our ultimate goal is to check if the commutator of two SUSY transformations on the spinor is of the same form as what we obtained for the scalar field in Eq. (6.61). To compare, we need to rewrite Eq. (6.91) in the form of something times $\partial_\mu \chi_a$, i.e., we want the index a to be on χ. We can move the index using the following identity, valid for any three spinors α, β and γ:

$$\boxed{\alpha_a(\beta \cdot \gamma) + \beta_a(\gamma \cdot \alpha) + \gamma_a(\alpha \cdot \beta) = 0} \tag{6.92}$$

Note that the spinor dot products are commuting quantities, so we are free to move them around. For example, $\alpha_a(\beta \cdot \gamma) = (\beta \cdot \gamma)\alpha_a$.

EXERCISE 6.6
Prove Eq. (6.92). *Hint*: Expand out all the terms in terms of the components of the spinors, and check explicitly that the identity is satisfied for both $a = 1$ and $a = 2$.

Obviously, to apply Eq. (6.92) to Eq. (6.91), we simply have to define $\partial_\mu \chi \equiv \gamma$. We then obtain [using Eq. (6.90)]

$$\begin{aligned}
\delta_\beta \delta_\zeta \chi_a &= -(\partial_\mu \chi) \cdot \alpha \, \beta_a - \alpha \cdot \beta \, \partial_\mu \chi_a \\
&= -(\partial_\mu \chi) \cdot \alpha \, \beta_a - \beta \cdot \alpha \, \partial_\mu \chi_a \\
&= -\left((\partial_\mu \chi)^T(-i\sigma^2)\sigma^\mu \sigma^2 \zeta^*\right) \beta_a - \left(\beta^T(-i\sigma^2)\sigma^\mu \sigma^2 \zeta^*\right) \partial_\mu \chi_a \\
&= i\left((\partial_\mu \chi)^T \sigma^2 \sigma^\mu \sigma^2 \zeta^*\right)\beta_a + i\left(\beta^T \sigma^2 \sigma^\mu \sigma^2 \zeta^*\right) \partial_\mu \chi_a
\end{aligned}$$

where in the second line we have rewritten $\alpha \cdot \beta$ as $\beta \cdot \alpha$ only so that the transpose would be applied to β, not to the complicated α [defined in Eq. (6.90)]. Note that the expressions inside the two parentheses in the last line are commuting quantities. We can freely move β_a or $\partial_\mu \chi_a$ through these parentheses without picking up any minus sign.

Using Eq. (A.5) and then Eq. (A.36), we can simplify the result greatly:

$$\delta_\beta \delta_\zeta \chi_a = i\left((\partial_\mu \chi)^T \sigma^2 \sigma^\mu \sigma^2 \zeta^*\right)\beta_a + i\left(\beta^T \sigma^2 \sigma^\mu \sigma^2 \zeta^*\right)\partial_\mu \chi_a$$

$$= i\left((\partial_\mu \chi)^T \bar\sigma^{\mu T} \zeta^*\right)\beta_a + i\left(\beta^T \bar\sigma^{\mu T} \zeta^*\right)\partial_\mu \chi_a$$

$$= -i\left(\zeta^\dagger \bar\sigma^\mu \partial_\mu \chi\right)\beta_a - i\left(\zeta^\dagger \bar\sigma^\mu \beta\right)\partial_\mu \chi_a$$

The second term on the right-hand side is what we wanted, something proportional to $\partial_\mu \chi_a$, but the first term will cause us some grief.

Obviously, by switching ζ and β, we get

$$\delta_\zeta \delta_\beta \chi_a = -i\left(\beta^\dagger \bar\sigma^\mu \partial_\mu \chi\right)\zeta_a - i\left(\beta^\dagger \bar\sigma^\mu \zeta\right)\partial_\mu \chi_a$$

so that

$$\left(\delta_\beta \delta_\zeta - \delta_\zeta \delta_\beta\right)\chi_a = -i\left(\zeta^\dagger \bar\sigma^\mu \partial_\mu \chi\right)\beta_a + i\left(\beta^\dagger \bar\sigma^\mu \partial_\mu \chi\right)\zeta_a$$

$$- i\left(\zeta^\dagger \bar\sigma^\mu \beta - \beta^\dagger \bar\sigma^\mu \zeta\right)\partial_\mu \chi_a \qquad (6.93)$$

Note that the last term is identical to the transformation we had for the scalar field (6.61) with ϕ replaced by χ! Unfortunately, the first two terms of Eq. (6.93) spoil the fun.

Note that if $i\bar\sigma^\mu \partial_\mu \chi = 0$, the two unwanted terms drop out. But this is nothing other than the Weyl equation! In other words, if the spinor satisfies its equation of motion (i.e., if it is on-shell), then we recover the algebra we had obtained previously using the transformations of the scalar field. However, for an off-shell spinor, one *cannot* write the algebra of the SUSY and Poincaré charges in terms of those same charges, as we did in Section 6.5. The conclusion therefore is that the algebra on the spinor field closes only on-shell.

What we would like to do now is to modify the theory so that the algebra on the spinor field will close off-shell, i.e., even if the spinor does not obey its equations of motion. Why do we want that? This is a subtle issue. There are SUSY theories that have no off-shell formulations, i.e., whose algebra closes only on-shell, so it would not be the end of the world if we would not modify our toy theory to close off-shell. However, when the algebra does not close offshell, SUSY transformations become nonlinear (they contain products of the fields) and depend on the coupling constants present in the theory. This makes it much more difficult to design lagrangians that are invariant under SUSY. We will therefore modify the theory to enforce that the algebra closes off-shell.

6.7 Introduction of an Auxiliary Field

There is a simple argument requiring no calculations that we could have used to predict that the SUSY algebra would not close off-shell. SUSY requires the number of bosonic and fermionic degrees of freedom to be equal (this will be demonstrated in the next chapter). A Weyl spinor has two complex components and so four degrees of freedom when it is off-shell. On-shell, the equation of motion imposes two constraints, reducing the number of degrees of freedom to two. A complex scalar field has two degrees of freedom, so we see that *on-shell*, the number of bosonic and fermionic degrees of freedom match. However, off-shell, the number of bosonic and fermionic degrees of freedom do not match anymore. This explains why the SUSY algebra only closes on-shell if we have a single complex scalar field and a Weyl spinor.

But the argument also suggests a solution. We need to add a bosonic field that will provide two degrees of freedom off-shell but no on-shell degrees of freedom! A field that has no on-shell degrees of freedom is called an *auxiliary field*, so what we need is to add to the theory a complex scalar auxiliary field, commonly called F. How do we get the field F to have no on-shell degrees of freedom? The trick is to make sure that the equation of motion for this field is $F^\dagger = F = 0$! This will clearly be the case if we add a term in the lagrangian that contains a product of F and $F^\dagger$; i.e., we do *not* include a kinetic term for F.

The simplest real term depending on F and $F^\dagger$ is $FF^\dagger$. So our lagrangian now takes the form

$$\mathcal{L} = \partial_\mu \phi \partial^\mu \phi^\dagger + \chi^\dagger i \bar{\sigma}^\mu \partial_\mu \chi + FF^\dagger \tag{6.94}$$

Note that despite being a scalar field, the field F is of dimension 2! This is something important to remember, so let's emphasize it:

$$\boxed{[F] = 2}$$

What should we take for the transformation of F under SUSY? Since it has a different dimension than ϕ, we cannot use the same transformation. We want a transformation that is linear in the SUSY parameter ζ and linear in one of the fields. Recall that ζ has dimension $-1/2$, so it must be multiplied by something of dimension $5/2$ (because F has a dimension equal to 2). This means that the transformation must contain the spinor field χ, so our starting point is

$$\delta F \simeq \zeta \chi$$

Since the dimensions of χ and ζ add up to 1, we need to include something with a dimension equal to 1. We don't want to use the scalar field to avoid a nonlinear transformation. What else do we have with dimension equal to 1? A derivative, of course! So let's try

$$\delta F \simeq \zeta \partial_\mu \chi$$

which has the right dimension. But, of course, it does not have the right Lorentz transformation because F is Lorentz-invariant. We need to include something that is dimensionless and yet carries a Lorentz index. Obviously, we want either σ^μ or $\bar{\sigma}^\mu$. We know, from the Dirac equation [see Eq. (2.26)] that it is the matrices $\bar{\sigma}^\mu$ that must be applied to a left-chiral spinor to yield an expression with well-defined Lorentz properties. So let's consider

$$\simeq \zeta \bar{\sigma}^\mu \partial_\mu \chi$$

Unfortunately, this still does not have the correct behavior under Lorentz transformations. From Eq. (2.26) again, we know that $\bar{\sigma}^\mu \partial_\mu \chi$ transforms like a *right-chiral spinor*. On the other hand, ζ is a left-chiral spinor. So we must recall how to combine a left- and right-chiral spinor to form a Lorentz invariant. The answer is that we must contract the hermitian conjugate of either spinor with the other spinor [see Eq. (2.42)]. It will be easier here to take the hermitian conjugate of ζ and contract it with $\bar{\sigma}^\mu \partial_\mu \chi$, so we finally take

$$\delta F = K \zeta^\dagger \bar{\sigma}^\mu \partial_\mu \chi$$

where K is some constant that we will determine shortly. This has the correct dimensions and Lorentz properties.

By taking the dagger, we get

$$\delta F^\dagger = K^* (\partial_\mu \chi^\dagger) \bar{\sigma}^\mu \zeta$$

What we must now find out is whether it is possible to add the term $FF^\dagger$ without spoiling the invariance of the lagrangian under SUSY. The variation of $FF^\dagger$ is

$$\delta(FF^\dagger) = K \zeta^\dagger \bar{\sigma}^\mu (\partial_\mu \chi) F^\dagger + K^* (\partial_\mu \chi^\dagger) \bar{\sigma}^\mu \zeta F \qquad (6.95)$$

which is not a total derivative, so we are in trouble. What we need to do is to modify the SUSY transformation of either ϕ or χ in such a way that the extra terms generated will cancel Eq. (6.95). Note that the variation of the spinor kinetic

Supersymmetry Demystified

energy is

$$\delta(\chi^{\dagger} i \bar{\sigma}^{\mu} \partial_{\mu} \chi) = (\delta \chi^{\dagger}) i \bar{\sigma}^{\mu} \partial_{\mu} \chi + \chi^{\dagger} i \bar{\sigma}^{\mu} \partial_{\mu}(\delta \chi) \qquad (6.96)$$

which looks a lot like Eq. (6.95)! To make the similarity more apparent, let's rewrite Eq. (6.95) in the form (after doing an integration by parts on the second term)

$$\delta(FF^{\dagger}) = (K^* \zeta F)^{\dagger} \bar{\sigma}^{\mu} \partial_{\mu} \chi - \chi^{\dagger} \bar{\sigma}^{\mu} \partial_{\mu}(K^* \zeta F) \qquad (6.97)$$

It is now obvious by comparing Eqs. (6.96) and (6.97) that if we take

$$\delta \chi = -i K^* \zeta F + \text{previous transformation}$$

the variation of the kinetic term of the spinor field will cancel out the variation of $FF^{\dagger}$ (modulo a total derivative).

We see that the value of the constant K is left completely arbitrary. The convention is to make the new term in the transformation of χ as simple as possible, namely,

$$-i K^* \zeta F \equiv \zeta F$$

which fixes $K = -i$.

Our final result is that the lagrangian

$$\boxed{\mathcal{L} = \partial_{\mu} \phi \partial^{\mu} \phi^{\dagger} + \chi^{\dagger} i \bar{\sigma}^{\mu} \partial_{\mu} \chi + FF^{\dagger}} \qquad (6.98)$$

is invariant (up to total derivatives, as always) under the following SUSY transformations:

$$\boxed{\begin{aligned} \delta \phi &= \zeta \cdot \chi \\ \delta \chi &= -i \sigma^{\mu} (i \sigma^2 \zeta^*) \, \partial_{\mu} \phi + F \zeta \\ \delta F &= -i \zeta^{\dagger} \bar{\sigma}^{\mu} \partial_{\mu} \chi \end{aligned}} \qquad (6.99)$$

6.8 Closure of the Algebra

Let's not lose track of the original motivation for introducing an auxiliary field, which was to ensure that the algebra closes off-shell for all the fields! What we now need to show is that for all three fields ϕ, χ, and F, the commutator of two SUSY transformations gives the same result, without recourse to any equation of motion.

We should start by repeating the calculation that led to Eq. (6.61) for the scalar field. Since we have modified the transformation of the spinor, there is no guarantee, at first sight, that we will recover the same result. However, it's easy to check that the new term present in the transformation of χ does not change the result of Eq. (6.61). Not only that, but now we do obtain the same result for all three fields without using the equations of motion; i.e., we get (see Exercise 6.7)

$$\delta_\beta \delta_\zeta X - \delta_\zeta \delta_\beta X = -i\left(\zeta^\dagger \bar{\sigma}^\mu \beta - \beta^\dagger \bar{\sigma}^\mu \zeta\right) \partial_\mu X \tag{6.100}$$

where X stands for any of the three fields ϕ, χ, or F.

This finally proves that the introduction of the auxiliary field allows closure of the algebra off-shell on all the fields of the theory. And the consequence of this is that the algebra we found for the SUSY charges holds in general.

EXERCISE 6.7
Prove that Eq. (6.100) is satisfied by all three fields ϕ, χ, and F.

6.9 Quiz

1. Do the supercharges commute with the momentum operator? Do they commute with the Lorentz generators?

2. What is the definition of an auxiliary field?

3. Explain how we could have predicted that the SUSY algebra would not close off-shell before the introduction of an auxiliary field, even before doing any calculation.

4. Without looking at the text, write down the most general transformation law of the auxiliary field F that is linear in the fields, of the right dimension, and with the correct Lorentz property. Of course, it must contain the SUSY parameter ζ. Recall that F is of dimension 2.

5. How many supercharges did we need to introduce (count two charges related by hermitian conjugation as independent). Why did we need to introduce that many?

CHAPTER 7

Applications of the SUSY Algebra

We have spent a considerable amount of energy working out the supersymmetry (SUSY) algebra, so it better be useful to something! Indeed, it proves to be a powerful tool for establishing general properties of supersymmetric theories without the need to ever write down any lagrangian! We will work out several such general properties in this chapter, but it seems like a good idea to first review how some of this works in the familiar context of the Poincaré group.

7.1 Classification of States Using the Algebra: Review of the Poincaré Group

Let's briefly review some basic facts about the representations of the Poincaré group (no SUSY here) on one-particle states. The details were worked out by Wigner in 1939[52] in what has become a classic paper. Here, we simply highlight some key results.

Casimir operators are operators that commute with all the generators of the group. These operators are of key importance in building representations because their eigenvalues can be used to classify the representations of the group. One Casimir operator of the Poincaré group is the squared momentum operator $P^\mu P_\mu$. This obviously commutes with the momentum generators, and the proof that it commutes with the Lorentz generators as well is straightforward. The eigenvalue of $P_\mu P^\mu$ is m^2, which is therefore one of the parameters used to classify one-particle states.

The second Casimir operator is not as obvious. It is built out of the so-called Pauli-Lubanski operator W_μ, defined as

$$W^\mu \equiv \frac{1}{2}\epsilon^{\mu\nu\rho\sigma} M_{\rho\sigma} P_\nu \tag{7.1}$$

where $\epsilon^{\mu\nu\rho\sigma}$ is the totally antisymmetric Levi-Civita symbol in four dimensions. We will use the normalization

$$\epsilon_{0123} = 1$$

which implies

$$\epsilon^{0123} = -1$$

With this normalization, it is clear that

$$\epsilon^{0ijk} = -\epsilon^{ijk} \tag{7.2}$$

where ϵ^{ijk} is the usual three-dimensional Levi-Civita symbol. Note that we automatically have

$$W^\mu P_\mu = \epsilon^{\mu\nu\rho\sigma} M_{\rho\sigma} P_\nu P_\mu$$

$$= 0 \tag{7.3}$$

which is obvious because $P_\nu P_\mu$ is symmetric under the exchange of those two indices, whereas the Levi-Civita symbol is antisymmetric. It is not as obvious that

$P_\mu W^\mu$ is also equal to zero (recall that P_μ and W^μ are operators), but it turns out to be true, as you will show in Exercise 7.1. Therefore,

$$[P_\mu, W^\mu] = 0 \qquad (7.4)$$

EXERCISE 7.1
Prove that $P_\mu W^\mu = 0$.

With some effort, one can show that $W_\mu W^\mu$ also commutes with all the generators of the Poincaré group, so that's our second Casimir operator.

To uncover the physical signification of the Pauli-Lubanski operator, consider first massive particles. In that case, we can choose to work in the rest frame of the particle, where its four-momentum is simply

$$p^\mu = p_\mu = (m, \vec{0}) \qquad (7.5)$$

We will be careful to distinguish the operator P^μ from its eigenvalues, which we will write as p^μ.

For the state with four-momentum given by Eq. (7.5), we then have

$$P_\mu |p\rangle = P_0 |p\rangle = m |p\rangle \qquad (7.6)$$

Let us now apply the Pauli-Lubanski operator to this state:

$$W^\mu |p\rangle = \frac{1}{2} \epsilon^{\mu\nu\rho\sigma} M_{\rho\sigma} P_\nu |p\rangle$$

$$= \frac{1}{2} \epsilon^{\mu 0 \rho\sigma} M_{\rho\sigma} P_0 |p\rangle$$

$$= \frac{m}{2} \epsilon^{\mu 0 \rho\sigma} M_{\rho\sigma} |p\rangle$$

Because of the antisymmetry of the Levi-Civita symbol, all three indices μ, ρ, and σ must be spacelike, so we conclude that

$$W^0 |p\rangle = 0 \qquad (7.7)$$

and

$$W^i|p\rangle = \frac{m}{2}\epsilon^{i0jk} M_{jk}|p\rangle$$

$$= -\frac{m}{2}\epsilon^{0ijk} M_{jk}|p\rangle$$

$$= \frac{m}{2}\epsilon^{ijk} M_{jk}|p\rangle$$

where we have used Eq. (7.2). It turns out that the operator $\frac{1}{2}\epsilon^{ijk} M_{jk}$ is simply the total angular momentum of the particle, i.e.,

$$\frac{1}{2}\epsilon^{ijk} M_{jk} = L^i + S^i \tag{7.8}$$

EXERCISE 7.2

The differential operator representation $\hat{M}_{\mu\nu}$ of the Lorentz generators acting on a spinor is defined as

$$\exp\left(\frac{i}{2}\omega^{\mu\nu}\hat{M}_{\mu\nu}\right)\chi \equiv \chi + \delta_\omega\chi$$

with $\delta_\omega\chi$ given in Eq. (6.81). This immediately leads to

$$\hat{M}_{\mu\nu} = \sigma_{\mu\nu} + i(x_\mu\partial_\nu - x_\nu\partial_\mu)$$

$$= \sigma_{\mu\nu} + x_\mu\hat{P}_\nu - x_\nu\hat{P}_\mu$$

Use this to prove Eq. (7.8). *HINT:* You may assume that the two sides of Eq. (7.8) are applied to a one-particle state of definite momentum so that you can replace the operators $\hat{P}_\mu$ and $\hat{P}_\nu$ by the actual four-momenta p_μ and p_ν.

We therefore have

$$W^i|p\rangle = m(S^i + L^i)|p\rangle \tag{7.9}$$

which, in the rest frame, reduces to

$$W^i|p\rangle = mS^i|p\rangle$$

So the Casimir operator $W_\mu W^\mu$ acting on this state gives

$$W_\mu W^\mu|p\rangle = W^0 W^0|p\rangle - W^i W^i|p\rangle = -m^2\vec{S}^2|p\rangle = -m^3 s(s+1)|p\rangle \tag{7.10}$$

We now see that the second property that can be used to specify the massive representations of the Lorentz group is the spin (or total angular momentum, if we are not working in the rest frame). Let's classify all the possible states of a massive particle at rest. For this, we need a complete set of commuting observables. In addition to the four-momentum squared (whose eigenvalue is just m^2) and the square of the spin, we also may use the component of the spin along the z axis, S_z, because this operator obviously commutes with $P_\mu P^\mu$ and with $\vec{S}^2$ (a fact familiar from quantum mechanics). Note that

$$W^3 = m(L_z + S_z) \tag{7.11}$$

which, in the rest frame of the particle, reduces to $W^3 = m S_z$.

Therefore, massive states are labeled in their rest frame by three quantum numbers: p, s, and s_z, with

$$P^\mu |p, s, s_z\rangle = m|p, s, s_z\rangle$$
$$W_\mu W^\mu |p, s, s_z\rangle = -m^2 \vec{S}^2 |p, s, s_z\rangle$$
$$= -m^2 s(s+1)|p, s, s_z\rangle$$
$$\frac{W^3}{m}|p, s, s_z\rangle = S_z |p, s, s_z\rangle = s_z |p, s, s_z\rangle$$

As we all know from quantum mechanics, s_z may take $2s + 1$ values, ranging from $-s$ to $+s$ in integer steps.

Now consider massless particles. In this case, we obviously cannot go to the rest frame of the particle, but we may choose to work in a frame where the four-momentum is given by

$$p^\mu = (E, 0, 0, E) \tag{7.12}$$

We will denote the corresponding state by $|p\rangle_0$. We then have

$$P^\mu |p\rangle_0 = (P^0, P^1, P^2, P^3)|p\rangle_0$$
$$= (E, 0, 0, E)|p\rangle_0 \tag{7.13}$$

or

$$P^1 |p\rangle_0 = P^2 |p\rangle_0 = 0$$
$$P^0 |p\rangle_0 = P^3 |p\rangle_0 = E|p\rangle_0$$

We can use these results to write

$$P^\mu |p\rangle_0 = (P^3, 0, 0, P^3)|p\rangle_0 \tag{7.14}$$

The reason why this is a useful thing to do will become clear in a moment.

It turns out that like $P^\mu P_\mu$, $W^\mu W_\mu$ gives zero when acting on a massless one-particle state. The proof of this result is a bit subtle. Obviously, we cannot simply let m go to zero in Eq. (7.10) because that equation had been obtained with the assumption that we could work in the rest frame of the particle. A detailed proof can be found in Wigner's original paper[52] or in Section 2.7.2 of Ref. 28.

The fact that $W_\mu W^\mu$ gives zero when applied to a one-particle massless state, together with Eq. (7.3), implies that when acting on such a state, the operator W^μ is proportional to the four-momentum operator

$$W^\mu |p\rangle_0 = h\, P^\mu |p\rangle_0 \tag{7.15}$$

(the meaning of the constant h will be elucidated shortly). To prove Eq. (7.15), consider

$$\begin{aligned} W^\mu P_\mu |p\rangle_0 &= (W^0 P^0 - W^1 P^1 - W^2 P^2 - W^3 P^3)|p\rangle_0 \\ &= (W^0 P^0 - W^3 P^3)|p\rangle_0 \\ &= E(W^0 - W^3)|p\rangle_0 \end{aligned}$$

where we have used Eqs. (7.13) and (7.14). But this must be zero because of Eq. (7.3), so we conclude that

$$W^0 |p\rangle_0 = W^3 |p\rangle_0 \tag{7.16}$$

Using this result, we have

$$\begin{aligned} W_\mu W^\mu |p\rangle_0 &= (W^0 W^0 - W^1 W^1 - W^2 W^2 - W^3 W^3)|p\rangle_0 \\ &= -(W^1 W^1 + W^2 W^2)|p\rangle_0 \end{aligned} \tag{7.17}$$

Since $W^\mu W_\mu$ is zero when acting on a massless state, we obtain

$$W^1 |p\rangle_0 = W^2 |p\rangle_0 = 0 \tag{7.18}$$

Therefore, when acting on the state, the action of the operator W^μ reduces to

$$W^\mu |p\rangle_0 = (W^3, 0, 0, W^3)|p\rangle_0 \qquad (7.19)$$

From Eq. (7.4) we know that W^3 commutes with P^3 (because it commutes with P_3), so we may take $|p\rangle_0$ to be a common eigenstate of these two operators. This implies that we may write

$$W^3 |p\rangle_0 = h\, P^3 |p\rangle_0 \qquad (7.20)$$

Now we see why we wrote Eq. (7.14) in that form. By comparing with Eq. (7.19), we now may conclude that

$$W^\mu |p\rangle_0 = h\, P^\mu |p\rangle_0 \qquad (7.21)$$

as we had set out to prove.

From Eq. (7.13) we see that the action of W^μ on the state is

$$W^\mu |p\rangle_0 = (hE, 0, 0, h\,E)|p\rangle_0 \qquad (7.22)$$

To find out what the constant h represents, we need to work out explicitly the left-hand side of Eq. (7.22). Let's pick $\mu = 3$. The derivation that led to Eq. (7.9) is still valid if we simply replace $P_0|p\rangle = m|p\rangle$ with $P_0|p\rangle_0 = E|p\rangle_0$. We then obtain

$$W^3 |p\rangle_0 = E(S_z + L_z)|p\rangle_0$$
$$= E\, s_z|p\rangle_0$$

where we have assumed that there is no orbital angular momentum. Using this in Eq. (7.22) implies that h is simply the z component of the spin, i.e.,

$$h = s_z \qquad (7.23)$$

Since our state has its momentum along the z direction, we may, of course, write

$$h = \vec{s} \cdot \hat{p} \qquad (7.24)$$

which is nothing other than the helicity of the particle. By the way, the helicity is often denoted by λ, but we prefer to use h because we will use λ to represent a left-chiral spinor that appears in supersymmetric gauge theories.

We have obtained this result by considering $\mu = 3$ in Eq. (7.22), but we get the same result if we pick $\mu = 0$, as you are invited to confirm in Exercise 7.3.

In contrast with massive states, massless representations of the Lorentz group are therefore completely specified *by only two numbers*: their energy (which then specifies their four-momentum) and their helicity h! The reason we don't have $2s + 1$ states, as for massive particles, lies in Eq. (7.18), which reveals that we do not have any ladder operators to change the z component of the spin. Therefore, there is only one spin state!

Does this mean that massless particles can be observed with only one helicity? Well, not quite. Recall that a physical theory also must respect *CPT* invariance. It turns out that *CPT* conjugate states have opposite helicities, so a physical theory necessarily must contain two states, with helicities h and $-h$. The states with the intermediate values of s_z that we encounter for massive particles therefore are absent for massless particles. This is the reason the photon is observed with helicities $+1$ and -1 or that the graviton is predicted to have helicities $+2$ and -2 only.

We will now repeat this exercise in the context of SUSY.

EXERCISE 7.3
Prove that we again obtain $h = \vec{s} \cdot \hat{p}$ if we set $\mu = 0$ in Eq. (7.22).

7.2 Effects of the Supercharges on States

We will now use the SUSY algebra to determine what states are needed to build a supersymmetric theory. Of course, we have worked out one such theory explicitly, the Wess-Zumino model, but we would like to see how the results we obtained there can be generalized.

We will only consider the classification of massless supersymmetric states. These are the only states of relevance if we are interested in the minimal supersymmetric extension of the standard model because in this theory, as is the case in the standard model itself, all masses are generated through spontaneous symmetry breaking.

As we saw in the previous section, massless states are identified by their four-vector and their helicity h. To emphasize this, we will change our notation from $|p\rangle_0$ to $|p, h\rangle$. What we want to now is to work out the effect of the supercharges on this state.

The supercharges commute with the four-momentum operator but not with the Lorentz generators, so they can change the helicity of states but not their four-momentum. Our first step, then, is to compute the commutator of the supercharges with W^μ. Actually, we really only need the commutator with W^0 because of Eqs. (7.16) and (7.18).

We will first compute $[Q_a, W^0]$ and come back to the hermitian conjugates of the supercharges later. Using Eqs. (7.1) and (6.83), this is given by

$$[Q_a, W^0] = \frac{1}{2}\epsilon^{0\nu\rho\sigma} [Q_a, M_{\rho\sigma} P_\nu]$$

$$= \frac{1}{2}\epsilon^{0\nu\rho\sigma} [Q_a, M_{\rho\sigma}] P_\nu$$

$$= \frac{1}{2}\epsilon^{0\nu\rho\sigma} (\sigma_{\rho\sigma})_a^{\ b} Q_b P_\nu$$

where we have used the fact that the supercharges commute with the momentum operator. We will drop the matrix indices a and b in the next few steps to alleviate the notation; we just need to keep in mind that on the right-hand side, Q is a column vector on which the Pauli matrices are acting.

The only components of P_ν that do not vanish when applied to a massless state are P_0 and P_3. If we take $\nu = 0$, we get zero because of the Levi-Civita symbol. This leaves us with

$$[Q, W^0] = \frac{1}{2}\epsilon^{03\rho\sigma} \sigma_{\rho\sigma} Q P_3$$

$$= \frac{1}{2}\epsilon^{03ij} \sigma_{ij} Q P_3$$

$$= -\frac{1}{2}\epsilon^{3ij} \sigma_{ij} Q P_3 \tag{7.25}$$

In Exercise 7.2 we showed that

$$\epsilon^{ijk}\sigma_{jk} = \sigma^i$$

which allows us to write

$$-\frac{1}{2}\epsilon^{3ij} \sigma_{ij} Q P_3 = -\frac{1}{2}\sigma^3 Q P_3$$

Reinstating the matrix indices, our final result is therefore is

$$[Q_a, W^0] = -\frac{1}{2}(\sigma^3)_a^{\ b} Q_b P_3 \tag{7.26}$$

This is also the commutator of Q_a with the operator W_0 because $W^0 = W_0$ with our choice of metric. The only elements of σ^3 that are nonzero are

$$(\sigma^3)^1_1 = -(\sigma^3)^2_2 = 1 \tag{7.27}$$

Using this in Eq. (7.26) we obtain

$$[Q_1, W_0] = -\frac{1}{2}Q_1 P_3 \tag{7.28}$$

and

$$[Q_2, W_0] = \frac{1}{2}Q_2 P_3 \tag{7.29}$$

Now let's find out what this is telling us. Recall that massless states satisfy Eq. (7.21). In particular, the effect of applying W_0 is simply

$$W_0|p, h\rangle = h\, E|p, h\rangle \tag{7.30}$$

We'd like to find out what is the effect of applying the supercharges Q_1 and Q_2 on $|p, h\rangle$. Consider first

$$Q_1|p, h\rangle \tag{7.31}$$

It's clear that this state has four-momentum p_μ because the supercharges commute with the momentum operator:

$$P_\mu Q_1|p, h\rangle = Q_1 P_\mu|p, h\rangle = Q_1 p_\mu|p, h\rangle = p_\mu\big(Q_1|p, h\rangle\big)$$

Now let's determine the helicity of Eq. (7.31). Using a trick familiar from quantum mechanics, we write

$$W_0(Q_1|p, h\rangle) = [W_0, Q_1]|p, h\rangle + Q_1 W_0|p, h\rangle$$

$$= \frac{1}{2}Q_1 E|p, h\rangle + Q_1 h\, E|p, h\rangle$$

$$= E\left(h + \frac{1}{2}\right)Q_1|p, h\rangle$$

The end result is

$$W_0(Q_1|p, h)) = E\left(h + \frac{1}{2}\right)Q_1|p, h\rangle$$

which, from Eq. (7.30), tells us that

$$Q_1|p, h\rangle = \left|p, h + \frac{1}{2}\right\rangle$$

Therefore *the action of the supercharge Q_1 is to raise the helicity by one-half*. Note that this implies that supercharges transform bosons into fermions and vice versa, as expected.

From Eq. (7.29) it is clear that Q_2 has the opposite effect of Q_1; i.e., *the supercharge Q_2 lowers the helicity by one-half*:

$$Q_2|p, h\rangle = \left|p, h - \frac{1}{2}\right\rangle$$

It is natural to expect that the hermitian conjugate $Q_1^\dagger$ and $Q_2^\dagger$ will have the opposite effect of Q_1 and Q_2. One indeed can show that this is the case using the commutator (6.86). The calculation is almost identical to the one we followed from Eq. (7.25) to Eq. (7.29) except that it involves supercharges with upper dotted indices. Since Chapter 3 was not prerequisite material for this chapter, we will only quote the final result, leaving the derivation to Exercise 7.4, which you are invited to do when you have become familiar with the van der Waerden notation. The end result is

$$\left[Q_1^\dagger, W_0\right] = \frac{1}{2}Q_1^\dagger P_3$$

$$\left[Q_2^\dagger, W_0\right] = -\frac{1}{2}Q_2^\dagger P_3 \tag{7.32}$$

By comparing with Eqs. (7.28) and (7.29), we see that, as expected, $Q_1^\dagger$ and $Q_2^\dagger$ lower and raise the helicity by one-half, respectively.

EXERCISE 7.4
Once you have become acquainted with the van der Waerden notation of Chapter 3, prove Eq. (7.32) using Eq. (6.86).

7.3 The Massless SUSY Multiplets

Now that we know the action of the supercharges on one-particle states, we will use the information to build irreducible representations of SUSY, also called *SUSY multiplets* or *supermultiplets* for short. To find these, we will work out the minimal set of states connected by the application of the supercharges. These states will represent the particle content necessary to build supersymmetric theories. Again, we restrict ourselves to massless states.

Our starting point is the SUSY algebra, Eqs. (6.70), (6.71), and (6.72). Using once again a frame in which

$$P^1|p, h\rangle = P^2|p, h\rangle = 0$$
$$P^3|p, h\rangle = P^0|p, h\rangle = E|h\rangle \tag{7.33}$$

we have

$$\sigma^\mu P_\mu|p, h\rangle = \left(P^0\sigma^0 - P^3\sigma^3\right)|p, h\rangle = \begin{pmatrix} 0 & 0 \\ 0 & 2P^0 \end{pmatrix}|p, h\rangle \tag{7.34}$$

Recall the anticommutation relations (6.72):

$$\{Q_a, Q_b^\dagger\} = (\sigma^\mu)_{ab} P_\mu \tag{7.35}$$

Using this and Eq. (7.34), we conclude that

$$\{Q_1, Q_1^\dagger\}|p, h\rangle = \{Q_1, Q_2^\dagger\}|p, h\rangle = \{Q_2, Q_1^\dagger\}|p, h\rangle = 0 \tag{7.36}$$

whereas

$$\{Q_2, Q_2^\dagger\}|p, h\rangle = 2P^0|p, h\rangle \tag{7.37}$$

The first relation in Eq. (7.36) is particularly useful. Its expectation value in a state $|p, h\rangle$ is

$$\langle p, h|\{Q_1, Q_1^\dagger\}|p, h\rangle = \langle p, h|Q_1 Q_1^\dagger|p, h\rangle + \langle p, h|Q_1^\dagger Q_1|p, h\rangle$$
$$= \left\|Q_1|p, h\rangle\right\|^2 + ||Q_1^\dagger|p, h\rangle||^2$$
$$= 0$$

Since we are working with physical states whose norm is positive definite, this implies that

$$Q_1|p, h\rangle = Q_1^\dagger|p, h\rangle = 0$$

We therefore only need to concern ourselves with the supercharges Q_2 and $Q_2^\dagger$.

What general statements can we make about the states forming a supermultiplet? They will all have the same four-momentum because the supercharges commute with the momentum operator, but they will, of course, have different helicities. It is clear that each supermultiplet has a state of minimum helicity because Q_2 is the only operator at our disposal that lowers the helicity, and $Q_2^2 = 0$. Let us then start with the state in a supermultiplet with the lowest helicity, which we will denote by

$$|p, h_{min}\rangle$$

By definition, acting with Q_2 on this state gives zero. On the other hand, acting with $Q_2^\dagger$ gives

$$Q_2^\dagger|p, h_{min}\rangle = \left|p, h_{min} + \frac{1}{2}\right\rangle$$

We cannot generate states of higher helicity because $(Q_2^\dagger)^2 = 0$. So this is it! A massless supermultiplet contains only two states of the same momentum but with helicities differing by half a unit (and therefore, one state is a boson, whereas the second is a fermion).

To be precise, our result applies to the so-called $N = 1$ SUSY, which involves only one Weyl spinor of charges Q (and its hermitian conjugate $Q^\dagger$). One can write down a supersymmetric with any number N of spinor charges, and these are often referred to as *extended supersymmetries*. We will briefly discuss extended supersymmetries at the end of this chapter.

As far as $N = 1$ SUSY is concerned, we therefore need only two states to build a theory. But it would not be physical because it would not be *CPT*-invariant. Recall that *CPT* changes the sign of the helicity, so the two members of a massless supermultiplet cannot be *CPT* conjugate states. A physical supersymmetric theory then must contain a supermultiplet plus its *CPT* conjugate, for a total of four states. To be precise, here we are counting only physical, on-shell states.

In the next section we will discuss in more details the massless supermultiplets. We will not discuss massive representations. The analysis is only slightly more involved, but for our purposes, only the massless representations will matter. Of course, except for the photon and the gluon, all particles are massive, but in the minimal supersymmetric standard model, we build the theory using massless multiplets and then incorporate their masses through the spontaneous symmetry breaking of the $SU(2)_L \times U(1)_Y$ group, as is done in the standard model.

7.4 Massless SUSY Multiplets and the MSSM

In this section we discuss only the multiplets of $N = 1$ SUSY. Let us first consider a state with $h_{\min} = 0$, which corresponds to a scalar field. Then SUSY requires the presence of an $h = 1/2$ state as well, which is a right-handed Weyl spinor. Since for a massless spinor handedness and chirality are equivalent, this is also a right-chiral spinor. *CPT* invariance further requires the introduction of a second $h = 0$ state as well as an $h = -1/2$ state. The two $h = 0$ states can be assembled into a complex scalar field, whereas the two fermion states correspond to a left-chiral Weyl spinor and its antiparticle. Of course, it is arbitrary which state is called the particle and which is the antiparticle, so we could just as well talk of a right-chiral particle together with its left-chiral antiparticle. In any case, the end result is that we obtain a scalar field plus a Weyl spinor and their antiparticles.

This type of multiplet is referred to as a *chiral multiplet* because the spin 1/2 fermion appears in only one chiral state (with the antifermion having the opposite chirality, of course). This is obviously the multiplet appearing in our model in Chapter 5. Note that the states we discuss here are the on-shell, physical states. Off shell, there are, of course, more degrees of freedom, but our discussion pertains only to physical states.

In the minimal supersymmetric standard model (MSSM) all the known fermions are taken to be members of chiral multiplets. Therefore, each fermion is paired up with a spin 0 supersymmetric partner. The accepted notation is to add a prefix *s* to the names of the known fermions to designate their spin 0 partners. So each fermion is paired up with a *sfermion*. The quarks are paired up with *squarks*, and leptons are paired up with *sleptons*. This terminology is even used with individual particle names, the scalar partner of the electron being called the *selectron*, whereas the supersymmetric partner of the top quark is referred to as the *stop*! This whimsical notation sometimes causes problems (The *scharm*, anyone? What about the *sstrange*?).

The Higgs field is also part of a chiral multiplet, being, of course, the scalar component. Actually, the Higgs content of the MSSM is more complex than in the standard model, as we will discuss in detail in Chapter 15, and requires the introduction of *two* $SU(2)$ doublets of complex Higgs fields, as opposed to the single doublet of the standard model. In any case, all these scalar fields belong to chiral multiplets and therefore have spinor superpartners. The convention for the names of spin 1/2 supersymmeric partners of standard model bosons is to add the suffix *ino* at the end of the name. The spin $1/2$ partners of the Higgs fields therefore are referred to as *Higgsinos*.

Next, let's consider supermultiplets with $h_{min} = 1/2$. SUSY requires a second state with $h = 1$. *CPT* then requires states with $h = -1$ and $h = -1/2$. We can think again of the two fermion states as a left-chiral Weyl spinor together with its antiparticle. The states $h = \pm 1$ obviously correspond to the two helicity states of a massless vector boson. This type of multiplet is therefore referred as a *vector multiplet* or sometimes as a gauge multiplet when the vector particle is a gauge field.

All the gauge bosons of the standard model belong to vector multiplets. Following the convention described earlier, their spin $1/2$ superpartners are generically called *gauginos*. More specifically, the superpartner of the photon, gluons, and weak-interaction bosons are, respectively, called the *photino*, the *gluinos*, the *winos*, and the *zino*. I am not making this up!

For $h_{min} = 1$, the four states have $h = \pm 1, \pm 3/2$. Again, we have the two helicity states of a massless vector particle. As for the massless spin $3/2$ state, it is sometimes referred to as a *Rarita-Schwinger field*.

For $h_{min} = 3/2$, we get four states with helicities $h = \pm 3/2, \pm 2$. The massless spin 2 particle is, of course, the famous graviton, and its spin $3/2$ partner is called the *gravitino*. These are, of course, crucial elements of supergravity (SUGRA) models and string theory.

7.5 Two More Important Results

We have seen how the algebra can be used to construct the massless irreducible representations of $N = 1$ SUSY without ever writing a lagrangian! We will now prove two additional important results using only the algebra.

First, we will prove that the number of boson states must be equal to the number of fermion states in a supermultiplet. The proof will be valid for massive as well as massless representations and for any value of N.

The first step is to introduce the fermion number operator N_f, which is defined, as its name suggests, to give the number of fermions present in a state when it is applied to that state. Now consider the operator $(-1)^{N_f}$. Applied to a bosonic state, this gives $+1$, and applied to a fermionic state, it gives -1. If we sum the expectation value of this operator over all the states belonging to a supermultiplet, we will get for an answer the number of bosons minus the number of fermions appearing in this multiplet. Let's carry out this calculation.

First, since the supercharges change a fermion into a boson, or vice versa, we have the obvious relation

$$(-1)^{N_f} Q_a = -Q_a (-1)^{N_f} \tag{7.38}$$

The same relation holds if we replace Q_a by $Q_a^\dagger$. This implies that the product of two supercharges commutes with $(-1)^{N_f}$.

The trick now is to sum the expectation value of

$$(-1)^{N_f}\{Q_a, Q_b^\dagger\}$$

over all the states belonging to a supermultiplet. In other words, we will take the *trace* of this operator over the members of a supermultiplet. A few simple manipulations then give

$$\text{Tr}\big((-1)^{N_f}\{Q_a, Q_b^\dagger\}\big) = \text{Tr}\big((-1)^{N_f} Q_a Q_b^\dagger + (-1)^{N_f} Q_b^\dagger Q_a\big)$$

$$= \text{Tr}\big((-1)^{N_f} Q_a Q_b^\dagger + Q_b^\dagger Q_a (-1)^{N_f}\big)$$

$$= \text{Tr}\big(-Q_a(-1)^{N_f} Q_b^\dagger + Q_b^\dagger Q_a (-1)^{N_f}\big)$$

where we have used Eq. (7.38) in the last line. Now we use the invariance of the trace under cyclic permutation of the arguments, $\text{Tr}(ABC) = \text{Tr}(CAB)$, to obtain

$$\text{Tr}\big(-Q_a(-1)^{N_f} Q_b^\dagger + Q_b^\dagger Q_a (-1)^{N_f}\big) = \text{Tr}\big(-Q_b^\dagger Q_a(-1)^{N_f} + Q_b^\dagger Q_a (-1)^{N_f}\big)$$

$$= 0$$

On the other hand, using the anticommutator (6.72), we also have that

$$\text{Tr}\big((-1)^{N_f}\{Q_a, Q_b^\dagger\}\big) = \text{Tr}\big((-1)^{N_f} \sigma^\mu P_\mu\big)$$

Since we have proved that the left-hand side vanishes identically, independent of the four-momentum, we must have

$$\text{Tr}(-1)^{N_f} = 0$$

in a SUSY supermultiplet, which implies that the number of fermions must be equal to the number of bosons. This is a general result and a well-known property of supersymmetric theories.

As a second proof, we will work out a condition on the expectation value of the hamiltonian in supersymmetric theories. This leads to a result that is a hallmark of SUSY, a result that will play a significant role in our discussion of SUSY breaking in Chapter 14.

The hamiltonian is the operator P^0. To isolate it in Eq. (6.72), we may use a simple trick. The trick becomes more obvious if we expand out the sum over Lorentz indices in Eq. (6.72):

$$\{Q_a, Q_b^\dagger\} = (\sigma^0)_{ab} H - (\sigma^i)_{ab} P^i$$

where σ^0 is simply the 2×2 identity matrix. What sets apart σ^0 from the Pauli matrices is that its trace is not zero. We therefore can isolate the hamiltonian by taking the trace of both sides, which means summing over a and b with the condition $a = b$. This leads to

$$\{Q_1, Q_1^\dagger\} + \{Q_2, Q_2^\dagger\} = 2H$$

Consider now calculating the expectation value of the two sides in an arbitrary state $|\psi\rangle$:

$$\langle\psi|Q_1 Q_1^\dagger|\psi\rangle + \langle\psi|Q_1^\dagger Q_1|\psi\rangle + \langle\psi|Q_2 Q_2^\dagger|\psi\rangle + \langle\psi|Q_2^\dagger Q_2|\psi\rangle = 2\langle\psi|H|\psi\rangle$$

$$(7.39)$$

Each term on the left-hand side is semipositive definite because it represents the norm squared of a state. For example,

$$\langle\psi|Q_1^\dagger Q_1|\psi\rangle = \|Q_1|\psi\rangle\|^2$$

Since the left-hand side of Eq. (7.39) is semipositive definite, we have

$$\langle\psi|H|\psi\rangle \geq 0$$

If we consider the special case of the state being the vacuum, then Eq. (7.39) gives the vacuum expectation value (vev) of the hamiltonian:

$$\langle 0|Q_1 Q_1^\dagger|\psi\rangle + \langle 0|Q_1^\dagger Q_1|0\rangle + \langle 0|Q_2 Q_2^\dagger|0\rangle + \langle 0|Q_2^\dagger Q_2|0\rangle = 2\langle 0|H|0\rangle \quad (7.40)$$

and we find again that

$$\langle 0|H|0\rangle \geq 0 \tag{7.41}$$

with the inequality saturated if the supercharges annihilate the vacuum. If they do not annihilate the vacuum, then we have a spontaneous breaking of SUSY, and we get a strict inequality, i.e.,

$$\langle 0|H|0\rangle > 0 \tag{7.42}$$

Therefore, a necessary and sufficient condition for SUSY to be spontaneously broken is that the vacuum expectation value of the hamiltonian is nonzero (and necessarily positive). We will come back to this issue in Chapter 14.

7.6 The Algebra in Majorana Form

Many references give the SUSY algebra in Majorana form instead of the Weyl form we just worked out. We will now show how to translate all the commutation relations involving the supercharges into the Majorana language.

The first step is to group the supercharges into a Majorana spinor. Recall from Eq. (4.20) that a Majorana spinor can be built out of a left-chiral Weyl spinor χ and the hermitian conjugate $\chi^\dagger$ via

$$\Psi_M = \begin{pmatrix} i\sigma^2 \chi^{\dagger T} \\ \chi \end{pmatrix}$$

We therefore can assemble the supercharges into a *Majorana supercharge*

$$Q_M \equiv \begin{pmatrix} i\sigma^2 Q^{\dagger T} \\ Q \end{pmatrix}$$

which, written explicitly, is

$$Q_M = \begin{pmatrix} Q_2^\dagger \\ -Q_1^\dagger \\ Q_1 \\ Q_2 \end{pmatrix} \tag{7.43}$$

Let us define the components $Q_{M1}, Q_{M2}, \ldots$ through

$$Q_M \equiv \begin{pmatrix} Q_{M1} \\ Q_{M2} \\ Q_{M3} \\ Q_{M4} \end{pmatrix}$$

Then *all* the anticommutators among the supercharges can be shown to be reproduced by the single expression

$$\{Q_{Ma}, \bar{Q}_{Mb}\} = (\gamma^\mu)_{ab} P_\mu \tag{7.44}$$

Here, the bar over Q_M represents the usual Dirac adjoint, so $\bar{Q}_{Mb}$ stands for $(Q_M^\dagger \gamma^0)_b$.

If we define the matrices $\gamma_{\mu\nu}$ by

$$\gamma_{\mu\nu} \equiv \frac{i}{4}(\gamma_\mu \gamma_\nu - \gamma_\nu \gamma_\mu)$$

then the commutation relation of the supercharges with the Lorentz generators are

$$[Q_{Ma}, M_{\mu\nu}] = (\gamma_{\mu\nu})_{ab} Q_{Mb}$$

It is not difficult to show that this expression reproduces both Eqs. (6.83) and (6.86) using the fact that $\gamma_{\mu\nu}$ may be written as

$$\gamma_{\mu\nu} = \begin{pmatrix} \bar{\sigma}_{\mu\nu} & 0 \\ 0 & \sigma_{\mu\nu} \end{pmatrix} \tag{7.45}$$

We will now obtain the SUSY charges from the supersymmetric currents, following Eq. (6.24). But first let's review the general approach.

EXERCISE 7.5
Verify that Eq. (7.44) reproduces the anticommutators (6.70), (6.71), and (6.72).

7.7 Obtaining the Charges from Symmetry Currents

To see explicitly how to construct the currents, consider the theory of a single complex scalar field. If the field is varied by an amount $\delta\phi$, the lagrangian varies by

$$\delta\mathcal{L}(\phi, \phi^\dagger) = \frac{\partial\mathcal{L}}{\partial\phi}\delta\phi + \frac{\partial\mathcal{L}}{\partial(\partial_\mu\phi)}\delta(\partial_\mu\phi) + \frac{\partial\mathcal{L}}{\partial\phi^\dagger}\delta\phi^\dagger + \frac{\partial\mathcal{L}}{\partial(\partial_\mu\phi^\dagger)}\delta(\partial_\mu\phi^\dagger)$$

This is simply an application of the chain rule from ordinary multivariable calculus (a similar trick is used to determine the equations of motion of the fields).

If this transformation is a symmetry of the theory, it leaves the action invariant, which, as we have discussed previously, implies either that the lagrangian density is invariant or that it changes by a total derivative. To be general, let's assume that it changes by a total derivative, which we will write as $\partial_\mu K^\mu$, with K^μ some function of the fields. We then have

$$\frac{\partial \mathcal{L}}{\partial \phi}\delta\phi + \frac{\partial \mathcal{L}}{\partial(\partial_\mu\phi)}\partial_\mu(\delta\phi) + \frac{\partial \mathcal{L}}{\partial \phi^\dagger}\delta\phi^\dagger + \frac{\partial \mathcal{L}}{\partial(\partial_\mu\phi^\dagger)}\partial_\mu(\delta\phi^\dagger) = \partial_\mu K^\mu \qquad (7.46)$$

where we have used $\delta(\partial_\mu\phi) = \partial_\mu(\delta\phi)$ and $\delta(\partial_\mu\phi^\dagger) = \partial_\mu(\delta\phi^\dagger)$. Now, assuming that ϕ and $\phi^\dagger$ obey the equations of motion, we have

$$\frac{\partial \mathcal{L}}{\partial \phi} = \partial_\mu\left(\frac{\partial \mathcal{L}}{\partial(\partial_\mu\phi)}\right) \qquad (7.47)$$

and

$$\frac{\partial \mathcal{L}}{\partial \phi^\dagger} = \partial_\mu\left(\frac{\partial \mathcal{L}}{\partial(\partial_\mu\phi^\dagger)}\right) \qquad (7.48)$$

Substituting Eqs. (7.47) and (7.48) into Eq. (7.46), we get

$$\partial_\mu\left[\frac{\partial \mathcal{L}}{\partial(\partial_\mu\phi)}\delta\phi + \frac{\partial \mathcal{L}}{\partial(\partial_\mu\phi^\dagger)}\delta\phi^\dagger\right] = \partial_\mu K^\mu \qquad (7.49)$$

The quantity in brackets is called the *Noether current* j^μ, i.e.,

$$j^\mu \equiv \frac{\partial \mathcal{L}}{\partial(\partial_\mu\phi)}\delta\phi + \frac{\partial \mathcal{L}}{\partial(\partial_\mu\phi^\dagger)}\delta\phi^\dagger$$

so Eq. (7.49) may be written as

$$\partial_\mu j^\mu = \partial_\mu K^\mu$$

Therefore, the quantity

$$\mathcal{J}^\mu \equiv (j^\mu - K^\mu) \qquad (7.50)$$

is conserved, in the sense that it satisfies

$$\partial_\mu \mathcal{J}^\mu = 0$$

Of course, in a theory where there are several fields, such as the supersymmetric theory we built in Chapters 5 and 6, we have to add terms such as Eqs. (7.47) and (7.48) for each field. We will illustrate this later in this chapter.

Now we get to the key point. The claim is that the charges generating the transformations are given by the spatial integral of the zeroth component of the currents:

$$Q = \int d^3x \, \mathcal{J}^0 \tag{7.51}$$

We won't prove that the charges are indeed given in general by Eq. (7.51) (see any textbook on quantum field theory for a derivation).

As an aside, the Q arc often described as *conserved* charges because they are time-independent.

As already emphasized, it is important to keep in mind that the currents, and hence the charges constructed via the currents, are functions of the quantum fields appearing in the theory, so they are themselves quantum field operators. Once the charges are found using Eq. (7.51), their commutation (or anticommutation) relations can be worked out explicitly using the equal time commutation/anticommutation of the fields.

Let's now apply this to SUSY!

7.8 Explicit Supercharges as Quantum Field Operators

Now let us build the SUSY current and find explicit expressions for the supercharges in terms of the fields of the theory.

In the case of SUSY, Eq. (7.51) is replaced by

$$Q \cdot \zeta + \bar{Q} \cdot \bar{\zeta} = \int d^3x \, \mathcal{J}^0_{\text{SUSY}}$$

We work with the free supersymmetric lagrangian, which we will repeat here:

$$\mathcal{L} = \partial_\mu \phi \partial^\mu \phi^\dagger + \chi^\dagger i \bar{\sigma}^\mu \partial_\mu \chi + F F^\dagger$$

First, let's find the SUSY Noether currents. In order to do this, we need

$$\frac{\partial \mathcal{L}}{\partial(\partial_\mu \phi)}\delta\phi = (\partial^\mu \phi^\dagger)\delta\phi \tag{7.52}$$

$$= (\partial_\mu \phi^\dagger)\chi \cdot \zeta \tag{7.53}$$

$$\frac{\partial \mathcal{L}}{\partial(\partial_\mu \chi)}\delta\chi = \chi^\dagger i\bar{\sigma}^\mu \delta\chi \tag{7.54}$$

$$= \chi^\dagger \bar{\sigma}^\mu (\partial_\nu \phi)\sigma^\nu i\sigma^2 \zeta^* \tag{7.55}$$

$$\frac{\partial \mathcal{L}}{\partial(\partial_\mu \phi^\dagger)}\delta\phi^\dagger = (\partial^\mu \phi)\bar{\chi} \cdot \bar{\zeta}$$

On the other hand,

$$\frac{\partial \mathcal{L}}{\partial(\partial_\mu F)} = \frac{\partial \mathcal{L}}{\partial(\partial_\mu F^\dagger)} = \frac{\partial \mathcal{L}}{\partial(\partial_\mu \chi^\dagger)} = 0$$

The Noether current is therefore

$$j^\mu = (\partial_\mu \phi^\dagger)\chi \cdot \zeta + (\partial_\nu \phi)\chi^\dagger \bar{\sigma}^\mu \sigma^\nu i\sigma^2 \zeta^* + (\partial^\mu \phi)\bar{\chi} \cdot \bar{\zeta}$$

To find $\partial_\mu K^\mu$, we simply have to add up the total derivatives we dropped when we calculated the supersymmetric transformation of our lagrangian. These total derivatives can be found by comparing Eqs. (5.14) and (5.15) to the equations preceding them, giving us (recall that we showed that the constant C appearing in those equations was equal to $-i$)

$$\partial_\mu K^\mu = \partial_\mu \left(\bar{\chi} \cdot \bar{\zeta}\partial^\mu \phi - i(\partial_\nu \phi^\dagger)\zeta^T i\sigma^2 \sigma^\nu i\bar{\sigma}^\mu \chi + (\partial^\mu \phi^\dagger)\zeta \cdot \chi \right) \tag{7.56}$$

where we have used $\chi^\dagger i\sigma^2 \zeta^* = \bar{\chi} \cdot \bar{\zeta}$ and $\zeta^T i\sigma^2 \chi = -\zeta \cdot \chi$. So K^μ is simply the expression inside the parenthesis. Using the fact that the order in spinor dot products does not matter, the supersymmetric current is simply

$$\mathcal{J}^\mu_{\text{SUSY}} = j^\mu - K^\mu$$

$$= (\partial_\nu \phi)\chi^\dagger \bar{\sigma}^\mu \sigma^\nu i\sigma^2 \zeta^* - (\partial_\nu \phi^\dagger)\zeta^T i\sigma^2 \sigma^\nu \bar{\sigma}^\mu \chi \tag{7.57}$$

We can then use this to find the charges:

$$\zeta \cdot Q + \bar{Q} \cdot \bar{\zeta} = \zeta^T(-i\sigma^2)Q + Q^\dagger i\sigma^2 \zeta^*$$

$$= \int d^3x\left((\partial_\nu\phi)\chi^\dagger\bar{\sigma}^0\sigma^\nu i\sigma^2\zeta^* - (\partial_\nu\phi^\dagger)\zeta^T i\sigma^2\sigma^\nu\bar{\sigma}^0\chi\right)$$

On the left-hand side we chose to write $\bar{\zeta} \cdot \bar{Q}$ as $\bar{Q} \cdot \bar{\zeta}$. This was simply in order to get ζ^* to appear on the right of the σ^2 matrix, as in the current.

The matrix $\bar{\sigma}^0$ is simply the identity matrix. Since the infinitesimal parameters ζ and ζ^* may be treated as independent, we can read off the charges Q and $Q^\dagger$. We find

$$Q = \int d^3x(\partial_\nu\phi^\dagger(x))\sigma^\nu\chi(x) \tag{7.58}$$

and

$$Q^\dagger = \int d^3x\,\chi^\dagger(x)\partial_\nu\phi(x)\sigma^\nu \tag{7.59}$$

which is simply the hermitian conjugate of Q.

This is what we were looking for: explicit expressions for the supercharges as quantum field operators. It is then easy to verify explicitly that these charges indeed generate the SUSY transformations of the fields given in Eq. (6.52). This will be left as an exercise. All we will need are the equal time commutators and anticommutators familiar from quantum field theory:

$$\left[\phi(\vec{x}, t), \dot{\phi}^\dagger(\vec{y}, t)\right] = i\delta^3(\vec{x} - \vec{y})$$

$$\{\chi_a(\vec{x}, t), \chi_b^\dagger(\vec{y}, t)\} = \delta_{ab}\delta^3(\vec{x} - \vec{y}) \tag{7.60}$$

where a dot indicates a derivative with respect to time. All other commutators and anticommutators are equal to zero. For example, $[\phi, \dot{\phi}] = 0$.

The commutators of the supercharges with the Poincaré generators also can be verified, but this requires that we first work out P_μ and $M_{\mu\nu}$ as quantum field operators. We will not pursue this here because, although straightforward, the calculation is a bit lengthy and is in any case similar to usual quantum field theory calculations. The interested reader will find the detailed calculation in Ref. 31.

EXERCISE 7.6
Obtain the commutators (6.52) using the quantum field operator representations of the supercharges (7.58) and (7.59) and the equal time commutators (7.60).

7.9 The Coleman-Mandula No-Go Theorem

A book on SUSY would be incomplete if it did not mention one of the most famous *no-go theorems* of particle physics. Now that we have worked a bit with charges and their algebra, the theorem will be easier to understand.

Since the discovery of relativity, it has been clear that particle states should belong to representations of the Poincaré group. On the other hand, it is natural to also group particles into multiplets of some additional "internal" symmetry group, and this viewpoint, of course, has been very successful in the construction of gauge theories. An obvious question to ask is whether such a symmetry group actually could "mix" nontrivially with the Poincaré group. To make this rather vague statement more precise, the question is whether the generators of the internal symmetry group could have nonzero commutators with the generators of the Poincaré group.

Let us write the charges (or *generators*) of the internal symmetry group as T^a and their algebra as

$$[T^a, T^b] = i f^{abc} T^c \tag{7.61}$$

The question is whether it is possible to write down some physically meaningful theory where the charges T^a do *not* commute with the generators of the Poincaré group $M_{\mu\nu}$ and P_μ (another way to put it is to say that the T^a should carry some Lorentz index). This is an interesting question because such a theory then would mix spacetime transformations with "internal" transformations and would lead to a highly nontrivial extension of gauge theories.

What Coleman and Mandula proved in 1967[11] (a derivation also can be found in[47]) was that for an algebra of the form of Eq. (7.61) and under a few very reasonable conditions, e.g., that elastic scattering amplitudes must be analytic functions of the Mandelstam variables and that plane waves must in general scatter, the answer is a resounding *no*! In other words, the generators T^a cannot carry spacetime indices. Exercise 7.7 illustrates the theorem in the simple case of a symmetric second-rank tensor charge $S_{\mu\nu} = S_{\nu\mu}$.

EXERCISE 7.7

Consider a symmetric tensor $S_{\mu\nu}$ that is conserved. Then its eigenvalue will be conserved in a collision. Consider the scattering process $a + b \rightarrow a + b$. Denote the four-momenta of the particles p_i^a, p_i^b, p_f^a, and p_f^b. Show that the conservation of the eigenvalue of $S_{\mu\nu}$ together with the conservation of total momentum implies that either there is no scattering at all, $p_f^a = p_i^a$ and $p_f^b = p_i^b$, or that the particles exchange their momenta, $p_f^a = p_i^b$ and $p_f^b = p_i^a$. No other direction of scattering is possible! This is clearly unphysical.

Hint: Write the most general eigenvalue for this operator for a one-particle state. Then use that the corresponding eigenvalue for a two-particle state is simply the sum of the one-particle state eigenvalues.

This seemed to be the kiss of death for the idea of building theories mixing spacetime transformations and internal symmetries nontrivially. One might think that physicists would have been eager to find some way to circumvent this no-go theorem as soon as it came out, but this was not how SUSY was first discovered. We won't get into the history of SUSY here; interested readers are invited to read the first chapter of Ref. 47 for a very informative historical summary.

Of course, SUSY *does* circumvent the theorem in a very simple way: The supercharges do not obey a set of *commutation* relations as in Eq. (7.61) but instead obey *anticommutation* relations. Another way to carry the same information is that supercharges transform as a spinor representation of the Lorentz group instead of transforming as vector (or more generally, tensor) representations.

The theorem is still useful because it tells us that no other symmetry with charges obeying commutation rules could mix nontrivially with spacetime transformations. But are the supercharge anticommutators we wrote down the most general possible ones? The next section addresses this question.

7.10 The Haag-Lopuszanski-Sohnius Theorem and Extended SUSY

The discovery of SUSY raised another obvious question: Is the algebra we wrote down the most general algebra that evades Coleman and Mandula's theorem? The answer was provided by Haag, Lopuszanski, and Sohnius in 1975,[23] and the answer again was a resounding no. Not only did they show that other possibilities exist, but they also worked out the most general algebra possible.

To write it down, we first introduce a set of N copies of the supercharges, i.e.,

$$Q_1^I, Q_2^I, Q_1^{\dagger I}, Q_2^{\dagger I}$$

where I is an extra label ranging from 1 to N. The SUSY we have considered so far in this book corresponds to $N = 1$ and therefore is referred to as $N = 1$ *supersymmetry*. The most general algebra satisfied by these charges is

$$\{Q_a^I, Q_b^J\} = (-i\sigma^2)_{ab} Z^{IJ}$$

$$\{Q_a^I, Q_b^{J\dagger}\} = (\sigma^\mu)_{ab} \delta^{IJ} P_\mu$$

$$\{Q_a^{I\dagger}, Q_b^{J\dagger}\} = (-i\sigma^2)_{ab}(Z^{IJ})^* \tag{7.62}$$

where the Z_{IJ} are called the *central charges* of the algebra (a central charge refers to a quantity which commutes with all the generators of an algebra) and always can be made antisymmetric, $Z^{IJ} = -Z^{JI}$. In the case of $N = 1$ supersymmetry, the unique central charge Z^{11} is therefore necessarily zero, and the algebra Eq. (7.62) reduces to what we obtained earlier in this chapter.

Extended SUSY is beyond the scope of this book, but one important point to make is that from a purely phenomenological point of view, only $N = 1$ is attractive, at least in theories describing physics at the low energies that will be accessible to the Large Hadron Collider (LHC). The reason is that starting with $N = 2$, supersymmetric theories are necessarily nonchiral; i.e., they must treat left- and right-chiral states on equal footing. For $N = 2$, the chiral supermultiplet contains both a right- and a left-chiral fermion, with $h = 1/2$ and $h = -1/2$ (as well as two states with $h = 0$). But the members of a supermultiplet all transform the same way under a gauge symmetry, so for $N = 2$ (and any larger value of N as well), the two chiral states must be treated on an equal footing by any gauge force. But this conflicts with what we know about weak force of the standard model, which treats differently left- and right-chiral states; hence only $N = 1$ SUSY is relevant to phenomenology. This is true if we think of the MSSM as some effective field theory to be used with energies on the order of a few TeV. A more fundamental theory could have a very different structure, and who knows how symmetry may be broken down to what we observe at small energies relative to the Planck scale. But even if we set aside the issue of chirality, other problems arise, if we increase too much the value of N because it does not seem possible to write down consistent quantum field theories containing fields with spin larger than 2.

More details on extended SUSY may be found in Refs. 8 and 47.

7.11 Quiz

1. What is the effect of the supercharge Q_1 on the four-momentum and helicity of a state?

2. For $N = 1$, how many states are required to form a SUSY multiplet?

3. What is extended SUSY?

4. What can we say about the vacuum expectation value of the hamiltonian in a supersymmetric theory?

5. How does SUSY evade the Coleman-Mandula no-go theorem?

CHAPTER 8

Adding Interactions: The Wess-Zumino Model

So far we have considered a rather uninteresting theory: All the fields are massless, and even more boring, there are no interactions at all. What we would like to do now is to see if we can include masses and interactions that will preserve supersymmetry (SUSY). The answer is obviously yes because SUSY would never have become such a hot topic if interactions could not be included. However, as you surely expect, this will prove to be a highly nontrivial undertaking.

In writing down possible interaction terms, we will follow certain rules:

- *We will not consider terms of dimension above 4.* This requires some explanation. Of course, any term in a lagrangian density is of dimension 4. What
  we mean here is that *if we ignore all the constants*, the dimension must not
  be more than 4.

  Consider for example the four-fermion interaction $\chi \cdot \chi \, \bar{\chi} \cdot \bar{\chi}$. Since the
  dimension of fermion fields is $3/2$, this term has dimension 6 and must be

rejected. Note that if we insisted on including this term in our lagrangian density, we would have to divide it by a constant having the dimension of energy squared:

$$\mathcal{L}_{\text{int}} = \frac{1}{\mu^2} \chi \cdot \chi \, \bar{\chi} \cdot \bar{\chi} \tag{8.1}$$

where μ would be a free parameter with dimension of energy. Therefore, another way of saying that we reject terms of dimension more than 4 is that we will not consider any term in the lagrangian that contains a dimensionful constant in the denominator.

The reason for this condition is that any interaction of dimension 5 or higher leads to a nonrenormalizable theory. There is nothing intrinsically wrong with nonrenormalizable theories if they are treated as effective field theories, but we won't consider these in this book.

- All the terms we add must, of course, be Lorentz-invariant. This obviously implies that any upper (contravariant) Lorentz index must be contracted with a corresponding lower (covariant) Lorentz index. But this is not all. We also must ensure that spinors are combined into Lorentz-invariant combinations. This is where all our work of Chapter 2 will come in handy.

- The lagrangian must be real, $\mathcal{L}^\dagger = \mathcal{L}$, in order to have *CPT* invariance.

- And finally, of course, we will insist that the lagrangian is invariant (up to total derivatives) under the SUSY transformations of the fields, which we repeat here for convenience:

$$\delta\phi = \zeta \cdot \chi = \zeta^T(-i\sigma^2)\chi$$
$$\delta\phi^\dagger = \bar{\chi} \cdot \bar{\zeta} = \chi^\dagger i\sigma^2 \zeta^*$$
$$\delta\chi = -i\sigma^\mu(\partial_\mu\phi)i\sigma^2\zeta^* + F\zeta$$
$$\delta F = -i\zeta^\dagger \bar{\sigma}^\mu \partial_\mu \chi \tag{8.2}$$

8.1 A Supersymmetric Lagrangian With Masses and Interactions

Let's begin. We want to write new terms built out of the fields ϕ, χ, F, and their hermitian conjugates that will respect our four criteria. Consider first interactions that depend only on ϕ and its dagger $\phi^\dagger$. To keep things general, let's just write this

interaction as

$$\mathcal{L}_1 = \mathcal{G}(\phi, \phi^\dagger) \tag{8.3}$$

This is automatically Lorentz-invariant because ϕ is a scalar field. We use a calligraphic G instead of $\mathcal{F}$ for this function in order to avoid possible confusion with the auxiliary field F. To satisfy the criterion of renormalizability, we may only consider functions of at most four scalar fields. Without loss of generality, we may take $\mathcal{G}$ to be real, $\mathcal{G}^\dagger = \mathcal{G}$ (if it weren't, we would simply add to it its hermitian conjugate and rename the sum $\mathcal{G}$).

What else can we write down? Well, we can consider some arbitrary function of ϕ times the auxiliary field F. We will write this term as

$$W_1(\phi, \phi^\dagger)F \tag{8.4}$$

The notation W_1 is a standard one. The reason for the subscript 1 is that we are dealing with a single set of fields ϕ, χ, and F. Later on we will extend our discussion to consider several sets of fields, and the 1 will be replaced by an index i.

The lagrangian Eq. (8.4) is automatically Lorentz-invariant because both ϕ and F are scalar fields. For this to be of dimension 5 or less, W must contain the product of at most two ϕ or $\phi^\dagger$ fields. However, Eq. (8.4) is not real. To get a real lagrangian, we must add to Eq. (8.4) its hermitian conjugate, so we will consider instead

$$\mathcal{L}_2 = W_1 F + W_1^\dagger F^\dagger \tag{8.5}$$

Note that there is no reality condition imposed on W_1.

Next, we also can consider a general function of the scalar fields multiplying a Lorentz-invariant combination of the spinor field χ. We have seen that a possible Lorentz invariant made of χ is $\chi \cdot \chi$, so we write this new contribution to the lagrangian as

$$-\frac{1}{2} W_{11}(\phi, \phi^\dagger)\chi \cdot \chi \tag{8.6}$$

Once again, we are using a standard notation for the name of the function of the scalar field, and the factor of $-1/2$ is purely conventional and chosen to simplify a later result. There is no a priori relation between W_{11} and W_1, although we will later find that they must indeed be closely related in order for the theory to be supersymmetric.

Again, Eq. (8.6) is not real, so we need to consider instead the real combination [using Eq. (A.44)]:

$$\mathcal{L}_3 = -\frac{1}{2} W_{11} \chi \cdot \chi - \frac{1}{2} W_{11}^\dagger (\chi \cdot \chi)^\dagger$$

$$= -\frac{1}{2} W_{11} \chi \cdot \chi - \frac{1}{2} W_{11}^\dagger \bar{\chi} \cdot \bar{\chi} \tag{8.7}$$

There is no reality condition imposed on W_{11}.

Next, we might think about considering terms containing a function of the auxiliary field F times $\chi \cdot \chi$, but this is necessarily of dimension 5 or higher. For the same reason, it is impossible to write any term that contains ϕ (or $\phi^\dagger$) and two auxiliary fields or to write terms containing all three types of fields (these start at dimension 6).

Denoting by $\mathcal{L}_0$ the free supersymmetric lagrangian obtained in Chapter 6, the most general lagrangian that is real, renormalizable, and Lorentz-invariant is therefore

$$\mathcal{L}_{\text{total}} = \mathcal{L}_0 + \mathcal{L}_1 + \mathcal{L}_2 + \mathcal{L}_3$$

$$= \partial_\mu \phi \partial^\mu \phi^\dagger + \chi^\dagger i \bar{\sigma}^\mu \partial_\mu \chi + F F^\dagger + \mathcal{G} + W_1 F + W_1^\dagger F^\dagger$$

$$- \frac{1}{2} W_{11} \chi \cdot \chi - \frac{1}{2} W_{11}^\dagger \bar{\chi} \cdot \bar{\chi} \tag{8.8}$$

where one must keep in mind that $\mathcal{G}$, W_1, and W_{11} are functions of ϕ and $\phi^\dagger$. We did not include any coefficients in front of the new terms because they can be absorbed into a redefinition of $\mathcal{G}$, W_1, and W_{11}.

Our goal now is to see if we can find expressions for those three functions of the scalar fields such that the total lagrangian is supersymmetric. We already know that $\mathcal{L}_0$ is invariant under SUSY so we only need to impose that $\mathcal{L}_1 + \mathcal{L}_2 + \mathcal{L}_3$ must be supersymmetric. From now on we will call the sum of these three terms $\mathcal{L}_{\text{int}}$ because they contain interactions between the different fields, i.e.,

$$\mathcal{L}_1 + \mathcal{L}_2 + \mathcal{L}_3 \equiv \mathcal{L}_{\text{int}} \tag{8.9}$$

Let us write the variation of $\mathcal{L}_{\text{int}}$ under SUSY explicitly:

$$\delta \mathcal{L}_{\text{int}} = \delta \mathcal{G} + \delta(W_1 F) + \delta(W_1^\dagger F^\dagger)$$

$$- \frac{1}{2} \delta(W_{11} \chi \cdot \chi) - \frac{1}{2} \delta(W_{11}^\dagger \bar{\chi} \cdot \bar{\chi}) \tag{8.10}$$

At first sight, it may seem like we will need to write down explicit expressions for W_1, W_{11}, and $\mathcal{G}$ before we can calculate the supersymmetric variation of our lagrangian, but this won't be necessary. For example, we can simply write the total variation of $\mathcal{G}(\phi, \phi^\dagger)$ under a SUSY transformation as

$$\delta \mathcal{G} = \frac{\partial \mathcal{G}}{\partial \phi} \delta \phi + \frac{\partial \mathcal{G}}{\partial \phi^\dagger} \delta \phi^\dagger \qquad (8.11)$$

using the rule familiar from calculus. We therefore get, using Eq. (8.2),

$$\delta \mathcal{G} = \frac{\partial \mathcal{G}}{\partial \phi} \zeta \cdot \chi + \frac{\partial \mathcal{G}}{\partial \phi^\dagger} \bar{\chi} \cdot \bar{\zeta} \qquad (8.12)$$

We can, of course, use the same trick to write down the variation of W_1 and W_{11}. For example, we have

$$\begin{aligned}
\delta(W_1 F) &= (\delta W_1)\, F + W_1 \delta F \\
&= \left(\frac{\partial W_1}{\partial \phi} \delta \phi + \frac{\partial W_1}{\partial \phi^\dagger} \delta \phi^\dagger \right) F + W_1 \delta F \\
&= \frac{\partial W_1}{\partial \phi} \zeta \cdot \chi F + \frac{\partial W_1}{\partial \phi^\dagger} \bar{\zeta} \cdot \bar{\chi} F - i W_1 \zeta^\dagger \bar{\sigma}^\mu \partial_\mu \chi \qquad (8.13)
\end{aligned}$$

The variation of $W_1^\dagger F^\dagger$ is obtained simply by taking the hermitian conjugate of this expression.

On the other hand, we find that

$$\begin{aligned}
-\frac{1}{2} \delta(W_{11} \chi \cdot \chi) &= -\frac{1}{2} \delta W_{11}\, \chi \cdot \chi - W_{11}\, \chi \cdot \delta \chi \\
&= -\frac{1}{2} \left(\frac{\partial W_{11}}{\partial \phi} \delta \phi + \frac{\partial W_{11}}{\partial \phi^\dagger} \delta \phi^\dagger \right) \chi \cdot \chi - W_{11}\, \chi \cdot \delta \chi \\
&= \quad -\frac{1}{2} \frac{\partial W_{11}}{\partial \phi} \zeta \cdot \chi\, \chi \cdot \chi - \frac{1}{2} \frac{\partial W_{11}}{\partial \phi^\dagger} \bar{\zeta} \cdot \bar{\chi}\, \chi \cdot \chi \\
&\qquad - i\, W_{11}\, \chi^T i\sigma^2 (\partial_\mu \phi)\, \sigma^\mu i\sigma^2 \zeta^* - W_{11}\, F\, \chi \cdot \zeta \qquad (8.14)
\end{aligned}$$

and again, the variation of $-W_{11}^\dagger \bar{\chi} \cdot \bar{\chi}/2$ is simply the hermitian conjugate of the preceding.

Before writing out Eq. (8.10) in gory detail, let us make a key observation: The first two terms (containing $\mathcal{G}$) cannot be canceled by anything arising from the last four terms (containing W_1 and W_{11}). This is clear because the variation of the terms in W_1 and W_{11} contains either a derivative, an auxiliary field F, or three spinors χ. But none of these appear in the terms in $\mathcal{G}$. Therefore, if we want the theory to be supersymmetric, $\mathcal{G}$ must be a constant, which always can be chosen to be zero, i.e.,

$$\boxed{\mathcal{G} = 0} \tag{8.15}$$

This leaves the more manageable form [using Eqs. (8.13) and (8.14)]

$$\delta\mathcal{L}_{\text{int}} = \delta(W_1 F) - \frac{1}{2}\delta(W_{11}\,\chi\cdot\chi) + \text{ h.c.}$$

$$= \underbrace{\frac{\partial W_1}{\partial\phi}\,\zeta\cdot\chi\,F}_{\text{I}} + \underbrace{\frac{\partial W_1}{\partial\phi^\dagger}\,\bar{\zeta}\cdot\bar{\chi}\,F}_{\text{II}} \underbrace{-iW_1\,\zeta^\dagger\bar{\sigma}^\mu\partial_\mu\chi}_{\text{III}} - \frac{1}{2}\underbrace{\frac{\partial W_{11}}{\partial\phi}\,\zeta\cdot\chi\,\chi\cdot\chi}_{\text{IV}}$$

$$- \frac{1}{2}\underbrace{\frac{\partial W_{11}}{\partial\phi^\dagger}\,\bar{\zeta}\cdot\bar{\chi}\,\chi\cdot\chi}_{\text{V}} \underbrace{-i\,W_{11}\,\chi^T i\sigma^2\,(\partial_\mu\phi)\,\sigma^\mu i\sigma^2\,\zeta^*}_{\text{VI}} - \underbrace{W_{11}F\chi\cdot\zeta}_{\text{VII}} + \text{h.c.}$$

where we have labeled the terms with roman numerals in order to make it easier to refer to them later. We did not write down the hermitian conjugate expressions explicitly because all we really need is to make the terms shown here cancel each other; if they cancel out, their hermitian conjugate will cancel out as well. Thus our task is now to see if there is any choice of the potentials W_1 and W_{11} such that the terms shown cancel each other out (modulo total derivatives, as always).

First, let us concentrate on the two terms that contain four spinors. Term IV is the only term that contains $\zeta\cdot\chi\,\chi\cdot\chi$. However, this expression is identically zero because it necessarily contains the square of a Grassmann number, so we get no constraint on $\partial W_{11}/\partial\phi$.

On the other hand, the quantity $\bar{\zeta}\cdot\bar{\chi}\,\chi\cdot\chi$, which appears in term V, is not zero, and since there are no other terms that could possibly cancel this one, we must impose

$$\boxed{\frac{\partial W_{11}}{\partial\phi^\dagger} = 0} \tag{8.16}$$

This is a result that deserves to be emphasized: The potential W_{11} *depends only on* ϕ, *not on* $\phi^\dagger$. A common way to express this is to say that W_{11} is a *holomorphic function* of ϕ. This expression comes from complex analysis, where a complex function that depends only on the complex variable z and not on $\bar{z}$ is said to be holomorphic.

Next, let us turn our attention to II. There are no other terms of this form, so it must be identically zero, which implies that W_1 is also holomorphic in ϕ:

$$\boxed{\frac{\partial W_1}{\partial \phi^\dagger} = 0} \tag{8.17}$$

This leaves us with four terms (plus their hermitian conjugate). Two of these terms contain derivatives, so they must cancel each other (or vanish separately). These two terms are III and VI:

$$\mathrm{III} + \mathrm{VI} = -i\, W_1\, \zeta^\dagger \bar{\sigma}^\mu \partial_\mu \chi - i\, W_{11}\, \chi^T\, i\sigma^2\, (\partial_\mu \phi)\, \sigma^\mu\, i\sigma^2\, \zeta^*$$

In the second term, we can move the $\partial_\mu \phi$ in front and then use Eq. (A.5) to simplify to

$$\mathrm{III} + \mathrm{VI} = -i\, W_1\, \zeta^\dagger \bar{\sigma}^\mu \partial_\mu \chi + i\, (\partial_\mu \phi)\, W_{11}\, \chi^T \bar{\sigma}^{\mu T}\, \zeta^*$$

which, using Eq. (A.36), may be written as

$$\mathrm{III} + \mathrm{VI} = -i\, W_1\, \zeta^\dagger \bar{\sigma}^\mu \partial_\mu \chi - i\, (\partial_\mu \phi)\, W_{11}\, \zeta^\dagger\, \bar{\sigma}^\mu\, \chi$$

In order to have the same spinor product in both terms, let us do an integration by parts on the first term (keep in mind that W_1 is a function of ϕ, so it is a spacetime-dependent quantity):

$$\mathrm{III} + \mathrm{VI} = i\, (\partial_\mu W_1)\, \zeta^\dagger \bar{\sigma}^\mu \chi - i\, (\partial_\mu \phi)\, W_{11}\, \zeta^\dagger \bar{\sigma}^\mu \chi$$

Now we set this equal to zero and obtain the condition

$$\partial_\mu W_1 = (\partial_\mu \phi)\, W_{11} \tag{8.18}$$

You see now why a factor of $-1/2$ has been included in Eq. (8.6); it was chosen in order to make the relation between W_1 and W_{11} as simple as possible.

Equation (8.18) provides a relation between W_1 and W_{11} that can be made more explicit by noting that, through an application of the chain rule,

$$\partial_\mu W_1 = \frac{\partial W_1}{\partial \phi}\, \partial_\mu \phi \tag{8.19}$$

which, when used in Eq. (8.18), yields

$$W_{11} = \frac{\partial W_1}{\partial \phi} \tag{8.20}$$

We have canceled all the terms except terms I and VII, i.e.,

$$I + VII = \frac{\partial W_1}{\partial \phi} \zeta \cdot \chi \, F - W_{11} \, F \, \chi \cdot \zeta$$

$$= \left(\frac{\partial W_1}{\partial \phi} - W_{11} \right) F \, \zeta \cdot \chi \tag{8.21}$$

where we have used $\zeta \cdot \chi = \chi \cdot \zeta$. Obviously, this vanishes because of Eq. (8.20)! We therefore have constructed a supersymmetric interaction lagrangian

$$\mathcal{L}_{\text{int}} = W_1(\phi) \, F - \frac{1}{2} \, W_{11}(\phi) \, \chi \cdot \chi + \text{h.c.}$$

$$= W_1(\phi) \, F - \frac{1}{2} \frac{\partial W_1(\phi)}{\partial \phi} \, \chi \cdot \chi + \text{h.c.} \tag{8.22}$$

where we made explicit the fact that W_1 and W_{11} may only depend on ϕ.

To have a renormalizable lagrangian, W_1 must be at most a quadratic function of the scalar field. The most general form is therefore

$$W_1(\phi) = m\phi + \frac{1}{2} y\phi^2 + C \tag{8.23}$$

where m has dimension of a mass, y is dimensionless, and the factor of $1/2$ is purely conventional.

The model we considered in this section is extremely simple. In the next section we will generalize to a set of n multiplets with fields ϕ_i, χ_i, and F_i, where i runs from 1 to n. Most of what we did here will apply *verbatim*, but there will be a few nontrivial complications owing to the presence of several fields. We will see that it proves convenient to introduce a new function of the scalar field denoted by $\mathcal{W}(\phi)$ and related to W_1 by

$$W_1 \equiv \frac{\partial \mathcal{W}}{\partial \phi} \tag{8.24}$$

which, using Eq. (8.20), implies

$$W_{11} = \frac{\partial^2 \mathcal{W}}{\partial \phi^2}$$

$\mathcal{W}$ is called the *superpotential* because it also will appear as a function of the superfields that will be introduced in Chapter 11. The superpotential corresponding to the W_1 given in Eq. (8.23) is obviously

$$\mathcal{W} = \frac{1}{2}\, m\, \phi^2 + \frac{1}{6}\, y\, \phi^3 + C\, \phi + f(\phi^\dagger) \qquad (8.25)$$

In terms of the superpotential, the interaction lagrangian (8.22) is given by

$$\boxed{\mathcal{L}_{\text{int}} = \frac{\partial \mathcal{W}}{\partial \phi}\, F - \frac{1}{2}\frac{\partial^2 \mathcal{W}}{\partial \phi^2}\, \chi \cdot \chi + \text{h.c.}} \qquad (8.26)$$

The function $f(\phi^\dagger)$ appearing in the superpotential plays no role because only derivatives of $\mathcal{W}$ with respect to ϕ appear in the lagrangian, so we may drop it and take the superpotential to be holomorphic in ϕ. The linear term in ϕ usually is not included but plays a role in the spontaneous symmetry breaking of SUSY, as we will see in Chapter 14.

8.2 A More General Lagrangian

We now generalize the free lagrangian to a set of n copies of the fields. Such a generalization will be useful when we consider the minimal supersymmetric standard model (MSSM). The obvious generalization of the lagrangian is

$$\mathcal{L}_{WZ} = \partial_\mu \phi_i^\dagger\, \partial^\mu \phi_i + \chi_i^\dagger\, i\bar{\sigma}^\mu\, \partial_\mu \chi_i + F_i^\dagger\, F_i + \left(W_i\, F_i - \frac{1}{2}\, W_{ij}\, \chi_i \cdot \chi_j + \text{h.c.} \right)$$

$$(8.27)$$

where the indices i and j are summed from 1 to n. We won't require that this index appear in the upper and lower positions to be summed over because it is not a spacetime index. The functions W_i and W_{ij} are a priori arbitrary functions of all the scalar fields ϕ_i and their hermitian conjugate

$$W_i = W_i(\phi_1, \phi_1^\dagger, \dots, \phi_n, \phi_n^\dagger) \qquad (8.28)$$

and similarly for W_{ij}.

Supersymmetry Demystified

We take the fields of each multiplet to transform exactly as in Eq. (8.2):

$$\delta\phi_i = \zeta \cdot \chi_i = -\zeta^T i\sigma^2 \chi_i$$

$$\delta\phi_i^\dagger = \bar{\chi}_i \cdot \bar{\zeta} = \chi_i^\dagger i\sigma^2 \zeta^*$$

$$\delta\chi_i = -i\sigma^\mu i\sigma^2 \zeta^* \partial_\mu \phi_i + F_i \zeta$$

$$\delta F_i = -i\,\zeta^\dagger \bar{\sigma}^\mu \partial_\mu \chi_i \tag{8.29}$$

Note that even though there are n fields of each type in the lagrangian, there is still only one spinor ζ appearing in the transformations.

The variation of the functions W_i is generalized to

$$\delta W_i(\phi, \phi^\dagger) = \frac{\partial W_i}{\partial \phi_k} \delta\phi_k + \frac{\partial W_i}{\partial \phi_k^\dagger} \delta\phi_k^\dagger \tag{8.30}$$

$$= \frac{\partial W_i}{\partial \phi_k} \zeta \cdot \chi_k + \frac{\partial W_i}{\partial \phi_k^\dagger} \bar{\zeta} \cdot \bar{\chi}_k \tag{8.31}$$

with a similar expression for the variation of W_{ij}.

We therefore have

$$\delta\big(W_i(\phi, \phi^\dagger) F_i + \text{h.c.}\big) = \frac{\partial W_i}{\partial \phi_k} \zeta \cdot \chi_k F_i + \frac{\partial W_i}{\partial \phi_k^\dagger} \bar{\zeta} \cdot \bar{\chi}_k F_i + W_i \,\delta F_i + \text{h.c.}$$

$$= \frac{\partial W_i}{\partial \phi_k} \zeta \cdot \chi_k F_i + \frac{\partial W_i}{\partial \phi_k^\dagger} \bar{\zeta} \cdot \bar{\chi}_k F_i$$

$$-i\,W_i \zeta^\dagger \bar{\sigma}^\mu \partial_\mu \chi_i + \text{h.c.} \tag{8.32}$$

and, similarly,

$$\delta\left(-\frac{1}{2} W_{ij} \chi_i \cdot \chi_j + \text{h.c.}\right) = -\frac{1}{2} \frac{\partial W_{ij}}{\partial \phi_k} \zeta \cdot \chi_k\, \chi_i \cdot \chi_j - \frac{1}{2} \frac{\partial W_{ij}}{\partial \phi_k^\dagger} \bar{\zeta} \cdot \bar{\chi}_k\, \chi_i \cdot \chi_j$$

$$-\frac{1}{2} W_{ij} \delta(\chi_i \cdot \chi_j) + \text{h.c.} \tag{8.33}$$

We are looking for conditions on W_i and W_{ij} such that the sum of Eqs. (8.32) and (8.33) vanishes, modulo total derivatives. We will focus on the terms written out explicitly and not worry about their hermitian conjugates because if they cancel out, their hermitian conjugates will too.

Terms with Four Spinors

Consider the first term of Eq. (8.33):

$$-\frac{1}{2}\frac{\partial W_{ij}}{\partial \phi_k}\, \zeta \cdot \chi_k \; \chi_i \cdot \chi_j \qquad (8.34)$$

There is no other term of this form. When the three indices are equal, $i = j = k$, it is identically zero because it contains the square of spinor components. What happens when the indices are different? Do we have to set W_{ij} equal to a constant? Actually, this would be too restrictive. To see this, consider identity (A.53). In order to use this identity, we must keep in mind that the indices i, j, and k are summed over. Consider, for example, the three terms

$$\frac{\partial W_{12}}{\partial \phi_3}\zeta \cdot \chi_3 \; \chi_1 \cdot \chi_2 + \frac{\partial W_{31}}{\partial \phi_2}\zeta \cdot \chi_2 \; \chi_3 \cdot \chi_1 + \frac{\partial W_{23}}{\partial \phi_1}\zeta \cdot \chi_1 \; \chi_2 \cdot \chi_3 \qquad (8.35)$$

Identity (A.53) tells us that

$$\zeta \cdot \chi_3 \; \chi_1 \cdot \chi_2 + \zeta \cdot \chi_2 \; \chi_3 \cdot \chi_1 + \zeta \cdot \chi_1 \; \chi_2 \cdot \chi_3 = 0 \qquad (8.36)$$

so the three terms of Eq. (8.35) will cancel out at the condition that

$$\frac{\partial W_{12}}{\partial \phi_3} = \frac{\partial W_{31}}{\partial \phi_2} = \frac{\partial W_{23}}{\partial \phi_1} \qquad (8.37)$$

It's clear that if we assume cyclic symmetry of $\partial W_{ij}/\partial \phi_k$ under the permutation of the three indices i, j, and k, we will have automatic cancellation of Eq. (8.34), all the terms canceling in groups of three.

As for the second term of Eq. (8.33), there is no identity to save us this time, so we must impose

$$\frac{\partial W_{ij}}{\partial \phi_k^{\dagger}} = 0 \qquad (8.38)$$

that is, W_{ij} must be a holomorphic function of the ϕ_k.

In order for $W_{ij}\,\chi_i \cdot \chi_j$ to be of dimension 4 or less, the most general holomorphic W_{ij} we can write down is

$$W_{ij} = m_{ij} + y_{ijk}\,\phi_k \qquad (8.39)$$

If we impose that $\partial W_{ij}/\partial \phi_k$ be invariant under cyclic permutation of the indices, we must take the couplings y_{ijk} to have cyclic symmetry, whereas there is no required symmetry on the constants m_{ij}. The end result is that the part of $\mathcal{L}_{\text{int}}$ containing W_{ij} is given by

$$-\frac{1}{2}\, m_{ij}\, \chi_i \cdot \chi_j - \frac{1}{2}\, y_{ijk}\, \phi_k\, \chi_i \cdot \chi_j \tag{8.40}$$

But because $\chi_i \cdot \chi_j$ is symmetric under the exchange of i and j, we can always make m_{ij} symmetric too. Thus, from now on, we take m_{ij} to be symmetric. With this choice, W_{ij} is symmetric under the exchange of its indices.

Now comes the key trick: We can automatically make W_{ij} symmetric if we write it as the second derivative of another function of the scalar fields, i.e.,

$$\boxed{W_{ij} \equiv \frac{\partial^2 \mathcal{W}}{\partial \phi_i \partial \phi_j}} \tag{8.41}$$

where $\mathcal{W}$ is, of course, the superpotential we encountered previously. Its most general form is

$$\mathcal{W} = \frac{1}{2}\, m_{ij}\, \phi_i \phi_j + \frac{1}{6}\, y_{ijk}\, \phi_i \phi_j \phi_k + c_{ij}\, \phi_i^\dagger \phi_j + c_i\, \phi_i + f(\phi_i^\dagger) \tag{8.42}$$

where f is an arbitrary function of the $\phi_i^\dagger$ (at most cubic). We will show below that the constants c_{ij} actually must be equal zero. The term linear in ϕ cannot be ruled out, and it is actually considered in certain forms of SUSY breaking.

Last Term of Eq. (8.33)

This term is

$$-\frac{1}{2}\, W_{ij}\, \delta(\chi_i \cdot \chi_j) = \frac{1}{2}\, W_{ij}\, \delta\big(\chi_i^T i\sigma^2 \chi_j\big)$$

$$= \frac{1}{2}\, W_{ij} \left(-\big(i\sigma^\mu i\sigma^2 \zeta^* \partial_\mu \phi_i\big)^T i\sigma^2\, \chi_j + F_i\, \zeta^T i\sigma^2\, \chi_j \right.$$

$$\left. - \chi_i^T i\sigma^2 i\sigma^\mu i\sigma^2 \zeta^* \partial_\mu \phi_j + F_j\, \chi_i^T i\sigma^2 \zeta \right) \tag{8.43}$$

Now we use identity (A.36) applied to the first term. For an arbitrary matrix A, this identity allows us to write

$$(A\zeta)^T (i\sigma^2\chi) = -(i\sigma^2\chi)^T A\zeta = -\chi^T (i\sigma^2)^T A\zeta = \chi^T i\sigma^2 A\zeta \qquad (8.44)$$

Therefore, the first term may be written as

$$\left(i\sigma^\mu i\sigma^2 \zeta^* \partial_\mu\phi_i\right)^T i\sigma^2 \chi_j = i \chi_j^T i\sigma^2 \sigma^\mu i\sigma^2 \zeta^* \partial_\mu\phi_i \qquad (8.45)$$

Substituting this into Eq. (8.43), we get

$$-\frac{1}{2} W_{ij}\,\delta(\chi_i \cdot \chi_j) = -\frac{i}{2} W_{ij}\, \chi_j^T i\sigma^2 \sigma^\mu i\sigma^2 \zeta^* \partial_\mu\phi_i + \frac{W_{ij}}{2} F_i\, \zeta^T i\sigma^2 \chi_j$$
$$-\frac{i}{2} W_{ij}\, \chi_i^T i\sigma^2 \sigma^\mu i\sigma^2 \zeta^* \partial_\mu\phi_j + \frac{W_{ij}}{2} F_j\, \chi_i^T i\sigma^2 \zeta \qquad (8.46)$$

We see that the first and third terms are identical (since the indices i and j are summed over).

It turns out that the second and fourth terms of Eq. (8.46) are also identical. This is obvious if we recall that

$$\zeta^T i\sigma^2 \chi_j = -\zeta \cdot \chi_j \qquad (8.47)$$

whereas $\chi_i^T i\sigma^2 \zeta = -\chi_i \cdot \zeta$ and $\zeta \cdot \chi = \chi \cdot \zeta$.

Using this, we obtain

$$-\frac{1}{2} W_{ij}\,\delta(\chi_i \cdot \chi_j) = -i\, W_{ij}\, \chi_j^T i\sigma^2 \sigma^\mu i\sigma^2 \zeta^* \partial_\mu\phi_i - W_{ij}\, F_i\, \zeta \cdot \chi_j \qquad (8.48)$$

This can be simplified even further using Eq. (A.5) and then Eq. (A.36) to write

$$\chi_j^T \sigma^2 \sigma^\mu \sigma^2 \zeta^* = \chi_j^T (\bar{\sigma}^\mu)^T \zeta^*$$
$$= -\zeta^\dagger \bar{\sigma}^\mu \chi_j$$

so that Eq. (8.48) simplifies to

$$-\frac{1}{2} W_{ij}\,\delta(\chi_i \cdot \chi_j) = -i\, W_{ij}\, \zeta^\dagger \bar{\sigma}^\mu \chi_j\, \partial_\mu\phi_i - W_{ij}\, F_i\, \zeta \cdot \chi_j \qquad (8.49)$$

Remaining Terms

We are left with the terms of Eqs. (8.32) and (8.49):

$$-i\, W_{ij}\, \zeta^\dagger\, \bar\sigma^\mu\, \chi_j\, \partial_\mu\phi_i - W_{ij}\, F_i\, \zeta\cdot\chi_j$$

$$+ \frac{\partial W_i}{\partial\phi_k}\, \zeta\cdot\chi_k\, F_i + \frac{\partial W_i}{\partial\phi_k^\dagger}\, \bar\zeta\cdot\bar\chi_k\, F_i - i\, W_i\zeta^\dagger\bar\sigma^\mu\partial_\mu\chi_i \qquad (8.50)$$

All these must cancel out, modulo total derivatives.

Consider first the terms that contain derivatives acting on some field, which are the first and last terms. These are very similar. To show even more clearly their relationship, we may use the fact that

$$W_{ij} = \frac{\partial^2 W}{\partial\phi_i\,\partial\phi_j} \qquad (8.51)$$

to write

$$W_{ij}\,\partial_\mu\phi_i = \frac{\partial^2 W}{\partial\phi_i\,\partial\phi_j}\,\partial_\mu\phi_i = \partial_\mu\left(\frac{\partial W}{\partial\phi_j}\right) \qquad (8.52)$$

So the sum of the first and last terms of Eq. (8.50) is

$$-i\,\partial_\mu\left(\frac{\partial W}{\partial\phi_j}\right)\zeta^\dagger\,\bar\sigma^\mu\,\chi_j - i\, W_i\,\zeta^\dagger\,\bar\sigma^\mu\,\partial_\mu\chi_i \qquad (8.53)$$

This is a total derivative at the condition that

$$\boxed{W_i = \frac{\partial W}{\partial\phi_i}} \qquad (8.54)$$

We see that both W_{ij} and W_i are given in terms of the superpotential [see Eq. (8.41].

Now we are left with three terms in Eq. (8.50):

$$-W_{ij}\, F_i\, \zeta\cdot\chi_j + \frac{\partial W_i}{\partial\phi_k}\, \zeta\cdot\chi_k\, F_i + \frac{\partial W_i}{\partial\phi_k^\dagger}\, \bar\zeta\cdot\bar\chi_k\, F_i \qquad (8.55)$$

Note that Eqs. (8.54) and (8.41) imply that

$$W_{ij} = \frac{\partial W_i}{\partial\phi_j} \qquad (8.56)$$

so the first two terms of Eq. (8.55) cancel out. We are left with the last term. Since we can't cancel this with anything and it is not a total derivative, W_i must be independent of $\phi^\dagger$, i.e.,

$$\frac{\partial W_i}{\partial \phi_j^\dagger} = 0 \tag{8.57}$$

If we plug the most general form of the superpotential Eq. (8.42) into Eq. (8.54), we get

$$W_i = m_{ij}\,\phi_j + \frac{1}{2}\,y_{ijk}\,\phi_j\,\phi_k + c_{ij}\phi_j^\dagger + c_i \tag{8.58}$$

We see that Eq. (8.57) implies that $c_{ij} = 0$.

Since the lagrangian contains only derivatives of the superpotential with respect to the fields ϕ_i, we also may set the function $f(\phi_i^\dagger) = 0$ in Eq. (8.42) so that the most general superpotential is the holomorphic function

$$\mathcal{W} = \frac{1}{2}\,m_{ij}\,\phi_i\,\phi_j + \frac{1}{6}\,y_{ijk}\,\phi_i\,\phi_j\,\phi_k + c_i\,\phi_i \tag{8.59}$$

And we are done! We have obtained a supersymmetric interacting theory. We will summarize our results in the next section.

8.3 The Full Wess-Zumino Lagrangian

We have found that the following lagrangian, referred to as the *Wess-Zumino lagrangian*, is supersymmetric:

$$
\begin{aligned}
\mathcal{L}_{WZ} &= \partial_\mu \phi_i^\dagger \partial^\mu \phi_i + \chi_i^\dagger i\bar\sigma^\mu \partial_\mu \chi_i + F_i^\dagger F_i + \left(W_i F_i - \frac{1}{2} W_{ij}\chi_i \cdot \chi_j + \text{h.c.} \right) \\
&= \partial_\mu \phi_i^\dagger \partial^\mu \phi_i + \chi_i^\dagger i\bar\sigma^\mu \partial_\mu \chi_i + F_i^\dagger F_i + \left(\frac{\partial \mathcal{W}}{\partial \phi_i} F_i - \frac{1}{2} \frac{\partial^2 \mathcal{W}}{\partial \phi_j\,\partial \phi_i} \chi_i \cdot \chi_j + \text{h.c.} \right)
\end{aligned}
\tag{8.60}
$$

where we have used Eqs. (8.54) and (8.41).

Let us now eliminate the auxiliary fields. Using the equation of motion for F_i, $\partial \mathcal{L}/\partial F_i = 0$, we get

$$F_i = -W_i = -\frac{\partial \mathcal{W}}{\partial \phi_i}$$

On the other hand, the equation of motion for $F_i^\dagger$ gives

$$F_i = -W_i^\dagger = -\left(\frac{\partial W}{\partial \phi_i}\right)^\dagger$$

so the two terms containing the auxiliary fields combine to form

$$F_i^\dagger F_i + \frac{\partial W}{\partial \phi_i} F_i + \left(\frac{\partial W}{\partial \phi_i}\right)^\dagger F_i^\dagger = -\left|\frac{\partial W}{\partial \phi_i}\right|^2 \tag{8.61}$$

Substituting this expression in the lagrangian, we get a different representation for the Wess-Zumino lagrangian:

$$\boxed{\mathcal{L}_{WZ} = \partial_\mu \phi_i^\dagger \partial^\mu \phi_i + \chi_i^\dagger i \bar{\sigma}^\mu \partial_\mu \chi_i - \left|\frac{\partial W}{\partial \phi_i}\right|^2 - \frac{1}{2}\left(\frac{\partial^2 W}{\partial \phi_j \, \partial \phi_i} \chi_i \cdot \chi_j + \text{h.c.}\right)}$$

$$\tag{8.62}$$

We see that the potential of the scalar fields is given by (recall $\mathcal{L} = K - V$)

$$V(\phi_i) = \left|\frac{\partial W}{\partial \phi_i}\right|^2 \tag{8.63}$$

Here, we are including the mass terms of the scalar field in the potential.

Using the general superpotential Eq. (8.59), we have

$$W_i = m_{ij}\, \phi_j + \frac{1}{2}\, y_{ijk}\, \phi_j\, \phi_k + c_i \tag{8.64}$$

and $W_{ij} = m_{ij} + y_{ijk}\phi_k$ so that the lagrangian is

$$\mathcal{L}_{WZ} = \partial_\mu \phi_i^\dagger \partial^\mu \phi_i + \chi_i^\dagger i \bar{\sigma}^\mu \partial_\mu \chi_i - \left|m_{ij}\phi_j + \frac{y_{ijk}}{2}\,\phi_j\phi_k + c_i\right|^2$$

$$- \frac{1}{2}\left(m_{ij}\, \chi_i \cdot \chi_j + y_{ijk}\,\phi_k\, \chi_i \cdot \chi_j + \text{h.c.}\right)$$

The coefficients c_i are usually set to zero, although they play a role in SUSY breaking, as we will discuss in Chapter 14.

We see that the spinor and scalar masses are equal as expected. Also, we see that there are three types of interactions: a cubic scalar interaction, a quartic scalar interaction, and a scalar-fermion interaction whose coupling constants are related in a very specific way in order to ensure invariance under SUSY.

8.4 The Wess-Zumino Lagrangian in Majorana Form

In Chapter 9 we will carry out explicit calculations in the Wess-Zumino model. Although it is, of course, possible to carry out calculations with Weyl spinors (see in particular Ref. 13), most references use Majorana spinors for calculations of scattering amplitudes, as well as real scalar fields. We will therefore rewrite the Wess-Zumino model in terms of Majorana spinors and real fields. Let's consider for simplicity the case of a single multiplet, for which (setting the linear term $c\phi$ in the superpotential equal to zero)

$$\mathcal{W} = \frac{m}{2}\,\phi^2 + \frac{y}{6}\,\phi^3 \qquad W_1 = m\,\phi + \frac{y}{2}\,\phi^2 \qquad W_{11} = m + y\,\phi \qquad (8.65)$$

If we don't get rid of the auxiliary fields, the lagrangian is

$$\mathcal{L}_{WZ} = \partial_\mu \phi^\dagger\,\partial^\mu \phi + \chi^\dagger\,i\bar{\sigma}^\mu\,\partial_\mu \chi + FF^\dagger + m\phi\,F + \frac{1}{2}\,y\phi^2\,F + m\,\phi^\dagger\,F^\dagger$$

$$+ \frac{y}{2}\,(\phi^\dagger)^2\,F^\dagger - \frac{m}{2}\,\left(\chi\cdot\chi + \bar{\chi}\cdot\bar{\chi}\right) - \frac{y}{2}\,\left(\phi\chi\cdot\chi + \phi^\dagger\,\bar{\chi}\cdot\bar{\chi}\right) \quad (8.66)$$

where we have taken the parameters m and y to be real. If we use the equations of motion of F and $F^\dagger$ to get rid of them, we get

$$\mathcal{L}_{WZ} = \partial_\mu \phi^\dagger\,\partial^\mu \phi + \chi^\dagger\,i\bar{\sigma}^\mu\,\partial_\mu \chi - m^2\,\phi\phi^\dagger - \frac{my}{2}\,\left(\phi^2\,\phi^\dagger + \phi\,\phi^{\dagger 2}\right) - \frac{y^2}{4}\,(\phi\phi^\dagger)^2$$

$$- \frac{m}{2}\,\left(\chi\cdot\chi + \bar{\chi}\cdot\bar{\chi}\right) - \frac{y}{2}\left(\phi\,\chi\cdot\chi + \phi^\dagger\,\bar{\chi}\cdot\bar{\chi}\right) \qquad (8.67)$$

It is conventional to rewrite the complex scalar field in terms of two real scalars conventionally named A and B according to

$$\phi \equiv \frac{1}{\sqrt{2}}(A + iB) \qquad (8.68)$$

Using this and the identities of Sections 4.7 and 4.11, we find that the free part of the lagrangian can be written as

$$\frac{1}{2} \partial_\mu A \, \partial^\mu A - \frac{1}{2} m^2 A^2 + \frac{1}{2} \partial_\mu B \, \partial^\mu B - \frac{1}{2} m^2 B^2 + \frac{1}{2} \overline{\Psi}_M (i \gamma^\mu \partial_\mu - m) \Psi_M \tag{8.69}$$

whereas the interactions are

$$\mathcal{L}_{\text{int}} = L_{I1} + L_{I2} + L_{I3} + L_{I4} \tag{8.70}$$

with

$$L_{I1} \equiv -\frac{1}{2} g^2 (A^2 + B^2)^2$$

$$L_{I2} \equiv -m \, g \, (A^3 + A B^2)$$

$$L_{I3} \equiv -g \, A \, \overline{\Psi}_M \Psi_M$$

$$L_{I4} \equiv i \, g \, B \, \overline{\Psi}_M \gamma_5 \Psi_M \tag{8.71}$$

where we have defined

$$g \equiv \frac{y}{\sqrt{8}} \tag{8.72}$$

8.5 Quiz

1. What is the scalar field potential in terms of the superpotential in the Wess-Zumino model?

2. What do we mean when we say that the superpotential is holomorphic in the scalar fields?

3. What is the relation between W_i, W_{ij}, and the superpotential $\mathcal{W}$?

4. Using the answer of the third question and the fact that the lagrangian contains a term $W_i F_i$, what is the maximum number of scalar fields that may enter the superpotential?

5. Write down the most general superpotential for a single scalar field.

CHAPTER 9

Some Explicit Calculations

In this chapter, we will consider the Wess-Zumino lagrangian with only one multiplet. It turns out that most calculations in the literature are carried out using the Majorana form of this lagrangian, given in the last section of Chapter 8.

Our only purpose in this chapter is to exhibit explicitly the cancellation of all quadratic divergences in the Wess-Zumino model in a few simple amplitudes. Some logarithmic divergences do *not* cancel out, and a more detailed analysis would show that these all can be canceled by a common wavefunction renormalization for both the scalar and fermion field. However, owing to a lack of space, we will not go into this in this book.

Calculations of scattering processes with Majorana spinors involve a few subtleties not present in calculations with the more familiar Dirac spinors. In addition, because our theory contains cubic interactions of the scalar field, we will encounter so-called tadpole diagrams that may not be familiar to some readers. For these reasons, the Feynman rules and Feynman diagrams for calculations with the Wess-Zumino lagrangian probably would look unfamiliar. We have therefore made the choice to not throw at you the Feynman rules for the Wess-Zumino model but instead to go back to the fundamental formula for the calculations of n-point functions in the canonical quantization approach.

9.1 Refresher About Calculations of Processes in Quantum Field Theory

Recall the the fundamental quantities of interest in quantum field theory are the so-called n-point functions, which, if we consider for simplicity the theory of a single real scalar field, take the form

$$\langle \Omega | T\left(\phi(x)\phi(y)\ldots\right)|\Omega\rangle = \frac{\langle 0|T\left(\phi_I(x)_I\phi_I(y)\ldots e^{\{i\int d^4z\mathcal{L}_{\text{int}}[\phi_I(z)]\}}\right)|0\rangle}{\langle 0|T\left(e^{\{id^4z\mathcal{L}_{\text{int}}[z]\}}\right)|0\rangle} \tag{9.1}$$

where T denotes the time-ordering operator, and the dots stand for some product of fields that depends on the process we are interested in (we will see some explicit examples soon). The brackets and braces are used instead of parentheses only to make the expressions easier to read. The state $|\Omega\rangle$ is the vacuum of the full (interacting) theory, whereas $|0\rangle$ is the vacuum state of the free (noninteracting) theory. The index I on the fields on the right mean that they are taken in the interaction picture, so their time evolution is governed by the free hamiltonian. $\mathcal{L}_{\text{int}}$ is the interaction part of the lagrangian (the free part representing the kinetic and mass terms). The division by $\langle 0| \exp(i \int \mathcal{L}_{\text{int}})|0\rangle$ has for only effect to remove all disconnected diagrams from the calculation.

These n-point functions are the fundamental quantities in perturbative quantum field theory. The amplitude of any process can be calculated in terms of the n-point functions using the Lehmann-Symanzik-Zimmermann reduction formula, and from the amplitude, one can calculate physical observables such as cross sections and decay rates. In this chapter, we will consider only n-point functions.

In case the use of this equation is not fresh in your memory, let's do a quick example. We will simply illustrate *how* the equation is used to calculate processes, leaving the proofs of *why* it works to textbooks on quantum field theory.

Consider, then, the simplest interacting field theory one can write down: the famous $\lambda\phi^4$ theory:

$$\mathcal{L} = \frac{1}{2}\partial^\mu\phi\partial_\mu\phi - \frac{1}{2}m^2\phi^2 - \lambda\phi^4 \tag{9.2}$$

We consider a real scalar field for simplicity (this explains why there is a factor of 1/2 in the kinetic and mass terms, a factor that is absent when the field is complex). For this theory, the interaction lagrangian is

$$\mathcal{L}_{\text{int}} = -\lambda\phi^4 \tag{9.3}$$

Let's look at the 2-point function, in other words, the propagator:

$$\langle\Omega|T\left(\phi(x)\phi(y)\right)|\Omega\rangle = \frac{\langle 0|T\left(\phi(x)\phi(y)e^{-i\int d^4z\,\lambda\phi^4(z)}\right)|0\rangle}{\langle 0|T\left(e^{-i\int d^4z\,\lambda\phi^4(z)}\right)|0\rangle} \tag{9.4}$$

Let's focus on the numerator (again, the effect of the denominator is simply to remove all disconnected diagrams). To order λ^0, we simply set the exponential equal to 1 and get

$$\langle 0|T\left(\phi(x)\phi(y)\right)|0\rangle \equiv \langle 0|\phi(x)\phi(y)|0\rangle$$

which is simply the free propagator, often denoted $D(x - y)$. This can be calculated explicitly using the expansion of ϕ in terms of creation and annihilation operators. The Wick contraction symbol means that all creation operators have to be moved to the right of the annihilation operators. The result (see any textbook on quantum field theory) is that $D(x - y)$ is the Fourier transform of the familiar scalar field propagator in momentum space, i.e.,

$$D(x - y) = \int \frac{d^4k}{(2\pi)^4} e^{-ik\cdot(x-y)} \frac{i}{k^2 - m^2 + i\epsilon}$$

$$\equiv \int \frac{d^4k}{(2\pi)^4} e^{-ik\cdot(x-y)}\, D(k) \tag{9.5}$$

where $D(k)$ is the momentum space lowest-order propagator.

For a less trivial example, consider the order λ correction to the propagator from the expansion of the exponential in the numerator of Eq. (9.4) to that order. We will call this correction $D_1(x - y)$ because it is of order λ^1:

$$D_1(x - y) \equiv -i\lambda\langle 0|T\left(\phi(x)\phi(y)\int d^4z\,\phi^4(z)\right)|0\rangle \tag{9.6}$$

Now we must contract all fields in pairs. One possibility is obviously

$$-i\lambda \int d^4z \, \langle 0|\phi(x)\phi(y)\phi(z)\phi(z)\phi(z)\phi(z)|0\rangle \tag{9.7}$$

however, this corresponds to a disconnected Feynman diagram. We therefore ignore this contribution [because it is canceled by a term arising from the denominator of Eq. (9.4)].

In order to get a connected diagram, we obviously must contract $\phi(x)$ with one of the four $\phi(z)$ and $\phi(y)$ with one of the remaining three $\phi(z)$. The remaining two $\phi(z)$ must be contracted together. There are $4 \times 3 = 12$ ways to do this, so we get

$$D_1(x - y) = -12i\lambda \int d^4z \langle 0|\phi(x)\phi(y)\phi(z)\phi(z)\phi(z)\phi(z)|0\rangle \tag{9.8}$$

There is no difference between contractions drawn below or above the fields, the two forms are used only for the sake of clarity. Now we simply replace each Wick contraction by the scalar propagator having for argument the difference of the spacetime points of the two fields and remove the ket and bra:

$$D_1(x - y) = -12i\lambda \int d^4z \, D(x - z)D(y - z)D(z - z) \tag{9.9}$$

which corresponds to Figure 9.1. The integral over three scalar propagators with these spacetime coordinates will occur many times, so let's have a closer look at it. Using Eq. (9.5), we have

$$\int d^4z \, D(x - z)D(y - z)D(z - z)$$

$$= \int d^4z \int \frac{d^4p}{(2\pi)^4} \frac{ie^{-ip\cdot(x-z)}}{p^2 - m^2 + i\epsilon} \int \frac{d^4k}{(2\pi)^4} \frac{ie^{-ik\cdot(y-z)}}{k^2 - m^2 + i\epsilon} \int \frac{d^4q}{(2\pi)^4} \frac{ie^{-iq\cdot(z-z)}}{q^2 - m^2 + i\epsilon}$$

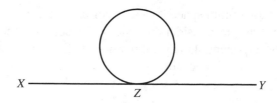

Figure 9.1   Diagram corresponding to Eq. (9.9).

The exponentials can be combined, and we can do the integration over z:

$$\int d^4z e^{-ip\cdot x - ik\cdot y}\, e^{iz\cdot(p+k)} = e^{-ip\cdot x - ik\cdot y}(2\pi)^4\delta^4(p+k)$$

$$= e^{-ip\cdot(x-y)}(2\pi)^4\delta^4(p+k) \qquad (9.10)$$

where in the last step we have set $k = -p$ in the exponential because this is enforced by the Dirac delta. We now can use the delta function to do trivially the integration over k which leads us to

$$\int d^4z\, D(x-z)D(y-z)D(z-z)$$

$$= \int \frac{d^4p}{(2\pi)^4} e^{-ip\cdot(x-y)}\frac{i}{p^2-m^2+i\epsilon}\frac{i}{p^2-m^2+i\epsilon}\int \frac{d^4q}{(2\pi)^4}\frac{i}{q^2-m^2+i\epsilon}$$

$$= \int \frac{d^4p}{(2\pi)^4} e^{-ip\cdot(x-y)}\, D(p)\, D(p)\int \frac{d^4q}{(2\pi)^4}\frac{i}{q^2-m^2+i\epsilon} \qquad (9.11)$$

If we Fourier transform to momentum space, we obtain

$$\left[\int d^4z\, D(x-z)D(y-z)D(z-z)\right]\Bigg|_{FT} = D(p)\, D(p)\, I_d \qquad (9.12)$$

where we have defined the quadratically divergent integral

$$I_d \equiv \int \frac{d^4q}{(2\pi)^4}\frac{i}{q^2-m^2+i\epsilon} \qquad (9.13)$$

To explicitly show the structure of the divergences arising from this integral, let us carry out the integration. The integrand has poles at $q_0 = \pm\sqrt{\vec{q}^2+m^2}\mp i\epsilon$. With a choice of appropriate contour, it is straightforward to show that

$$\int \frac{d^4q}{(2\pi)^4}\frac{i}{q^2-m^2+i\epsilon} = \frac{1}{2}\int \frac{d^3\vec{q}}{(2\pi)^3}\frac{1}{\sqrt{\vec{q}^2+m^2}} \qquad (9.14)$$

Using an explicit cutoff on the magnitude of the three-momentum, we get

$$
I_d = \frac{1}{4\pi^2} \int_0^\Lambda dq \frac{q^2}{\sqrt{q^2 + m^2}}
$$

$$
= \frac{1}{8\pi^2} \left[\Lambda^2 \sqrt{1 + \frac{m^2}{\Lambda^2}} - m^2 \ln\left(\frac{\Lambda}{m}\right) - m^2 \ln\left(1 + \sqrt{1 + \frac{m^2}{\Lambda^2}}\right) \right]
$$

$$
\approx \frac{1}{8\pi^2} \left[\Lambda^2 - m^2 \ln\left(\frac{\Lambda}{m}\right) + m^2 \times \text{finite piece} \right] \tag{9.15}
$$

This shows that the integral contains a quadratically divergent term that is *independent of the mass of the particle* as well as a logarithmic divergence and a finite contribution that both depend on the mass. These observations will be important when we discuss SUSY breaking in Chapter 14.

When it comes to calculations of physical processes, the quantity of relevance (to use in the LSZ reduction formula) is actually the *amputated* momentum space propagator that is obtained from Eq. (9.12) by dropping the two propagators $D(p)$ associated with the external lines:

$$
\left[\int d^4z\, D(x - z)D(y - z)D(z - z) \right]\Big|_{FT}^{AM} = I_d \tag{9.16}
$$

Therefore, the amputated one-loop propagator is

$$
D_1^{AM}(p) = -12\, i\, \lambda\, I_d \tag{9.17}
$$

Quadratic divergences are ubiquitous in theories with scalar fields. The beauty of SUSY is that it rids us of these nasty divergences in a very clever manner, as we will soon see in simple examples.

9.2 Propagators

Before doing any calculations in the Wess-Zumino model, we need the propagators of a Majorana fermion. Since we will use Greek subscripts to indicate the components of the Majorana spinors, these will be represented by Ψ^M instead of Ψ_M to make the notation less cumbersome.

As for a Dirac spinor, we have

$$\langle 0| T \left(\Psi_\alpha^M(x) \overline{\Psi}_\beta^M(y) \right) |0 \rangle = \int \frac{d^4 k}{(2\pi)^4} \, e^{-ik \cdot (x-y)} S_{\alpha\beta}(k) \qquad (9.18)$$

with

$$S_{\alpha\beta}(k) \equiv i \, \frac{(\slashed{k} + m)_{\alpha\beta}}{k^2 - m^2 + i\epsilon} \qquad (9.19)$$

However, unlike Dirac spinors, there are two additional propagators coming from the contractions of $\Psi^M \Psi^M$ and $\overline{\Psi}^M \overline{\Psi}^M$. These propagators are identically zero for a Dirac spinor because they contain the vacuum expectation value of the product of a creation and an annihilation operator for the fermion and the antifermion, which is zero because those are two distinct states. The corresponding expressions for a Majorana fermion are not zero because it is its own antiparticle.

It is easy to work out these two extra propagators from the result in Eq. (9.18) if we recall that for a Majorana fermion [see Eq. (4.2)],

$$(\Psi^M)^c = \Psi^M \qquad (9.20)$$

where the charge-conjugation operation is defined as

$$\Psi^c = C \, \overline{\Psi}^T \qquad (9.21)$$

and the matrix C, as given in Section 4.1, is equal to

$$C = \begin{pmatrix} i\sigma^2 & 0 \\ 0 & -i\sigma^2 \end{pmatrix} \qquad (9.22)$$

Note that $C^2 = -\mathbf{1}$, which implies that $C^{-1} = -C$. We obviously also have $C^T = -C$, so $C^T = C^{-1}$. These relations will be very useful when we calculate scattering amplitudes.

Showing the spinor indices explicitly, Eqs. (9.20) and (9.21) translate to

$$\Psi_\alpha^M = C_{\alpha\beta} \, (\overline{\Psi})_\beta^M \qquad (9.23)$$

Consider the propagator

$$\langle 0| T \left(\Psi_\alpha^M(x) \Psi_\beta^M(y) \right) |0 \rangle \qquad (9.24)$$

What we want to do is to rewrite this in a form similar to Eq. (9.18) using Eq. (9.23). Obviously, we simply have to replace Ψ_β^M in Eq. (9.24) with $C_{\beta\gamma}(\overline{\Psi})_\gamma^M$, which allows us to express the propagator (9.24) in terms of Eq. (9.18):

$$\langle 0|T\left(\Psi_\alpha^M(x)\Psi_\beta^M(y)\right)|0\rangle = \int \frac{d^4k}{(2\pi)^4}\, e^{-ik\cdot(x-y)}\, C_{\beta\gamma}\, S_{\alpha\gamma}(k)$$

However, this is not the most convenient form. When we will calculate diagrams, we will want to calculate the traces of the matrices acting on the spinors, and for this reason, it is more useful to have the index γ in the C matrix to appear in the first position, which we can do simply by using $C_{\beta\gamma} = C_{\gamma\beta}^T$, to finally get

$$\langle 0|T\left(\Psi_\alpha^M(x)\Psi_\beta^M(y)\right)|0\rangle = \int \frac{d^4k}{(2\pi)^4}\, e^{-ik\cdot(x-y)}\, S_{\alpha\gamma}(k)\, C_{\gamma\beta}^T \tag{9.25}$$

This way, the matrices C^T and S form a matrix product equal to SC^T.

Following the same approach, it is easy to show that

$$\langle 0|T\left(\overline{\Psi}_\alpha^M(x)\overline{\Psi}_\beta^M(y)\right)|0\rangle = \int \frac{d^4k}{(2\pi)^4}\, e^{-ik\cdot(x-y)} C_{\alpha\gamma}^T S_{\gamma\beta}(k) \tag{9.26}$$

When drawing Feynman diagrams involving Majorana spinors (and Weyl spinors too), one has to distinguish among the three different types of fermion propagators. This is done by drawing on the fermion lines two arrows flowing in opposite directions (either toward one another or away from one another) to represent the two new types of propagators, Eqs. (9.25) and (9.26), in addition to the usual propagator, Eq. (9.18), which is represented, as usual, with a single arrow. We won't introduce this notation for the very few calculations we will do. We will simply use the same diagram to represent all three types of propagators. The double-arrow notation is very useful for drawing general conclusions about the types of counterterms that can be generated by a given lagrangian. A simple illustration of this notation for Majorana spinors can be found in Section 1.3 of Ref. 8.

Let us note that we could, of course, carry out all the calculations using two-component Weyl spinors instead of four-component Majorana spinors. In Ref. 13, a detailed explanation of the Feynman rules and diagrammatic notation for Weyl spinors (which also involves the double-arrow notation) is presented, as well as a large number of detailed calculations.

The propagator of a scalar field is, of course, the usual

$$\langle 0|T\big(A(x)A(y)\big)|0\rangle = \int \frac{d^4k}{(2\pi)^4}\, e^{-ik\cdot(x-y)} D_A(k) \qquad (9.27)$$

where

$$D_A(k) \equiv \frac{i}{k^2 - m^2 + i\epsilon} \qquad (9.28)$$

Note that we don't put any label on the masses because all the particles have the same mass.

For the rest of this chapter we will not include a label M on the Majorana spinors. It will be understood that all spinors Ψ will represent Majorana spinors, not Dirac spinors.

9.3 One Point Function

As a warm-up exercise, consider the one-point function of the A field in the Wess-Zumino model:

$$\langle \Omega|A(x)|\Omega\rangle \qquad (9.29)$$

This turns out to be much simpler than the zero-point (vacuum energy) calculation. If we work to order g in the coupling constant, the calculation involves only three simple diagrams. Despite its simplicity, this is still an interesting calculation to work through because it will exhibit clearly the clever way in which a supersymmetric theory takes care of the quadratic divergences.

To be explicit, the one-point function is given by

$$\langle \Omega|T\big(A(x)\big)|\Omega\rangle = \frac{\langle 0|T\big(A(x)e^{i\int d^4z\mathcal{L}'(z)}\big)|0\rangle}{\langle 0|T\big(e^{i\int d^4z\mathcal{L}'(z)}\big)|0\rangle} \qquad (9.30)$$

where the interaction lagrangian of the Wess-Zumino model is given by the sum of the four terms of Eq. (8.71). Note in particular that L_{I1} is of order g^2, whereas the other three interactions are of order g.

To lowest order (order g^0), we get

$$\langle\Omega|T\big(A(x)\big)|\Omega\rangle = \langle 0|A(x)|0\rangle = 0 \tag{9.31}$$

because $A(x)$ is a sum of single creation and annihilation operators. The first nonzero contribution is of order g and is given by

$$\langle\Omega|T\big(A(x)\big)|\Omega\rangle\big|_g = \int d^4z\langle 0|T\{iA(x)(L_{12}(z)+L_{13}(z))\}|0\rangle$$

$$= -ig\int d^4z\langle 0|T\Big(\underbrace{mA(x)A^3(z)}_{\text{I}}+\underbrace{mA(x)A(z)B^2(z)}_{\text{II}}$$

$$+\underbrace{A(x)A(z)\overline{\Psi}\Psi}_{\text{III}}\Big) \tag{9.32}$$

Note that the interaction L_{14} does not contribute because it does not contain any A field to pair up with $A(x)$. We haven't written the denominator of Eq. (9.30) (whose purpose is to remove the disconnected diagrams) because it is equal to one at the order we are working.

In the first term of Eq. (9.32), we have three ways to contract $A(x)$ with one of the $A(z)$ and only one way to contract the two remaining $A(z)$ which gives us a factor of 3. The first term is therefore

$$\text{I} = -3igm\int d^4z\, D_A(x-z)D_A(z-z)$$

$$= -3igm\int d^4z\int\frac{d^4q}{(2\pi)^4}\int\frac{d^4k}{(2\pi)^4}\frac{ie^{-iq\cdot(x-z)}}{q^2-m^2+i\epsilon}\frac{ie^{-ik\cdot(z-z)}}{k^2-m^2+i\epsilon}$$

$$= -3igm\int d^4z\int\frac{d^4q}{(2\pi)^4}\int\frac{d^4k}{(2\pi)^4}\frac{ie^{-iq\cdot(x-z)}}{q^2-m^2+i\epsilon}\frac{i}{k^2-m^2+i\epsilon} \tag{9.33}$$

which is illustrated in Figure 9.2.

Carrying out the integral over z gives $(2\pi^4)\delta^4(q)$, which allows us to trivially integrate over q, finally giving us

$$\text{I} = -\frac{3ig}{m}\int\frac{d^4k}{(2\pi)^4}\frac{1}{k^2-m^2+i\epsilon} = -3\frac{g}{m}I_d \tag{9.34}$$

with I_d given in Eq.(9.15).

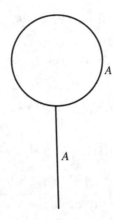

Figure 9.2   Diagram corresponding to Eq. (9.33).

Now look at the second term of Eq. (9.32). There is only one way to contract the fields, so this contribution is simply

$$\text{II} = -igm \int d^4z \, D_a(x - z) D_B(z - z)$$

$$= -\frac{ig}{m} \int \frac{d^4k}{(2\pi)^4} \frac{1}{k^2 - m^2 + i\epsilon} = -\frac{g}{m} I_d \tag{9.35}$$

with the corresponding diagram illustrated in Figure 9.3.

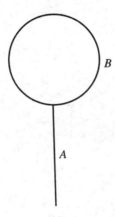

Figure 9.3   Diagram corresponding to Eq. (9.35).

Supersymmetry Demystified

Finally, there is the third term of Eq. (9.32). It is equal to

$$
\begin{aligned}
\text{III} &= -ig \int d^4z \, D_A(x-z) \, \langle 0|T\left(\overline{\Psi}_\alpha(z)\,\Psi_\alpha(z)\right)|0\rangle \\
&= ig \int d^4z \, D_A(x-z) \, \langle 0|T\left(\Psi_\alpha(z)\overline{\Psi}_\alpha(z)\right)|0\rangle \\
&= ig \int d^4z \, D_A(x-z) \, \text{Tr}\left(S(z-z)\right) \\
&= ig \int d^4z \int \frac{d^4q}{(2\pi)^4} \int \frac{d^4k}{(2\pi)^4} \frac{ie^{-iq\cdot(x-z)}}{q^2-m^2+i\epsilon} \frac{i\,\text{Tr}(\slashed{k}+m)}{k^2-m^2+i\epsilon} \\
&= 4igm \int d^4z \int \frac{d^4q}{(2\pi)^4} \int \frac{d^4k}{(2\pi)^4} \frac{ie^{-iq\cdot(x-z)}}{q^2-m^2+i\epsilon} \frac{i}{k^2-m^2+i\epsilon} \\
&= 4i\frac{g}{m} \int d^4z \int \frac{d^4k}{(2\pi)^4} \frac{1}{k^2-m^2+i\epsilon} \\
&= 4\frac{g}{m} I_d
\end{aligned}
\tag{9.36}
$$

where we have used the usual trace identities for gamma matrices (which are summarized in Appendix A). The corresponding Feynman diagram is shown in Figure 9.4.

The three terms in Eqs. (9.33), (9.35), and (9.36) are seen to cancel out exactly. More specifically, the contribution from the fermion loop cancels out exactly the contribution from the two scalar loops.

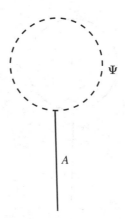

Figure 9.4 Diagram corresponding to Eq. (9.36).

Note that this works only because the same coupling constant g appears in both L_{12} and L_{13} and because the three particles have the same mass, which is also equal to the parameter m appearing as a coupling constant in L_{12}.

9.4 Propagator of the *B* Field to One Loop

We consider now

$$\langle \Omega | T\big(B(x)B(y)\big)|\Omega\rangle = \frac{\langle 0|T\big(B(x)B(y)e^{i\int d^4z \mathcal{L}'(z)}\big)|0\rangle}{\langle 0|T\big(e^{i\int d^4z \mathcal{L}'(z)}\big)|0\rangle} \tag{9.37}$$

Let us recall that the effect of the denominator is to remove the disconnected diagrams from the perturbation expansion.

The lowest-order (order g^0) contribution is simply the free-particle propagator $D_B(x-y)$. There is no distinction between the free propagators of the A and B fields because they have the same mass. However, we will still distinguish the two by calling them D_A and D_B because it will be useful to distinguish the contributions of the two fields when we discuss SUSY spontaneous symmetry breaking in Chapter 14.

The only way to have a nonzero result in the expansion of the exponential in the numerator is to have an even number of A and B fields (if there is an odd number of fields, there will always be a creation or annihilation operator that will be unpaired, which, when sandwiched between the vacuum, will give zero). Because of this, we can see that there is no order g contributions.

$\mathcal{O}(g^2)$ contribution comes from the L_{I1} term in the *first* order of the expansion of the exponential, i.e.,

$$\int d^4z \langle 0|T\big(iB(x)B(y)L_{I1}(z)\big)|0\rangle \tag{9.38}$$

and six contributions from the second order of the expansion of the exponential, i.e.,

$$\int d^4z \int d^4w \,\langle 0|T\left\{ B(x)B(y)\left[-\frac{1}{2}L_{12}(z)L_{12}(w) - \frac{1}{2}L_{13}(z)L_{13}(w) \right.\right.$$
$$\left.\left. -\frac{1}{2}L_{14}(z)L_{14}(w) - L_{12}(z)L_{13}(w) - L_{12}(z)L_{14}(w) - L_{13}(z)L_{14}(w) \right]\right\}|0\rangle \tag{9.39}$$

A Few More Useful Integrals

The integral in Eq. (9.13) will be useful several times in this chapter. There are two more integrals that will show up repeatedly. We will evaluate them here to alleviate the presentation in the next few sections.

One of these integrals is

$$\int d^4z\, d^4w\, D(x-z)D(y-w)D(w-z)D(w-z)$$

$$= \int d^4z\, d^4w \left[\int \frac{d^4p}{(2\pi)^4} \frac{ie^{-ip\cdot(x-z)}}{p^2 - m^2 + i\epsilon} \int \frac{d^4q}{(2\pi)^4} \frac{ie^{-iq\cdot(y-w)}}{q^2 - m^2 + i\epsilon} \right.$$

$$\left. \int \frac{d^4k}{(2\pi)^4} \frac{ie^{-ik\cdot(w-z)}}{k^2 - m^2 + i\epsilon} \int \frac{d^4l}{(2\pi)^4} \frac{ie^{-il\cdot(w-z)}}{l^2 - m^2 + i\epsilon} \right]$$

Doing the z integration gives a factor $(2\pi)^4\delta^4(p+k+l)$. Integrating then over l, over w, and finally, over q, we finally get

$$\int \frac{d^4p}{(2\pi)^4} e^{-ip\cdot(x-y)} \frac{i}{p^2 - m^2 + i\epsilon} \frac{i}{p^2 - m^2 + i\epsilon}$$

$$\times \left[\int \frac{d^4k}{(2\pi)^4} \frac{i}{k^2 - m^2 + i\epsilon} \frac{i}{(p+k)^2 - m^2 + i\epsilon} \right] \qquad (9.40)$$

whose amputated Fourier transform is

$$\left[\int d^4z\, d^4w\, D(x-z)D(y-w)D(w-z)D(w-z) \right]\bigg|_{FT}^{AM} \equiv I_{\log} \qquad (9.41)$$

where $I_{\log}$ is the logarithmically divergent integral between brackets in Eq. (9.40).

The second integral we will often need is

$$\int d^4z\, d^4w\, D(x-z)\, D(y-z)\, D(z-w)\, D(w-w)$$

$$= \int d^4z\, d^4w \left[\int \frac{d^4p}{(2\pi)^4} \frac{ie^{-ip\cdot(x-z)}}{p^2 - m^2 + i\epsilon} \int \frac{d^4q}{(2\pi)^4} \frac{ie^{-iq\cdot(y-z)}}{q^2 - m^2 + i\epsilon} \right.$$

$$\left. \int \frac{d^4k}{(2\pi)^4} \frac{ie^{-ik\cdot(z-w)}}{k^2 - m^2 + i\epsilon} \int \frac{d^4l}{(2\pi)^4} \frac{ie^{-il\cdot(w-w)}}{l^2 - m^2 + i\epsilon} \right] \qquad (9.42)$$

Doing the z integration, we get a factor $(2\pi)^4\delta^4(p+q-k)$ that we may use to integrate over k. Integrating then over w and finally over q, we get

$$-\frac{i}{m^2}\int\frac{d^4p}{(2\pi)^4}\frac{ie^{-ip\cdot(x-y)}}{p^2-m^2+i\epsilon}\left[\int\frac{d^4l}{(2\pi)^4}\frac{i}{l^2-m^2+i\epsilon}\right]\frac{i}{p^2-m^2+i\epsilon} \quad (9.43)$$

so the Fourier transform of Eq. (9.42) is (after amputating the external lines)

$$\left[\int d^4z\,d^4w\,D(x-z)D(y-z)D(z-w)D(w-w)\right]\Bigg|_{FT}^{AM}=-\frac{i}{m^2}I_d \quad (9.44)$$

with I_d defined in Eq. (9.13).

Now we are well equipped to calculate the one-loop corrections to the B propagator.

Contribution from L_{I1}

Consider first the contribution from L_{I1} [see Eqs. (9.38) and (8.71)]. It is

$$\int d^4z\langle 0|T\left\{B(x)B(y)\left[-\frac{ig^2}{2}B(z)^4-ig^2A(z)^2B(z)^2-\frac{ig^2}{2}A(z)^4\right)\right]\right\}|0\rangle$$

$$(9.45)$$

In the first term, there are four ways to contract $B(x)$ with one of the $B(z)$. Then there are three ways to contract $B(y)$ with one of the remaining $B(z)$. The two $B(z)$ left over must be contracted together. We therefore get

$$-6ig^2\left[\int d^4z\,D_B(x-z)D_B(y-z)D_B(z-z)\right]\Bigg|_{FT}^{AM}=-6ig^2I_d \quad (9.46)$$

where we have used Eq. (9.12). This integral corresponds to Figure 9.5.

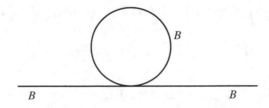

Figure 9.5 Diagram corresponding to Eq. (9.46).

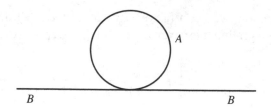

Figure 9.6 Diagram corresponding to Eq. (9.47).

In the case of the second term of Eq. (9.45), there are two ways to contract $B(x)$ with a $B(z)$, leaving only one possible contraction for $B(y)$. The two A fields must be contracted together, so we get

$$-2ig^2\left[\int d^4z\, D_B(x-z)D_B(y-z)D_A(z-z)\right]_{FT}^{AM} = -2ig^2 I_d \qquad (9.47)$$

which is depicted in Figure 9.6.

Now consider the third term of Eq. (9.45). The A fields cannot be contracted with the external lines, so this produces a disconnected diagram that we must drop. The total contribution of L_{I1} is therefore the sum of Eqs. (9.46) and (9.47):

$$\boxed{\text{Contribution from } L_{I1} = -8ig^2 I_d} \qquad (9.48)$$

with I_d defined in Eq. (9.13).

Contribution from L_{I2}^2

This contribution is given by [see Eqs. (9.39) and (8.71)]

$$-\frac{m^2g^2}{2}\int d^4w\, d^4z\, \langle 0|T\left(B(x)B(y)A^3(z)A^3(w)\right)|0\rangle$$

$$-\frac{m^2g^2}{2}\int d^4w\, d^4z\, \langle 0|T\left(B(x)B(y)A(z)A(w)B^2(z)B^2(w)\right)|0\rangle$$

$$-m^2g^2\int d^4w\, d^4z\, \langle 0|T\left(B(x)B(y)A^3(w)A(z)B^2(z)\right)|0\rangle \qquad (9.49)$$

The first term produces only disconnected diagrams, so we may ignore it.

Consider the second term. We will have two choices. We may contract $B(x)$ and $B(y)$ with B fields at different spacetime points, or we may contract both of them

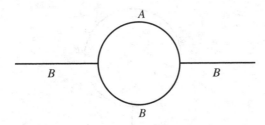

Figure 9.7 Diagram corresponding to Eq. (9.50).

with B fields at the same point. Let's start by considering contracting them with B fields at different points.

We can contract $B(x)$ with either one of the $B(z)$ or one of the $B(w)$. Let's choose to contract $B(x)$ with one of the two $B(z)$ so we get a factor of two. Now, the remaining $B(z)$ must be contracted with one of the two $B(w)$ (otherwise, we get a disconnected diagram), and there are two ways to do this. Finally, the remaining $B(w)$ must be contracted with $B(y)$, and the two A fields must be contracted together. Thus, in all, there are four combinations. But if we would have started by contracting $B(x)$ with one of the $B(w)$ instead, we obviously would have obtained another four combinations identical to the first four, so there is a total of eight combinations. We finally get the diagram of Figure 9.7, which corresponds to the expression

$$-4m^2g^2\left[\int d^4z d^4w\, D_B(x-z)D_B(y-w)D_B(w-z)D_A(w-z)\right]\Bigg|_{FT}^{AM}$$

$$= -4m^2g^2 I_{\log} \tag{9.50}$$

using Eq. (9.41). Since this is only logarithmically divergent, we will ignore this contribution (in a complete calculation, this divergence is taken care of by the renormalization of the wavefunction).

Now consider the other possibility: contracting both $B(x)$ and $B(y)$ with a B field at the same spacetime point. Let's say that we contract both with $B^2(z)$. There are two ways to do this. Then the remaining B fields at w must be contracted together, whereas the two A fields must be contracted together as well. But we could have started by contracting $B(x)$ and $B(y)$ with $B^2(w)$. Thus the overall factor of this diagram is 4. We therefore get the diagram of Figure 9.8, which is represented by the integral

$$-2m^2g^2\left[\int d^4z d^4w\, D_B(x-z)D_B(y-z)D_A(z-w)D_B(w-w)\right]\Bigg|_{FT}^{AM}$$

$$= 2ig^2 I_d \tag{9.51}$$

where we have used Eq. (9.44).

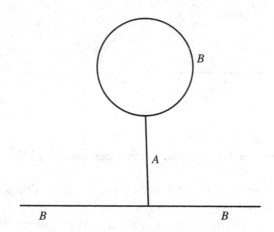

Figure 9.8 Diagram corresponding to Eq. (9.51).

Third Term of L_{I2}^2

Consider now the third and last term in Eq. (9.49). To get a connected diagram, $B(x)$ must be contracted with a $B(z)$ (two ways), and $B(y)$ obviously must be contracted with the remaining $B(z)$. $A(z)$ must be contracted with one of the $A(w)$ (three ways), and the remaining two $A(w)$ must be contracted together (see Figure 9.9). This gives an overall factor of 6, and we get for this diagram

$$-6m^2g^2\left[\int d^4z\,d^4w\,D_B(x-z)D_B(y-z)D_A(z-w)D_A(w-w)\right]\Bigg|_{FT}^{AM} = 6ig^2I_d$$
(9.52)

using again Eq. (9.44).

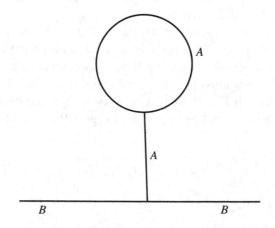

Figure 9.9 Diagram corresponding to Eq. (9.52).

Contribution from $L_{I2}\,L_{I3}$

This contribution is given by [see Eqs. (9.39) and (8.71)]

$$-\int d^4w\,d^4z\langle 0|T\big(B(x)B(y)\{mg^2A(z)\overline{\Psi}(z)\Psi(z)\big(A^3(w)+A(w)B^2(w)\big)\}\big)|0\rangle$$

(9.53)

The A^3 term gives a disconnected diagram. In the case of the second term, we must contract both $B(x)$ and $B(y)$ with the $B^2(w)$. There are two ways to do this. The $A(z)$ obviously must be contracted with $A(w)$, and the $\overline{\Psi}$ must be contracted with the Ψ so that the overall factor is simply 2, and we get [the change of sign is due to the need to move Ψ to the left of $\overline{\Psi}$ in order to use Eq.(9.18)]

$$2mg^2\int d^4z\,d^4w\,D_B(x-w)D_B(y-w)D_A(z-w)\mathrm{Tr}[S(z-z)]\qquad(9.54)$$

which is represented by Figure 9.10 (we use dashed lines to represent fermion propagators).

The trace of the fermion propagator involves calculating (using k for the four-momentum of the fermion)

$$\mathrm{Tr}\left(\frac{i}{\not{k}-m+i\epsilon}\right)=i\,\mathrm{Tr}\left(\frac{\not{k}+m}{k^2-m^2+i\epsilon}\right)$$

$$=\frac{4im}{k^2-m^2+i\epsilon}\qquad(9.55)$$

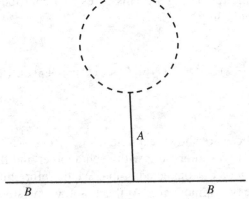

Figure 9.10   Diagram corresponding to Eq. (9.54). The dashed line is the fermion propagator.

This is simply $4m$ times the expression we would have in a boson propagator, so we may write

$$\text{Tr}[S(z - z)] = 4m D(z - z) \tag{9.56}$$

Equation (9.54) is therefore

$$8g^2m^2 \left[\int d^4z d^4w \, D_B(x - w) D_B(y - w) D_A(z - w) D(z - z) \right]\Big|_{FT}^{AM} = -8ig^2 I_d \tag{9.57}$$

where once more we have used Eq. (9.44).

Contribution from L_{I4}^2

The integral is [see Eqs. (9.39) and (8.71)]

$$\int d^4w d^4z \langle 0|T \left\{ B(x)B(y) \left(\frac{g^2}{2} B(z)\overline{\Psi}(z)\gamma_5 \Psi(z) \, B(w)\overline{\Psi}(w)\gamma_5 \Psi(w) \right) \right\} |0\rangle \tag{9.58}$$

Obviously, $B(x)$ must be contracted with either $B(w)$ or $B(z)$, and $B(y)$ must be contracted with the remaining B. This gives two contributions. We get, after contracting the B fields only,

$$g^2 \int d^4w d^4z D_B(x - z) D_B(y - w) \langle 0|T \left(\overline{\Psi}_\alpha(z)\gamma_{5\alpha\beta} \Psi_\beta(z) \, \overline{\Psi}_\delta(w)\gamma_{5\delta\rho} \Psi_\rho(w) \right) |0\rangle \tag{9.59}$$

where the spinor indices are shown explicitly.

If we contract the two spinors at z with each another and the two spinors at w with each other, we get a disconnected graph. We therefore must contract the two spinors at w with the two spinors at z. At this point, we must keep in mind that we are dealing with Majorana fermions, so we have more possibilities than for a Dirac

fermion. Keeping this in mind, we get

$$\langle 0|T\left(\overline{\Psi}_\alpha(z)\gamma_{5\alpha\beta}\Psi_\beta(z)\,\overline{\Psi}_\delta(w)\gamma_{5\delta\rho}\Psi_\rho(w)\right)|0\rangle$$

$$= \langle 0|T\left(\overline{\Psi}_\alpha(z)\gamma_{5\alpha\beta}\Psi_\beta(z)\,\overline{\Psi}_\delta(w)\gamma_{5\delta\rho}\Psi_\rho(w)\right)|0\rangle$$

$$+ \langle 0|T\left(\overline{\Psi}_\alpha(z)\gamma_{5\alpha\beta}\Psi_\beta(z)\,\overline{\Psi}_\delta(w)\gamma_{5\delta\rho}\Psi_\rho(w)\right)|0\rangle \quad (9.60)$$

Consider the second term first. This is a term we also would get for Dirac fermions. Recall that in the fermion propagator (9.18), the Ψ appears to the left of the Dirac conjugate field $\overline{\Psi}$, so we must move $\Psi_\rho(w)$ to the left of $\overline{\Psi}_\alpha(z)$. Every time we move a fermion field passed a Ψ or a $\overline{\Psi}$, we pick up a minus sign, so we get

$$\langle 0|T\left(\overline{\Psi}_\alpha(z)\gamma_{5\alpha\beta}\Psi_\beta(z)\,\overline{\Psi}_\delta(w)\gamma_{5\delta\rho}\Psi_\rho(w)\right)|0\rangle$$

$$= -\langle 0|T\left(\Psi_\rho(w)\overline{\Psi}_\alpha(z)\gamma_{5\alpha\beta}\Psi_\beta(z)\overline{\Psi}_\delta(w)\gamma_{5\delta\rho}\right)|0\rangle \quad (9.61)$$

Now we can replace the contractions by fermion propagators, which gives us that the second term of Eq. (9.60) is finally equal to

$$-S_{\rho\alpha}(w-z)\gamma_{5\alpha\beta}S_{\beta\delta}(z-w)\gamma_{5\delta\rho} = -\mathrm{Tr}\big(S(w-z)\gamma_5 S(z-w)\gamma_5\big) \quad (9.62)$$

We obtain a trace of the matrices because all the indices appear next to one another except for the index ρ, which appears at the very beginning and at the very end.

Let us now turn our attention to the first term of Eq. (9.60). This is a contribution that we would not have if we were considering Dirac fermions. In order to use the propagators (9.25) and (9.26), we need to move the two Ψ next to one another and the two $\overline{\Psi}$ next to one another as well. In principle, we could use any order between the two Ψ and between the two $\overline{\Psi}$, but the calculation will be much simpler if we arrange the order so that we get a trace. Unfortunately, we can't quite make it work. For example, if we move the $\overline{\Psi}_\delta(w)$ to the left of $\overline{\Psi}_\alpha(z)$ and $\Psi_\rho(w)$ to the right of $\Psi_\beta(z)$, we get

$$\langle 0|T\left(\overline{\Psi}_\alpha(z)\gamma_{5\alpha\beta}\Psi_\beta(z)\,\overline{\Psi}_\delta(w)\gamma_{5\delta\rho}\Psi_\rho(w)\right)|0\rangle$$

$$= \langle 0|T\left(\overline{\Psi}_\delta(w)\overline{\Psi}_\alpha(z)\gamma_{5\alpha\beta}\Psi_\beta(z)\Psi_\rho(w)\gamma_{5\delta\rho}\right)|0\rangle \quad (9.63)$$

We see that the indices are not in order: There is a mismatch because of the indices of the γ_5 matrix on the far right. The way out is of course to reverse the indices of that matrix. This can be done by using the identity (valid for any matrix)

$$\gamma_{5\delta\rho} = (\gamma_5^T)_{\rho\delta} \tag{9.64}$$

However, because $\gamma_5^T = \gamma_5$, we see that we are free to switch the order of its indices,

$$\gamma_{5\delta\rho} = \gamma_{5\rho\delta} \tag{9.65}$$

which allows us to write Eq. (9.63) as

$$\langle 0 | T \left(\overline{\Psi}_\delta(w) \overline{\Psi}_\alpha(z) \gamma_{5\alpha\beta} \Psi_\beta(z) \Psi_\rho(w) \gamma_{5\rho\delta} \right) | 0 \rangle$$
$$= C_{\delta\gamma}^T S_{\gamma\alpha}(w-z) \gamma_{5\alpha\beta} S_{\beta\mu}(z-w) C_{\mu\rho}^T \gamma_{5\rho\delta}$$
$$= \text{Tr}\left(C^T S(w-z) \gamma_5 S(z-w) C^T \gamma_5 \right)$$

Now, it turns out that there is always a way to get rid of the C^T matrices. The trick is to use that $C^T = -C = C^{-1}$ so that $(C^T)^2 = -1$. In order to use this, though, we need to get the two C_T matrices next to one another. We will do this in two steps. First, we use the invariance of the trace under cyclic permutations, $\text{Tr}(ABCD) = \text{Tr}(DABC)$ to write

$$\text{Tr}\left(C^T S(w-z) \gamma_5 S(z-w) C^T \gamma_5 \right) = \text{Tr}\left(S(w-z) \gamma_5 S(z-w) C^T \gamma_5 C^T \right) \tag{9.66}$$

Next, we use the fact that γ_5 and C^T commute; i.e., $\gamma_5 C^T = C^T \gamma_5$ (which can be easily checked using their explicit representations), so we finally obtain

$$\text{Tr}\left(S(w-z) \gamma_5 S(z-w) C^T \gamma_5 C^T \right) = \text{Tr}\left(S(w-z) \gamma_5 S(z-w) C^T C^T \gamma_5 \right)$$
$$= -\text{Tr}\left(S(w-z) \gamma_5 S(z-w) \gamma_5 \right) \tag{9.67}$$

which is equal to what we had obtained for the second term of Eq. (9.60) [see also Eq. (9.62)]. Therefore, Eq. (9.60) is equal to

$$\langle 0 | T \left(\overline{\Psi}_\alpha(z) \gamma_{5\alpha\beta} \Psi_\beta(z) \overline{\Psi}_\delta(w) \gamma_{5\delta\rho} \Psi_\rho(w) \right) | 0 \rangle = -2\text{Tr}\left(S(w-z) \gamma_5 S(z-w) \gamma_5 \right) \tag{9.68}$$

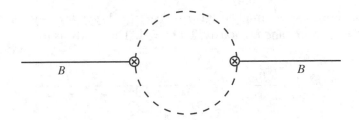

Figure 9.11   Diagram corresponding to Eq. (9.69). The vertices drawn as a cross inside a circle represent interactions containing a γ_5 matrix.

Plugging this back into the expression for the diagram, i.e., Eq. (9.59), we get that the contribution of L_{14}^2 is

$$-2g^2 \int d^4w\, d^4z\, D_B(x-z) D_B(y-w) \text{Tr}\big(S(z-w)\gamma_5 S(w-z)\gamma_5\big) \qquad (9.69)$$

which is shown in Figure 9.11 and corresponds to the integral

$$-2g^2 \int d^4w\, d^4z \int \frac{d^4p}{(2\pi)^4} \frac{ie^{-ip\cdot(x-z)}}{p^2-m^2+i\epsilon} \int \frac{d^4q}{(2\pi)^4} \frac{ie^{-iq\cdot(y-w)}}{q^2-m^2+i\epsilon}$$

$$\int \frac{d^4k}{(2\pi)^4} \int \frac{d^4l}{(2\pi)^4} \text{Tr}\left(\frac{ie^{-ik\cdot(z-w)}}{\slashed{k}-m+i\epsilon}\gamma_5 \frac{ie^{-il\cdot(w-z)}}{\slashed{l}-m+i\epsilon}\gamma_5\right) \qquad (9.70)$$

Doing the integration over z gives a factor $(2\pi)^4\delta^4(p-k+l)$, which we may use to integrate over d^4l. We get

$$-2g^2 \int d^4w \int \frac{d^4p}{(2\pi)^4} \frac{ie^{-ip\cdot x}}{p^2-m^2+i\epsilon} \int \frac{d^4q}{(2\pi)^4} \frac{ie^{-iq\cdot(y-w)}}{q^2-m^2+i\epsilon}$$

$$\int \frac{d^4k}{(2\pi)^4} e^{ip\cdot w} \text{Tr}\left(\frac{i}{\slashed{k}-m+i\epsilon}\gamma_5 \frac{i}{\slashed{k}-\slashed{p}-m+i\epsilon}\gamma_5\right) \qquad (9.71)$$

Doing the integral over w and then over q, we get

$$2g^2 \int \frac{d^4p}{(2\pi)^4} e^{-ip\cdot(x-y)} \frac{i}{p^2-m^2+i\epsilon} \frac{i}{p^2-m^2+i\epsilon}$$

$$\int \frac{d^4k}{(2\pi)^4} \text{Tr}\left\{\frac{(\slashed{k}+m)\gamma_5(\slashed{k}-\slashed{p}+m)\gamma_5}{(k^2-m^2)[(k-p)^2-m^2]}\right\}$$

To carry out the traces, we simply need to use $(\gamma_5)^2 = 1$, $\gamma_5 \gamma^\mu = -\gamma^\mu \gamma_5$, $\text{Tr}(\not{q}) = 0$, $\text{Tr}(\not{q}\gamma_5\not{b}\gamma_5) = -4a \cdot b$, and obviously, $\text{Tr}(\mathbf{1}) = 4$. This leads us to

$$8g^2 \int \frac{d^4p}{(2\pi)^4} e^{-ip\cdot(x-y)} \frac{i}{p^2 - m^2 + i\epsilon} \frac{i}{p^2 - m^2 + i\epsilon}$$

$$\times \int \frac{d^4k}{(2\pi)^4} \frac{k \cdot p - k^2 + m^2}{(k^2 - m^2)[(k-p)^2 - m^2]} \tag{9.72}$$

This contains a quadratic divergence, but it is not easy to isolate it in that form. A standard trick is to write the numerator as

$$k \cdot p - k^2 + m^2 = \frac{-(p-k)^2 + k^2 + p^2 - 2k^2 + 2m^2}{2}$$

$$= \frac{-(p-k)^2 + m^2 - k^2 + m^2 + p^2}{2} \tag{9.73}$$

so that the integral is

$$-4g^2 \int \frac{d^4p}{(2\pi)^4} e^{-ip\cdot(x-y)} \frac{i}{p^2 - m^2 + i\epsilon} \frac{i}{p^2 - m^2 + i\epsilon}$$

$$\times \int \frac{d^4k}{(2\pi)^4} \left\{ \frac{1}{(k-p)^2 - m^2} + \frac{1}{k^2 - m^2} - \frac{p^2}{(k^2 - m^2)[(k-p)^2 - m^2]} \right\}$$

In this form, it is clear that the quadratic divergence is isolated in the first two terms. Ignoring the last term and making the change of variable $k \to k + p$ in the first integral, we can finally see that the quadratically divergent contribution of this diagram is

$$-8g^2 \left[\int \frac{d^4p}{(2\pi)^4} e^{-ip\cdot(x-y)} \frac{i}{p^2 - m^2 + i\epsilon} \frac{i}{p^2 - m^2 + i\epsilon} \int \frac{d^4k}{(2\pi)^4} \frac{1}{k^2 - m^2} \right]\Big|_{FT}^{AM}$$

$$= 8ig^2 I_d \tag{9.74}$$

As an aside, the change of variable we just performed is fine as long as we are only interested in isolating the quadratic divergent contribution of the diagram. If we were interested in extracting the finite contribution, we could not be as cavalier.

The Remaining Terms

There are still three terms we haven't calculated: the terms in L_{13}^2, $L_{12}L_{14}$, and $L_{13}L_{14}$. However, none of these terms contribute connected diagrams to the one-loop B propagator as you can easily verify.

9.5 Putting It All Together

Let us summarize our results. We will group the contributions into two groups: the tadpole diagrams and the nontadpole diagrams. Our goal is to demonstrate that quadratic divergences cancel out, so we don't take into consideration the diagram of Figure 9.7 which contains a logarithmic divergence.

For the tadpole diagrams, we found

$$A \text{ tadpole} = 6ig^2 I_d$$

$$B \text{ tadpole} = 2ig^2 I_d$$

$$\text{Fermion tadpole} = -8ig^2 I_d$$

$$\text{Sum} = 0 \tag{9.75}$$

The results for the nontadpole diagrams are

$$A \text{ loop} = -2ig^2 I_d$$

$$B \text{ loop} = -6ig^2 I_d$$

$$\text{Fermion loop} = 8ig^2 I_d$$

$$\text{Sum} = 0 \tag{9.76}$$

As advertised, all quadratic divergences cancel out!

The choice of grouping the contributions into tadpole versus nontadpole diagrams to show the cancellation may seem arbitrary in this example. For example, we could have grouped instead the fermion and scalar contributions separately and still obtained a cancellation. However, there is an accidental symmetry here that obscures the structure of the cancellations. For this reason, it is instructive to calculate as well the one-loop quadratic divergent contributions to the A propagator. We will only quote the results here, which you are invited to double-check (all the tricks needed to do the calculation have been introduced in the calculation of the B propagator).

The one-loop quadratic divergent contributions to the propagator of A are, for the tadpole diagrams,

$$A \text{ tadpole} = 18ig^2 I_d$$

$$B \text{ tadpole} = 6ig^2 I_d$$

$$\text{Fermion tadpole} = -24ig^2 I_d$$

$$\text{Sum} = 0 \tag{9.77}$$

and for the non-tadpole diagrams, they are

$$A \text{ loop} = -6ig^2 I_d$$

$$B \text{ loop} = -2ig^2 I_d$$

$$\text{Fermion loop} = 8ig^2 I_d$$

$$\text{Sum} = 0 \tag{9.78}$$

Again, there is one diagram that is only logarithmically divergent and therefore was not included.

By looking at the results for the A propagator, it is now clear that the tadpole diagrams cancel separately from the nontadpole contributions. These cancellations are highly nontrivial. They obviously require a very subtle "conspiracy" between all the parameters appearing in the lagrangian. Obviously, SUSY imposes very stringent conditions on these parameters (not only at tree level, as we will discuss in the next section). This is striking when we realize that there are three particles and seven interaction terms but only one mass and one coupling constant in the theory! Thus there are only two parameters, where we could have, in principle, 10 different parameters (three masses and seven coupling constants).

However, if we are only interested in having the quadratic divergences cancel, it turns out that SUSY offers more constraints than what is necessary. For example, consider leaving the dimensionless coupling constants in all the interactions the same, as in SUSY, but let's assign different symbols to the four different parameters with a dimension of mass: the masses of the three particles m_A, m_B, and m_Ψ and the dimensionful coupling constant that appears in the interaction L_{12}, which we will denote by m_c.

The important point to notice is that the scalar masses m_A and m_B do not appear at all in the quadratically divergent corrections to the propagator! On the other hand, if we go over the calculation of the contribution from the fermion tadpole, we can see that this diagram is actually proportional to m_c/m_Ψ, and therefore, for the quadratic divergences to cancel, it is essential that $m_c = m_\Psi$.

The conclusion is that we can change the masses of the scalar particles without disrupting the cancellation of quadratic divergences! This would clearly break SUSY because the fermion and the scalar would no longer be degenerate in mass, but it would not spoil the cancellation of the quadratic divergences. Of course, we have only shown this for the very specific case of the one-loop corrections to the B propagator. But a more detailed analysis shows that it is in fact valid to all orders of perturbation theory and for all processes!

In addition, we also may modify some of the coupling constants of the various interactions without spoiling the cancellations of the quadratic divergences. Thus, as long as we are motivated only by the desire to cancel these divergences, we have actually more leeway with the parameters of the theory than SUSY allows. We could therefore add terms to the lagrangian that would break explicitly the invariance under SUSY but which would preserve the cancellation of quadratic divergences in the theory. Such terms are said to break SUSY *softly*. We will list all the possible terms that break SUSY softly in Chapter 14. They will play a crucial role in Chapters 15 and 16 because they are required to break SUSY in the MSSM.

9.6 A Note on Nonrenormalization Theorems

Unfortunately, we don't have enough space to explore more in-depth the perturbative properties of supersymmetric theories. The very simple examples we did had for their only goal to give you a peek at the types of cancellations of ultraviolet divergences that occur owing to SUSY.

However, this very short chapter does not do justice to the power of SUSY. Given more space, we would discover not only that all quadratic divergences cancel to all orders of perturbation theory but also that the only renormalization required in the Wess-Zumino model is a common wavefunction renormalization of all the fields (which involves only a logarithmic divergence). The mass m and the coupling constant g *do not get renormalized at all* to any order in perturbation theory!

The theorems demonstrating that certain parameters in supersymmetric theories do not get renormalized at all are known as *nonrenormalization theorems*. A proof using a diagrammatic approach was worked out in Ref. 22. Seiberg[42] has derived the same result using a very nifty trick that makes the proof extremely simple. The basic idea is that the parameters of the theory, such as m and g in the Wess-Zumino model, are treated as the expectation values of some new fields. Of course, these are not physical fields but rather simply mathematical tricks to carry out the proof and therefore are sometimes referred to as *spurious fields* or *spurions*. The introduction of these spurious fields enhances the symmetry of the lagrangian, introducing an additional $U(1)$ symmetry. Using the superspace approach that we will cover in later chapters and the holomorphicity of the potential, it is a simple matter to prove that some interactions cannot get renormalized to any order in perturbation theory. A simple illustration of this technique can be found in Section 3 of Ref. 43, and more details are given in Section 8.2 of Ref. 8.

Let us also add that calculations of Feynman diagrams in SUSY are simplified greatly by the use of so-called supergraphs. These graphs involve the propagators (or *superpropagators*) of the superfields that we will introduce in Chapter 11. The end result is that the effect of all the field members of a supermultiplet are taken into account simultaneously instead of having many diagrams with the various fields. This obviously simplifies the calculations greatly, and is a powerful tool to build directly into the Feynman diagrams the highly nontrivial relations between the various terms imposed by SUSY. The supergraph approach therefore not only simplifies calculations but also facilitates the proof of many results, including nonrenormalization theorems. This is unfortunately beyond the scope of this book, but if you are interested, you can find a nice introduction in Ref. 47 and a more technical presentation in Ref. 49.

9.7 Quiz

1. What is the difference between the propagator of a Dirac fermion and the propagator of a Majorana fermion?

2. How many different parameters are there in the Wess-Zumino model (for a single multiplet)?

3. Are supersymmetric theories completely free of all divergences?

4. Give one example of a term we could add to the lagrangian that would break SUSY but which would not spoil the cancellation of quadratic divergences.

5. What is meant by saying that an interaction breaks SUSY softly?

CHAPTER 10

Supersymmetric Gauge Theories

The only theory we have built so far contained a spinor and a scalar field. If we are to apply supersymmetry (SUSY) to real life, we need to deal with gauge theories. As we saw in Section 7.4, a vector (massless) supermultiplet contains states with helicities $h = 1/2$ and $h = 1$. Together with the corresponding *CPT* conjugates, we therefore have four states with $h = \pm 1/2, \pm 1$. The particle content is therefore a massless fermion and a massless spin 1 particle. If we consider an abelian $U(1)$ gauge theory, this is all the particle content we need to construct the supersymmetric version because there is only one gauge field. For nonabelian gauge theories, we will, of course, need to introduce as many fermions as there are gauge fields, and these fermions will have to transform in the same representation as the gauge fields. But let's not get ahead of ourselves, and let's start by building a supersymmetric abelian gauge theory. By the way, the term *gauge multiplet* is often used to denote a vector multiplet, and we will use both terms interchangeably.

10.1 Free Supersymmetric Abelian Gauge Theory

We consider a $U(1)$ free gauge theory where the superpartner of the photon field A_μ will be a left-chiral Weyl spinor that we will denote by λ. The spin 1/2 superpartners of gauge bosons are generically referred to as *gauginos*. The gaugino for the supersymmetric version of quantum electrodynamics (QED) is also referred to as a *photino*.

Our starting point for the lagrangian is therefore a sum of kinetic terms for the photon and the spinor

$$\mathcal{L} = -\frac{1}{4} F_{\mu\nu} F^{\mu\nu} + i\lambda^\dagger \bar{\sigma}^\mu \partial_\mu \lambda \tag{10.1}$$

But this is incomplete because, as in the case of the Wess-Zumino model, we will need to introduce an auxiliary scalar field in order for the SUSY algebra to close off-shell. We will set aside this issue for now and come back to it in the next section.

The members of a SUSY multiplet all must transform the same way under any gauge group acting on them. Since the photon field is neutral, we take the photino to be neutral as well so we don't need to replace the derivative ∂_μ acting on it by a covariant derivative, and the lagrangian (10.1) is gauge-invariant as it is. Note that when we say that the photon is neutral, this is a statement about the gauge invariance of the field strength $F_{\mu\nu}$, not of A_μ (which, of course, is not gauge-invariant). In nonabelian gauge theories, the field strength is *not* invariant, which is why the gauge bosons are "charged," i.e., interact directly (at tree level) with one another. In supersymmetric extensions of nonabelian gauge theories, the gauginos also will be charged under the gauge group.

Because there is no interaction between the spinor and the vector field, this is a free theory. We will include interactions later.

Now consider the SUSY transformations of the fields. The gauge field A^μ is real (in the sense $A^{\mu\dagger} = A^\mu$), so its transformation must be real as well. Let us try

$$\delta A^\mu = \zeta^\dagger \bar{\sigma}^\mu \lambda + \lambda^\dagger \bar{\sigma}^\mu \zeta \tag{10.2}$$

where we had to add the hermitian conjugate to make the transformation real. Since the dimension of A^μ is equal to 1, and since λ has a dimension of 3/2, we get that the SUSY parameter ζ is of dimension $-1/2$ as before.

EXERCISE 10.1
Show that Eq. (10.2) is real.

Now consider the transformation of the photino. There is a new constraint that we did not have previously: The transformation must be gauge-invariant. Thus we can't use $\delta\lambda \simeq A_\mu$. Instead, we start with the gauge-invariant combination

$$\delta\lambda \simeq C\zeta F_{\mu\nu}$$

where C is an arbitrary constant. Since $F_{\mu\nu}$ has dimension 2, the dimensions work out. However, the Lorentz properties are not all right, obviously. We need to contract the two Lorentz indices of the field strength without changing the dimensions of the right-hand side. We should use some combination of σ^μ and $\bar{\sigma}^\mu$ but we must be careful; we also must make sure that we end up with something that will transform like a left-chiral spinor. Recall that $P_\mu \bar{\sigma}^\mu \chi$ transforms like a right-chiral spinor, whereas $P_\mu \sigma^\mu \eta$ transforms like a left-chiral spinor. Therefore $P_\mu P_\nu \sigma^\mu \bar{\sigma}^\nu \chi$ transforms like a left-chiral spinor. So we see that the quantity

$$\delta\lambda = C F_{\mu\nu} \sigma^\mu \bar{\sigma}^\nu \zeta \tag{10.3}$$

transforms like a left-chiral spinor, as desired (note that $\bar{\sigma}^\nu$ must be applied on the spinor first).

Taking the hermitian conjugate of Eq. (10.3), we get

$$\delta\lambda^\dagger = C^* F_{\mu\nu} \zeta^\dagger \bar{\sigma}^\nu \sigma^\mu$$

Let's now find out if there is some value of C that will make the lagrangian invariant. Consider first the variation of the Maxwell lagrangian:

$$\begin{aligned}
\delta\left(-\frac{1}{4}F_{\mu\nu}F^{\mu\nu}\right) &= -\frac{1}{2}F_{\mu\nu}(\partial^\mu \delta A^\nu - \partial^\nu \delta A^\mu) \\
&= -F_{\mu\nu}\partial^\mu \delta A^\nu \\
&= -F_{\mu\nu}\partial^\mu(\zeta^\dagger \bar{\sigma}^\nu \lambda) - F_{\mu\nu}\partial^\mu(\lambda^\dagger \bar{\sigma}^\nu \zeta) \\
&= \underbrace{-F_{\mu\nu}\zeta^\dagger \bar{\sigma}^\nu \partial^\mu \lambda}_{\text{I}} \underbrace{-F_{\mu\nu}(\partial^\mu \lambda^\dagger)\bar{\sigma}^\nu \zeta}_{\text{II}}
\end{aligned} \tag{10.4}$$

where in the second step we used the antisymmetry of $F_{\mu\nu}$.

Now consider the variation of the spinor term in Eq. (10.1):

$$\delta(i\lambda^\dagger \bar{\sigma}^\mu \partial_\mu \lambda) = \underbrace{iC^* F_{\mu\nu}\zeta^\dagger \bar{\sigma}^\nu \sigma^\mu \bar{\sigma}^\rho \partial_\rho \lambda}_{\text{III}} + \underbrace{iC\lambda^\dagger \bar{\sigma}^\rho \sigma^\mu \bar{\sigma}^\nu \zeta \partial_\rho F_{\mu\nu}}_{\text{IV}}$$

Terms I and III must cancel each other (modulo total derivatives) because they are the only ones containing $\zeta^\dagger$. Likewise, the second and fourth terms must cancel out because they contain ζ.

Consider III first. We shall use identity (A.11), which we will repeat here for convenience:

$$\bar{\sigma}^\nu \sigma^\mu \bar{\sigma}^\rho = \eta^{\mu\nu}\bar{\sigma}^\rho - \eta^{\nu\rho}\bar{\sigma}^\mu + \eta^{\mu\rho}\bar{\sigma}^\nu - i\epsilon^{\nu\mu\rho\delta}\bar{\sigma}_\delta$$

We use this to write term III as

$$\text{III} = iC^* F_{\mu\nu}\zeta^\dagger(\eta^{\mu\nu}\bar{\sigma}^\rho - \eta^{\nu\rho}\bar{\sigma}^\mu + \eta^{\mu\rho}\bar{\sigma}^\nu - i\epsilon^{\nu\mu\rho\delta}\bar{\sigma}_\delta)\partial_\rho\lambda$$

The first piece is identically zero because $\eta^{\mu\nu}$ is symmetric in the two Lorentz indices, whereas $F_{\mu\nu}$ is antisymmetric. The term with the Levi-Civita symbol also vanishes, although this is not as obvious. To prove it, we do an integration by parts and rewrite

$$\epsilon^{\nu\mu\rho\delta}(\partial_\rho\lambda)F_{\mu\nu} = -\epsilon^{\nu\mu\rho\delta}\lambda\partial_\rho F_{\mu\nu}$$
$$= -\epsilon^{\nu\mu\rho\delta}\lambda\partial_\rho\partial_\mu A_\nu + \epsilon^{\nu\mu\rho\delta}\lambda\partial_\rho\partial_\nu A_\mu$$

Now we see that this is identically zero because both terms contain a quantity symmetric under the exchange of two indices (the two derivatives) times the antisymmetric Levi-Civita symbol.

We are left with

$$\text{III} = -iC^* F_{\mu\nu}\zeta^\dagger\bar{\sigma}^\mu\partial^\nu\lambda + iC^* F_{\mu\nu}\zeta^\dagger\bar{\sigma}^\nu\partial^\mu\lambda$$

If we relabel $\mu \leftrightarrow \nu$ in the first term and use the antisymmetry of the field strength, we may combine the two terms to get the simple expression

$$\text{III} = 2iC^* F_{\mu\nu}\zeta^\dagger\bar{\sigma}^\nu\partial^\mu\lambda \tag{10.5}$$

This will be canceled by the first term, namely,

$$\text{I} = -F_{\mu\nu}\zeta^\dagger\bar{\sigma}^\nu\partial^\mu\lambda$$

at the condition of choosing $C = i/2$.

Let's now look at IV. Using the same tricks we used for the third term, we find that it may be written as

$$\mathrm{IV} = 2iC(\partial^\nu \lambda^\dagger)\bar\sigma^\mu \zeta F_{\mu\nu}$$
$$= -2iC(\partial^\mu \lambda^\dagger)\bar\sigma^\nu \zeta F_{\mu\nu}$$

which cancels out II at the condition that $C = i/2$ once again.

The lagrangian (10.1) is therefore supersymmetric under the transformations

$$\delta A_\mu = \zeta^\dagger \bar\sigma^\mu \lambda + \lambda^\dagger \bar\sigma^\mu \zeta$$

$$\delta\lambda = \frac{i}{2}F_{\mu\nu}\sigma^\mu \bar\sigma^\nu \zeta$$

$$\delta\lambda^\dagger = -\frac{i}{2}F_{\mu\nu}\zeta^\dagger \bar\sigma^\nu \sigma^\mu$$

However, as noted at the beginning of this chapter, this is incomplete because the algebra does not close off-shell. Let's discuss this now.

10.2 Introduction of the Auxiliary Field

The next step would be to compute the commutator of two SUSY transformations on all the fields, as we did for the left-chiral multiplet of the Wess-Zumino model. We won't do the calculation here but only quote the final result that, again, the algebra does not close off-shell.

This should not be surprising if we count the fermionic and bosonic degrees of freedom in our theory. On-shell, A_μ has two degrees of freedom (the two transverse polarization states), and so does a left-chiral Weyl fermion, as we have mentioned several times. Off-shell, a vector field A_μ has three degrees of freedom, whereas a left-chiral Weyl fermion has four. Therefore, we need to introduce *one* extra bosonic degree of freedom (as opposed to the two needed for the chiral multiplet) in the form of a *real* scalar field that is conventionally denoted D. As with the photino and the field strength, we take D to be gauge-invariant.

We therefore add the following term to the lagrangian:

$$\mathcal{L}_{\mathrm{aux}} = \frac{1}{2}D^2$$

Like the auxiliary field F, D has dimension 2. For its variation, we may take the same expression we had for the variation of F, at the condition of adding the hermitian conjugate of that variation, because D is a real field. This gives

$$\delta D = -i\zeta^\dagger \bar{\sigma}^\mu \partial_\mu \lambda + i(\partial_\mu \lambda)^\dagger \bar{\sigma}^\mu \zeta$$

Note that the transformation is a total derivative. Interestingly, the transformation of the auxiliary field F is also a total derivative, (which is not a coincidence, as will be explained in Section 12.5). Since D is real, gauge invariant and transforms with a total derivative under SUSY, *we can include a term linear in D in the lagrangian.* We therefore include a term

$$\mathcal{L}_{FI} = \xi D \tag{10.6}$$

where the arbitrary constant ξ has the dimension of a mass squared because D is of dimension 2. This term is known as the *Fayet-Illiopoulos term* (hence the choice of subscript) and will play an important role in our discussion of SUSY spontaneous breaking in Chapter 14. Note that we could not have added a term linear in F to the Wess-Zumino lagrangian because F is not a real field.

Now consider the variation of $D^2/2$, i.e.,

$$\delta\left(\frac{1}{2}D^2\right) = -iD\zeta^\dagger \bar{\sigma}^\mu \partial_\mu \lambda + iD(\partial_\mu \lambda)^\dagger \bar{\sigma}^\mu \zeta$$

As before, to cancel these terms, we introduce an extra piece to the variation of the spinor field λ:

$$\delta\lambda = \text{previous term} + D\zeta$$
$$\delta\lambda^\dagger = \text{previous term} + D\zeta^\dagger \tag{10.7}$$

where we have used the fact that D is real.

One can check that the lagrangian is still invariant under SUSY (modulo total derivatives, as always) after including those modifications.

The complete supersymmetric lagrangian of a free abelian gauge multiplet is therefore

$$\boxed{\mathcal{L} = -\frac{1}{4}F_{\mu\nu}F^{\mu\nu} + i\lambda^\dagger \bar{\sigma}^\mu \partial_\mu \lambda + \frac{1}{2}D^2 + \xi D} \tag{10.8}$$

with the SUSY transformation of the fields given by

$$\delta A_\mu = \zeta^\dagger \bar\sigma^\mu \lambda + \lambda^\dagger \bar\sigma^\mu \zeta$$

$$\delta\lambda = \frac{i}{2} F_{\mu\nu} \sigma^\mu \bar\sigma^\nu \zeta + D\zeta$$

$$\delta D = -i\zeta^\dagger \bar\sigma^\mu \partial_\mu \lambda + i(\partial_\mu \lambda)^\dagger \bar\sigma^\mu \zeta \qquad (10.9)$$

Let's now consider nonabelian gauge theories.

10.3 Review of Nonabelian Gauge Theories

Before discussing supersymmetric nonabelian gauge theories, let's do a quick review of some basic facts about group theory. We will restrict our attention to semisimple groups, i.e., groups with no $U(1)$ invariant subgroup. The generators of a group obey an algebra of the form

$$[T^a, T^b] = if^{abc} T^c \qquad (10.10)$$

where the constants f^{abc} are called the *structure constants* of the group, and the indices run from 1 to the order of the group. A basis always may be chosen such that the structure constants are completely antisymmetric.

It is easy to check that for any generators A, B, and C, the following *Jacobi identity* holds:

$$\big[[A, B], C\big] + \big[[B, C], A\big] + \big[[C, A], B\big] = 0$$

This leads to the following identity for the structure constants:

$$f^{abe} f^{ecd} + f^{bce} f^{ead} + f^{cae} f^{ebd} = 0 \qquad (10.11)$$

The generators are traceless, but the trace of the product of two generators is in general nonzero. It is always possible to choose a basis such that

$$\mathrm{Tr}(T^a T^b) = C(R)\delta^{ab}$$

where the coefficient $C(R)$ depends on the representation used for the generators T^a and on their normalization. However, once the value of $C(R)$ for one particular representation is fixed, the values of $C(R)$ for all the other representations is automatically determined.

We will now focus on $SU(N)$ because these are the only groups [beside the trivial $U(1)$] that will be of interest to us in this book. For $SU(N)$, the number of generators is equal to $N^2 - 1$, so the indices in Eqs. (10.10) and (10.11) run from 1 to $N^2 - 1$.

We will be concerned with only two specific representations: the fundamental and adjoint representations. The fundamental representation is of dimension N (the generators are $N \times N$ matrices) and is the smallest irreducible representation. There is also an "antifundamental" representation related to the fundamental representation by complex conjugation. For $N = 2$, the fundamental and antifundamental representations are actually equivalent; i.e., they are related by a unitary transformation. On the other hand, for $N > 2$, the fundamental and antifundamental representations are not equivalent. We will see an explicit illustration of the difference between the two representations in the next section, where we discuss quantum chromodynamics (QCD).

It is customary to normalize the generators of the fundamental representation to

$$\text{Tr}(T_{\text{F}}^a T_{\text{F}}^b) = \frac{\delta^{ab}}{2} \tag{10.12}$$

As mentioned earlier, this then sets the normalization of all other representations.

For $SU(2)$, the structure constants are given by the Levi-Civita symbol $f^{ijk} = \epsilon^{ijk}$. We won't make any distinction between upper and lower indices for the $SU(N)$ indices. The generators of the fundamental representation are taken to be half the Pauli matrices:

$$T_{\text{F}}^a = \frac{\sigma^a}{2} \qquad \text{for } SU(2) \tag{10.13}$$

where the σ^a are just the usual Pauli matrices. These obey the normalization condition (10.12).

The structure constants for $SU(3)$ can be found in all books on particle physics (often with typos, unfortunately), and we won't repeat them here (to avoid writing them with typos!). The generators of the fundamental representation are half the Gell-Mann matrices:

$$T_{\text{F}}^a = \frac{\lambda^a}{2} \qquad \text{for } SU(3)$$

Do not confuse the Gell-Mann matrices with the components of the photino λ introduced earlier in this chapter!

In the standard model, all the fermions, as well as the Higgs fields, transform in the fundamental and antifundamental representations of the gauge groups.

The adjoint representation is also of particular interest because it is the representation in which the gauge field strengths transform. The adjoint representation can be obtained from the structure constants by defining

$$\left(T^a_{\text{AD}}\right)^{bc} \equiv -if^{abc} \tag{10.14}$$

On the left-hand side, the index i labels the generator with which we are dealing, whereas the jk denote the specific entry of that generator. Clearly, the dimension of these matrices is $N^2 - 1$, the same as the number of generators. Using the identity (10.11), one can show that these matrices indeed obey the commutation relations (10.10), which proves that Eq. (10.14) is indeed a valid representation. With the normalization (10.12) for the fundamental representation, the adjoint representation satisfies

$$\text{Tr}\left(T^a_{\text{AD}} T^b_{\text{AD}}\right) = N\delta^{ab} \tag{10.15}$$

Consider now a set of N fields ψ_i (which we take to be fermions but which could be bosons as well, such as the Higgs doublet of the standard model) that transform in the fundamental representation of $SU(N)$. They can be assembled into an N-component column vector ψ. This means that under a gauge transformation, they transform as

$$\psi \rightarrow \psi' = \exp\left[igT^a_{\text{F}}\Lambda^a(x)\right]\psi$$
$$\equiv U\psi \tag{10.16}$$

where g is the gauge coupling constant, and the $\Lambda^a(x)$ are a set of $N^2 - 1$ spacetime-dependent gauge parameters. In this chapter, the symbol U will represent the matrix implementing the gauge transformation; this should not be confused with the matrices U used in Chapter 6 for Lorentz and SUSY transformations.

We will suppress the x dependence of the gauge parameters in the following to alleviate the notation. It is also convenient to define

$$T^a_{\text{F}}\Lambda^a \equiv \Lambda$$

and to refer to Λ as *the* gauge parameter, even though it is really an N by N matrix containing a set of $N^2 - 1$ parameters.

We define the gauge covariant derivative on this multiplet to be

$$D_\mu \psi \equiv \left(\partial_\mu + i g A_\mu^a T_F^a\right)\psi \tag{10.17}$$

When we consider the standard model, we will use the notation $A_\mu^a \equiv W_\mu^i$ for the $SU(2)_L$ gauge fields and $A_\mu^a = G_\mu^a$ for the gluons.

We see that there is the same number of gauge fields as there are generators of the group, i.e., $N^2 - 1$. To simplify the notation, we also define

$$A_\mu^a T_F^a \equiv A_\mu \tag{10.18}$$

which we shall simply call *the* gauge field. The transformation of this gauge field is taken to be

$$A_\mu \to A_\mu' = U A_\mu U^\dagger + \frac{i}{g}(\partial_\mu U)U^\dagger \tag{10.19}$$

With this definition, the covariant derivative $D_\mu \psi$ transforms the same way as ψ itself:

$$D_\mu \psi \to D_\mu \psi' = U D_\mu \psi \tag{10.20}$$

For a set of Dirac spinors Ψ_i transforming in the fundamental representation, a gauge-invariant lagrangian is obtained simply by replacing ∂_μ by the gauge-covariant derivative in the Dirac lagrangian:

$$\mathcal{L} = \overline{\Psi}\left(i\gamma^\mu D_\mu - m\right)\Psi$$

where, again, Ψ is to be understood as a column vector containing the N Dirac spinors transforming in the fundamental representation.

We have a similar expression for Weyl spinors. Consider a set of N left-chiral Weyl spinors χ_i transforming in the fundamental representation. The gauge-invariant kinetic energy then is given by

$$\mathcal{L}_w = \chi^\dagger i \bar{\sigma}^\mu D_\mu \chi$$

As mentioned earlier, for $SU(N)$ with $N > 2$, there are two nonequivalent fundamental representations. One involves the generators T_F^a, and the second involves their complex conjugate T_F^{a*} (we then say that the group admits complex representations). It turns out that when we express the QCD lagrangian in terms of the left-chiral quark and antiquark Weyl spinors χ_q and $\chi_{\bar{q}}$, we find that they transform under the fundamental and antifundamental representations of $SU(3)$ (this will be discussed in more detail in the next section). This is noteworthy because when we will write down the minimal supersymmetric standard model (MSSM) in Chapter 15, it will be entirely in terms of left-chiral spinors, and it will be important to be able to check that the standard model lagrangian will be contained within the MSSM. We will get back to this point in the next section.

Be warned that there is much freedom in the choice of signs in Eqs. (10.19) and (10.17), and in the exponential of (10.16), the only constraint being the transformation property (10.20). Many different permutations of the three signs are found in the literature.

The field strength corresponding to a gauge field is

$$F_{\mu\nu} = \partial_\mu A_\nu - \partial_\nu A_\mu - ig[A_\mu, A_\nu] \tag{10.21}$$

where it is understood that each A is meant to be the matrix defined in Eq. (10.18). The commutator is, of course, zero in the case of an abelian gauge theory, and we then recover the familiar field strength of electromagnetism. Since the field strength is also a matrix in the Lie algebra, it may be expanded over the generators as

$$F_{\mu\nu} \equiv F_{\mu\nu}^a T_{\mathrm{F}}^a$$

Written in terms of components, Eq. (10.21) corresponds to

$$F_{\mu\nu}^a T_{\mathrm{F}}^a = \partial_\mu A_\nu^a T_{\mathrm{F}}^a - \partial_\nu A_\mu^a T_{\mathrm{F}}^a - ig A_\mu^b A_\nu^c \left[T_{\mathrm{F}}^b, T_{\mathrm{F}}^c \right]$$

Using Eq. (10.10), this leads to

$$F_{\mu\nu}^a = \partial_\mu A_\nu^a - \partial_\nu A_\mu^a + g f^{abc} A_\mu^b A_\nu^c$$

Using Eq. (10.19), one can show that the field strength transforms as

$$
\begin{aligned}
F_{\mu\nu} \rightarrow F_{\mu\nu}' &= U F_{\mu\nu} U^\dagger \\
&\approx F_{\mu\nu} + i[\Lambda, F_{\mu\nu}] \\
&= F_{\mu\nu}^a T_{\mathrm{F}}^a + i \Lambda^b F_{\mu\nu}^c \left[T_{\mathrm{F}}^b, T_{\mathrm{F}}^c \right] \\
&= F_{\mu\nu}^a T_{\mathrm{F}}^a - f^{abc} \Lambda^b F_{\mu\nu}^c T_{\mathrm{F}}^a
\end{aligned} \tag{10.22}
$$

from which we can read off the transformation of $F_{\mu\nu}^a$:

$$\left(F_{\mu\nu}^a \right)' = F_{\mu\nu}^a - f^{abc} \Lambda^b F_{\mu\nu}^c$$

Using the antisymmetry of the structure constants, this can be written as

$$\left(F_{\mu\nu}^a\right)' = F_{\mu\nu}^a + f^{bac}\Lambda^b F_{\mu\nu}^c$$

This step was made because now we can use Eq. (10.14) to write this as

$$\left(F_{\mu\nu}^a\right)' = F_{\mu\nu}^a + i\left(T_{AD}^b\right)^{ac}\Lambda^b F_{\mu\nu}^c$$

which is the infinitesimal limit of

$$F_{\mu\nu} = \exp\left(iT_{AD}^b\Lambda^b\right)F_{\mu\nu} \tag{10.23}$$

This shows that the field strength transforms in the adjoint representation! Note however that in Eq. (10.23), $F_{\mu\nu}$ is a column vector with $N^2 - 1$ entries, in contrast with the $F_{\mu\nu}$ of Eq. (10.22) which is an N by N matrix.

As we see, unlike the abelian case, the field strength of a nonabelian gauge theory is not gauge-invariant, and we cannot simply use $-F^{\mu\nu}F_{\mu\nu}/4$ as a kinetic term (of course we can't, it's a matrix!). On the other hand, the trace of this quantity *is* gauge-invariant, and the kinetic term is taken to be

$$\mathcal{L}_{kin} = -\frac{1}{2}\text{Tr}\left(F_{\mu\nu}F^{\mu\nu}\right)$$

The factor of $1/2$ is chosen so that the trace reduces to a sum of kinetic terms with conventional normalization, i.e.,

$$-\frac{1}{2}\text{Tr}\left(F_{\mu\nu}F^{\mu\nu}\right) = -\frac{1}{2}F_{\mu\nu}^a F_{\mu\nu}^b \text{Tr}\left(T_F^a T_F^b\right)$$

$$= -\frac{1}{4}F_{\mu\nu}^a F_{\mu\nu}^a \tag{10.24}$$

where Eq. (10.12) was used.

Note that the field strength $F_{\mu\nu}$ contains the coupling constant g, as we can see from Eq. (10.21). It is sometimes convenient to have all dependence on the coupling constant made explicit in the lagrangian. This can be done by rescaling the gauge field in the following way:

$$A_\mu \rightarrow \frac{A_\mu}{g}$$

in which case the field strength is rescaled to

$$F_{\mu\nu} \rightarrow \frac{F_{\mu\nu}}{g} \tag{10.25}$$

where the new field strength *does not contain the coupling constant* anymore. The kinetic energy of the gauge bosons then is

$$\mathcal{L}_{\text{kin}} = -\frac{1}{2g^2} \text{Tr}\left(F_{\mu\nu} F^{\mu\nu}\right) = -\frac{1}{4g^2} F^a_{\mu\nu} F^a_{\mu\nu} \tag{10.26}$$

EXERCISE 10.2

a. Verify that the generators of Eq. (10.13) obey the normalization condition [Eq. (10.12)].

b. Construct explicitly the matrices representing the generators, of $SU(2)$ in the adjoint representation, and verify that they obey Eqs. (10.10) and (10.15).

10.4 The QCD Lagrangian in Terms of Weyl Spinors

Since most of us have been taught the standard model in terms of Dirac spinors, it may be useful to pause and see what the lagrangian of a nonabelian gauge theory looks like in terms of Weyl spinors. We will use the example of QCD for a single quark species (i.e., a single flavor). This is particularly interesting because quarks and antiquarks belong to two nonequivalent representations of $SU(3)_c$, so it will be instructive to see how this works out in terms of Weyl spinors.

Let us then write the QCD lagrangian in terms of left-chiral Weyl spinors. The derivation is almost identical to the one we followed in Section 4.6 to write the QED lagrangian in terms of left-chiral spinors.

We start with the gauge-invariant lagrangian for a quark, i.e.,

$$\mathcal{L}_{\text{QCD}} = \overline{\Psi}_D \left(i \gamma^\mu D_\mu - m \right) \Psi_D$$

where Ψ_D actually represents a column vector of the three-color quark states in the fundamental representation **3**. To be more explicit, this is

$$\mathcal{L}_{\text{QCD}} = \overline{\Psi}_D \left(i \gamma^\mu \partial_\mu - g_s A_\mu - m \right) \Psi_D \tag{10.27}$$

where g_s is the strong coupling constant, and

$$A_\mu \equiv \frac{\lambda^a}{2} A_\mu^a$$

As noted in the preceding section, the λ^a are the eight Gell-Mann matrices. Of course, in the standard model, there are no explicit mass terms, all masses arising from spontaneous symmetry breaking.

What we now want to do is to write Eq. (10.27) in terms of the left-chiral quark and antiquark spinors χ_q and $\chi_{\bar{q}}$. We have already shown in Section 4.5 [see Eq. (4.14)] that

$$\overline{\Psi}_D \big(i\gamma^\mu \partial_\mu - m\big)\Psi_D = i\chi_{\bar{q}}^\dagger \bar{\sigma}^\mu \partial_\mu \chi_{\bar{q}} + i\chi_q^\dagger \bar{\sigma}^\mu \partial_\mu \chi_q - m(\chi_{\bar{q}} \cdot \chi_q + \bar{\chi}_q \cdot \bar{\chi}_{\bar{q}})$$

The two extra terms coupling the Weyl spinors to the gauge field can be obtained from Eq. (4.12) by replacing ∂_μ with $ig_s A_\mu$:

$$\text{Terms containing } A_\mu = -g_s \chi_{\bar{q}}^T \sigma^2 \sigma^\mu A_\mu \sigma^2 \chi_{\bar{q}}^{\dagger T} - g_s \chi_q^\dagger \bar{\sigma}^\mu A_\mu \chi_q$$

$$= g_s \chi_{\bar{q}}^\dagger A_\mu^T \bar{\sigma}^\mu \chi_{\bar{q}} - g_s \chi_q^\dagger \bar{\sigma}^\mu A_\mu \chi_q$$

where we have used Eqs. (A.5) and (A.36) on the first term. Note that the spin matrices $\bar{\sigma}^\mu$ and σ^μ commute with the gauge field matrix A^μ because the former acts on the spin degrees of freedom, whereas the latter acts in color space.

We see that there is a new twist compared with the situation we faced with the QED lagrangian: The transpose affects the gauge field A^μ because it is a matrix.

Using the fact that the Gell-Mann matrices are hermitian, we may write

$$A_\mu^T = A_\mu^a \frac{(\lambda^a)^T}{2}$$

$$= A_\mu^a \frac{(\lambda^a)^*}{2}$$

so the QCD lagrangian may be written

$$\mathcal{L}_{\text{QCD}} = i\chi_{\bar{q}}^\dagger \bar{\sigma}^\mu \left[\partial_\mu - \frac{i}{2} g_s A_\mu^a (\lambda^a)^* \right] \chi_{\bar{q}} + i\chi_q^\dagger \bar{\sigma}^\mu \left(\partial_\mu + ig_s A_\mu^a \frac{\lambda^a}{2} \right) \chi_q$$

$$-m(\chi_{\bar{q}} \cdot \chi_q + \bar{\chi}_q \cdot \bar{\chi}_{\bar{q}}) \tag{10.28}$$

The fact that a complex conjugation must be taken on the term containing the gauge field in the kinetic energy of the antiquark tells us that this field transforms in the representation $\bar{\mathbf{3}}$. As usual, we may introduce gauge-covariant derivatives D_μ acting on the quark fields and write Eq. (10.28) as

$$\mathcal{L}_{\text{QCD}} = i\chi_{\bar{q}}^\dagger \bar{\sigma}^\mu D_\mu \chi_{\bar{q}} + i\chi_q^\dagger \bar{\sigma}^\mu D_\mu \chi_q - m(\chi_{\bar{q}} \cdot \chi_q + \bar{\chi}_q \cdot \bar{\chi}_{\bar{q}})$$

with the understanding that when acting on a $\mathbf{3}$ representation, such as χ_e, the covariant derivative is

$$D_\mu = \partial_\mu + \frac{i}{2}g_s A_\mu^a \lambda^a$$

whereas, when acting on the representation $\bar{\mathbf{3}}$, it is

$$D_\mu = \partial_\mu - \frac{i}{2}g_s A_\mu^a (\lambda^a)^*$$

We see that writing the lagrangian in terms of the quark and antiquark Weyl spinors makes it explicit that they transform under the nonequivalent representations $\mathbf{3}$ and $\bar{\mathbf{3}}$.

10.5 Free Supersymmetric Nonabelian Gauge Theories

Now we are ready to describe free supersymmetric nonabelian gauge theories (free in the sense that we don't consider interaction terms beyond those dictated by gauge invariance). We know that in an $SU(N)$ gauge theory, there are $N^2 - 1$ gauge fields because that's the number of generators. To make the theory supersymmetric, we will therefore need the same number of left-chiral gauginos λ^a and auxiliary fields D^a. Our starting point for our lagrangian thus is

$$\mathcal{L} = -\frac{1}{4}F_{\mu\nu}^a F^{\mu\nu a} + i\lambda^{a\dagger}\bar{\sigma}^\mu D_\mu \lambda^a + \frac{1}{2}D^a D^a \qquad (10.29)$$

Be careful about not confusing the D_μ of the covariant derivative with the auxiliary fields D^a! And keep in mind that the index a on λ labels the $N^2 - 1$ gauginos forming the adjoint representation, not their spinor components.

Note that we cannot write a Fayet-Illiopolos term, i.e., a term linear in the auxiliary field, in the case of a nonabelian gauge theory because the D^a are not gauge-invariant.

As a first guess for the supersymmetric transformations that leave this lagrangian invariant (up to total derivatives, as always), it makes sense to try the transformations that we worked out in the abelian case, Eq. (10.9), applied separately to each of the $N^2 - 1$ fields $F_{\mu\nu}^a$, λ^a and D^a.

Unfortunately, things are not so simple. We have one more constraint now: The SUSY transformations must not conflict with gauge transformations. To see if there is such a conflict, we first need to know in what representation the gauginos and auxiliary field transform, if at all. To answer this, consider a generic field X (which could be bosonic or fermionic) whose transformation under SUSY is, symbolically,

$$X^a \rightarrow X^a + \delta_s X^a \tag{10.30}$$

If we want this transformation (together with the SUSY transformations of all the other fields in the theory) to be a symmetry of the lagrangian, it better still hold if we first perform a gauge transformation of the field X before applying it! And for this to be true, the two sides of Eq. (10.30) obviously must transform the same way under a gauge transformation.

Let us see what this implies. As our first guess for the transformation of each of the gauginos λ^a, let's take the exact same expression we had in the abelian case, namely,

$$\delta\lambda^a = \frac{i}{2}\sigma^\mu\bar{\sigma}^\nu\zeta F_{\mu\nu}^a + \zeta D^a \tag{10.31}$$

For this to be valid in the nonabelian case, we are forced to conclude that *the gauginos and the auxiliary fields must transform like the field strengths under gauge transformations, i.e., in the adjoint representation*. With that choice, the SUSY transformation (10.31) does not conflict with the gauge transformation, in the sense that supersymmetry and gauge transformations commute. Of course, it does not mean that the guess (10.31) is the correct SUSY transformation for the lagrangian (10.29); that's another issue. For now, we are simply making sure that what we'll try for the SUSY transformations does not conflict with the gauge symmetry.

Now let's see if the other transformations we had obtained in the abelian case cause any problem. Consider next the transformation of the auxiliary field. The transformation in the abelian case suggests that we take

$$\delta D^a = -i\zeta^\dagger\bar{\sigma}^\mu\partial_\mu\lambda^a + i(\partial_\mu\lambda^a)^\dagger\bar{\sigma}^\mu\zeta$$

Is that acceptable? Well, the auxiliary fields transform in the adjoint representation, as we just saw. Does the right-hand side transform the same way? The answer is clearly no, because of the derivatives acting on the gauginos, which will pick up a derivative of the gauge parameter on transformation. But we know how to fix that from elementary particle physics; we simply have to replace the derivative with a covariant derivative! However, since the gauginos transform in the adjoint representation, *we need to use this representation to define the covariant derivative acting on them.*

At this point, expressions become quite awkward if we show all the $SU(N)$ indices explicitly, so let's drop the indices on the gauginos λ. All we need to do is to think of each λ as being a column vector with $N^2 - 1$ entries. With this simplified notation, the covariant derivative acting on a gaugino is simply

$$D_\mu \lambda = \left(\partial_\mu + i g_s T^a_{\mathrm{AD}} A^a_\mu\right)\lambda$$

In case you are interested, with explicit $SU(N)$ indices, this equation becomes

$$D_\mu \lambda^b = \partial_\mu \lambda^b + i g_s \left(T^a_{\mathrm{AD}}\right)^{bc} A^a_\mu \lambda^c$$

Therefore, our guess for the transformation of the auxiliary fields is [hiding once more the $SU(N)$ indices]:

$$\delta D = -i \zeta^\dagger \bar{\sigma}^\mu D_\mu \lambda + i (D_\mu \lambda)^\dagger \bar{\sigma}^\mu \zeta$$

What about the SUSY transformation of the gauge fields A^a_μ? The transformation in the abelian case suggests that we take

$$\delta A^a_\mu = \zeta^\dagger \bar{\sigma}^\mu \lambda^a + \lambda^{a\dagger} \bar{\sigma}^\mu \zeta \tag{10.32}$$

However, at first sight, this seems completely wrong. The gauginos transform in the adjoint representation, whereas the gauge fields don't. But, there is no need to despair. Recall that what really matters is that the transformed field $A_\mu + \delta A_\mu$ has the same gauge transformation as A_μ itself. Let's have a closer look.

Consider contracting both sides of Eq. (10.32) with T^a_F. The matrix $A_\mu = A^a_\mu T^a_F$ transforms as in Eq. (10.19):

$$A_\mu \to A'_\mu = U A_\mu U^\dagger + \frac{i}{g}\left(\partial_\mu U\right)U^\dagger \tag{10.33}$$

whereas $\lambda = \lambda^a T^a_F$ transforms as

$$\lambda \to \lambda' = U \lambda U^\dagger$$

Since δA_μ contains only the gauginos, we also have

$$\delta A_\mu \rightarrow U \delta A_\mu U^\dagger \tag{10.34}$$

Using Eqs. (10.33) and (10.34), we find that the quantity $A_\mu + \delta A_\mu$ transforms as

$$A_\mu + \delta A_\mu \rightarrow U A_\mu U^\dagger + \frac{i}{g} \left(\partial_\mu U \right) U^\dagger + U \delta A_\mu U^\dagger$$

$$= U \left(A_\mu + \delta A_\mu \right) U^\dagger + \frac{i}{g} \left(\partial_\mu U \right) U^\dagger$$

which is precisely the gauge transformation of A_μ itself! This tells us that the SUSY transformation (10.32) does not conflict with the gauge symmetry.

To summarize, our first guesses for the SUSY transformations of the fields are

$$\delta \lambda = \frac{i}{2} \sigma^\mu \bar{\sigma}^\nu \zeta F_{\mu\nu} + \zeta D$$

$$\delta D = -i \zeta^\dagger \bar{\sigma}^\mu D_\mu \lambda + i (D_\mu \lambda)^\dagger \bar{\sigma}^\mu \zeta$$

$$\delta A_\mu = \zeta^\dagger \bar{\sigma}^\mu \lambda + \lambda^\dagger \bar{\sigma}^\mu \zeta$$

Remains the task of checking whether these transformations indeed leave the lagrangian (10.29) invariant. It turns out that they do! However, we won't prove this here because the demonstration is almost identical to the one we carried out in detail for the abelian case. The interested reader can find the details in Ref. 31. Instead, it is more interesting to turn our attention to supersymmetric interactions between vector and chiral multiplets.

10.6 Combining an Abelian Vector Multiplet With a Chiral Multiplet

Now we consider building supersymmetric theories that couple chiral and vector multiplets. We start with a free abelian gauge multiplet and a single free chiral multiplet. By *free chiral multiplet*, we mean that we will not include any self-interaction among the fields of the chiral multiplet or any mass terms for χ and ϕ. We will come back to the question of interactions of the chiral multiplet at the end of this section.

We want to build a supersymmetric theory in which the fields of the chiral multiplet interact with the fields of the vector multiplet. The simplest way to couple the two multiplets is to take the chiral multiplet to have a $U(1)$ charge. In other words, under the $U(1)$ gauge transformation, we take the fields of the chiral multiplet to transform as

$$X \Rightarrow \exp[iq\Lambda(x)]X \qquad (10.35)$$

where X stands for any of the three fields ϕ, χ, or F, and q is their common charge. As before, we take the spinor and auxiliary fields of the abelian gauge multiplet λ and D to be gauge-invariant, whereas the gauge field A_μ transforms as usual.

In order for the lagrangian of the chiral multiplet to be gauge-invariant, we must replace all the derivatives acting on the fields of that multiplet by gauge-covariant derivatives; i.e., we must make the replacement

$$\partial_\mu X \to D_\mu X = \left(\partial_\mu + iqA_\mu\right)X$$

where X stands again for any of the fields of the chiral multiplet. We therefore have

$$\mathcal{L} = (D_\mu\phi)^\dagger(D^\mu\phi) + i\chi^\dagger\bar{\sigma}^\mu D_\mu\chi + F^\dagger F - \frac{1}{4}F_{\mu\nu}F^{\mu\nu} + i\lambda^\dagger\bar{\sigma}^\mu\partial_\mu\lambda + \frac{1}{2}D^2 + \xi D$$
$$(10.36)$$

This is our starting point. Note that the gauge field A_μ is already coupled to the field of the chiral multiplet through the gauge-covariant derivative. We see that we get three new interaction terms arising from gauging the fields:

$$(\mathcal{L}_{\text{int}})_1 = iq\phi A^\mu\partial_\mu\phi^\dagger - iq\phi^\dagger A^\mu\partial_\mu\phi - q\chi^\dagger A_\mu\bar{\sigma}^\mu\chi \qquad (10.37)$$

which provide interactions between the gauge field A_μ and the fields of the chiral multiplet. We have included a label 1 because there will be a second interaction lagrangian.

To be as general as possible, we now look for gauge-invariant interactions combining the fields of the two multiplets that do not arise from gauging the derivatives. Since the fields of the chiral multiplets transform as Eq. (10.35), gauge-invariant combinations can be formed if we multiply any two of them as long as we apply a hermitian conjugate to one of them. Since our goal is to couple these fields with the fields $F_{\mu\nu}$, λ, and D, which have dimensions 2, 3/2, and 2, we need only to retain combinations of ϕ, χ, and F with dimension 5/2 at most. The only combinations of fields in a chiral multiplet that are gauge-invariant and of dimension less than or equal to 5/2 are

$$\phi^\dagger\phi \qquad \phi^\dagger\chi$$

and their hermitian conjugates. We now combine these terms with the gauge-invariant fields $F_{\mu\nu}$, λ, and D to get interactions. Keeping only terms of dimension 4 or less, we get (we are not worrying about matching indices and Lorentz invariance for now)

$$F_{\mu\nu}\phi^\dagger\phi \qquad D\phi^\dagger\phi \qquad \phi^\dagger\chi\lambda + \text{h.c.} \qquad \phi^\dagger\phi\lambda + \text{h.c.} \qquad (10.38)$$

The first two terms are of dimension 4, whereas the last two have dimension 7/2.

Now let's worry about Lorentz invariance. The first term contains two Lorentz indices that we must contract with something. We cannot use derivatives because the terms are already of dimension 4. The only dimensionless quantities with Lorentz indices that we have at our disposal are σ^μ and $\bar{\sigma}^\mu$, but since $F_{\mu\nu}\phi^\dagger\phi$ does not contain any spinor between which to sandwich the Pauli matrices, we are out of luck, and we must drop that term.

The last term in Eq. (10.38) has a similar problem. In order to get a Lorentz invariant, we must introduce a second spinor to dot with λ, which produces an expression of dimension 5, so we must drop that term as well.

In the third term of Eq. (10.38), the two spinors can be combined in a Lorentz-invariant way using the spinor dot products introduced in the Chapter 1, so our final form is

$$\phi^\dagger\chi \cdot \lambda + \phi\bar{\chi} \cdot \bar{\lambda} \qquad (10.39)$$

Recall that in spinor dot products, the order does not matter, which is why we did not list, say, $\phi^\dagger\lambda \cdot \chi$ as a different possibility.

The second term of Eq. (10.38) is already Lorentz-invariant, so it is fine as it stands. Thus we finally get the following interaction lagrangian

$$(\mathcal{L}_{\text{int}})_2 = c_1\left(\phi^\dagger\chi \cdot \lambda + \phi\bar{\chi} \cdot \bar{\lambda}\right) + c_2\phi^\dagger\phi D \qquad (10.40)$$

where c_1 and c_2 are are arbitrary constants to be determined. We have assumed that c_1 is real. Of course, there is no a priori reason why this should be the case, but it will simplify our calculation, and it will turn out to be general enough for building a supersymmetric lagrangian. For the same reason, we will also take c_2 to be real in the rest of our calculations.

Our total interaction lagrangian therefore is the sum of Eqs. (10.37) and (10.40).

EXERCISE 10.3

We have taken the fields of the chiral multiplet to be charged. We also could have considered them to be neutral. In that case, what would be the most general (renormalizable, gauge-invariant, and real) interaction lagrangian mixing fields of both multiplets?

The goal now is to see if it is possible to find values of c_1 and c_2 for which the theory will be supersymmetric. Of course, we know that the free lagrangians for the vector and chiral multiplets are supersymmetric, so we need only to worry about the interaction terms.

A priori, there is no reason to believe that the vector and chiral multiplets should both transform with a common SUSY parameter ζ. In fact, we will find that we *cannot* make the theory supersymmetric if we use a common parameter for both supermultiplets. So let's use ζ for the parameter of the transformation of the chiral multiplet and $a\zeta$ for the vector multiplet, where a is an arbitrary constant that we will take to be real (with hindsight, we will see that this was indeed sufficiently general). Then the transformation we take for the fields of the vector multiplet are simply

$$\delta A^\mu = a(\zeta^\dagger \bar\sigma^\mu \lambda + \lambda^\dagger \bar\sigma^\mu \zeta)$$

$$\delta\lambda = \frac{ia}{2} F_{\mu\nu} \sigma^\mu \bar\sigma^\nu \zeta + aD\zeta$$

$$\delta\lambda^\dagger = -\frac{ia}{2} F_{\mu\nu} \zeta^\dagger \bar\sigma^\nu \sigma^\mu + a\zeta^\dagger D$$

$$\delta D = -ia\zeta^\dagger \bar\sigma^\mu \partial_\mu \lambda + ia(\partial_\mu \lambda^\dagger)\bar\sigma^\mu \zeta$$

whereas the transformations of the field of the chiral multiplet have to be modified by replacing the partial derivatives by covariant derivatives, i.e.,

$$\delta\phi = \zeta \cdot \chi$$

$$\delta\phi^\dagger = \bar\chi \cdot \bar\zeta$$

$$\delta\chi = -i(D_\mu\phi)\sigma^\mu i\sigma^2 \zeta^* + F\zeta$$

$$\delta\chi^\dagger = -i(D_\mu\phi)^\dagger \zeta^T i\sigma^2 \sigma^\mu + F^\dagger \zeta^\dagger$$

$$\delta F = -i\zeta^\dagger \bar\sigma^\mu D_\mu\chi$$

$$\delta F^\dagger = i(D_\mu\chi)^\dagger \bar\sigma^\mu \zeta$$

These are our "guesses" for the SUSY transformations. It turns out that the transformation of the auxiliary field will have to be modified to make our theory supersymmetric.

Note that because of the covariant derivative, F and χ (and their hermitian conjugates) pick up an extra term compared with what we had previously, when the chiral multiplet had no charge. To be explicit, these are

$$\delta F = \text{previous terms} + q\zeta^\dagger \bar{\sigma}^\mu \chi A_\mu$$

$$\delta F^\dagger = \text{previous terms} + q\chi^\dagger \bar{\sigma}^\mu \zeta A_\mu$$

$$\delta\chi = \text{previous terms} + q\phi A_\mu \sigma^\mu i\sigma^2 \zeta^*$$

$$\delta\chi^\dagger = \text{previous terms} - q\phi^\dagger A_\mu \zeta^T i\sigma^2 \sigma^\mu$$

These will generate new terms in the variation of the terms in the *free* part of the lagrangian for the chiral multiplet contained in Eq. (10.36) that we did not have before. Let us use $(\delta F)_A$ to represent the new term appearing in the variation of F and similar expressions for the other fields, with a subscript A to remind us that this new term contains the gauge field A_μ. Then the new terms generated in the variation of the free part of the chiral multiplet lagrangian are simply

$$\delta(\mathcal{L}_{\text{free}}) = i(\delta\chi^\dagger)_A \bar{\sigma}^\mu \partial_\mu \chi + i\chi^\dagger \bar{\sigma}^\mu \partial_\mu (\delta\chi)_A + (\delta F^\dagger)_A F + F^\dagger (F)_A \quad (10.41)$$

Now consider the variation of the terms in the first interaction lagrangian, Eq. (10.37), under SUSY:

$$\delta(\mathcal{L}_{\text{int}})_1 = iq(\delta\phi)A_\mu \partial^\mu \phi^\dagger + iq\phi(\delta A_\mu)\partial^\mu \phi^\dagger + iq\phi A_\mu \partial^\mu(\delta\phi^\dagger)$$

$$- iq(\delta\phi^\dagger)A^\mu \partial_\mu \phi - iq\phi^\dagger(\delta A^\mu)\partial_\mu \phi - iq\phi^\dagger A^\mu \partial_\mu(\delta\phi)$$

$$- q(\delta\chi^\dagger)A_\mu \bar{\sigma}^\mu \chi - q(\delta A_\mu)\chi^\dagger \bar{\sigma}^\mu \chi - q\chi^\dagger A_\mu \bar{\sigma}^\mu(\delta\chi) \quad (10.42)$$

The variation of Eq. (10.40) is

$$\delta(\mathcal{L}_{\text{int}})_2 = c_1(\delta\phi^\dagger)\chi \cdot \lambda + c_1\phi^\dagger(\delta\chi) \cdot \lambda + c_1\phi^\dagger \chi \cdot (\delta\lambda)$$

$$+ c_1(\delta\phi)\bar{\lambda} \cdot \bar{\chi} + c_1\phi(\delta\bar{\lambda}) \cdot \bar{\chi} + c_1\phi\bar{\lambda} \cdot \delta\bar{\chi}$$

$$+ c_2(\delta\phi^\dagger)\phi D + c_2\phi^\dagger(\delta\phi)D + c_2\phi^\dagger \phi \delta D \quad (10.43)$$

The goal is to see if there is a choice of values for c_1, c_2, and a such that the sum of Eqs. (10.41), (10.42), and (10.43) cancel, modulo total derivatives. This is obviously a huge undertaking! We will not carry out the full calculation here but simply consider just enough terms to allow us to determine the values of these three constants. We also will find that one of the SUSY transformations must be modified.

Terms Linear in *D*

Let's focus on the terms linear in D that must cancel among themselves (modulo total derivatives, as usual). There are four such terms: the two terms that have explicit factors of D and the two terms generated by the variation of λ and its hermitian conjugate. These are

$$c_1 a(\phi^\dagger D \chi \cdot \zeta + \phi D \bar{\zeta} \cdot \bar{\chi}) + c_2(\phi D \bar{\chi} \cdot \bar{\zeta} + \phi^\dagger D \zeta \cdot \chi)$$

Of course, the terms in ϕ and the terms in $\phi^\dagger$ must cancel out separately. We see that the condition in both cases is

$$\boxed{c_1 a + c_2 = 0} \tag{10.44}$$

Terms with Four Spinors

Next, consider terms that are generated with four spinors. One source is from the next to last term in Eq. (10.42). The other two sources are from the first and fourth terms of Eq. (10.43). We get

$$-qa\chi^\dagger \bar{\sigma}^\mu \chi (\zeta^\dagger \bar{\sigma}_\mu \lambda + \lambda^\dagger \bar{\sigma}_\mu \zeta) + c_1 \bar{\chi} \cdot \bar{\zeta} \chi \cdot \lambda + c_1 \zeta \cdot \chi \bar{\lambda} \cdot \bar{\chi} \tag{10.45}$$

Using the identity (A.57), we have

$$\left(\chi^\dagger \bar{\sigma}^\mu \chi\right)(\zeta^\dagger \bar{\sigma}_\mu \beta) = 2(\bar{\chi} \cdot \bar{\zeta})(\chi \cdot \beta)$$

so Eq. (10.45) becomes

$$-2qa(\bar{\chi} \cdot \bar{\zeta})(\chi \cdot \lambda) - 2qa(\bar{\chi} \cdot \bar{\lambda})(\chi \cdot \zeta) + c_1(\bar{\chi} \cdot \bar{\zeta})(\chi \cdot \lambda) + c_1(\zeta \cdot \chi)(\bar{\lambda} \cdot \bar{\chi})$$
$$= (c_1 - 2qa)\big((\bar{\chi} \cdot \bar{\zeta})(\chi \cdot \lambda) + (\zeta \cdot \chi)(\bar{\lambda} \cdot \bar{\chi})\big)$$

where the parentheses around the spinor dot products have been included just to make the expression easier to read. We have used the fact that the order in spinor dot products does not matter; e.g., $\bar{\lambda} \cdot \bar{\chi} = \bar{\chi} \cdot \bar{\lambda}$.

We therefore get the condition

$$\boxed{c_1 - 2qa = 0} \tag{10.46}$$

Putting together Eqs. (10.44) and (10.46), we see that we can express both c_1 and c_2 in terms of a, i.e.,

$$\boxed{c_1 = 2qa \qquad c_2 = -2qa^2}$$

(10.47)

Terms with ϕ and a Derivative of ϕ

We have contributions of this form in the second and fifth terms of Eq. (10.42). We also generate terms like this from the variation of χ and $\chi^\dagger$ in the second and sixth terms of Eq. (10.43). Explicitly, these are [using the explicit form for the spinor dot products, Eq. (A.37)]

$$iqa(\phi\partial_\mu\phi^\dagger - \phi^\dagger\partial_\mu\phi)(\zeta^\dagger\bar{\sigma}^\mu\lambda + \lambda^\dagger\bar{\sigma}^\mu\zeta) + c_1\phi^\dagger(-i\partial_\mu\phi)(\sigma^\mu i\sigma^2\zeta^*)^T(-i\sigma^2)\lambda$$
$$-c_1\phi\lambda^\dagger(i\sigma^2)(i\partial_\mu\phi^\dagger\zeta^T i\sigma^2\sigma^\mu)^T$$

(10.48)

At first sight, it seems as if there are no other terms of this form. But we have to remember that terms related by integration by parts must be considered as being equivalent. Thus we also must include terms of the form $\phi\phi^\dagger$ times the derivative of other fields. There is indeed such a term: the last term of Eq. (10.43):

$$c_2\phi^\dagger\phi\delta D = -2iqa^3\phi^\dagger\phi[(\partial_\mu\lambda^\dagger)\bar{\sigma}^\mu\zeta - \zeta^\dagger\bar{\sigma}^\mu\partial_\mu\lambda]$$
$$= -2iqa^3\phi^\dagger\phi\partial_\mu(\lambda^\dagger\bar{\sigma}^\mu\zeta - \zeta^\dagger\bar{\sigma}^\mu\lambda)$$

(10.49)

where we have used $c_2 = -2qa^2$.

We must find if it's possible to get the terms in Eqs. (10.48) and (10.49) to cancel, modulo total derivatives. Adding the two terms and taking the transpose of the pieces in the last two terms of Eq. (10.48), we get

$$iqa(\phi\partial_\mu\phi^\dagger - \phi^\dagger\partial_\mu\phi)(\zeta^\dagger\bar{\sigma}^\mu\lambda + \lambda^\dagger\bar{\sigma}^\mu\zeta) - ic_1\phi^\dagger(\partial_\mu\phi)\zeta^\dagger i\sigma^2\sigma^{\mu T} i\sigma^2\lambda$$
$$+ic_1\phi(\partial_\mu\phi^\dagger)\lambda^\dagger i\sigma^2\sigma^{\mu T} i\sigma^2\zeta - 2iqa^3\phi^\dagger\phi\partial_\mu(\lambda^\dagger\bar{\sigma}^\mu\zeta - \zeta^\dagger\bar{\sigma}^\mu\lambda)$$

Using the identity (A.5) to simplify the two terms with σ^2 matrices, and setting $c_1 = 2qa$, we get a much simpler expression:

$$iqa(\phi\partial_\mu\phi^\dagger + \phi^\dagger\partial_\mu\phi)(\zeta^\dagger\bar{\sigma}^\mu\lambda - \lambda^\dagger\bar{\sigma}^\mu\zeta) - 2iqa^3\phi^\dagger\phi\partial_\mu(\lambda^\dagger\bar{\sigma}^\mu\zeta - \zeta^\dagger\bar{\sigma}^\mu\lambda)$$

(10.50)

There is no value of a for which this is zero (other than the trivial and uninteresting $a = 0$). However, it is not necessary for these terms to cancel out identically, all we need is for them to be equal to zero modulo total derivatives! Thus, if there is a value of a for which the whole thing is a total derivative, we have achieved our goal. Note that the first parentheses may be written as

$$(\phi \partial_\mu \phi^\dagger + \phi^\dagger \partial_\mu \phi) = \partial_\mu (\phi^\dagger \phi)$$

and then it becomes obvious that Eq. (10.50) is a total derivative at the condition that $iqa = 2iqa^3$. This finally gives us

$$\boxed{1 - 2a^2 = 0 \Rightarrow a = -\frac{1}{\sqrt{2}}}$$

where the choice of sign is purely conventional. Using this in Eq. (10.47), we finally get

$$\boxed{c_1 = -\sqrt{2}q \qquad c_2 = -q}$$

We have now fixed completely the only three parameters available to us. But there are still many many more terms to cancel out! It turns out that, quite remarkably, almost all terms do cancel out with the values of a, c_1, and c_2 we have determined. We won't carry out the entire calculation here, for lack of space (and of energy), but we will point out the single set of terms that do *not* cancel out and show how the problem is taken care of.

Terms in F

The problem appears in the terms linear in F or $F^\dagger$. Some of these terms are generated by the variation of χ and $\chi^\dagger$ in the seventh and last terms of Eq. (10.42) and in the second and sixth terms of Eq. (10.43):

$$-q F^\dagger \zeta^\dagger A_\mu \bar{\sigma}^\mu \chi - q F \chi^\dagger A_\mu \bar{\sigma}^\mu \zeta - \sqrt{2}q \phi^\dagger F \zeta \cdot \lambda - \sqrt{2}q F^\dagger \phi \bar{\lambda} \cdot \bar{\zeta} \qquad (10.51)$$

In addition, we have the two terms produced by the variation of $F^\dagger F$ in Eq. (10.41):

$$(\delta F^\dagger)_A F + F^\dagger (\delta F)_A = q A_\mu F^\dagger \zeta^\dagger \bar\sigma^\mu \chi + q A_\mu F \chi^\dagger \bar\sigma^\mu \zeta$$

We see that these terms cancel out the first two terms of Eq. (10.51). However, we are left with the last two terms of Eq. (10.51), which cannot be canceled with anything. The only way out is to introduce an extra term in the variation of F. We must impose

$$\delta F = -i\zeta^\dagger \bar\sigma^\mu D_\mu \chi + \sqrt{2} q \phi \bar\lambda \cdot \bar\zeta$$

where the first term is what we had previously. Of course, we also have

$$\delta F^\dagger = i(D_\mu \chi)^\dagger \bar\sigma^\mu \zeta + \sqrt{2} q \phi^\dagger \lambda \cdot \zeta$$

With this modification and with our choices of c_1, c_2, and a, *all* the terms in Eqs. (10.41), (10.42), and (10.43) cancel out. Amazing!

Final Result for the Coupling of a Chiral with an Abelian Gauge Multiplet

The full lagrangian that couples a chiral multiplet with an abelian multiplet is therefore

$$\mathcal{L} = (D_\mu \phi)^\dagger (D^\mu \phi) + i\chi^\dagger \bar\sigma^\mu D_\mu \chi + F^\dagger F - \frac{1}{4} F_{\mu\nu} F^{\mu\nu} + i\lambda^\dagger \bar\sigma^\mu \partial_\mu \lambda$$

$$+ \frac{1}{2} D^2 + \xi D - \sqrt{2} q (\phi^\dagger \chi \cdot \lambda + \text{h.c.}) - q\phi^\dagger \phi D$$

Note the interesting fact that the coupling constants in the interactions terms are all proportional to the electric charge, a relation imposed by SUSY. In non-SUSY theories, there would be no way to generate this type of constraint other than putting it in by hand.

The fields transform as

$$\delta A^\mu = -\frac{1}{\sqrt{2}}(\zeta^\dagger \bar{\sigma}^\mu \lambda + \lambda^\dagger \bar{\sigma}^\mu \zeta)$$

$$\delta \lambda = -\frac{i}{2\sqrt{2}} F_{\mu\nu} \sigma^\mu \bar{\sigma}^\nu \zeta - \frac{1}{\sqrt{2}} D\zeta$$

$$\delta D = \frac{i}{\sqrt{2}} \zeta^\dagger \bar{\sigma}^\mu \partial_\mu \lambda - \frac{i}{\sqrt{2}} (\partial_\mu \lambda^\dagger) \bar{\sigma}^\mu \zeta$$

$$\delta \phi = \zeta \cdot \chi$$

$$\delta \chi = -i(D_\mu \phi) \sigma^\mu i \sigma^2 \zeta^* + F\zeta$$

$$\delta F = -i\zeta^\dagger \bar{\sigma}^\mu D_\mu \chi + \sqrt{2} q \phi \bar{\lambda} \cdot \bar{\zeta} \tag{10.52}$$

The lagrangian (10.54) does not contain any interaction among the fields of the chiral multiplet. We have seen in Section 8.1 [see Eq. (8.26)] that the interaction lagrangian of a single chiral multiplet, which is written entirely in terms of the superpotential $\mathcal{W}$, is given by

$$\mathcal{L}_W = \frac{\partial \mathcal{W}}{\partial \phi} F - \frac{1}{2} \frac{\partial^2 \mathcal{W}}{\partial \phi^2} \chi \cdot \chi + \text{h.c.}$$

But since all the members of the chiral multiplet have the same charge, there is no superpotential $\mathcal{W}$ for which this expression is gauge-invariant (recall that the superpotential depends only on ϕ, not on its hermitian conjugate). Therefore, we must set $\mathcal{W}$ equal to zero to respect gauge invariance.

If we allow for the presence of several chiral multiplets, as in Section 8.2, then it becomes possible to write down gauge-invariant interactions between them at the condition that there are some combinations of the charges that add up to zero. The interaction lagrangian can be read off from Eq. (8.60):

$$\mathcal{L}_W = \left(\frac{\partial \mathcal{W}}{\partial \phi_i} F_i - \frac{1}{2} \frac{\partial^2 \mathcal{W}}{\partial \phi_j \partial \phi_i} \chi_i \cdot \chi_j + \text{h.c.} \right) \tag{10.53}$$

If we include this term and allow for the presence of several left-chiral multiplets, we obtain the following generalization:

$$
\begin{aligned}
\mathcal{L} =& (D_\mu \phi_i)^\dagger \left(D^\mu \phi_i\right) + i \chi_i^\dagger \bar{\sigma}^\mu D_\mu \chi_i + F_i^\dagger F_i \\
& - \frac{1}{4} F_{\mu\nu} F^{\mu\nu} + i\lambda^\dagger \bar{\sigma}^\mu \partial_\mu \lambda + \frac{1}{2} D^2 + \xi D - \sqrt{2} q \times \left(\phi_i^\dagger \chi_i \cdot \lambda + \text{h.c.}\right) \\
& - q \phi_i^\dagger \phi_i D + \left(\frac{\partial \mathcal{W}}{\partial \phi_i} F_i - \frac{1}{2} \frac{\partial^2 \mathcal{W}}{\partial \phi_j \partial \phi_i} \chi_i \cdot \chi_j + \text{h.c.}\right)
\end{aligned}
$$

$$(10.54)$$

10.7 Eliminating the Auxiliary Fields

Eliminating the auxiliary fields F_i leads to the replacement

$$
F_i^\dagger F_i + \left(\frac{\partial \mathcal{W}}{\partial \phi_i} F_i + \text{h.c.}\right) = -\left|\frac{\partial \mathcal{W}}{\partial \phi_i}\right|^2
$$

Now let's turn our attention to the auxiliary field D. Can we also eliminate it for a general supersymmetric abelian gauge theory? The answer is yes, and it is actually easier than eliminating the auxiliary fields F_i. The solution to the equation of motion $\delta \mathcal{L}/\delta D = 0$ is

$$
D = q_i \phi_i^\dagger \phi_i - \xi
$$

where the ξ comes from the Fayet-Illiopoulos term. Plugging this back into the D dependent terms in the lagrangian, we get

$$
-q_i \phi_i^\dagger \phi_i D + \frac{1}{2} D^2 + \xi D = -\frac{1}{2} \left(\sum_i q_i \phi_i^\dagger \phi_i - \xi\right)^2 \qquad (10.55)
$$

where we wrote the summation symbol in the first term to make it clear that the sum is performed before squaring the expression.

We see that this represents *another contribution to the potential of the scalar fields*. After elimination of the auxiliary fields using their equations of motion, the total lagrangian therefore can be written as

$$\mathcal{L} = \mathcal{L}_{F_i = D = 0} - \left| \frac{\partial \mathcal{W}}{\partial \phi_i} \right|^2 - \frac{1}{2} \left(\sum_i q_i \phi_i^\dagger \phi_i - \xi \right)^2 \qquad (10.56)$$

This means that the scalar field potential is simply

$$\boxed{ V(\phi_i) = \left| \frac{\partial \mathcal{W}}{\partial \phi_i} \right|^2 + \frac{1}{2} \left(\sum_i q_i \phi_i^\dagger \phi_i - \xi \right)^2 } \qquad (10.57)$$

To be precise, these terms not only contain the interactions of the scalar fields, but they also contain their masses as well.

Equation (10.57) will play a crucial role in our discussion of SUSY spontaneous symmetry breaking in Chapter 14.

10.8 Combining a Nonabelian Gauge Multiplet With a Chiral Multiplet

As in the abelian case, we assign the fields of the chiral multiplet (ϕ, χ, and F) to some representation of the gauge group, the usually choice being the fundamental or antifundamental representations. These representations of $SU(N)$ are N-dimensional, so we must take N copies of each of the fields of the chiral multiplet. We then replace the derivatives acting on the these fields with covariant derivatives. For example, if we assemble N scalar fields transforming in the fundamental representation ϕ^a into a column vector ϕ, we have

$$D_\mu \phi = \partial_\mu \phi + i g A_\mu^a T_{\text{F}}^a \phi$$

As we saw in Section 10.5, this is a shorthand for

$$D_\mu \phi^b = \partial_\mu \phi^b + i g A_\mu^a \left(T_{\text{F}}^a \right)^{bc} \phi^c$$

We have, of course, a similar expression for the covariant derivative acting on the chiral spinors fields χ^a.

If we do not consider self-interactions of the chiral multiplet (i.e., if we set the superpotential equal to zero), the simplest lagrangian containing interactions between a chiral and nonabelian gauge multiplets is

$$\mathcal{L} = \left(D_\mu \phi^b\right)^\dagger (D^\mu \phi^b) + i(\chi^b)^\dagger \bar{\sigma}^\mu D_\mu \chi^b + (F^b)^\dagger F^b$$

$$- \frac{1}{4} F^a_{\mu\nu} F^{\mu\nu}_a + i(\lambda^a)^\dagger \bar{\sigma}^\mu (D_\mu \lambda)^a + \frac{1}{2} D^a D^a$$

$$\equiv (D_\mu \phi)^\dagger (D^\mu \phi) + i\chi^\dagger \bar{\sigma}^\mu D_\mu \chi + F^\dagger F - \frac{1}{2} \text{Tr}\left(F_{\mu\nu} F^{\mu\nu}\right) + i\lambda^\dagger \bar{\sigma}^\mu D_\mu \lambda + \frac{1}{2} D^2$$

where the index b runs from 1 to N, whereas the index a runs from 1 to $N^2 - 1$. This is the nonabelian generalization of Eq. (10.36). As mentioned earlier, there is no Fayet-Illiopoulos term in the nonabelian case because such a term is not gauge-invariant.

This lagrangian contains interactions between the gauge fields of the gauge multiplet and the fields χ and ϕ through the covariant derivatives, but we also can add extra gauge-invariant interactions between the fields of the two multiplets, as we did in the abelian case.

The fields of the vector multiplet are all in the adjoint representation, so what we need is to build gauge-invariant expressions out of fields transforming in the fundamental and the adjoint representations. From Section 10.3, we know that the transformation of a field belonging to the adjoint representation, e.g., $F_{\mu\nu}$, is given by [see Eqs. (10.22) and (10.23)]

$$F_{\mu\nu} \rightarrow U F_{\mu\nu} U^\dagger$$

where $F_{\mu\nu} = F^a_{\mu\nu} T^a_{\rm F}$. The other fields of the vector multiplet, fields λ and D, transform the same way. On the other hand, the fields of the chiral multiplet transform in the fundamental representation, i.e.,

$$\phi \rightarrow U\phi$$

and similarly for χ and F. Here, it is understood that ϕ actually stands for a column vector with N entries ϕ^b. We therefore have the following 27 gauge-invariant combinations available:

$$(\phi, \chi, F)^\dagger (F_{\mu\nu}, \lambda, D)(\phi, \chi, F)$$

Nine terms are real, whereas the remaining 18 terms form 9 pairs related by hermitian conjugation.

However, the dimension of most of these terms is too large. Recall that the fields of the gauge muliplet $F^a_{\mu\nu}$, D^a, and λ^a have dimensions 2, 2, and 3/2, respectively, whereas the fields F^b, ϕ^b, and χ^b have dimensions 2, 1, and 3/2. If we don't worry about Lorentz invariance for now, the available gauge-invariant terms of dimension 4 or less are

$$\phi^\dagger F_{\mu\nu}\phi \qquad \phi^\dagger D\phi \qquad \phi^\dagger \lambda\chi + \text{h.c.} \qquad \phi^\dagger \lambda\phi + \text{h.c.} \qquad (10.58)$$

These are the obvious nonabelian generalization of Eq. (10.38), and the exact same arguments that we used after that equation show once more that the only Lorentz invariants of dimension 3 or less are

$$\mathcal{L}_{\text{int}} = c_1 \phi^\dagger \lambda \cdot \chi + c_1 \phi \bar{\lambda} \cdot \bar{\chi} + c_2 \phi^\dagger D\phi$$

where c_2 must be real, and we have taken c_1 to be real as well. Without showing all the matrix indices, it may be a bit clearer to show explicitly the matrix structure of this expression by writing

$$\mathcal{L}_{\text{int}} = c_1 \big(\phi^\dagger T_{\text{F}}^a \chi\big) \cdot \lambda^a + c_1 \big(\phi T_{\text{F}}^a \bar{\chi}\big) \cdot \bar{\lambda}^a + c_2 \big(\phi^\dagger T_{\text{F}}^a \phi\big) D^a$$

This is, of course, the nonabelian generalization of Eq. (10.40). If we do show all the gauge indices, we have

$$\mathcal{L}_{\text{int}} = c_1 \big[(\phi^b)^\dagger \big(T_{\text{F}}^a\big)^{bc} \chi^c\big] \cdot \lambda^a + c_1 \big[(\phi)^b \big(T_{\text{F}}^a\big)^{bc} \bar{\chi}^c\big] \cdot \bar{\lambda}^a + c_2 \big[(\phi^b)^\dagger \big(T_{\text{F}}^a\big)^{bc} \phi^c\big] D^a$$

Now we should repeat the steps of Section 10.6 to find out if there is a choice of c_1 and c_2 for which the theory is supersymmetric. We won't present the calculation here because it is almost identical to what we did in Section 10.6. The final result turns out to be that the theory is SUSY invariant at the same condition that we found for the abelian case, namely,

$$c_1 = -\sqrt{2}g \qquad \text{and} \qquad c_2 = -g$$

The complete lagrangian of a chiral multiplet interacting with a nonabelian gauge multiplet is finally

$$\begin{aligned}
\mathcal{L} =& (D_\mu\phi)^\dagger(D^\mu\phi) + i(\chi)^\dagger\bar{\sigma}^\mu D_\mu\chi + (F)^\dagger F - \frac{1}{2}\text{Tr}\big(F_{\mu\nu}F^{\mu\nu}\big) + i(\lambda)^\dagger\bar{\sigma}^\mu D_\mu\lambda \\
& + \frac{1}{2}D^2 - \sqrt{2}g\phi^\dagger\lambda\cdot\chi - \sqrt{2}g\phi\bar{\lambda}\cdot\bar{\chi} - g\phi^\dagger D\phi
\end{aligned}$$

$$(10.59)$$

As when we combined a chiral and abelian gauge multiplet, the transformation of the nonabelian gauge multiplet must be considered with ζ replaced by $-\zeta/\sqrt{2}$, and we also must add a term to the transformation of the auxiliary field F:

$$\delta F = \text{previous terms} + \sqrt{2}g\phi T_{\mathrm{F}}^a \bar{\zeta} \cdot \bar{\lambda}^a$$

Let's see what happens if we eliminate the auxiliary fields D^a. The equation of motion for D^a gives

$$D^a = g\left(\phi^\dagger T_{\mathrm{F}}^a \phi\right)$$

Plugging this back into the terms containing D, we get

$$\frac{1}{2}D^2 - g\phi^\dagger D\phi = \frac{1}{2}D^a D^a - g\phi^\dagger D^a T_{\mathrm{F}}^a \phi$$

$$= -\frac{g^2}{2}\left(\phi^\dagger T_{\mathrm{F}}^a \phi\right)\left(\phi^\dagger T^a \phi\right)$$

which gives a contribution to the scalar potential equal to

$$V_D(\phi) = \frac{g^2}{2}\left(\phi^\dagger T_{\mathrm{F}}^a \phi\right)\left(\phi^\dagger T_{\mathrm{F}}^a \phi\right) \tag{10.60}$$

If there are several left-chiral multiplets labeled by an index i, we get instead

$$V_D(\phi_i) = \frac{g^2}{2}\left(\phi_i^\dagger T_{\mathrm{F}}^a \phi_i\right)\left(\phi_j^\dagger T_{\mathrm{F}}^a \phi_j\right)$$

where the index i labels the different scalar fields, each of which is a multiplet in the fundamental representation. In other words, each ϕ_i must be thought of as a column vector with N entries. Thus the complete scalar potential in this case is

$$\boxed{V(\phi_i) = \left|\frac{\partial\mathcal{W}}{\partial\phi_i}\right|^2 + \frac{g^2}{2}\left(\phi_i^\dagger T_{\mathrm{F}}^a \phi_i\right)\left(\phi_j^\dagger T_{\mathrm{F}}^a \phi_j\right)} \tag{10.61}$$

10.9 Quiz

1. In what representation does the photino transform?

2. When we combine a vector multiplet with a chiral multiplet, what are the changes that must be made to the SUSY transformations of the component fields?

3. What is the Fayet-Illiopoulos term?

4. What is the gauge-covariant derivative for an antiquark left-chiral Weyl spinor?

5. Why is the auxiliary field of a vector multiplet a real scalar field, whereas the auxiliary field for a chiral multiplet is a complex scalar field?

CHAPTER 11

Superspace Formalism

So far we have constructed supersymmetric theories the "hard way": by writing the most general renormalizable lagrangians that are real and Lorentz-invariant and then tweaking them to see if we could make them invariant under supersymmetry (SUSY). There is an easier way to obtain supersymmetric theories: the so-called superspace approach, pioneered by Salam and Strathdee.[41] We now discuss this approach.

The basic idea is the following: We have seen that in the case of spacetime symmetries, such as invariance under translation or rotation, the charges can be written as differential operators instead of functions of the fields. At first glance, it might seem that this idea can't be applied to SUSY because the supersymmetric transformations of the fields yield other fields at the same spacetime point. However, we have discovered that the anticommutation of two supersymmetric transformations of a field produces the same field evaluated at a shifted spacetime point! Indeed, the anticommuator of two SUSY charges gives the four-momentum operator P^μ, whose effect is to translate fields in spacetime. This is a key aspect of SUSY: It is a symmetry that mixes in a nontrivial way an internal symmetry with spacetime symmetries.

This relation between the supercharges and spacetime symmetries suggests the possibility of setting up the SUSY charges as differential operators acting on an extended spacetime which is referred to as superspace. We will make this idea more explicit in this chapter.

Until now, we have not used the van der Waerden notation described in Chapter 3; all Weyl spinors were expressed in terms of their lower, undotted components. However, calculations in superspace would become amazingly cumbersome if we would insist on sticking to spinors with lower undotted indices only. Besides, most SUSY references make extensive use of the van der Waerden notation, so it's a good idea to get used to manipulating expressions written in that notation. If you haven't read Chapter 3 yet, grab a cup of coffee, put on some relaxing music, and go over it before coming back to the next section.

11.1 The Superspace Coordinates

For this idea to work, a nontrivial step first must be taken. To understand this, let's go back to the simple example of the charges P_μ. The corresponding unitary transformation U is [see Eq. (6.18)]

$$U(a) \equiv \exp(i a^\mu P_\mu)$$

This operator is what we sometimes called U_a in Chapter 6. The notation $U(a)$ is preferable here because we will be considering transformations depending on several parameters, and it becomes awkward to write the unitary transformations with several subscripts.

The quantity a^μ is obviously the infinitesimal four-vector parameterizing the transformation. But there is something special about this four-vector: It is a *displacement* four-vector which appears in the *argument* of the transformed field, as is clear if we write the transformed field explicitly [see Eqs. (6.8) and (6.15)]:

$$\phi(x') = U(a)\,\phi(x)\,U^\dagger(a)$$
$$= \phi(x^\mu + a^\mu) \tag{11.1}$$

In particular, this implies that if we start with the field evaluated at the spacetime origin (in some given frame) $\phi(0)$, we can generate the field at any position $\phi(x)$ by applying a transformation U with parameter $a_\mu = x_\mu$:

$$\phi(x) = U(x)\,\phi(0)\,U^\dagger(x)$$
$$= \exp(i x^\mu P_\mu)\,\phi(0)\,\exp(-i x^\mu P_\mu) \tag{11.2}$$

We would like to be able to view the SUSY charges as producing a translation of the fields too. The parameters that are dotted with the supercharges (ζ and $\bar{\zeta}$) are Grassmann quantities, which we now want to treat as displacement vectors. Therefore, we must extend the spacetime dependence of the fields to include a dependence on four *Grassmann coordinates* in addition to the usual four spacetime coordinates. This extended spacetime is what is referred to as *superspace*. In this context, superspace must not be seen as a physical space but simply as a mathematical trick that simplifies the construction of supersymmetric theories, as we will see.

The extra four Grassmann coordinates are commonly denoted by θ_1, θ_2, and their hermitian conjugates. Since the SUSY parameter ζ is a left-chiral Weyl spinor, it is natural to take θ_1 and θ_2 to form a left-chiral Weyl spinor that we will denote simply by θ:

$$\theta \equiv \begin{pmatrix} \theta_1 \\ \theta_2 \end{pmatrix}$$

We also define

$$\bar{\theta} = \begin{pmatrix} \bar{\theta}^{\dot{1}} \\ \bar{\theta}^{\dot{2}} \end{pmatrix}$$

Recall that, by definition,

$$\bar{\theta}^{\dot{1}} = (\theta^1)^\dagger \quad \text{and} \quad \bar{\theta}^{\dot{2}} = (\theta^2)^\dagger$$

But the Grassmann coordinates are not quantum fields, so we may replace the dagger acting on the components by a simple complex conjugation, i.e.,

$$(\theta^1)^\dagger = \theta^{1*} \quad \text{and} \quad (\theta^2)^\dagger = \theta^{2*}$$

We now introduce *superfields*, which are functions of all eight coordinates x_μ, θ, and $\bar{\theta}$. Consider a general superfield

$$\Phi(x, \theta, \bar{\theta})$$

that we take to be a scalar under Lorentz transformations (later we will encounter superfields that are not scalars).

Now we will apply a SUSY transformation to this superfield. Our SUSY transformation will be a function of the infinitesimal parameters a_μ, ζ, and $\bar{\zeta}$, as we have

used so far. Thus let us write this as

$$U(a, \zeta, \bar{\zeta}) \, \Phi(x, \theta, \bar{\theta}) \, U^{\dagger}(a, \zeta, \bar{\zeta}) = \Phi(x', \theta', \bar{\theta}') \qquad (11.3)$$

This is our starting point. Our goal is to obtain representations for the SUSY charges as differential operators acting in superspace.

This is a nontrivial exercise. In principle, what we would like to do is to follow the steps we took in Chapter 6 to work out the differential operator representations of the Poincaré generators from the transformations of the coordinates:

$$\text{Coordinate transformations} \implies \text{charges as differential operators}$$

Unfortunately, we don't know how the superspace coordinates transform under SUSY, only the SUSY algebra. What can we do? The trick is that we can actually invert the step described in Eq. (6.33); i.e., we can go from the algebra directly to the transformation of the coordinates:

$$\text{Algebra} \implies \text{coordinate transformations}$$

Once we have the coordinate transformations, it is a simple matter to find the differential operator representation of the charges.

Let us first see how this works in the simple context of ordinary translations.

11.2 Example of Spacetime Translations

Our goal, then, is to figure out how coordinates transform under translations using only the algebra of the components of the momentum operator as input. Of course, the approach here will be overkill when applied to simple translations, but it is nevertheless instructive to see the technique applied in a familiar and simple context before forging ahead with SUSY.

Let us go back to the action of the four-momentum charges P^{μ} on a scalar field shown in Eqs. (11.1) and (11.2). Note that P^{μ} here is a quantum field operator, not a differential operator. We cannot freely move the P^{μ} through the quantum field $\phi(x)$, but we are free to move a^{μ} or x^{μ} through the charges P^{μ} with impunity.

Now, let us write again Eq. (11.1) but use Eq. (11.2) for $\phi(x)$, giving

$$U(a) \, U(x) \, \phi(0) \, U^{\dagger}(x) \, U^{\dagger}(a) = \phi(x')$$

The trick is to realize that we can write the transformed field as arising from *a single transformation* produced by the operator $U(x')$:

$$U(a)\, U(x)\, \phi(0)\, U^\dagger(x)\, U^\dagger(a) \equiv U(x')\, \phi(0)\, U^\dagger(x')$$

which yields

$$U(a)\, U(x) = U(x')$$

Writing the unitary operators in terms of the charges P^μ, we have

$$\exp(i a^\nu P_\nu)\, \exp(i x^\mu P_\mu) = \exp(i x'^\mu P_\mu) \tag{11.4}$$

In order to relate x' to x and a, we must simply combine the exponentials on the left-hand side using the Baker-Campbell-Hausdorff formula:

$$\exp(A)\exp(B) = \exp\!\left(A + B + \frac{1}{2}[A,\, B] + \cdots\right)$$

Using this for the left-hand side of Eq. (11.4) and setting equal the arguments of the exponents, we finally get

$$i\,(a^\mu + x^\mu)\, P_\mu - \frac{1}{2} a^\nu x^\mu\, [P_\nu,\, P_\mu] + \cdots = i\, x'^\mu\, P_\mu$$

This is what we were seeking: a way to determine the transformed coordinates in terms of the old coordinates using only the algebra of the charges. Of course, in the case of the momentum operator, the algebra is trivial, i.e., $[P^\mu,\, P^\nu] = 0$, which yields

$$x'^\mu = x^\mu + a^\mu$$

Once we have established how the coordinates are transformed by the symmetry operation, it is simple to find the differential operator representation of the four-momentum as we did in Chapter 6. The only subtle point to keep in mind is that the action of the differential operator is defined through [see Eqs. (6.25) and (6.26)]

$$\exp\!\left(-i\, a^\mu\, \hat{P}_\mu\right) \phi(x) = \phi(x + \delta x) \tag{11.5}$$

From here it is trivial to obtain $\hat{P}_\mu = i\, \partial_\mu$.

We will now repeat these steps in the case of supersymmetric transformations applied to superfields.

11.3 Supersymmetric Transformations of the Superspace Coordinates

We now apply the approach of the preceding section to SUSY. Our starting point is Eq. (11.3). The SUSY equivalent of Eq. (11.2) is obviously

$$\Phi(x, \theta, \bar{\theta}) = U(x, \theta, \bar{\theta}) \, \Phi(0) \, U^{\dagger}(x, \theta, \bar{\theta}) \tag{11.6}$$

On the other hand, we have, by definition,

$$\Phi(x', \theta', \bar{\theta}') = U(x', \theta', \bar{\theta}') \, \Phi(0) \, U^{\dagger}(x', \theta', \bar{\theta}') \tag{11.7}$$

Inserting Eqs. (11.6) and (11.7) into Eq. (11.3), we are led to

$$U(a, \zeta, \bar{\zeta}) \, U(x, \theta, \bar{\theta}) = U(x', \theta', \bar{\theta}') \tag{11.8}$$

This is what we are going to use to figure out how the shifted coordinates are related to the original coordinates after a SUSY transformation with parameters a, ζ, and $\bar{\zeta}$.

It might seem that we are being a bit masochistic here. One might be tempted to conclude that in analogy with the result for translations in real space $x' = x + a$, the corresponding result here simply should be $x' = x + a$, $\theta' = \theta + \zeta$, and $\bar{\theta}' = \bar{\theta} + \bar{\zeta}$. However, the situation is more subtle than this because we can, using σ^{μ}, build quantities out of θ and ζ that will transform like four-vectors, so the transformation of x could turn out to contain more than just a. In fact, we know that the transformation of x *has* to be modified by SUSY because we have already established that the commutator of two SUSY transformations yields a shift in spacetime. Therefore, x' will depend on θ and ζ in a nontrivial way.

The next step is to write down explicit expressions for U in terms of the charges. What should we take for the supersymmetric equivalent of $a^{\mu} P_{\mu}$ or $x^{\mu} P_{\mu}$? The role of the momentum operator is played by the supercharges Q and $\bar{Q}$, whereas the coordinates x^{μ} should be replaced by θ and $\bar{\theta}$. Since we need Lorentz invariants to use in the exponential of U, the equivalent of $x^{\mu} P_{\mu}$ is

$$Q \cdot \theta + \bar{Q} \cdot \bar{\theta}$$

On the other hand, the role of the a^μ is played by ζ and $\bar{\zeta}$, so the equivalent of $a^\mu P_\mu$ is

$$Q \cdot \zeta + \bar{Q} \cdot \bar{\zeta}$$

Since the commutator of two supersymmetric transformations contains the generators P_μ, we need to consider both translations and supersymmetric transformations together to work out the transformations of the coordinates in superspace.

However, when we get to the point of actually writing out U, we face a difficulty: There are several possible choices for U that are not equivalent. For example, we could choose

$$U(x, \theta, \bar{\theta}) = \exp(ix \cdot P) \, \exp(i\theta \cdot Q) \, \exp(i\bar{\theta} \cdot \bar{Q}) \qquad (11.9)$$

but we could as well choose

$$U(x, \theta, \bar{\theta}) = \exp(i\, x \cdot P) \, \exp(i\, \bar{\theta} \cdot \bar{Q}) \, \exp(i\, \theta \cdot Q) \qquad (11.10)$$

or

$$U(x, \theta, \bar{\theta}) = \exp(i\, x \cdot P + i\, \bar{Q} \cdot \bar{\theta} + i\, Q \cdot \theta) \qquad (11.11)$$

These three choices are not equivalent because the charges Q and $\bar{Q}$ do not commute (however, $\hat{P}^\mu$ commutes with everything, so other choices with $a \cdot P$ in different positions are all equivalent to one of the three expressions given above). There is no right or wrong choice here, but the details of the following steps depend on which expression we pick. It turns out that the rest of the discussion is simpler if we use Eq. (11.9) or Eq. (11.10), but most references on SUSY use the more symmetric Eq. (11.11) as a starting point. We will follow the majority of the literature and consider Eq. (11.11).

Note that the operator (11.11) is unitary. To show this, we first need to evaluate $U^\dagger$:

$$U^\dagger = \exp\!\left(-i\, x \cdot P - i\, (\bar{Q} \cdot \bar{\theta})^\dagger - i\, (Q \cdot \theta)^\dagger\right)$$
$$= \exp\!\left(-i\, x \cdot P - i\, Q \cdot \theta - i\, \bar{Q} \cdot \bar{\theta}\right)$$

where we have used the fact that P_μ is hermitian as well as the identities (A.44) to apply the hermitian conjugate on the spinor dot products. It is now clear that $U^\dagger U = U^\dagger U = 1$ because the arguments of the exponentials in U and $U^\dagger$ differ

only by a sign. Note that for this to work, it is crucial that U contains both $Q \cdot \theta$ and $\bar{Q} \cdot \bar{\theta}$! If we had only one of the two terms, U would no longer be unitary.

Now that we have an explicit expression for U in terms of the charges, we can rewrite Eq. (11.8) as

$$\exp(i\,a \cdot P + i\,\bar{\zeta} \cdot \bar{Q} + i\,\zeta \cdot Q) \exp(i\,x \cdot P + i\,\bar{\theta} \cdot \bar{Q} + i\,\theta \cdot Q)$$
$$= \exp(i\,x' \cdot P + i\,\bar{\theta}' \cdot \bar{Q} + i\,\theta' \cdot Q) \tag{11.12}$$

We will now spend some time to calculate the left-hand side of Eq. (11.12). Using the Baker-Campbell-Hausdorff formula, it becomes

$$\exp\Bigl(i\,(x + a) \cdot P + i\,(\bar{\theta} + \bar{\zeta}) \cdot \bar{Q} + i\,(\theta + \zeta) \cdot Q$$
$$-\frac{1}{2} [\bar{\zeta} \cdot \bar{Q} + \zeta \cdot Q, \, \bar{\theta} \cdot \bar{Q} + \theta \cdot Q] + \cdots \Bigr) \tag{11.13}$$

where we have used the fact that $\hat{P}^\mu$ commutes with all charges to simplify the commutator term. This looks like a nightmare to calculate (well, an infinite number of nightmares), but luckily, we will see that the expansion terminates pretty quickly.

The commutators we need actually have been calculated in Chapter 6 using all components with lower, undotted indices, which forced us to carry around a bunch of σ^2 matrices. We can calculate the commutators *much* more efficiently now using the notation of Chapter 3. Consider first the commutator $[\zeta \cdot Q, \theta \cdot Q]$ (recall that the order in spinor dot products does not matter). Expanding, we get

$$[\zeta \cdot Q, \theta \cdot Q] = [\zeta^a Q_a, \, \theta^b Q_b]$$
$$= \zeta^a Q_a \theta^b Q_b - \theta^b Q_b \zeta^a Q_a$$
$$= -\zeta^a \theta^b \{Q_a, Q_b\}$$
$$= 0 \tag{11.14}$$

where we have used that the supercharges anticommute, [Eq. (6.70)].

Similarly,

$$[\bar{Q} \cdot \bar{\zeta}, \bar{Q} \cdot \bar{\theta}] = [\bar{Q}_{\dot{a}} \bar{\zeta}^{\dot{a}}, \, \bar{Q}_{\dot{b}} \bar{\theta}^{\dot{b}}]$$
$$= -\bar{\zeta}^{\dot{a}} \bar{\theta}^{\dot{b}} \{\bar{Q}_{\dot{a}}, \bar{Q}_{\dot{b}}\}$$
$$= 0 \tag{11.15}$$

Next, we turn our attention to

$$[\zeta \cdot Q, \bar{Q} \cdot \bar{\theta}] = [\zeta^a Q_a, \bar{Q}_{\dot{b}} \bar{\theta}^{\dot{b}}] \tag{11.16}$$

Let's pause to make a comment before we proceed with the calculation of this expression. It is frequent to see in SUSY expressions the same letter appearing both dotted and undotted. For example, we could have written Eq. (11.16) as

$$[\zeta \cdot Q, \bar{Q} \cdot \bar{\theta}] = [\zeta^a Q_a, \bar{Q}_{\dot{a}} \bar{\theta}^{\dot{a}}] \tag{11.17}$$

In such expressions, it is implicit that the index $\dot{a}$ is totally unrelated to the index a, so Eq. (11.17) is completely equivalent to Eq. (11.16). In this book we will avoid using a letter and the dot of the same letter in any given expression only to make equations easier to read. It is indeed easier to tell apart b and $\dot{c}$ than b and $\dot{b}$, especially given that these appear as small indices.

Proceeding with the calculation, we find that

$$[\zeta^a Q_a, \bar{Q}_{\dot{b}} \bar{\theta}^{\dot{b}}] = \zeta^a \bar{\theta}^{\dot{b}} \{Q_a, \bar{Q}_{\dot{b}}\}$$

$$= \zeta^a \bar{\theta}^{\dot{b}} (\sigma^\mu)_{a\dot{b}} P_\mu$$

using Eq. (6.72)

$$= \zeta \sigma^\mu \bar{\theta} P_\mu \tag{11.18}$$

From this, we get $[\bar{Q} \cdot \bar{\xi}, \theta \cdot Q]$ for free (or almost). We simply have to write

$$[\bar{Q} \cdot \bar{\xi}, \theta \cdot Q] = -[\theta \cdot Q, \bar{Q} \cdot \bar{\xi}]$$

and use Eq. (11.18) with ζ and θ switched to get

$$[\bar{Q} \cdot \bar{\xi}, \theta \cdot Q] = -\theta \sigma^\mu \bar{\xi} P_\mu \tag{11.19}$$

The key results are the commutators (11.14), (11.15), (11.18), and (11.19). Now we can use them to evaluate Eq. (11.13). Note that since the commutators of Q and $Q^\dagger$ yield the momentum operator P^μ, which commutes with everything, all the higher-order commutators represented by the dots in Eq. (11.13) vanish! So Eq. (11.13) is simply

$$\exp\left(i (x + a) \cdot P + i (\bar{\theta} + \bar{\xi}) \cdot \bar{Q} + i (\theta + \zeta) \cdot Q - \frac{1}{2} \zeta \sigma^\mu \bar{\theta} P_\mu + \frac{1}{2} \theta \sigma^\mu \bar{\xi} P_\mu \right)$$

$$\tag{11.20}$$

Recall that this is the left-hand side of Eq. (11.12). Setting this equal to the right-hand side of Eq. (11.12), we obtain

$$\exp\left(i\,(x+a)\cdot P + i\,(\bar\theta+\bar\zeta)\cdot\bar Q + i\,(\theta+\zeta)\cdot Q - \frac{1}{2}\zeta\sigma^\mu\bar\theta\,P^\mu + \frac{1}{2}\theta\sigma^\mu\bar\zeta\,P^\mu\right)$$

$$= \exp\left(i\,\theta'\cdot Q + i\,\bar\theta'\cdot\bar Q + i\,x'\cdot P\right)$$

from which we can finally read off the transformations of the superspace coordinates:

$$\boxed{\begin{aligned} x' &= x + a + \frac{i}{2}\zeta\sigma^\mu\bar\theta - \frac{i}{2}\theta\sigma^\mu\bar\zeta \\ \theta' &= \theta + \zeta \\ \bar\theta &= \bar\theta + \bar\zeta \end{aligned}}$$

(11.21)

As expected, the transformation of x acquires extra terms that mix ordinary space with the fermionic coordinates of superspace.

Using Eq. (A.50), the transformation of x also could be written as

$$x' = x + a - \frac{i}{2}\theta\bar\sigma^\mu\zeta + \frac{i}{2}\bar\zeta\bar\sigma^\mu\theta$$

(11.22)

Now we are ready to write the SUSY charges as differential operators acting in superspace.

11.4 Introduction to Superfields

Before going any further, it is important to take a short pause to review some important facts about the calculus of Grassmann variables. Although this may seem a bit intimidating at first, it turns out that Grassmann calculus is *much* easier than calculus involving real or complex variables. As one might guess, this comes from the fact that the square of any Grassmann quantity is zero.

Before discussing superfields, let's start with something simpler. Consider an arbitrary (but commuting) function F of a single Grassmann variable, say, θ_1. Let us now look at the Taylor expansion of this function around $\theta_1 = 0$. Usual Taylor expansions (with respect to real or complex variables) contain an infinite number of terms, but here, since $\theta_1^2 = 0$, the expansion terminates after only two terms!

$$F(\theta_1) = c_0 + c_1\,\theta_1$$

(11.23)

where we really mean an equality, not an approximation, and the coefficients c_0 and c_1 are some complex numbers. If the function depends on two Grassmann variables θ_1 and θ_2, then the Taylor expansion clearly terminates after four terms:

$$F(\theta_1, \theta_2) = c_0 + c_1\,\theta_1 + c_2\,\theta_2 + c_3\,\theta_1\,\theta_2$$

There is no need to write a term proportional to $\theta_2\theta_1$ because this term could be combined with the term in $\theta_1\,\theta_2$ after using $\theta_2\,\theta_1 = -\theta_1\,\theta_2$.

Now we add a twist: We make F a function of both Grassmann variables and spacetime and call it a *superfield* $\mathcal{F}$ (in this book, we will use calligraphic letters to represent most superfields). Then the Taylor expansion with respect to the Grassmann variables obviously is given by

$$\mathcal{F}(x, \theta_1, \theta_2) = A(x) + B(x)\,\theta_1 + C(x)\,\theta_2 + D(x)\,\theta_1\,\theta_2 \qquad (11.24)$$

Note that in this expression, $B(x)$ and $C(x)$ are *Grassmann* functions, whereas $A(x)$ and $D(x)$ are ordinary (i.e., commuting) functions. Of course, our interest is in quantum field theory, so we promote the functions $A(x) \cdots D(x)$ to quantum fields.

$\mathcal{F}$ is not the most general superfield we can construct because we have chosen it to be independent of $\bar{\theta}$. Exercise 11.2 invites you to work out the expansion of the most general superfield, which is a function of x, θ, and $\bar{\theta}$.

We will be mostly concerned with superfields that are invariant under Lorentz transformations, but even in that case, it is not straigthforward to infer how the four fields appearing in Eq. (11.24) transform. It is preferable instead to write the expansion in terms of the Weyl spinor θ, i.e.,

$$\mathcal{F}(x, \theta) = \phi(x) + \theta \cdot \chi(x) + \frac{1}{2}\theta \cdot \theta F(x) \qquad (11.25)$$

where the factor of $1/2$ in the last term is purely conventional and chosen to make this resemble usual Taylor expansions. As an aside, note that many authors rescale $\theta \to \sqrt{2}\theta$ so that they write the expansion as

$$\mathcal{F}(x, \theta) = \phi(x) + \sqrt{2}\theta \cdot \chi(x) + \theta \cdot \theta\, F(x)$$

This introduces various powers of $\sqrt{2}$ with respect to our convention.

If the field $\mathcal{F}$ is a scalar under Lorentz transformations, we can tell right away that $\phi(x)$ is a scalar field and that $\chi(x)$ is a left-chiral spinor field, because θ is itself a left-chiral Weyl spinor. We also can tell that $F(x)$ is a scalar field because the quantity $\theta \cdot \theta$ is invariant under Lorentz transformations. Note that the number of independent fields in Eqs. (11.24) and (11.25) is the same because $\chi(x)$ is a two-component spinor.

The fields $\phi(x)$, $\chi(x)$, and $F(x)$ are often referred to as the *component fields* of the superfield $\mathcal{F}$.

EXERCISE 11.1

Find the relations between the fields appearing in Eq. (11.24), and the fields $\phi(x)$, $\chi_1(x)$, $\chi_2(x)$, and $F(x)$ appearing in Eq. (11.25).

EXERCISE 11.2

Consider a general scalar superfield $\mathcal{S}(x, \theta, \bar{\theta})$. Write down the most general Taylor expansion of this superfield [not in terms of components as in Eq. (11.24) but in the index-free notation, as in Eq. (11.25)]. *Hint*: You should find nine terms. Don't forget that you may use σ^μ (but then you will not need $\bar{\sigma}^\mu$; do you see why?).

11.5 Aside on Grassmann Calculus

Besides Taylor expansions, we also will need to differentiate and integrate with respect to Grassmann variables. Consider first taking derivatives. Of course, we define

$$\frac{\partial \theta_1}{\partial \theta_1} = 1$$

Note that we treat θ_1, θ_2, $\bar{\theta}_1$, and $\bar{\theta}_2$ as four independent variables, so the derivative of any of those four quantities with respect to any of the other three gives zero. For example,

$$\frac{\partial \theta_2}{\partial \theta_1} = \frac{\partial \bar{\theta}_1}{\partial \theta_1} = \frac{\partial \bar{\theta}_2}{\partial \theta_1} = 0$$

On the other hand, the variables with upper indices must be viewed as functions of the variables with lower indices, not as independent variables. Indeed, recall that as we saw in Chapter 3 [see Eqs. (3.13) and (3.16)],

$$\theta^1 = \theta_2 \qquad \theta^2 = -\theta_1$$

$$\bar{\theta}^{\dot{1}} = \bar{\theta}_{\dot{2}} \qquad \bar{\theta}^{\dot{2}} = -\bar{\theta}_{\dot{1}} \tag{11.26}$$

This means that, for example,

$$\frac{\partial \theta^1}{\partial \theta_1} = 0$$

whereas

$$\frac{\partial \theta^2}{\partial \theta_1} = -1$$

Note that the relations (11.26) allow us to also rewrite derivatives with respect to variables with upper indices in terms of derivatives with respect to variables with lower indices, and vice versa. For example,

$$\frac{\partial}{\partial \theta^1} = \frac{\partial}{\partial \theta_2}$$

and

$$\frac{\partial}{\partial \theta^2} = -\frac{\partial}{\partial \theta_1}$$

Consider now taking the derivative of something less trivial, let's say

$$\frac{\partial}{\partial \theta_1} (\theta_2 \, \theta_1)$$

We may apply the product rule on the condition that we keep in mind that the derivative is a Grassmann quantity so that we pick up a minus sign when passing it through θ_2:

$$\frac{\partial}{\partial \theta_1} (\theta_2 \, \theta_1) = \left(\frac{\partial \theta_2}{\partial \theta_1}\right) \theta_1 - \theta_2 \left(\frac{\partial \theta_1}{\partial \theta_1}\right)$$

Since θ_2 is independent of θ_1, we simply get

$$-\theta_2 \left(\frac{\partial \theta_1}{\partial \theta_1}\right) = -\theta_2$$

Of course, we could have simply moved θ_2 in front of the derivative because it is a constant as far as the derivative with respect to θ_1 is concerned, to get

$$\frac{\partial}{\partial \theta_1} \theta_2 \, \theta_1 = -\theta_2 \frac{\partial}{\partial \theta_1} \theta_1 = -\theta_2$$

Particularly useful will be the identities mentioned in the following exercise.

EXERCISE 11.3
Prove

$$\frac{\partial}{\partial \theta^a} \theta \cdot \theta = 2\theta_a$$

and

$$\frac{\partial}{\partial \theta_a} \theta \cdot \theta = -2\theta^a$$

These formulas will prove to be very useful in the next section.

Now let's turn our attention to integration of Grassmann variables. Here, integration is not related to the limit of a sum and has some properties quite different from conventional Riemann integrals. It is better thought of as an abstract algebraic operation defined in terms of its properties. As such, a different symbol than an integral sign probably would be more appropriate. The integration over Grassmann variables originally was defined by Berezin[7] and is therefore sometimes referred to as *Berezin integration*.

So how do we define this "integration" operation? First, we impose that it be a linear operation:

$$\int d\theta_1 \left[k_1 \, F(\theta_1) + k_2 \, G(\theta_1) \right] = k_1 \int d\theta_1 \, F(\theta_1) + k_2 \int d\theta_1 \, G(\theta_1) \qquad (11.27)$$

where k_1 and k_2 are arbitrary complex numbers.

Now comes the key definition. We know that an arbitrary function of θ_1 has the expansion given in Eq. (11.23). We *define* the integration of $F(\theta_1)$ to give for result the coefficient of the term linear in θ_1:

$$\int d\theta_1 \, F(\theta_1) \equiv c_1$$

Using Eqs. (11.23) and (11.27), this means that

$$c_0 \int d\theta_1 + c_1 \int d\theta_1 \, \theta_1 = c_1$$

which implies

$$\int d\theta_1 = 0 \qquad \int d\theta_1 \, \theta_1 = 1$$

Note that "integrating" $F(\theta_1)$ with respect to θ_1 gives exactly the same result as differentiating it with respect to θ_1! This is not what we are used to in conventional calculus (except for the function e^x!).

Now consider integration with respect to two variables, θ_1 and θ_2. We define the "measures" to be Grassmann quantities, like the θ_a, so we impose

$$\{d\theta_a, d\theta_b\} = \{d\theta_a, \theta_b\} = \{\theta_a, \theta_b\} = 0$$

This means that $\int d\theta_2 \, \theta_1 = -\theta_1 \int d\theta_2$ for example.

As an example, consider

$$\int d\theta_1 \int d\theta_2 \, \theta_1 = - \int d\theta_2 \left(\int d\theta_1 \theta_1 \right)$$

$$= - \int d\theta_2$$

$$= 0$$

Clearly, we also have

$$\int d\theta_1 \int d\theta_2 \, \theta_2 = 0$$

More interesting is

$$\int d\theta_1 \int d\theta_2 \, \theta_1 \, \theta_2 = - \int d\theta_1 \, \theta_1 \int d\theta_2 \, \theta_2$$

$$= -1$$

Note that if we integrate the superfield (11.24) over both of its Grassmann variables, we obtain

$$\int d\theta_1\, d\theta_2\, \mathcal{F}(x, \theta_1, \theta_2) = -D(x)$$

so the integration *projects out the component proportional to the two Grassmann variables*. The fact that we can use an integration to project out terms like this will play a key role in the rest of this book.

Let's define the measure $d^2\theta$ to represent

$$d^2\theta \equiv d\theta_1\, d\theta_2$$

in that order. With this definition,

$$\int d^2\theta\, \theta \cdot \theta = 2 \int d^2\theta\, \theta_2\theta_1 = 2$$

so that

$$\boxed{\frac{1}{2} \int d^2\theta\, \theta \cdot \theta = 1} \tag{11.28}$$

a relation we will use very often.

We obviously also can integrate over the barred variables $\bar{\theta}_{\dot{a}}$. All the preceding results hold as well for those variables. However, we will define

$$d^2\bar{\theta} \equiv -d\bar{\theta}_{\dot{1}}\, d\bar{\theta}_{\dot{2}}$$

instead. This choice is made so that

$$\int d^2\bar{\theta}\, \bar{\theta} \cdot \bar{\theta} = 2 \int d^2\bar{\theta}\, \bar{\theta}_{\dot{1}}\, \bar{\theta}_{\dot{2}} = 2$$

and therefore,

$$\boxed{\frac{1}{2} \int d^2\bar{\theta}\, \bar{\theta} \cdot \bar{\theta} = 1}$$

Of course, we also have

$$\boxed{\frac{1}{4} \int d^2\bar{\theta}\, d^2\theta\, \theta \cdot \theta\, \bar{\theta} \cdot \bar{\theta} = 1}$$

(11.29)

which we will also need often. Note that when we integrate any other combination of θ and $\bar{\theta}$, we get zero. For example, if we integrate a general superfield $S(x, \theta, \bar{\theta})$, we get

$$\int d^2\bar{\theta}\, d^2\theta\, S(x, \theta, \bar{\theta}) = \text{coefficient of the term } \frac{\theta \cdot \theta\, \bar{\theta} \cdot \bar{\theta}}{4} \text{ in } S(x, \theta, \bar{\theta})$$

Similarly, Eq. (11.28) implies

$$\int d^2\theta\, S(x, \theta, \bar{\theta}) = \text{coefficient of the term } \frac{\theta \cdot \theta}{2} \text{ in } S(x, \theta, \bar{\theta})$$

11.6 The SUSY Charges as Differential Operators

We will keep using S to denote a general superfield. The expression analogous to Eq. (11.5) in the case of a translation plus a supersymmetric transformation is

$$\exp\left(-i\, a^\mu \hat{P}_\mu - i\, \zeta \cdot Q - i\, \bar{\zeta} \cdot \bar{Q}\right) S(x, \theta, \bar{\theta}) = S(x + \delta x, \theta + \delta\theta, \bar{\theta} + \delta\bar{\theta})$$

(11.30)

Now we just need to plug in the transformed coordinates (11.21) into the superfield and Taylor expand. This gives

$$\begin{aligned}
S(x + \delta x, \theta + \delta\theta, \bar{\theta} + \delta\bar{\theta}) \\
\approx S(x, \theta, \bar{\theta}) + \delta x^\mu \partial_\mu S(x, \theta, \bar{\theta}) + \delta\theta^a\, \partial_a S(x, \theta, \bar{\theta}) + \delta\bar{\theta}_{\dot{a}} \partial^{\dot{a}} S(x, \theta, \bar{\theta}) \\
= S + \left(a^\mu + \frac{i}{2} \zeta\sigma^\mu\bar{\theta} - \frac{i}{2}\theta\sigma^\mu\bar{\zeta}\right) \partial_\mu S + \zeta^a \partial_a S + \bar{\zeta}_{\dot{a}} \bar{\partial}^{\dot{a}} S
\end{aligned}$$

(11.31)

where we have suppressed the $(x, \theta, \bar{\theta})$ dependence of S to ease the notation a bit and where the partial derivatives with respect to the Grassmann variables are defined by

$$\partial_a \mathcal{S} \equiv \frac{\partial \mathcal{S}}{\partial \theta^a} \qquad \bar{\partial}^{\dot{a}} \mathcal{S} \equiv \frac{\partial \mathcal{S}}{\partial \bar{\theta}_{\dot{a}}}$$

Note that, as usual, an upper position of the index on the partial derivative denotes the derivative with respect to a variable with a lower index, and vice versa.

Note that in Eq. (11.31) we have chosen the undotted indices to go diagonally downward and the dotted indices to go diagonally upward, following the convention introduced in Chapter 3. We can write Eq. (11.31) in a more compact form if we use the notation for spinor dot products and define

$$\zeta^a \partial_a \mathcal{S} \equiv \zeta \cdot \partial \mathcal{S}$$
$$\bar{\zeta}_{\dot{a}} \bar{\partial}^{\dot{a}} \mathcal{S} \equiv \bar{\zeta} \cdot \bar{\partial} \mathcal{S}$$

in which case Eq. (11.31) is given by

$$\mathcal{S}(x + \delta x, \theta + \delta\theta, \bar{\theta} + \delta\bar{\theta}) \approx \mathcal{S} + \left(a^\mu + \frac{i}{2} \zeta \sigma^\mu \bar{\theta} - \frac{i}{2} \theta \sigma^\mu \bar{\zeta} \right) \partial_\mu \mathcal{S}$$

$$+ \zeta \cdot \partial \mathcal{S} + \bar{\zeta} \cdot \bar{\partial} \mathcal{S} \tag{11.32}$$

Expanding to first order the exponential in Eq. (11.30) and setting this equal to Eq. (11.32), we get

$$(-i a^\mu \hat{P}_\mu - i \zeta \cdot \mathcal{Q} - i \bar{\zeta} \cdot \bar{\mathcal{Q}}) \mathcal{S} = \left(a^\mu + \frac{i}{2} \zeta \sigma^\mu \bar{\theta} - \frac{i}{2} \theta \sigma^\mu \bar{\zeta} \right) \partial_\mu \mathcal{S}$$

$$+ \zeta \cdot \partial \mathcal{S} + \bar{\zeta} \cdot \bar{\partial} \mathcal{S} \tag{11.33}$$

We can read off the differential operators $\hat{P}$, $\mathcal{Q}$, and $\mathcal{Q}^\dagger$ by separately setting equal the terms in a^μ, ζ^a, and $\bar{\zeta}_{\dot{a}}$. If we look at the terms proportional to a^μ, we get

$$\boxed{\hat{P}_\mu = i \, \partial_\mu} \tag{11.34}$$

as we had before, in the absence of SUSY.

Recall that $\zeta \sigma^\mu \bar{\theta}$ stands for $\zeta^a (\sigma^\mu)_{a\dot{b}} \bar{\theta}^{\dot{b}}$. Keeping only the terms containing ζ in Eq. (11.33) and writing out explicitly all the indices, we get

$$-i\zeta^a \mathcal{Q}_a \mathcal{S} = \frac{i}{2} \zeta^a (\sigma^\mu)_{a\dot{b}} \bar{\theta}^{\dot{b}} \partial_\mu \mathcal{S} + \zeta^a \partial_a \mathcal{S}$$

which implies

$$\boxed{\mathcal{Q}_a = i\,\partial_a - \frac{1}{2}\,(\sigma^\mu)_{ab}\,\bar{\theta}^{\dot{b}}\,\partial_\mu} \qquad (11.35)$$

Using Eq. (11.34), this also could be written as

$$\mathcal{Q}_a = i\,\partial_a + \frac{i}{2}\,(\sigma^\mu)_{ab}\,\bar{\theta}^{\dot{b}}\,\hat{P}_\mu$$

The terms in $\bar{\zeta}$ in Eq. (11.33) give the following relation:

$$-i\,\bar{\zeta}_{\dot{a}}\,\bar{\mathcal{Q}}^{\dot{a}} = -\frac{i}{2}\theta\sigma^\mu\bar{\zeta}\,\partial_\mu + \bar{\zeta}_{\dot{a}}\bar{\partial}^{\dot{a}} \qquad (11.36)$$

Here, we have a slight problem. Recall that $\theta\sigma^\mu\bar{\zeta}$ stands for $\theta^a(\sigma^\mu)_{a\dot{b}}\bar{\zeta}^{\dot{b}}$, so the ζ does not appear with the same indices in all the terms, preventing us from isolating $\bar{\mathcal{Q}}^{\dot{a}}$. We are saved by Eq. (A.50), which allows us to write

$$\theta\sigma^\mu\bar{\zeta} = -\bar{\zeta}\bar{\sigma}^\mu\theta = -\bar{\zeta}_{\dot{a}}\,(\bar{\sigma}^\mu)^{\dot{a}b}\,\theta_b$$

Using this in Eq. (11.36), we get

$$-i\,\bar{\zeta}_{\dot{a}}\,\bar{\mathcal{Q}}^{\dot{a}} = \frac{i}{2}\,\bar{\zeta}_{\dot{a}}\,(\bar{\sigma}^\mu)^{\dot{a}b}\,\theta_b\partial_\mu + \bar{\zeta}_{\dot{a}}\bar{\partial}^{\dot{a}}$$

which yields

$$\boxed{\bar{\mathcal{Q}}^{\dot{a}} = i\,\bar{\partial}^{\dot{a}} - \frac{1}{2}\,(\bar{\sigma}^\mu)^{\dot{a}b}\,\theta_b\partial_\mu} \qquad (11.37)$$

However, we also would like to have the corresponding expression with a lower dotted index, namely, $\bar{\mathcal{Q}}_{\dot{a}}$, because we then could check that we recover the algebra of the charges derived in Chapter 6, which was written entirely in terms of components with lower indices. We can proceed in two ways. We can use the ϵ symbol introduced in Section 3.4 to lower the index of $\bar{\mathcal{Q}}^{\dot{a}}$, or we may go back to Eq. (11.36) and write all the terms with the parameter $\bar{\zeta}$ having an upper index. Either way, there are some subtleties in the calculation.

We will follow the second approach and leave the demonstration using the ϵ for Exercise 11.5. We'll begin by writing again Eq. (11.36) with all the indices explicitly shown:

$$-i\, \bar{\xi}_{\dot{a}}\, \bar{Q}^{\dot{a}} = -\frac{i}{2}\, \theta^{a}\, (\sigma^{\mu})_{a\dot{b}}\, \bar{\xi}^{\dot{b}}\, \partial_{\mu} + \bar{\xi}_{\dot{a}}\, \bar{\partial}^{\dot{a}} \tag{11.38}$$

To express the left-hand side in terms of the charges with lower dotted components, we use Eq. (A.43)

$$\bar{\xi}_{\dot{a}}\, \bar{Q}^{\dot{a}} = \bar{Q}_{\dot{a}}\, \bar{\xi}^{\dot{a}}$$

Using this in Eq. (11.38), we get

$$-i\, \bar{Q}_{\dot{a}}\, \bar{\xi}^{\dot{a}} = -\frac{i}{2}\, \theta^{a}(\sigma^{\mu})_{a\dot{b}}\, \bar{\xi}^{\dot{b}}\, \partial_{\mu} + \bar{\xi}_{\dot{a}}\, \bar{\partial}^{\dot{a}} \tag{11.39}$$

Before we can isolate the charge, we need to rewrite the last term with the index on $\bar{\xi}$ raised. This is a bit more tricky. The identity we need is

$$\bar{\xi}_{\dot{a}}\, \bar{\partial}^{\dot{a}} = \bar{\xi}^{\dot{a}}\, \bar{\partial}_{\dot{a}} \tag{11.40}$$

which you are invited to prove in Exercise 11.4. We then can move the $\bar{\xi}$ to the right of the derivative if we change the sign (because they are both Grassmann quantities):

$$\bar{\xi}^{\dot{a}}\, \bar{\partial}_{\dot{a}} = -\bar{\partial}_{\dot{a}}\, \bar{\xi}^{\dot{a}} \tag{11.41}$$

We can move ζ in front of the derivative because it is a constant spinor. Things would not be so simple if it had a $\bar{\theta}$ dependence. Note that the end result is

$$\boxed{\bar{\xi}_{\dot{a}}\, \bar{\partial}^{\dot{a}} = -\bar{\partial}_{\dot{a}}\, \bar{\xi}^{\dot{a}}}$$

which has the *opposite sign* to what Eq. (A.43) would predict! This is so because, as demonstrated in Exercise 11.4,

$$\partial_{2} = -\partial^{1} \qquad \partial_{1} = \partial^{2} \qquad \bar{\partial}_{\dot{2}} = -\bar{\partial}^{\dot{1}} \qquad \bar{\partial}_{\dot{1}} = \bar{\partial}^{\dot{2}}$$

and these relations have the *opposite sign* to the usual relations between Grassmann quantities with upper and lower indices shown in Eq. (11.26). Incidentally, a consequence of this is that while for components of spinors we have [see Eq. (A.31)]

$$\epsilon^{ab}\, \chi_b = \chi^a$$

$$\epsilon_{\dot{a}\dot{b}}\, \bar{\chi}^{\dot{b}} = \bar{\chi}_{\dot{a}}$$

for derivatives we have instead

$$\epsilon^{ab}\, \partial_b = -\partial^a$$

$$\epsilon_{\dot{a}\dot{b}}\, \bar{\partial}^{\dot{b}} = -\bar{\partial}_{\dot{a}} \tag{11.42}$$

Thus the ϵ introduces a minus sign when it is used to raise or lower indices of partial derivatives.

EXERCISE 11.4
Prove that

$$\zeta^a \partial_a = \zeta_a \partial^a$$

Then argue that this also implies the validity of Eq. (11.40).

Substituting Eq. (11.41) into Eq. (11.39), we get

$$-i\,\bar{Q}_{\dot{a}}\, \bar{\xi}^{\dot{a}} = -\frac{i}{2}\, \theta^a\, (\sigma^\mu)_{a\dot{b}}\, \bar{\xi}^{\dot{b}}\, \partial_\mu - \bar{\partial}_{\dot{a}}\, \zeta^{\dot{a}}$$

from which we finally obtain

$$\boxed{\bar{Q}_{\dot{a}} = -i\,\bar{\partial}_{\dot{a}} + \frac{1}{2}\, \theta^b (\sigma^\mu)_{b\dot{a}}\, \partial_\mu} \tag{11.43}$$

or, using $\hat{P}_\mu = i\partial_\mu$,

$$\bar{Q}_{\dot{a}} = -i\,\bar{\partial}_{\dot{a}} - \frac{i}{2}\, \theta^b\, (\sigma^\mu)_{b\dot{a}}\, \hat{P}_\mu$$

EXERCISE 11.5
Rederive Eq. (11.43) from Eq. (11.37) using ϵ to lower the indices. *Hint*: You will need Eqs. (A.27) and (11.42).

We have finally accomplished the goal we had set for ourselves: We have expressed the charges as differential operators acting on superfields (i.e., functions of superspace). As a nontrivial double-check, let us now compute the algebra of the charges, which should agree with what we found in Chapter 6 in Eqs. (6.70), (6.71), and (6.72), which we will repeat here using the van der Waerden notation:

$$\{Q_a, Q_b\} = 0$$

$$\{\bar{Q}_{\dot{a}}, \bar{Q}_{\dot{b}}\} = 0$$

$$\{Q_a, \bar{Q}_{\dot{b}}\} = (\sigma^\mu)_{a\dot{b}}\, P^\mu \tag{11.44}$$

The first two almost trivially follow from our operators (11.35) and (11.43). The only nontrivial point is that we must remember to pick up a sign when moving Grassmann quantities through one another. For example, consider the following anticommutator:

$$\{\partial_a, \bar{\theta}^{\dot{b}}\} \tag{11.45}$$

which appears in the anticommutator $\{Q_a, Q_b\}$. In computing anticommutators of charges represented by differential operators, we must imagine that the whole expression is applied to some test function of superspace, as we imagine that a function of x is present when we compute $[\hat{x}, \hat{p}]$ in quantum mechanics.

Expanding Eq. (11.45), we get

$$\{\partial_a, \bar{\theta}^{\dot{b}}\} = \partial_a\, \bar{\theta}^{\dot{b}} + \bar{\theta}^{\dot{b}}\, \partial_a$$

$$= (\partial_a\, \bar{\theta}^{\dot{b}}) - \bar{\theta}^{\dot{b}}\, \partial_a + \bar{\theta}^{\dot{b}}\, \partial_a$$

$$= (\partial_a\, \bar{\theta}^{\dot{b}})$$

$$= 0$$

where in the parenthesis the derivative is applied only on $\bar{\theta}^{\dot{b}}$, and this gives zero because the dotted upper components are independent of the undotted upper components (recall that $\partial_a = \partial/\partial\theta^a$). From this, the first anticommutator of Eq. (11.44) follows trivially. The second one is as straightforward.

For the third one, we need a bit more work, but not much. We need to compute

$$\{\partial_a, \theta^b\} = \partial_a\, \theta^b + \theta^b\, \partial_a$$

$$= (\partial_a\, \theta^b) - \theta^b\, \partial_a + \theta^b\, \partial_a$$

$$= \delta_a^b \tag{11.46}$$

where δ_a^b is simply the ordinary Kronecker delta, equal to 1 when $a = b$ and 0 when $a \neq b$. Similarly, we find

$$\{\bar{\partial}_{\dot{a}}, \bar{\theta}^{\dot{b}}\} = \delta_{\dot{a}}^{\dot{b}} \tag{11.47}$$

Again, this delta is simply the usual Kronecker delta.

Using these two results together with Eqs. (11.35) and (11.43) (after relabeling the indices to avoid summing the same letter twice), we find (using $\partial_\mu = -i\hat{P}_\mu$)

$$\{Q_a, \bar{Q}_{\dot{b}}\} = \left\{ i\partial_a + \frac{i}{2}(\sigma^\mu)_{a\dot{c}}\,\bar{\theta}^{\dot{c}}\,\hat{P}_\mu, \, -i\partial_{\dot{b}} - \frac{i}{2}\theta^d\,(\sigma^\mu)_{d\dot{b}}\,\hat{P}_\mu \right\}$$

$$= \frac{1}{2}\{\partial_a, \theta^d\}(\sigma^\mu)_{d\dot{b}}\,\hat{P}_\mu + \frac{1}{2}\{\bar{\theta}^{\dot{c}}, \partial_{\dot{b}}\}(\sigma^\mu)_{a\dot{c}}\hat{P}_\mu$$

where we have used the fact that the components of σ^μ are simply numbers, so we are free to move them around as we please. Using $\{A, B\} = \{B, A\}$, which applies to any operator, including Grassmann ones, and applying Eqs. (11.46) and (11.47), we find

$$\{Q_a, \bar{Q}_{\dot{b}}\} = \frac{1}{2}\delta_a^d(\sigma^\mu)_{d\dot{b}}\,\hat{P}_\mu + \frac{1}{2}\delta_{\dot{b}}^{\dot{c}}(\sigma^\mu)_{a\dot{c}}\,\hat{P}_\mu$$

$$= \frac{1}{2}(\sigma^\mu)_{a\dot{b}}\hat{P}_\mu + \frac{1}{2}(\sigma^\mu)_{a\dot{b}}\hat{P}_\mu$$

$$= (\sigma^\mu)_{a\dot{b}}\,\hat{P}_\mu$$

which is indeed the third relation of Eq. (11.44)!

11.7 Constraints and Superfields

We have almost reached the point where we can start harnessing the full power of the superspace approach. The basic idea is brilliant and yet rather simple. We have seen in Eqs. (11.30) and (11.31) how a superfield transforms under SUSY. On the other hand, we have seen in Section 11.4 how a superfield can be expanded in terms of a finite number of component fields that only depend on spacetime. In addition, those component fields have well-defined behavior under Lorentz transformations. They will play the role of the actual quantum fields appearing in supersymmetric lagrangian.

Putting now these two ideas together; i.e., substituting the expansion of a super-field in terms of component fields into the SUSY transformation [Eq. (11.30)], we will get in a single equation a picture of how *all* the fields making up the theory transform under supersymmetry!

This is a neat trick, but if this were the only motivation for superspace, it would be a bit of a let-down. After all, the main difficulty we had in constructing supersymmetric theories was not in writing down the SUSY transformations of the fields but in finding how to build lagrangians that are invariant under SUSY. This is actually where the superspace approach truly shines: We will see how, using a few devilish tricks, superfields make the construction of supersymmetric theories child's play! The full power of this approach will be demonstrated when we will be able to write down with almost no effort the SUSY-invariant interactions of the minimal supersymmetric standard model.

But we must first start with less lofty goals. Our first step will be to recover the Wess-Zumino lagrangian from the superfield approach. Unfortunately, we face a difficulty right from the start. The Wess-Zumino lagrangian involves two complex scalar fields and one left-chiral Weyl spinor field. However, a general superfield $S(x, \theta, \bar{\theta})$ has many more fields, as you found out if you solved Exercise 11.2. We *could* build a theory out of a general superfield, but we obviously would get something more complex than the Wess-Zumino lagrangian.

On the other hand, a superfield $\mathcal{F}$ that is a function of only x and θ contains just the correct number of fields [see Eq. (11.25)], which have in addition the right behavior under Lorentz transformations if we take the superfield to be a scalar (we also could have considered a superfield depending only on x and $\bar{\theta}$ instead). Unfortunately, it is *not* consistent with SUSY to consider a superfield depending only on x and θ. The reason is simple: If we apply a supersymmetric transformation to such a superfield, we end up with a transformed field $\mathcal{F}'$ that will depend on *all* the superspace coordinates.

To see this, let's look again at Eq. (11.32):

$$\mathcal{F}'(x, \theta, \bar{\theta}) = \mathcal{F}(x, \theta, \bar{\theta}) + \left(a^\mu + \frac{i}{2} \zeta \sigma^\mu \bar{\theta} - \frac{i}{2} \theta \sigma^\mu \bar{\zeta} \right) \partial_\mu \mathcal{F}(x, \theta, \bar{\theta})$$

$$+ \zeta \cdot \partial \mathcal{F}(x, \theta, \bar{\theta}) + \bar{\zeta} \cdot \bar{\partial} \mathcal{F}(x, \theta, \bar{\theta}) \qquad (11.48)$$

If we start with a superfield $\mathcal{F}$ depending only on x and θ, this simplifies to

$$\mathcal{F}'(x, \theta) = \mathcal{F}(x, \theta) + \left(a^\mu + \frac{i}{2} \zeta \sigma^\mu \bar{\theta} - \frac{i}{2} \theta \sigma^\mu \bar{\zeta} \right) \partial_\mu \mathcal{F}(x, \theta)$$

$$+ \zeta \cdot \partial \mathcal{F}(x, \theta)$$

We see the problem now. Even though our initial superfield $\mathcal{F}$ depends only on x and θ, the transformed superfield $\mathcal{F}'$ depends on $\bar{\theta}$ as well because this superspace coordinate appears explicitly in the transformation. What this tells us is that it is inconsistent with SUSY to consider a superfield that is a function of x and θ alone.

Let us now explain the problem again from a different angle. Considering a superfield $\mathcal{F}$ that does not depend on $\bar{\theta}$ is equivalent to starting from a general superfield $\mathcal{S}$ on which the following *constraint* has been applied:

$$\frac{\partial}{\partial \bar{\theta}^{\dot{a}}} \mathcal{S}(x, \theta, \bar{\theta}) = 0$$

so that $\mathcal{S}$ is actually a function of x and θ only, i.e.,

$$\mathcal{S}(x, \theta, \bar{\theta}) = \mathcal{F}(x, \theta) \tag{11.49}$$

By the way, the choice of differentiating with respect to the Grassmann variable with an upper index is only one of convenience. It makes life easier to have the same position of the index as in the derivatives that appear in the supercharges.

Now let's assume that we don't already know the answer and ask again if this constraint is consistent with SUSY transformations or, in other words, whether SUSY transformations *preserve* this constraint. To answer this, we must check whether, if we apply a SUSY transformation to a superfield $\mathcal{F}$ independent of $\bar{\theta}$, the SUSY-transformed field $\mathcal{F}'$ also will satisfy the same constraint:

$$\frac{\partial}{\partial \bar{\theta}^{\dot{a}}} \mathcal{F}' \overset{?}{=} 0$$

The left-hand side may be written more explicitly as

$$\frac{\partial}{\partial \bar{\theta}^{\dot{a}}} \mathcal{F}' = \frac{\partial}{\partial \bar{\theta}^{\dot{a}}} \left(1 - i\, a^{\mu} \hat{P}_{\mu} - i\, \zeta \cdot \mathcal{Q} - i\, \bar{\xi} \cdot \bar{\mathcal{Q}} \right) \mathcal{F}$$

$$= \frac{\partial}{\partial \bar{\theta}^{\dot{a}}} \mathcal{F} - i\, \frac{\partial}{\partial \bar{\theta}^{\dot{a}}} \left(a^{\mu} \hat{P}_{\mu} + \zeta \cdot \mathcal{Q} + \bar{\xi} \cdot \bar{\mathcal{Q}} \right) \mathcal{F}$$

$$= -i\, \frac{\partial}{\partial \bar{\theta}^{\dot{a}}} \left(a^{\mu} \hat{P}_{\mu} + \zeta \cdot \mathcal{Q} + i\, \bar{\xi} \cdot \bar{\mathcal{Q}} \right) \mathcal{F}$$

where in the last step we used the fact that by definition $\mathcal{F}$ obeys the constraint. Note that in the last line, the derivative with respect to $\bar{\theta}^{\dot{a}}$ acts on everything to its right, including the superfield.

Recall that we are checking whether or not this is equal to zero, which will tell us if the constraint is consistent with supersymmetric transformations. Now comes the crucial observation: If we can "pass" the $\partial / \partial \bar{\theta}^{\dot{a}}$ through the parentheses to apply it to $\mathcal{F}$, we will automatically get zero because $\mathcal{F}$ satisfies the constraint, by assumption. This is therefore the key point: *The constraint is consistent with SUSY if it commutes with*

$$a^\mu \hat{P}_\mu + \zeta \cdot Q + \bar{\zeta} \cdot \bar{Q}$$

The operator $\partial/\partial\bar{\theta}^{\dot{a}}$ obviously commutes with $a^\mu \hat{P}_\mu$, so we are left with checking whether the commutator

$$\left[\frac{\partial}{\partial\bar{\theta}^{\dot{a}}}, \zeta \cdot Q + \bar{\zeta} \cdot \bar{Q} \right]$$

vanishes. Recall that the spinor ζ is a constant spinor (independent of both the spacetime coordinates x and the Grassmann coordinates θ) so that we can pass the derivative through ζ or $\bar{\zeta}$ on the condition that we add a minus sign because both ζ and $\partial/\partial\theta$ are Grassmann quantities. This allows us to write the commutator as

$$\left[\frac{\partial}{\partial\bar{\theta}^{\dot{a}}}, \zeta^b Q_b + \bar{\zeta}_{\dot{b}} \bar{Q}^{\dot{b}} \right] = \frac{\partial}{\partial\bar{\theta}^{\dot{a}}} \left(\zeta^b Q_b + \bar{\zeta}_{\dot{b}} \bar{Q}^{\dot{b}} \right) - \left(\zeta^b Q_b + \bar{\zeta}_{\dot{b}} \bar{Q}^{\dot{b}} \right) \frac{\partial}{\partial\bar{\theta}^{\dot{a}}}$$

$$= -\zeta^b \left(\frac{\partial}{\partial\bar{\theta}^{\dot{a}}} Q_b + Q_b \frac{\partial}{\partial\bar{\theta}^{\dot{a}}} \right) - \bar{\zeta}_{\dot{b}} \left(\frac{\partial}{\partial\bar{\theta}^{\dot{a}}} \bar{Q}^{\dot{b}} + \bar{Q}^{\dot{b}} \frac{\partial}{\partial\bar{\theta}^{\dot{a}}} \right)$$

$$= -\zeta^b \left\{ \frac{\partial}{\partial\bar{\theta}^{\dot{a}}}, Q_b \right\} - \bar{\zeta}_{\dot{b}} \left\{ \frac{\partial}{\partial\bar{\theta}^{\dot{a}}}, \bar{Q}^{\dot{b}} \right\}$$

Since the parameters ζ_b and $\bar{\zeta}_{\dot{b}}$ are independent, *the constraint is consistent with SUSY on the condition that it anticommutes with the SUSY supercharges*:

$$\left\{ \frac{\partial}{\partial\bar{\theta}^{\dot{a}}}, Q_b \right\} \stackrel{?}{=} 0$$

$$\left\{ \frac{\partial}{\partial\bar{\theta}^{\dot{a}}}, \bar{Q}^{\dot{b}} \right\} \stackrel{?}{=} 0$$

Using the explicit representations of the SUSY charges (11.35) and (11.37), we see that the operator $\partial/\partial\bar{\theta}^{\dot{a}}$ does indeed anticommute with $\bar{Q}^{\dot{b}}$ but *not* with Q_b because the latter contains $\bar{\theta}$. The conclusion is that the constraint $\bar{\partial}_{\dot{a}}\mathcal{S} = 0$ is not preserved by a supersymmetric transformation and is therefore useless to us.

We already knew that the constraint $\partial\Phi/\partial\bar{\theta}^{\dot{a}} = 0$ is not preserved by SUSY. The reason we went through this whole discussion, pretending all along that we did not already know the answer, is that now we have a simple criterion to find out if a constraint on superfields is consistent with SUSY. Let's write such a constraint as

$$\bar{D}_{\dot{a}}\mathcal{S}(x,\theta,\bar{\theta}) = 0$$

where $\bar{D}_{\dot{a}}$ is some differential operator acting in superspace.[*] What we have discovered is that for this to be a valid constraint, the operator $\bar{D}_{\dot{a}}$ must anticommute with the SUSY supercharges:

$$\{\bar{D}_{\dot{a}}, \mathcal{Q}_b\} = 0 \tag{11.50}$$

$$\{\bar{D}_{\dot{a}}, \bar{\mathcal{Q}}^{\dot{b}}\} = 0 \tag{11.51}$$

What we now want to do is to construct an explicit operator $\bar{D}_{\dot{a}}$ (as a differential operator acting in superspace) that will satisfy those two conditions.

We know that the partial derivative with respect to $\bar{\theta}^{\dot{a}}$ does not work because only the second condition is fulfilled:

$$\left\{\frac{\partial}{\partial\bar{\theta}^{\dot{a}}}, \mathcal{Q}_b\right\} \neq 0 \tag{11.52}$$

$$\left\{\frac{\partial}{\partial\bar{\theta}^{\dot{a}}}, \bar{\mathcal{Q}}^{\dot{b}}\right\} = 0 \tag{11.53}$$

Still, let's use this operator as our starting point in constructing $\bar{D}_{\dot{a}}$. After all, it satisfies Eq. (11.53), so that's not a bad start. The simplest thing we can try is to add a term to the derivative that will ensure that Eq. (11.52) also will be satisfied. Of course, we don't want to spoil Eq. (11.53) by doing that, so the added term must by itself anticommute with $\bar{\mathcal{Q}}^{\dot{b}}$.

It is clear by looking at the definition of $\bar{\mathcal{Q}}^{\dot{b}}$ [Eq. (11.37)], that it anticommutes with θ but not with x^{μ} or $\bar{\theta}$. So let's add a term linear in θ to $\partial/\partial\bar{\theta}^{\dot{a}}$. It will make our life easier if we add a term proportional to θ^a (instead of θ_a) because the charge $\mathcal{Q}_b$ contains a derivative with respect to Grassmann variables with upper indices. So let's try

$$\bar{D}_{\dot{a}} = \frac{\partial}{\partial\bar{\theta}^{\dot{a}}} + C_{\dot{a}c}\,\theta^c \tag{11.54}$$

---

[*] The reason for choosing the operator to have a lower dotted index is simply based on convenience because it will simplify the next few steps. Of course, once we have the components with lower dotted indices, we can work out the components with any other type of indices.

where we have included the indices that the quantity C must necessarily carry. This operator satisfies Eq. (11.53) if we take $C_{\dot{a}c}$ to be independent of the Grassmann variables. Our goal now is to see if it is possible to choose an expression $C_{\dot{a}c}$ such that Eq. (11.54) anticommutes with $\mathcal{Q}_b$:

$$\left\{ \frac{\partial}{\partial \bar{\theta}^{\dot{a}}} + C_{\dot{a}c}\, \theta^c,\, \mathcal{Q}_b \right\} = 0$$

Using the explicit representation of $\mathcal{Q}_b$ [Eq. (11.35)], we get

$$\left\{ \frac{\partial}{\partial \bar{\theta}^{\dot{a}}} + C_{\dot{a}c}\, \theta^c,\, i\, \frac{\partial}{\partial \theta^b} + \frac{i}{2}\, (\sigma^\mu)_{b\dot{d}}\, \bar{\theta}^{\dot{d}}\, \hat{P}_\mu \right\} = 0$$

or

$$i\left\{ \frac{\partial}{\partial \bar{\theta}^{\dot{a}}},\, \frac{\partial}{\partial \theta^b} \right\} + \frac{i}{2}\left\{ \frac{\partial}{\partial \bar{\theta}^{\dot{a}}},\, (\sigma^\mu)_{b\dot{d}}\, \bar{\theta}^{\dot{d}}\, \hat{P}_\mu \right\} + i\left\{ C_{\dot{a}c}\, \theta^c,\, \frac{\partial}{\partial \theta^b} \right\}$$

$$+ \frac{i}{2}\left\{ C_{\dot{a}c}\, \theta^c,\, (\sigma^\mu)_{b\dot{d}}\, \bar{\theta}^{\dot{d}}\, \hat{P}_\mu \right\} = 0 \tag{11.55}$$

The first anticommutator is identically zero. The last anticommutator also vanishes on the condition that C has no spacetime dependence (so that it commutes with $\hat{P}_\mu$). This leaves two nontrivial anticommutators. The second and third anticommutators of Eq. (11.55) cancel out on the condition that (see Exercise 11.6)

$$C_{\dot{a}b} = -\frac{1}{2}\, (\sigma^\mu)_{b\dot{a}}\, \hat{P}_\mu$$

$$= -\frac{i}{2}\, (\sigma^\mu)_{b\dot{a}}\, \partial_\mu \tag{11.56}$$

EXERCISE 11.6

Prove that the second and third anticommutators of Eq. (11.55) cancel out if $C_{\dot{a}b}$ is given by Eq. (11.56). You may, of course, just plug in the result in Eq. (11.56), but it is more interesting (and rewarding) to obtain the answer directly from Eq. (11.55). In doing it this way, you may assume that $C_{\dot{a}b}$ does not depend on θ so that it commutes with $\partial/\partial\theta^b$ to pull it out of the third anticommutator. Of course, if that assumption had not worked, we would have had to consider a more general ansatz for $C_{\dot{a}b}$.

Using this in Eq. (11.54), we finally obtain

$$\bar{D}_{\dot{a}} = \frac{\partial}{\partial \bar{\theta}^{\dot{a}}} - \frac{i}{2} \theta^c (\sigma^\mu)_{c\dot{a}} \partial_\mu \qquad (11.57)$$

Of course, we could have started with a constraint containing a derivative with respect to θ^a instead of a derivative with respect to $\bar{\theta}^{\dot{a}}$. Imposing that this second constraint anticommutes with all the supercharges then leads to the expression

$$D_a = \frac{\partial}{\partial \theta^a} - \frac{i}{2} (\sigma^\mu)_{a\dot{b}} \bar{\theta}^{\dot{b}} \partial_\mu \qquad (11.58)$$

The two differential operators we just obtained can be used to write down constraints that are consistent with SUSY transformations. Our next step obviously is to look for the most general superfields that satisfy

$$\bar{D}_{\dot{a}} \, \mathcal{S}_{\text{LC}}(x, \theta, \bar{\theta}) = 0 \qquad (11.59)$$

These superfields are called *left-chiral superfields* for reasons that will become clear in Chapter 12, hence the subscript LC. The fields satisfying

$$D_a \, \mathcal{S}_{\text{RC}} = 0 \qquad (11.60)$$

are called *right-chiral superfields*. However, most supersymmetric theories, including the minimal supersymmetric standard model, do not use right-chiral superfields, so we will focus on left-chiral superfields in Chapter 12.

Using some of the tricks of Exercise 11.5, it is straightforward to raise the indices to obtain

$$D^a = -\frac{\partial}{\partial \theta_a} + \frac{i}{2} \bar{\theta}_{\dot{b}} (\bar{\sigma}^\mu)^{\dot{b}a} \partial_\mu$$

$$\bar{D}^{\dot{a}} = -\frac{\partial}{\partial \bar{\theta}_{\dot{a}}} + \frac{i}{2} (\bar{\sigma}^\mu)^{\dot{a}b} \theta_b \partial_\mu \qquad (11.61)$$

We may then define dot products between the constraints, i.e.,

$$D \cdot D \equiv D^a D_a$$

$$\bar{D} \cdot \bar{D} \equiv \bar{D}_{\dot{a}} \bar{D}^{\dot{a}}$$

that we will use later on. An important observation is that on any superfield $\mathcal{S}$, we have

$$D_a D \cdot D \mathcal{S} = D^a D \cdot D \mathcal{S} = 0$$

and

$$\bar{D}_{\dot{a}} \bar{D} \cdot \bar{D} \mathcal{S} = \bar{D}^{\dot{a}} \bar{D} \cdot \bar{D} \mathcal{S} = 0 \tag{11.62}$$

This follows from the fact that three constraints of the same type necessarily contain either a Grassmann variable squared or the second derivative with respect to a Grassmann variable.

Later we will need the anticommutation relations satisfied by the constraints. Obviously,

$$\{D_a, D_b\} = \{D^a, D_b\} = \{\bar{D}_{\dot{a}}, \bar{D}_{\dot{b}}\} = \{\bar{D}^{\dot{a}}, \bar{D}_{\dot{b}}\} = 0$$

whereas a short calculation gives

$$\{D_a, \bar{D}_{\dot{b}}\} = -i(\sigma^\mu)_{a\dot{b}}\, \partial_\mu \tag{11.63}$$

11.8 Quiz

1. What is the result of the integral $\int d\theta_1\, \theta_2$?
2. What is the derivative

$$\frac{\partial}{\partial \theta_1}\left(\theta^1 \theta_2\right)$$

   equal to?

3. Consider a general function of *three* Grassmann variables $F(\theta_1, \theta_2, \theta_3)$. If we expand in all three Grassmann variables, how many independent terms do we generate?

4. How many supercharges did we have to introduce?

5. What is the result of $\epsilon^{ab}\partial_b$?

CHAPTER 12

Left-Chiral Superfields

As mentioned at the end of Chapter 11, a superfield satisfying the constraint (11.59) is called a *left-chiral superfield*. In this chapter we will only discuss left-chiral superfields that are scalars under Lorentz transformations. We will denote these superfields by Φ. The reason why such a superfield is called *left-chiral* even though it is a scalar under Lorentz transformations is that one of its component fields is a left-chiral spinor, as we will see explicitly shortly. In Chapter 13 we will encounter a left-chiral superfield that is not a scalar.

12.1 General Expansion of Left-Chiral Superfields

One obvious observation we can make about left-chiral superfields is that since the constraint does not involve the coordinates θ_a, they can be arbitrary functions of that variable. Of course, a superfield depending only on θ would automatically be left-chiral, but we want to be as general as possible by introducing a dependence on x and $\bar{\theta}$ as well. The dependence of Φ on these two variables clearly must be related in a nontrivial manner.

To proceed, the trick is to construct a new coordinate, let's call it y^μ, that is in principle a function of both x^μ and $\bar{\theta}$ and that satisfies by itself the constraint equation

$$\bar{D}_{\dot{a}} y^\mu(x^\mu, \bar{\theta}) = 0 \tag{12.1}$$

If we succeed in identifying such a coordinate, then *any* function of y^μ and θ will be a left-chiral superfield Φ because it will trivially satisfy

$$\bar{D}_{\dot{a}} \Phi(y^\mu, \theta) = 0$$

The solution of Eq. (12.1) is surprisingly simple. Since the differential operator $\bar{D}_{\dot{a}}$ contains derivatives with respect to x^μ and $\bar{\theta}^{\dot{a}}$ [see Eq. (11.57)], the simplest guess is to write y_μ as x_μ plus something proportional to $\bar{\theta}$. We therefore try the ansatz

$$y^\mu = x^\mu + \bar{\theta}^{\dot{b}} K^\mu_{\dot{b}} \tag{12.2}$$

where K is an unknown expression that we must now find. We have included the indices on K that it must carry for the equation to be consistent. Note that y^μ is an ordinary (i.e., commuting) variable, so $K^\mu_{\dot{b}}$ must be a fermionic quantity (which is also obvious from the fact that it carries a spinor index). Let us now assume that $K^\mu_{\dot{b}}$ does not depend on x_μ and $\bar{\theta}^{\dot{a}}$ so that $\bar{D}_{\dot{a}} K^\mu_{\dot{b}} = 0$. Of course, if this doesn't work, we will have to come back and drop this assumption. Note that we have chosen to place $\theta^{\dot{a}}$ to the left of K so that we won't have to worry about picking up a minus sign as we move the partial derivative $\bar{\partial}_{\dot{a}}$ through $K^\mu_{\dot{b}}$.

Now we impose Eq. (12.1). A quick calculation reveals that

$$\bar{D}_{\dot{a}} y^\mu = K^\mu_{\dot{a}} - \frac{i}{2} \theta^c (\sigma^\mu)_{c\dot{a}}$$

so we must take

$$K^\mu_{\dot{a}} = \frac{i}{2} \theta^c (\sigma^\mu)_{c\dot{a}}$$

Putting this back into Eq. (12.2), we finally get

$$y^\mu = x^\mu + \frac{i}{2} \bar{\theta}^{\dot{b}} \theta^c (\sigma^\mu)_{c\dot{b}}$$

We can move the $\bar{\theta}^{\dot{b}}$ to the far right to get

$$\boxed{y^\mu = x^\mu - \frac{i}{2}\theta\sigma^\mu\bar{\theta}}$$

(12.3)

The most general left-chiral superfield is therefore a general function of y_μ and θ, i.e.,

$$\Phi(y^\mu, \theta)$$

We will *choose* to take Φ to be a scalar under Lorentz transformations, although it is possible to write left-chiral fields with different Lorentz properties. We will meet such left-chiral superfields in Chapter 13.

It is straightforward to see that the most general right-chiral superfield [that obeys the constraint (11.60)] is a function of the variables $\bar{\theta}$ and z^μ with

$$z^\mu \equiv x^\mu + \frac{i}{2}\theta\sigma^\mu\bar{\theta}$$

Note that if we take the hermitian conjugate of a left-chiral superfield, we obtain

$$\Phi^\dagger(y^\mu, \theta) = \Phi(y^{\mu\dagger}, \bar{\theta}) = \Phi(z^\mu, \bar{\theta})$$

which is a right-chiral superfield! So we obtain the important result that for any left-chiral superfield Φ,

$$\boxed{D_a \Phi^\dagger = 0}$$

(12.4)

Let us now see what the component fields of a left-chiral superfield are. To find this out, we simply expand in the Grassmann variables appearing in the field, as we did in Section 11.4. We will do this in two steps. We first expand in θ only, leaving y^μ untouched. This, of course, gives us exactly the same result as Eq. (11.25), namely,

$$\Phi_{lc}(y, \theta) = \phi(y) + \theta \cdot \chi(y) + \frac{1}{2}\theta \cdot \theta F(y)$$

(12.5)

We see that the first field, ϕ, is a scalar because the superfield itself is taken to be a scalar. Since $\theta \cdot \theta$ is a Lorentz invariant, F is also a scalar field. Finally, since θ is a left-chiral spinor, χ also must be a left-chiral spinor. Those are the component fields of the left-chiral superfield Φ. The fact that one of the component fields of Φ

is a left-chiral spinor, whereas there is no right-chiral spinor, is the reason why the superfield is called a *left-chiral superfield*. Keep in mind, however, that Φ itself is a Lorentz scalar!

We are not quite done. We still haven't fully expanded in terms of the Grassmann variables because y^μ depends on them. We do that now. The Taylor expansion of the component fields is straightforward. Consider first the scalar field:

$$\phi(y) = \phi\left(x^\mu - \frac{i}{2}\theta\sigma^\mu\bar\theta\right)$$

$$= \phi(x) - \frac{i}{2}\theta\sigma^\mu\bar\theta\partial_\mu\phi(x) - \frac{1}{8}\theta\sigma^\mu\bar\theta\theta\sigma^\nu\bar\theta\partial_\mu\partial_\nu\phi(x) \qquad (12.6)$$

The last line is an exact result, not an approximation, because higher-order terms automatically vanish since they contain Grassmann variables squared. Note that $\theta\sigma^\mu\bar\theta$ is a commuting quantity, so we may move it as a block anywhere we want. The last term on the right of Eq. (12.6) can be simplified using Eq. (A.56), which allows us to write

$$\theta\sigma^\mu\bar\theta\theta\sigma^\nu\bar\theta\partial_\mu\partial_\nu\phi(x) = \frac{1}{2}\eta^{\mu\nu}\theta\cdot\theta\bar\theta\cdot\bar\theta\partial_\mu\partial_\nu\phi(x)$$

$$= \frac{1}{2}\theta\cdot\theta\bar\theta\cdot\bar\theta\partial^\nu\partial_\nu\phi(x)$$

$$\equiv \frac{1}{2}\theta\cdot\theta\bar\theta\cdot\bar\theta\Box\phi(x)$$

where $\Box$ is the d'Alembertian operator. Using this result, Eq. (12.6) can be written as

$$\phi(y) = \phi\left(x^\mu - \frac{i}{2}\theta\sigma^\mu\bar\theta\right)$$

$$= \phi(x) - \frac{i}{2}\theta\sigma^\mu\bar\theta\partial_\mu\phi(x) - \frac{1}{16}\theta\cdot\theta\bar\theta\cdot\bar\theta\Box\phi(x) \qquad (12.7)$$

The expansion of $\chi(y)$ is identical except that we need only to keep the first two terms in the expansion with respect to θ because χ is already dotted with θ. Thus we get

$$\chi_a(y) = \chi_a(x^\mu - \frac{i}{2}\theta\sigma^\mu\bar\theta) = \chi_a(x) - \frac{i}{2}\theta\sigma^\mu\bar\theta\partial_\mu\chi_a(x)$$

On the other hand, since F is multiplied by $\theta \cdot \theta$, we may replace $F(y)$ directly by $F(x)$ because all higher-order terms are automatically zero.

Substituting the expansions of $\phi(y)$, $\chi(y)$, and $F(y)$ into Eq. (12.5), we finally get that a left-chiral superfield has the following expansion in terms of its component fields:

$$
\begin{aligned}
\Phi = {}& \phi(x) - \frac{i}{2} \theta \sigma^\mu \bar\theta \partial_\mu \phi(x) - \frac{1}{16} \theta \cdot \theta \bar\theta \cdot \bar\theta \Box \phi(x) \\
& + \theta \cdot \chi(x) - \frac{i}{2} \theta \sigma^\mu \bar\theta \theta \cdot \partial_\mu \chi(x) + \frac{1}{2} F(x) \theta \cdot \theta
\end{aligned}
\tag{12.8}
$$

Recall that $\theta \cdot \partial_\mu \chi$ stands for $\theta^a \cdot \partial_\mu \chi_a$. Obviously, a left-chiral superfield does not contain all the terms that the most general superfields have (see Exercise 11.2).

The fact that the constraint we have used is preserved by supersymmetry (SUSY) tells us that applying an arbitrary SUSY transformation to Eq. (12.8) will generate a result that will be of the same form; i.e., there will be no new type of terms produced (e.g., no terms with $\bar\theta \cdot \bar\theta$ and no θ will be generated by a SUSY transformation).

We see that a general left-chiral superfield contains fields with the same Lorentz transformation properties as the ones we needed to build the Wess-Zumino lagrangian: a left-chiral spinor field and two complex scalar fields. However, to really show that we can identify the components fields of Φ with the fields of the Wess-Zumino lagrangian, we need to show that they transform the same way under SUSY. We will prove that this is indeed the case, a result that we anticipated by using the same names for the three fields that we used in Chapters 5 and 6. The beauty of the superfield formalism is that we will recover the SUSY transformations of all three component fields from the single transformation of the superfield.

12.2 SUSY Transformations of the Component Fields

By definition, the SUSY-transformed superfield Φ' can be written in terms of the transformed component fields as

$$
\begin{aligned}
\Phi' = {}& \phi'(x) - \frac{i}{2} \theta \sigma^\mu \bar\theta \partial_\mu \phi'(x) - \frac{1}{16} \theta \cdot \theta \bar\theta \cdot \bar\theta \Box \phi'(x) \\
& + \theta \cdot \chi'(x) - \frac{i}{2} \theta \sigma^\mu \bar\theta \theta \cdot \partial_\mu \chi'(x) + \frac{1}{2} F'(x) \theta \cdot \theta
\end{aligned}
\tag{12.9}
$$

All we did was rewrite Eq. (12.8) with a prime on all the fields.

Now we calculate the explicit SUSY-transformed superfield using Eq. (11.48). The effect of $\hat{P}_\mu$ is simply to translate all the field in spacetime by $x^\mu \to x^\mu + a^\mu$, so we will only consider the effect of the supercharges. We have, from Eq. (11.32),

$$\Phi' = \Phi + \frac{i}{2}\left(\zeta\sigma^\mu\bar{\theta} - \theta\sigma^\mu\bar{\zeta}\right)\partial_\mu\Phi + \zeta\cdot\partial\Phi + \bar{\zeta}\cdot\bar{\partial}\Phi \qquad (12.10)$$

So now we have two different expressions for the SUSY-transformed field. Equation (12.9) gives us the transformed superfield in terms of the transformed component fields $\phi'(x)$, $\chi'(x)$, and $F'(x)$. On the other hand, the right-hand side of Eq. (12.10) also gives the transformed superfield in terms of the infinitesimal parameters ζ, $\bar{\zeta}$, and the original fields $\phi(x)$, $\chi(x)$, and $F(x)$ [after using Eq. (12.8) for Φ]. Setting the two results equal to one another will give us the SUSY transformations of all the component fields.

The first step in evaluating the right-hand side of Eq. (12.10) is to calculate the partial derivatives of Φ with respect to x^μ, θ^a, and $\bar{\theta}^{\dot{a}}$. The derivative with respect to x^μ is straightforward:

$$\partial_\mu\Phi = \partial_\mu\phi - \frac{i}{2}\theta\sigma^\nu\bar{\theta}\partial_\mu\partial_\nu\phi - \frac{1}{16}\theta\cdot\theta\,\bar{\theta}\cdot\bar{\theta}\partial_\mu\partial^\nu\partial_\nu\phi + \theta\cdot\partial_\mu\chi$$

$$-\frac{i}{2}\theta\sigma^\nu\bar{\theta}\theta\cdot\partial_\mu\partial_\nu\chi + \frac{1}{2}\theta\cdot\theta\,\partial_\mu F \qquad (12.11)$$

The derivatives with respect to the Grassmann variables require some care. We get

$$\frac{\partial\Phi}{\partial\theta^a} = -\frac{i}{2}(\sigma^\mu)_{ab}\bar{\theta}^b\,\partial_\mu\phi - \frac{1}{8}\theta_a\bar{\theta}\cdot\bar{\theta}\Box\phi + \chi_a - \frac{i}{2}(\sigma^\mu)_{ab}\bar{\theta}^b\theta\cdot\partial_\mu\chi$$

$$-\frac{i}{2}(\partial_\mu\chi_a)\theta\sigma^\mu\bar{\theta} + \theta_a F \qquad (12.12)$$

where we have used the result of Exercise 11.3 for the derivative of $\theta\cdot\theta$. On the other hand, we find [after using identity (A.50)]

$$\frac{\partial\Phi}{\partial\bar{\theta}_{\dot{a}}} = \frac{i}{2}(\bar{\sigma}^\mu)^{\dot{a}b}\theta_b\partial_\mu\phi - \frac{1}{8}\theta\cdot\theta\bar{\theta}^{\dot{a}}\,\Box\phi + \frac{i}{2}(\bar{\sigma}^\mu)^{\dot{a}b}\theta_b\theta\cdot\partial_\mu\chi \qquad (12.13)$$

What we must do now is to substitute Eqs. (12.11), (12.12), and (12.13) into the right-hand side of Eq. (12.10). Note that all terms with three θ or three $\bar{\theta}$ automatically vanish because they necessarily contain the square of a Grassmann

coordinate. This leaves us with

$$\Phi' = \Phi + \frac{i}{2}\left(\zeta\sigma^\mu\bar\theta - \theta\sigma^\mu\bar\zeta\right)\partial_\mu\phi + \frac{1}{4}\left(\zeta\sigma^\mu\bar\theta - \theta\sigma^\mu\bar\zeta\right)\theta\sigma^\nu\bar\theta\partial_\mu\partial_\nu\phi$$

$$+ \frac{i}{2}\left(\zeta\sigma^\mu\bar\theta - \theta\sigma^\mu\bar\zeta\right)\theta\cdot\partial_\mu\chi + \frac{1}{4}\zeta\sigma^\mu\bar\theta\theta\sigma^\nu\bar\theta\theta\cdot\partial_\mu\partial_\nu\chi + \frac{i}{4}\zeta\sigma^\mu\bar\theta\partial_\mu F\,\theta\cdot\theta$$

$$- \frac{i}{2}\zeta\sigma^\mu\bar\theta\partial_\mu\phi - \frac{1}{8}\zeta\cdot\theta\bar\theta\cdot\bar\theta\Box\phi + \zeta\cdot\chi - \frac{i}{2}\zeta\sigma^\mu\bar\theta\theta\cdot\partial_\mu\chi - \frac{i}{2}\zeta\cdot\partial_\mu\chi\theta\sigma^\mu\bar\theta$$

$$+ \zeta\cdot\theta F + \frac{i}{2}\bar\zeta\bar\sigma^\mu\theta\theta\partial_\mu\phi - \frac{1}{8}\theta\cdot\theta\bar\zeta\cdot\bar\theta\Box\phi + \frac{i}{2}\bar\zeta\bar\sigma^\mu\theta\theta\cdot\partial_\mu\chi \qquad (12.14)$$

All we have to do now is to replace Φ and Φ' by Eqs. (12.8) and (12.9), and we will be able to read off the transformations of the fields (after a bit of work, of course). The trick is to realize that terms with different powers of the Grassmann variables θ^a and $\bar\theta^{\dot a}$ are independent.

For example, consider the terms with no Grassmann variables appearing on the two sides of Eq. (12.14). Setting them equal, we find the following result:

$$\phi' = \phi + \zeta\cdot\chi \qquad (12.15)$$

which is exactly the transformation law for ϕ that we obtained in Chapter 6 [see Eq. (6.99)]!

Now consider the terms that are linear in θ and have no $\bar\theta$. We get

$$\theta\cdot\chi' = \theta\cdot\chi - \frac{i}{2}\theta\sigma^\mu\bar\zeta\partial_\mu\phi + \zeta\cdot\theta F + \frac{i}{2}\bar\zeta\bar\sigma^\mu\theta\partial_\mu\phi$$

Using Eq. (A.49), we may write $\bar\zeta\bar\sigma^\mu\theta = -\theta\sigma^\mu\bar\zeta$, which shows that the last term is actually equal to the second term, giving us (showing now explicitly all the indices):

$$\theta^a\chi'_a = \theta^a\chi_a - i\,\theta^a\,(\sigma^\mu)_{ab}\,\bar\zeta^{\dot b}\,\partial_\mu\phi + \theta^a\,\zeta_a\,F$$

which yields

$$\chi'_a = \chi_a - i\,(\sigma^\mu)_{ab}\,\bar\zeta^{\dot b}\partial_\mu\phi + \zeta_a\,F \qquad (12.16)$$

This is indeed the same as in Eq. (6.99), although it is not obvious at first sight. Recall that in Chapter 6, all the spinors had lower undotted indices, so we need to express the ζ spinor on right-hand side of Eq. (12.16) in terms of its lower undotted components. Using $\bar{\zeta}^{\dot{b}} = (i\sigma^2)^{\dot{b}\dot{c}}\bar{\zeta}_{\dot{c}}$ [see Eq. (A.31)] and $\bar{\zeta}_{\dot{c}} = \zeta_c^*$, we can write our result as

$$\chi' = \chi - i\sigma^\mu (i\sigma^2) \zeta^* \partial_\mu \phi + \zeta F$$

with the convention now that the indices on all the spinors are downstairs and undotted. This is exactly our result in Eq. (6.99), including the term proportional to the auxiliary field F in the transformation of the spinor.

Now let's consider the terms containing two Grassmann variables θ^a and no $\bar{\theta}$ in Eq. (12.14). These are

$$\frac{1}{2}F'\theta \cdot \theta = \frac{1}{2}F\theta \cdot \theta - \frac{i}{2}\theta\sigma^\mu\bar{\zeta}\theta \cdot \partial_\mu\chi + \frac{i}{2}\bar{\zeta}\bar{\sigma}^\mu\theta\theta \cdot \partial_\mu\chi$$

The two last terms can be combined using again $\bar{\zeta}\bar{\sigma}^\mu\theta = -\theta\sigma^\mu\bar{\zeta}$, giving us

$$\frac{1}{2}F'\theta \cdot \theta = \frac{1}{2}F\theta \cdot \theta - i\theta\sigma^\mu\bar{\zeta}\theta \cdot \partial_\mu\chi$$

The last term can be rewritten with the two Grassmann variables dotted into one another using the identity (12.47) (which will be derived in Exercise 12.2):

$$\frac{1}{2}F'\theta \cdot \theta = \frac{1}{2}F\theta \cdot \theta + \frac{i}{2}\theta \cdot \theta(\partial_\mu\chi)\sigma^\mu\bar{\zeta}$$

from which we can read off the transformation of the auxiliary field, i.e.,

$$F' = F + i(\partial_\mu\chi)\sigma^\mu\bar{\zeta}$$

Using once again Eq. (A.49), we may write

$$(\partial_\mu\chi)\sigma^\mu\bar{\zeta} = -\bar{\zeta}\bar{\sigma}^\mu\partial_\mu\chi$$

which leads to

$$\delta F = -i\bar{\zeta}\bar{\sigma}^\mu\partial_\mu\chi \qquad (12.17)$$

in agreement with Eq. (6.99).

Let us pause and summarize what we have accomplished in this section. By comparing the terms with no Grassmann variables, with a single Grassmann variable θ^a, and with two Grassmann variables $\theta^a\theta^b$ on the two sides of Eq. (12.14), we have established the transformation properties of the fields $\phi(x)$, $\chi(x)$, and $F(x)$ under

SUSY and have found that they agree with the transformations we established in Chapters 5 and 6.

However, we are clearly not done. We should consider all the other terms that appear in Eq. (12.14). Now that we have established the transformation laws of all the fields, there is no leeway left: We should show that with the transformation laws we just established, *all* the terms on the two sides of Eq. (12.14) agree. This is obviously quite labor-intensive, and we won't pursue the demonstration here, but rest assured that it does work. The courageous reader is invited to complete the calculation.

12.3 Constructing SUSY Invariants Out of Left-Chiral Superfields

The key reason for the introduction of superfields is that they make the construction of SUSY-invariant theories very easy. We will now see how this works in the case of left-chiral superfields.

The starting point is the seemingly innocuous observation that the $F(x)$ field transforms as a total derivative under SUSY. Since the $F(x)$ field is the coefficient of the $\theta \cdot \theta / 2$ term in the expansion of Φ, we can phrase this by saying: *The coefficient of the $\theta \cdot \theta / 2$ term in the expansion of a left-chiral superfield $\Phi(x, \theta, \bar{\theta})$ transforms as a total derivative under SUSY.*

We saw in the section on Grassmann calculus that it is possible to "project out," i.e., to isolate, the coefficient of $\theta \cdot \theta / 2$ of an arbitrary function of the Grassmann variables simply by integrating over $d^2\theta$, so we can say that

$$\int d^2\theta \, \Phi(x, \theta, \bar{\theta}) \tag{12.18}$$

transforms as a total derivative under SUSY. Recall that $d^2\theta$ is a shorthand for $d\theta_1 d\theta_2$.

The fact that Eq. (12.18) transforms as a total derivative tells us that it can be used as a term in a SUSY-invariant lagrangian density. In other words, the quantity

$$\int d^4x \, d^2\theta \, \Phi(x, \theta, \bar{\theta}) \tag{12.19}$$

is *invariant* under a SUSY transformation.

In itself, Eq. (12.19) is not interesting because it just gives the field $F(x)$, not interactions or kinetic terms. What we would like to do is to introduce SUSY-invariant interactions and kinetic terms using superfields.

The next key observation is that *the product of any number of left-chiral superfields is itself a chiral superfield*. What we mean by this is that the product of left-chiral superfields can be written as a superfield that has the same expansion in the Grassmann variables *and* that transforms as a single left-chiral superfield under SUSY.

This will become more clear if we carry out the explicit proof in the special case of the product of two left-chiral superfields. To be general, we will consider the product of two distinct superfields that we will call Φ_i and Φ_j. This is preferable to using Φ_1 and Φ_2, which might lead to confusion when we start writing out the Weyl spinor fields because χ_1 and χ_2 could be mistaken for the components of the spinors.

If we go back to writing left-chiral superfields in terms of the variables y^μ introduced in Eq. (12.3), we have

$$\Phi_i(y, \theta) = \phi_i(y) + \theta \cdot \chi_i(y) + \frac{1}{2} \theta \cdot \theta F_i(y) \tag{12.20}$$

and

$$\Phi_j(y, \theta) = \phi_j(y) + \theta \cdot \chi_j(y) + \frac{1}{2} \theta \cdot \theta F_j(y) \tag{12.21}$$

Evidently, the product of those two superfields can be cast in the form of a left-chiral superfield, i.e., with the expansion

$$\Phi_i \Phi_j \equiv \Phi_{eq} = \phi_{eq}(y) + \theta \cdot \chi_{eq}(y) + \frac{1}{2} \theta \cdot \theta F_{eq}(y) \tag{12.22}$$

The three fields ϕ_{eq}, χ_{eq}, and F_{eq} are obviously functions of the fields appearing in Φ_i and Φ_j. A short calculation reveals that

$$\phi_{eq}(y) = \phi_i(y)\phi_j(y)$$
$$\chi_{eq}(y) = \phi_i(y)\, \chi_j(y) + \phi_j(y)\, \chi_i(y)$$
$$F_{eq}(y) = \phi_i(y)\, F_j(y) + \phi_j(y)\, F_i(y) - \chi_i(y) \cdot \chi_j(y) \tag{12.23}$$

where we have used Eq. (A.54).

On the other hand, it is clear that multiplying an arbitrary superfield with a left-chiral superfield does not yield a left-chiral superfield. As a special case of this, consider multiplying a left-chiral superfield by its hermitian conjugate $\Phi^{\dagger}$:

$$\Phi^{\dagger}\Phi = \left[\phi^{\dagger}(y) + \bar{\theta} \cdot \bar{\chi}(y) + \frac{1}{2}\bar{\theta} \cdot \bar{\theta} F^{\dagger}(y)\right]\left[\phi(y) + \theta \cdot \chi(y) + \frac{1}{2}\theta \cdot \theta F(y)\right]$$

Obviously, the result is *not* a left-chiral superfield. For example, it contains a function of y times $\bar{\theta} \cdot \bar{\theta}$, which is absent in a left-chiral superfield.

Going back to the product of two left-chiral superfields $\Phi_i\Phi_j$, it is important to realize that it is not sufficient to show that it has the same Grassmann expansion as a single left-chiral superfield. To really prove that the result, Eq. (12.22), is a bona fide left-chiral superfield, we must prove that under SUSY, the fields ϕ_{eq}, χ_{eq}, and F_{eq} transform the same way as the components of a left-chiral superfield. In other words, we need to show that the SUSY transformations of these three fields are identical to transformations (12.15), (12.16), and (12.17). This turns out to be an easy proof if we use supercharges to implement SUSY transformations. In that language, what we need to check is whether the following relation holds:

$$\left[\exp(-i\,\zeta \cdot \mathcal{Q} - i\,\bar{\zeta} \cdot \bar{\mathcal{Q}})\,\Phi_i\right]\left[\exp(-i\,\zeta \cdot \mathcal{Q} - i\,\bar{\zeta} \cdot \bar{\mathcal{Q}})\,\Phi_j\right]$$
$$\overset{?}{=} \exp(-i\,\zeta \cdot \mathcal{Q} - i\,\bar{\zeta} \cdot \bar{\mathcal{Q}})\,(\Phi_i\,\Phi_j)$$

Let us define

$$-i\,\zeta \cdot \mathcal{Q} - i\,\bar{\zeta} \cdot \bar{\mathcal{Q}} \equiv \mathcal{O}$$

Then, to first order in ζ, what we need to check is whether

$$\Phi_i\,(\mathcal{O}\Phi_j) + (\mathcal{O}\Phi_i)\,\Phi_j \overset{?}{=} \mathcal{O}(\Phi_i\,\Phi_j) \qquad (12.24)$$

This is actually trivially true because the supercharges $\mathcal{Q}$ and $\bar{\mathcal{Q}}$ are linear operators, and the left-chiral superfields are bosonic quantities. These two facts allow us to write

$$\mathcal{Q}(\Phi_i\,\Phi_j) = (\mathcal{Q}\Phi_i)\,\Phi_j + \Phi_i\,(\mathcal{Q}\Phi_j)$$

and similarly for $\bar{\mathcal{Q}}$. Therefore, Eq. (12.24) is indeed valid, and this completes the proof that the product of two left-chiral superfield does transform like a single left-chiral superfield under SUSY. It is then clear, by induction, that the product of any number of left-chiral superfields also transforms like a left-chiral superfield.

Now that we have convinced ourselves that multiplying left-chiral superfields produces new superfields of the same nature, the obvious question is: Why should we care? To answer this, let's go back to the observation that the auxiliary field F transforms as a total derivative under SUSY. This means that *in the product of any number of left-chiral superfields, the term proportional to $\theta \cdot \theta$ necessarily transforms as a total derivative under a SUSY transformation.*

Let's define the *F term* of any function of superfields (not necessarily left-chiral) as the coefficient of the $\theta \cdot \theta/2$ term. The factor of one-half is included because then the F term can be written very simply in terms of integrations over Grassmann variables. Let's write a function of superfields as $\mathcal{F}(\mathcal{S}_1, \mathcal{S}_2, \ldots)$, where the $\mathcal{S}_i$ are general superfields. We use the following notation to represent the F term of that function:

$$\int d^2\theta \, \mathcal{F}(\mathcal{S}_1, \mathcal{S}_2, \ldots) \equiv \mathcal{F}(\mathcal{S}_1, \mathcal{S}_2, \ldots)\Big|_F \qquad (12.25)$$

Thus we use the notation $|_F$ to mean "extraction of the F term."

Expression (12.25) is obviously a nontrivial combination of the component fields in the superfields $\mathcal{S}_1, \mathcal{S}_2, \ldots$ and is *not* in general invariant under SUSY. However, if $\mathcal{F}$ is a *holomorphic function of left-chiral superfields*, i.e., a function of the Φ_i and not of the $\Phi_i^\dagger$, then it is itself a left-chiral superfield. In that case, its F term transforms as a total derivative under SUSY, and therefore, Eq. (12.25) is a supersymmetric lagrangian density!

In the case of a holomorphic function of left-chiral superfields, we will use the notation $\mathcal{W}$ instead of $\mathcal{F}$. The conclusion then is that

$$\mathcal{W}(\Phi_1, \Phi_2, \ldots)\Big|_F \qquad (12.26)$$

is a supersymmetric lagrangian density (i.e., it transforms with a total derivative under SUSY).

The function $\mathcal{W}$ is, of course, the superpotential we introduced in Chapter 8 as a function of the scalar fields ϕ_i. Here, it is a function of the *superfields* Φ_i, which is the origin of the name *superpotential*. Now we understand the reason for the requirement of holomorphicity we had obtained in a roundabout way earlier: If we multiply by the hermitian conjugate $\Phi_i^\dagger$ of any of the superfields, the superpotential will no longer be a left-chiral superfield, and its F term will not be a supersymmetric lagrangian density.

It's hard to overemphasize the power of the result that Eq. (12.26) is a supersymmetric lagrangian. We can take any number of left-chiral superfields, multiply them and extract the F term of the result, and we will *automatically* have created a SUSY-invariant theory! Compared with the efforts we had to put in to build the simplest supersymmetric theories in Chapters 5, 6, and 8, this is child's play! The ease it brings to the construction of supersymmetric theories is one of the reasons that make the superspace approach to SUSY so enticing.

We will soon show how we can use this observation to recover some of the SUSY-invariant lagrangians we built in previous chapters. We will not reproduce all of them this way, but luckily, only a slight generalization of the trick will be needed to construct all the supersymmetric theories we have found previously at a fraction of the effort! Moreover, we will be able to construct more complex supersymmetric lagrangians, including the minimal supersymmetric standard model (MSSM), with very little work.

Before doing an explicit example, two points deserve to be emphasized. First, although the trick just described gives us a way to write down an infinite number of supersymmetric theories, the fact that we want to consider only renormalizable theories will greatly restrict the choices available to us, as we will discuss shortly.

The second point is that the F term of an *arbitrary* superfield obviously does not, in general, transform as a total derivative. For example, the F term of $\Phi^\dagger \Phi$ does not transform as a total derivative because this combination of fields is not itself a chiral superfield, as we discussed previously.

This raises the question of whether it is possible to extract a supersymmetric lagrangian from a product of arbitrary superfields. The answer is yes, and the way to do this will provide the generalization to which we hinted earlier. We will come back to this later in this chapter.

Let us now construct explicitly a supersymmetric theory using our newfound approach. The simplest (nontrivial) case is the F term of the product of two left-chiral superfields $\Phi_i \Phi_j$. We multiplied two left-chiral superfields in Eq. (12.23), but the result was expressed in terms of the variable y, and what we need is the F term of $\Phi_i(x, \theta, \bar{\theta})\Phi_j(x, \theta, \bar{\theta})$. However, a second of thought shows that *as long as we are interested only in picking up the F term, we simply may set $y = x$ directly in the product of left-chiral superfields*. In other words,

$$\left[\Phi_i(y, \theta)\, \Phi_j(y, \theta) \ldots \right]\big|_F = \left[\Phi_i(y = x, \theta)\, \Phi_j(y = x, \theta) \ldots \right]\big|_F \qquad (12.27)$$

This result is very useful because the expansion of left-chiral superfields in terms of y and θ is much simpler than the expansion in terms of x, θ, and $\bar{\theta}$!

EXERCISE 12.1
Prove Eq. (12.27). *Hint*: Using the relation between y^μ and x^μ, the proof is trivial.

Using Eqs. (12.23) and (12.27), we therefore have

$$(\Phi_i\, \Phi_j)\big|_F = \phi_i(x)\, F_j(x) + \phi_j(x)\, F_i(x) - \chi_i(x) \cdot \chi_j(x) \qquad (12.28)$$

The key point is that this quantity transforms as a total derivative under SUSY, so it is a supersymmetric lagrangian density!

If this expression seems vaguely familiar, there is a good reason: These terms appear as part of the Wess-Zumino lagrangian derived in Chapter 8. We still need more work before recovering the complete Wess-Zumino lagrangian using this approach, but for now, note how easy it was to obtain a set of SUSY-invariant interactions compared with the long calculations we had to go through in Chapter 8. All we needed to do this time was to multiply two left-chiral superfields and extract the F term of the result!

We have a supersymmetric action, but we must not forget the other conditions that a viable lagrangian must have. In particular, it must be a real quantity. As it stands, the lagrangian density in Eq. (12.28) is not real (recall that ϕ and F are complex scalar fields), so we must add its hermitian conjugate to get a real lagrangian.

In addition, Eq. (12.28) does not have the correct dimension for an action. Before correcting this, let's take a moment to open a parenthesis on the dimension of left-chiral superfields. The explicit expansion of such a superfield is (going back to the variable y for a moment)

$$\Phi = \phi(y) + \chi(y) \cdot \theta + \frac{1}{2} \theta \cdot \theta F(y)$$

The superfield therefore must have the same dimension as a scalar field, which is 1. Recall that the auxiliary field F has dimension 2, *which shows that the Grassmann variables θ_1 and θ_2 have dimension $-1/2$* . On the other hand, since

$$\frac{1}{2} \int d^2\theta \, \theta \cdot \theta = 1$$

we see that *the Grassmann infinitesimals $d\theta_1$ and $d\theta_2$ have dimension $1/2$!* This is in stark contrast with ordinary variables, for which the infinitesimal element has the same dimension as the variable itself (e.g., the dimensions of dx and x are the same).

Now let us look at

$$\int d^2\theta \, \Phi_i \, \Phi_j \Big|_F$$

As we discussed previously, the superfields have dimension 1, so $\Phi_i \, \Phi_j$ has dimension 2. But the projection of the F term, which involves integrating over $d^2\theta$, *increases the dimension by 1*, so the F term $\int d^2\theta \Phi_i \, \Phi_j$ has dimension 3. We therefore have to multiply it by a constant m having dimension of mass in order to obtain a lagrangian density with the correct dimension.

If we allow the presence of n left-chiral superfields, the most general supersymmetric real lagrangian that is constructed out of the product of two such superfields is

$$\frac{1}{2} m_{ij} \Phi_i \, \Phi_j |_F + \text{h.c.} = \frac{1}{2} m_{ij} \left(\phi_i \, F_j + \phi_j \, F_i - \chi_i \cdot \chi_j \right) + \text{h.c.} \qquad (12.29)$$

where the sum over the indices i and j goes from 1 to the number of superfields, and the constants m_{ij} have the dimension of mass. Because the superfields commute (they are bosonic quantities), the product $\Phi_i \, \Phi_j$ is symmetric under the exchange of the indices i and j, which allows us to choose the coefficients m_{ij} to also be symmetric, which is why we chose to introduce a factor of $1/2$.

So this is it! We have constructed a nontrivial interaction lagrangian density that is invariant, up to total derivatives, under SUSY. And the way we did it was to simply multiply two left-chiral superfields and extract the F term. As mentioned earlier, we have in this way recovered *some* of the terms that appeared in the Wess-Zumino lagrangian. This should not be too surprising because left-chiral superfields have the same field content as the Wess-Zumino model, so it was to be expected that constructing SUSY-invariant actions out of these superfields should lead us back to that model.

However, there is something even more interesting to mention. The expression on the left-hand side of Eq. (12.29) is exactly the same form as the first term we included in the superpotential $\mathcal{W}(\phi)$ in Chapter 8! Except that, obviously, in Chapter 8 the superpotential was a function of the scalar fields $\phi_i(x)$ and $\phi_j(x)$, not of superfields. This is, of course, not accidental, and the connection will be explained in more detail shortly.

We have only recovered some of the interactions present in the Wess-Zumino model. To obtain the other ones, the next obvious thing to try is the F term of the product of three left-chiral superfields:

$$\Phi_i \, \Phi_j \, \Phi_k \Big|_F = \int d^2\theta \, \Phi_i \, \Phi_j \, \Phi_k$$

If we write explicitly only the terms containing two factors of θ, we have

$$\Phi_i \, \Phi_j \, \Phi_k = \frac{1}{2}\theta \cdot \theta \left(\phi_i \, \phi_j \, F_k + \phi_j \, \phi_k \, F_i + \phi_i \, \phi_k \, F_j \right) + \phi_i \, \theta \cdot \chi_j \, \theta \cdot \chi_k$$

$$+ \phi_j \, \theta \cdot \chi_k \, \theta \cdot \chi_i + \phi_k \, \theta \cdot \chi_i \, \theta \cdot \chi_j + \cdots$$

$$= \frac{1}{2}\theta \cdot \theta \left(\phi_i \, \phi_j \, F_k + \phi_j \, \phi_k \, F_i + \phi_i \, \phi_k \, F_j - \phi_i \, \chi_j \cdot \chi_k - \phi_j \, \chi_k \cdot \chi_i \right.$$

$$\left. - \phi_k \, \chi_i \cdot \chi_j \right) + \cdots$$

where we have again made use of Eq. (A.54), and the dots represent terms linear in or independent of θ. Therefore, the F term is simply

$$\Phi_i \, \Phi_j \, \Phi_k \Big|_F = \phi_i \, \phi_j \, F_k + \phi_j \, \phi_k \, F_i + \phi_i \, \phi_k \, F_j$$
$$- \phi_i \, \chi_j \cdot \chi_k - \phi_j \, \chi_k \cdot \chi_i - \phi_k \, \chi_i \cdot \chi_j$$

Note that the whole expression is symmetric under the exchange of any two indices (as we expected because the chiral superfields commute). The most general combination of three chiral superfields therefore is given by

$$\frac{1}{6} \, y_{ijk} \, \Phi_i \, \Phi_j \, \Phi_k \Big|_F \tag{12.30}$$

where the dimensionless coefficients y_{ijk} can be taken to be symmetric under the exchange of any two indices for the same reason that we could take the coefficients m_{ij} to be symmetric (and the division by $3! = 6$ has the same origin as the division by 2 accompanying m_{ij}). Again, we also must add the hermitian conjugate to get a real lagrangian.

We see that Eq. (12.30) contains all the interactions terms of the Wess-Zumino lagrangian that were not present in the F term of the product of two superfields! And notice that, once more, this is precisely the second term of the superpotential obtained in Chapter 8 but now expressed in terms of the left-chiral superfields.

All the interactions of the Wess-Zumino lagrangian therefore are contained in the simple expression

$$\boxed{\mathcal{L}_{\text{int}}(\text{w-z}) = \left(\frac{1}{2} \, m_{ij} \, \Phi_i \, \Phi_j + \frac{1}{6} \, y_{ijk} \, \Phi_i \, \Phi_j \, \Phi_k + \text{h.c.} \right) \Big|_F} \tag{12.31}$$

We understand now why this potential is called the *superpotential*: It is a function of the superfields that contains all the supersymmetric interactions terms.

I hope you are impressed by how easy it is to construct SUSY-invariant interactions using superfields!

Are there other SUSY-invariant lagrangians we can generate this way? Of course, since we can multiply an arbitrary number of left-chiral superfields and extract the F term to get such lagrangians. However, remember that we only want to consider renormalizable lagrangians, for which the sum of the dimensions of the fields must not be larger than 4. This means that *we won't consider the product of more than three superfields*. At first, this might seem incorrect because a left-chiral superfield has dimension 1, so shouldn't we include products of four superfields as well?

But recall that the integration over $d^2\theta$ necessary to extract the F term increases the dimension by 1, so even though Φ^4 has dimension 4, *the F term of Φ^4 has dimension 5.*

12.4 Relation Between the Superpotential in Terms of Superfields and the Superpotential of Chapter 8

Let us now try to understand the relation between the superpotential we wrote here in terms of superfields and the superpotential we had in Chapter 8. There is obviously a connection because the two have exactly the same functional dependence. However, in Chapter 8 the superpotential was expressed in terms of the scalar component field ϕ, and even more significant, there also were derivatives of the superpotential appearing in the lagrangian. So what is the precise connection?

Consider calculating the F term of a superpotential $\mathcal{W}$ that is a function of n left-chiral superfields:

$$\mathcal{W}(\Phi_1, \Phi_2, \dots, \Phi_n)$$

To begin, assume that the superpotential is a monomial in the superfields, which contain at most one power of each superfield. In other words, $\mathcal{W}$ is simply

$$\mathcal{W}(\Phi_1, \Phi_2, \dots, \Phi_n) = \Phi_1 \Phi_2 \cdots \Phi_n \qquad (12.32)$$

Of course, some of the superfields can be omitted from the product, but none is raised to a power. It will be trivial to extend our discussion to an arbitrary polynomial later on.

To make the discussion more transparent, let us write one of the superfields, let's say Φ_i, in terms of its component fields:

$$\mathcal{W}(\Phi_1 \cdots \Phi_i \cdots \Phi_n) = \mathcal{W}\left(\Phi_1 \cdots \left(\phi_i + \chi_i \cdot \theta + \frac{1}{2} \theta \cdot \theta F_i \right) \cdots \Phi_n \right)$$

Now let's find out what will be the F term of this expression or, equivalently, what is the coefficient of $\theta \cdot \theta / 2$.

There are many contributions. One comes from keeping the auxiliary field F_i of Φ_i, in which case, since it already multiplies $\theta \cdot \theta$, we must only keep the ϕ

component of all the other superfields multiplying Φ_i. So this contribution is equal to

$$\mathcal{W}\left(\phi_1 \cdots \frac{1}{2}\theta \cdot \theta F_i \cdots \phi_n\right)$$

which may be written as

$$\mathcal{W}\left(\phi_1 \cdots 1 \cdots \phi_n\right)\frac{1}{2}\theta \cdot \theta F_i \tag{12.33}$$

In other words, we set all the superfields equal to their scalar component except Φ_i, which is set equal to 1, and we multiply the whole thing by the auxiliary field F_i and $\theta \cdot \theta/2$. But Eq. (12.33) is simply

$$\frac{\partial \mathcal{W}(\phi_1 \cdots \phi_n)}{\partial \phi_i}\frac{1}{2}\theta \cdot \theta F_i \tag{12.34}$$

where $\mathcal{W}$ is now seen as a function of the scalar components of the superfields (and ϕ_i is obviously not set equal to 1 in $\mathcal{W}$). If we extract the F term of this, we finally get

$$\mathcal{W}\left(\phi_1 \cdots \frac{1}{2}\theta \cdot \theta F_i \cdots \phi_n\right)\Big|_F = \frac{\partial \mathcal{W}(\phi_1 \cdots \phi_n)}{\partial \phi_i} F_i \tag{12.35}$$

To keep things simple, we have assumed that the superfield Φ_i appears linearly in the superpotential. However, it's clear that Eq. (12.35) remains valid no matter what the exponent of Φ_i is. If Φ_i appears raised to the power k, we will get the following k contributions:

$$k\,\mathcal{W}\left(\phi_1 \cdots 1 \cdots \phi_n\right)\frac{1}{2}\theta \cdot \theta F_i\, \phi_i^{k-1}$$

which is indeed still given by

$$\frac{\partial \mathcal{W}(\phi_1 \cdots \phi_n)}{\partial \phi_i}\frac{1}{2}\theta \cdot \theta F_i$$

because ϕ_i appears as ϕ_i^k in $\mathcal{W}$.

In this expression, the index i is *not* summed over because we have only considered replacing one of all the superfields by its F term. But, clearly, we also must include the contributions coming from keeping the F term of each of the other superfields, and this done simply by summing over all possible values of the index i.

We have only considered one type of contribution to the F term of the superpotential. There is a second source of F terms that comes from keeping the term linear in θ (and containing a Weyl spinor) from two different superfields. Let's then choose to keep the spinor terms from the superfields Φ_i, and Φ_j (with $i \neq j$), and of course, we set all the other superfields equal to their scalar components ϕ. In this case, we get a contribution equal to

$$
\mathcal{W}\Big(\phi_1 \cdots \theta \cdot \chi_i \cdots \theta \cdot \chi_j \cdots \phi_n\Big) = \mathcal{W}\Big(\phi_1 \cdots 1 \cdots 1 \cdots \phi_n\Big)\theta \cdot \chi_i \theta \cdot \chi_j
$$

$$
= -\mathcal{W}\Big(\phi_1 \cdots 1 \cdots 1 \cdots \phi_n\Big)\frac{1}{2}\theta \cdot \theta \, \chi_i \cdot \chi_j
$$

where we have used Eq. (A.54) for the last step. In terms of partial derivatives, this is obviously

$$
-\frac{1}{2}\,\theta \cdot \theta \, \frac{\partial^2 \mathcal{W}(\phi_1 \cdots \phi_n)}{\partial \phi_i \partial \phi_j}\chi_i \cdot \chi_j
$$

The F term of this expression is simply

$$
\mathcal{W}\Big(\phi_1 \cdots \theta \cdot \chi_i \cdots \theta \cdot \chi_j \cdots \phi_n\Big)\Big|_F = -\frac{\partial^2 \mathcal{W}(\phi_1 \cdots \phi_n)}{\partial \phi_i \partial \phi_j}\chi_i \cdot \chi_j \qquad (12.36)
$$

The indices i and j are not summed over because we have only considered the contribution from a specific (distinct) pair of superfields. Now we want to include the contributions from all pairs. First, let's still consider $i \neq j$. If we sum over all values of i and j while excluding $i = j$, we will double count each contribution because

$$
\frac{\partial^2 \mathcal{W}(\phi_1 \cdots \phi_n)}{\partial \phi_i \partial \phi_j} = \frac{\partial^2 \mathcal{W}(\phi_1 \cdots \phi_n)}{\partial \phi_j \partial \phi_i}
$$

We therefore must divide our total result by two, so that the generalization of Eq. (12.36) is

$$
-\frac{1}{2}\frac{\partial^2 \mathcal{W}(\phi_1 \cdots \phi_n)}{\partial \phi_i \partial \phi_j}\chi_i \cdot \chi_j \qquad (12.37)
$$

with now i and j summed over, excluding the case $i = j$. Now consider $i = j$ and assume again that Φ_i appears raised to the power k. Then it is clear that the number of combinations of the form $\chi_i \cdot \theta \, \chi_i \cdot \theta$ we can make is $k(k-1)/2$, so Eq. (12.37) is still valid for $i = j$!

We have exhausted the possible sources of F terms from a general superpotential. To summarize, the supersymmetric lagrangian obtained from the superpotential is given by the sum of Eqs. (12.35) and (12.37), to which we must add the hermitian conjugate to obtain a real lagrangian:

$$
\boxed{\mathcal{W}(\Phi_1 \cdots \Phi_n)\Big|_F = \frac{\partial \mathcal{W}(\phi_1 \cdots \phi_n)}{\partial \phi_i} F_i - \frac{1}{2} \frac{\partial^2 \mathcal{W}(\phi_1 \cdots \phi_n)}{\partial \phi_i \partial \phi_j} \chi_i \cdot \chi_j + \text{h.c.}}
$$

(12.38)

which is exactly the expressions we obtained in Chapter 8 [see Eq. (8.60)].

12.5 The Free Part of the Wess-Zumino Model

We have reconstructed the interactions appearing in the Wess-Zumino model, but the *free* part,

$$
\mathcal{L}_{\text{free}} = \partial_\mu \phi \, \partial^\mu \phi^\dagger + i \chi^\dagger \bar{\sigma}^\mu \partial_\mu \chi + F^\dagger F
$$

(12.39)

is missing. We have dropped the index i on the fields because the free part does not mix fields belonging to different superfields, so there is no possibility of confusion. In other words, we might as well consider a single superfield; if there are several of them, we simply add a lagrangian of the form of Eq. (12.39) for each one.

Let's repeat here the expression for a left-chiral superfield given in Eq. (12.8):

$$
\Phi = \phi(x) - \frac{i}{2} \theta \sigma^\mu \bar{\theta} \, \partial_\mu \phi(x) - \frac{1}{16} \theta \cdot \theta \bar{\theta} \cdot \bar{\theta} \Box \phi(x)
$$

$$
+ \theta \cdot \chi(x) - \frac{i}{2} \theta \sigma^\mu \bar{\theta} \theta \cdot \partial_\mu \chi(x) + \frac{1}{2} F(x) \theta \cdot \theta
$$

(12.40)

For the rest of this chapter, all the component fields will be functions of x, not of y. Keeping this in mind, we will stop writing explicitly the x dependence to alleviate the notation.

It is clear that we cannot get the free Wess-Zumino lagrangian by multiplying Φ with itself because it does not contain the hermitian conjugates of the fields. For example, we see that two terms in Eq. (12.40) contain a derivative of ϕ, but we don't have any term with a derivative of $\phi^\dagger$. To obtain such a term, we obviously need to consider the hermitian conjugate of Eq. (12.40):

$$\Phi^\dagger = \phi^\dagger + \frac{i}{2}\theta\sigma^\nu\bar\theta\,\partial_\nu\phi^\dagger - \frac{1}{16}\theta\cdot\theta\bar\theta\cdot\bar\theta\Box\phi^\dagger$$

$$+\bar\theta\cdot\bar\chi + \frac{i}{2}\theta\sigma^\nu\bar\theta\bar\theta\cdot\partial_\nu\bar\chi + \frac{1}{2}F^\dagger\bar\theta\cdot\bar\theta \qquad (12.41)$$

where we have used Eqs. (A.52) and (A.44) and have relabeled the dummy Lorentz index $\mu \to \nu$ so that it does not get confused with the Lorentz index appearing in Φ.

To get a kinetic term for ϕ, the obvious thing to try is the product $\Phi^\dagger\Phi$. Let us for now only write the terms containing factors of ϕ and $\phi^\dagger$ and two derivatives:

$$\Phi^\dagger\Phi = -\frac{1}{16}\theta\cdot\theta\bar\theta\cdot\bar\theta\left(\phi^\dagger\Box\phi + \phi\Box\phi^\dagger\right) + \frac{1}{4}\theta\sigma^\nu\bar\theta\theta\sigma^\mu\bar\theta\partial_\mu\phi\partial_\nu\phi^\dagger + \cdots \quad (12.42)$$

where we are not showing the x dependence of the fields to simplify the expression. Using Eq. (A.56), the last term can be simplified to

$$\frac{1}{4}\theta\sigma^\nu\bar\theta\theta\sigma^\mu\bar\theta\partial_\mu\phi\,\partial_\nu\phi^\dagger = \frac{1}{8}\theta\cdot\theta\bar\theta\cdot\bar\theta\eta^{\mu\nu}\partial_\mu\phi\,\partial_\nu\phi^\dagger$$

$$= \frac{1}{8}\theta\cdot\theta\bar\theta\cdot\bar\theta\partial^\mu\phi\,\partial_\mu\phi^\dagger$$

Substituting this into Eq. (12.42), we get

$$\Phi^\dagger\Phi = -\frac{1}{16}\theta\cdot\theta\bar\theta\cdot\bar\theta\left(\phi^\dagger\Box\phi + \phi\Box\phi^\dagger - 2\,\partial^\mu\phi\partial_\mu\phi^\dagger\right) + \cdots$$

Recall that this is a lagrangian density, so we are free to do integrations by parts. Doing integrations by parts on the first two terms, we see that all three terms have exactly the same form and therefore can be summed to

$$\Phi^\dagger\Phi = \frac{1}{4}\partial_\mu\phi\,\partial^\mu\phi^\dagger\theta\cdot\theta\bar\theta\cdot\bar\theta + \cdots \qquad (12.43)$$

where the dots represent all the terms that do not contribute to the the kinetic term of the field ϕ. We have thus established that *the kinetic term of ϕ appears as part of the coefficient of $\theta\cdot\theta\bar\theta\cdot\bar\theta/4$ in the product $\Phi^\dagger\Phi$!*

The coefficient of this combination of Grassmann variables is known as the *D term* because in the superfields describing gauge multiplets that we will consider in

Chapter 13, this coefficient happens to be the auxiliary field D that we encountered in Chapter 10.

We therefore have that the D term of $\Phi^\dagger \Phi$ is given by

$$\left.(\Phi^\dagger \Phi)\right|_D = \partial_\mu \phi \, \partial^\mu \phi^\dagger + \cdots$$

where the dots represent other terms to which we will soon turn our attention. Of course, this D term can be projected out using integrations over the Grassmann coordinates, using [see Eq. (11.29)]

$$\frac{1}{4} \int d^2\theta d^2\bar{\theta} \, \theta \cdot \theta \bar{\theta} \cdot \bar{\theta} = 1$$

which shows that the D term of $\Phi^\dagger \Phi$ is given by

$$\left.(\Phi^\dagger \Phi)\right|_D = \int d^2\theta \, d^2\bar{\theta} \Phi^\dagger \Phi$$

$$= \partial_\mu \phi \, \partial^\mu \phi^\dagger + \cdots \tag{12.44}$$

Note that projecting out the D term *increases the dimension by 2*. Indeed, $\Phi^\dagger \Phi$ is of dimension 2, but its D term is of dimension 4. This is, of course, obvious from the fact that $d\theta$ and $d\bar{\theta}$ are of dimension $1/2$.

Our result in Eq. (12.44) should make you wonder what other expressions are contained in the D term of $\Phi^\dagger \Phi$. Let's find out! Multiplying Eqs. (12.40) and (12.41), and writing only the terms containing two θ and two $\bar{\theta}$, we get

$$\Phi^\dagger \Phi = \frac{1}{4} \theta \cdot \theta \bar{\theta} \cdot \bar{\theta} \left(\partial_\mu \phi^\dagger \partial^\mu \phi + F F^\dagger\right) + \frac{i}{2} \theta \sigma^\mu \bar{\theta} \left(\theta \cdot \chi \bar{\theta} \cdot \partial_\mu \bar{\chi} - \bar{\theta} \cdot \bar{\chi} \theta \cdot \partial_\mu \chi\right) \tag{12.45}$$

where we have combined right away the ϕ term using our hard work from Eqs. (12.42) and (12.43), and we have relabeled a Lorentz index $\nu \to \mu$.

The last two terms can be combined after an integration by parts. Indeed, we may rewrite the first term in the second parentheses as

$$\theta \cdot \chi \, \bar{\theta} \cdot \partial_\mu \bar{\chi} = -\theta \cdot (\partial_\mu \chi) \bar{\theta} \cdot \bar{\chi}$$

so that Eq. (12.45) becomes

$$\Phi^\dagger \Phi = \frac{1}{4}\theta\cdot\theta\,\bar\theta\cdot\bar\theta\left(\partial_\mu\phi^\dagger\partial^\mu\phi + FF^\dagger\right) - i\,\theta\sigma^\mu\bar\theta\,\bar\theta\cdot\bar\chi\theta\cdot\partial_\mu\chi \qquad (12.46)$$

What we would like to do now is to rewrite the last term with the Grassmann variables appearing in an overall factor $\theta\cdot\theta\,\bar\theta\cdot\bar\theta$ so that we can easily read off the D term.

First, we will need the identity (see Exercise 12.2)

$$\theta\sigma^\mu\bar\lambda\theta\cdot\partial_\mu\chi = -\frac{1}{2}\theta\cdot\theta(\partial_\mu\chi)\sigma^\mu\bar\lambda \qquad (12.47)$$

to group together the two θ and where λ is a left-chiral spinor. In the same exercise, you are invited also to prove

$$\bar\theta\cdot\lambda(\partial_\mu\chi)\sigma^\mu\bar\theta = -\frac{1}{2}\bar\theta\cdot\bar\theta(\partial_\mu\chi)\sigma^\mu\lambda \qquad (12.48)$$

Using those two results, the last term of Eq. (12.46) is

$$-i\,\theta\sigma^\mu\bar\theta\bar\theta\cdot\bar\chi\theta\cdot\partial_\mu\chi = -\frac{i}{4}\theta\cdot\theta\bar\theta\cdot\bar\theta(\partial_\mu\chi)\sigma^\mu\bar\chi$$

EXERCISE 12.2
Prove Eqs. (12.47) and (12.48). *Hint*: Use Eqs. (A.54) and (A.55).

Finally,

$$\Phi^\dagger\Phi = \frac{1}{4}\theta\cdot\theta\bar\theta\cdot\bar\theta\left[\partial_\mu\phi^\dagger\,\partial^\mu\phi + FF^\dagger - i\,(\partial_\mu\chi)\sigma^\mu\bar\chi\right] + \cdots$$

where the dots now represent terms that do not contribute to the D term. Finally, we extract the D term to obtain

$$\Phi^\dagger\Phi\Big|_D = \partial_\mu\phi^\dagger\,\partial^\mu\phi + FF^\dagger - i\,(\partial_\mu\chi)\sigma^\mu\bar\chi \qquad (12.49)$$

This is not exactly the free part of the Wess-Zumino lagrangian of Chapter 8, where the spinor term was written as $i\,\chi^\dagger\bar\sigma^\mu\partial_\mu\chi$, which using the bar notation, is equivalent to $i\,\bar\chi\bar\sigma^\mu\partial_\mu\chi$. But using the identity (A.50) once again, we may write

$$-i\,(\partial_\mu\chi)\sigma^\mu\bar\chi = i\,\bar\chi\bar\sigma^\mu\partial_\mu\chi$$

so that Eq. (12.49) becomes

$$\Phi^\dagger \Phi \Big|_D = \partial_\mu \phi^\dagger \, \partial^\mu \phi + F F^\dagger + i \, \bar{\chi} \bar{\sigma}^\mu \partial_\mu \chi \qquad (12.50)$$

which is exactly the free part of the Wess-Zumino lagrangian!

12.6 Why Does It All Work?

We have just seen how amazingly simple it is to obtain the complete Wess-Zumino lagrangian in the superfield approach: The F term of the most general function of left-chiral superfields gives us the interaction terms, whereas the D term of $\Phi^\dagger \Phi$ gives us the free lagrangian.

This is all nice and easy, but it's only because of our hard work in Chapters 5, 6, and 8 that we know that these lagrangians are supersymmetric. And it was only after we worked out the transformation of the auxiliary field F and showed that it transforms with a total derivative that we could prove that the F term of the product of any number of left-chiral superfields also transforms with a total derivative.

What we need is some criterion to tell us how to extract supersymmetric lagrangians from combinations of superfields without having to check the invariance under SUSY by hand or to work out the SUSY transformations of the component fields explicitly. For example, we would like to understand why we need to extract the F term of a product of left-chiral superfields instead of keeping, say, the term independent of Grassmann variables. Or why is it that for $\Phi^\dagger \Phi$ we need the D term instead of the F term.

We will now see how we could have predicted these results from the very beginning, even before working out any explicit transformation of the component fields. Let's start with a single left-chiral superfield. Recall that the dimensions of the fields ϕ, χ, and F are

$$[\phi] = 1 \qquad [\chi] = \frac{3}{2} \qquad [F] = 2$$

Recall also that the infinitesimal parameter ζ of a SUSY transformation has dimension $-1/2$. The transformation of F contains one factor of ζ and therefore must be of the form

$$\delta F = \zeta \times \text{ something of dimension 5/2}$$

for the dimensions to match. The "something" must be a function that is linear in one of the other fields appearing in the theory. Clearly, we must use χ to get something with a half-integer dimension. And now the punch line: In order to get something of dimension 5/2 out of χ, we *have* to apply a partial derivative to it. Therefore, the variation of F has to be of the following form

$$\delta F \simeq \zeta \partial_\mu \chi$$
$$= \partial_\mu (\zeta \chi)$$

based only on considerations of dimensions. Of course, this is not complete; we need something else that has a Lorentz index and which is dimensionless in order to make the final result a scalar quantity, like F, and this something else turns out to be $\bar{\sigma}^\mu$, as we saw previously. For the present discussion, however, Pauli matrices are irrelevant, so we will ignore them.

The main point is that using only dimensional considerations, we could have predicted that F must transform as a total derivative. It is instructive to go through the same type of argument for the two other fields of the multiplet, ϕ and χ, and see that only F has to transform as a total derivative. For example, consider the transformation of χ. From dimensional analysis, we have

$$\delta\chi = \zeta \times \text{ something of dimension 2}$$

This time, we have two choices for the "something": We could use either $\partial_\mu \phi$ or F. So we expect the transformation to be a linear combination of these two terms, which is obviously not a total derivative. The reason that we have two possibilities now is that there is a field of lower dimension than χ, the scalar field ϕ, but also a field of higher dimension, the auxiliary field F.

We see that the key point that sets apart the transformation of the field F is that it is the field of *highest dimension*, so its transformation necessarily *must* contain another field of lesser dimension, which forces us to include a derivative in the transformation. The conclusion is that the field of highest dimension in a supermultiplet must transform as a total derivative!

Finally, since the product of any number of left-chiral superfields is itself a left-chiral superfield, we know that the F term of such a product necessarily transforms as a total derivative under SUSY. Consequently, *the F term of the product of any number of left-chiral superfields provides a supersymmetric lagrangian density* (which is invariant up to total derivatives).

Now, what about superfields, or products of superfields, that are not left-chiral? It remains true that the field with the highest dimension must transform as the total derivative of a field of lower dimension. For a general superfield, the field of highest dimension is the field that multiplies $\theta \cdot \theta\, \bar{\theta} \cdot \bar{\theta}/4$. In gauge superfields, this will be the auxiliary field D that we encountered in Chapter 10. By extension, we call the quantity that multiplies this particular combination of Grassmann variables in a general superfield the D *term* of that superfield.

Consequently, the D term of *any* type of superfield or of any combination of superfields also transforms as a total derivative. Therefore, *the D term of any combination of superfields is a supersymmetric lagrangian density*. This is how we could have known ahead of time that the D term of $\Phi^\dagger\Phi$ necessarily would give a valid candidate for a supersymmetric lagrangian, even without any of the work of the previous chapters. Isn't that neat?

An obvious question that you may be asking yourself is why don't we use the D term of the superpotential as well as its F term? After all, we have just argued that the D term of any product of superfields is a valid candidate for a supersymmetric lagrangian density! The answer is that the D term of a product of left-chiral superfields is uninteresting because it is necessarily a total derivative. We are not saying that it transforms with a total derivative but that it *is* a total derivative to start with, which vanishes if we carry out the spacetime integation. This can be seen explicitly in Eq. (12.8), where the D term is

$$-\frac{1}{4}\,\Box\phi(x)$$

Since the product of any number of left-chiral superfields is necessarily of the form of Eq. (12.8) with some equivalent fields ϕ_{eq}, χ_{eq}, and F_{eq}, we can tell that the D term of any product of left-chiral superfields is a total derivative. This is the reason we do not bother with the D term of the superpotential.

One last aside before moving on to the next chapter. We have seen that the D term of $\Phi^\dagger\Phi$ is already of dimension 4, so if we are only interested in renormalizable theories, we cannot consider D terms of any other combinations, such as $\Phi^\dagger\Phi^\dagger\Phi$. However, it is natural to think of the supersymmetric theories we have constructed so far as effective field theories valid only at a energies below some physical cutoff Λ (see Chapter 1). Indeed, we know that a theory without gravity cannot be truly fundamental and that reason alone indicates that any supersymmetric theory without gravity must be treated as an effective field theory (here, we are only discussing theories with global SUSY, i.e., with the SUSY parameter ζ spacetime independent, not supergravity theories, in which SUSY is gauged). In that spirit, one should include nonrenormalizable interactions.

Consider, then, an effective field theory whose field content is a set of left-chiral superfields $\Phi_i, \ldots, \Phi_n$. The most general supersymmetric lagrangian contains two types of terms: the F term of a superpotential, which is holomorphic in the fields Φ_i, and the D term of a function containing at least one hermitian conjugate of a superfield. This second function is referred to as the *Kähler potential* $\mathcal{K}(\Phi_i, \Phi_j^\dagger)$. In a renormalizable theory, the Kähler potential simply contains the sum of the kinetic terms, i.e.,

$$\mathcal{K}_{\text{ren}}(\Phi_i, \Phi_j^\dagger) = \Phi_i^\dagger \, \Phi_i \tag{12.51}$$

The topics of supergraviy and nonrenormalizable supersymmetric interactions are beyond the scope of this book, so we will not discuss the Kähler potential any further.

12.7 Quiz

1. Describe the component fields of a left-chiral superfield.
2. Why do we not consider the D term of the superpotential when building supersymmetric lagrangians?
3. What is the dimension of the D term of $\Phi^\dagger \Phi^\dagger \Phi$?
4. Why do we impose the superpotential to be a holomorphic function of left-chiral superfields?
5. What is the Kähler potential?

CHAPTER 13

Supersymmetric Gauge Field Theories in the Superfield Approach

Now, we would like to see how to construct gauge-invariant theories using super-fields. As a warm-up, we will start with abelian gauge theories.

13.1 Abelian Gauge Invariance in the Superfield Formalism

First, let's consider a left-chiral superfield Φ that carries a $U(1)$ charge q. What do we mean by this? Well, by direct analogy with the usual gauge transformation of charged particles, we will take this as meaning that the superfield acquires a phase

$$\Phi \to e^{2iq\Lambda}\Phi$$

under a gauge transformation, with q being the charge of the field and Λ the gauge parameter. This is of the same form as a standard $U(1)$ transformation except for the factor of 2, which is, of course, purely conventional and has been chosen because it will simplify a later result.

To build a gauge theory, we proceed as usual: We make the gauge parameter spacetime-dependent, i.e., $\Lambda = \Lambda(x)$, so it is now a field. But we are working in superspace, so it actually must be taken to be a *superfield*! However, it cannot be an arbitrary superfield because we obviously want the left-chiral superfield Φ to remain a left-chiral superfield after a gauge transformation. This means that we must take the superfield Λ to be a chiral superfield itself, so it has the expansion [see Eq. (12.8)]:

$$\Lambda = \phi_\Lambda - \frac{i}{2}\theta\sigma^\mu\bar{\theta}\partial_\mu\phi_\Lambda - \frac{1}{16}\theta\cdot\theta\,\bar{\theta}\cdot\bar{\theta}\Box\phi_\Lambda$$
$$+\theta\cdot\chi_\Lambda - \frac{i}{2}\theta\sigma^\mu\bar{\theta}\,\theta\cdot\partial_\mu\chi_\Lambda + \frac{1}{2}F_\Lambda\theta\cdot\theta$$

Even though this has the same expansion as the left-chiral superfields we have met so far, the dimensions of the fields are different. In order for the superfield Λ to appear in an exponential, it must be dimensionless. If we recall that each Grassmann variable has dimension $-1/2$, then the dimensions of the component fields are

$$[\phi_\Lambda] = 0 \qquad [\chi_\Lambda] = \frac{1}{2} \qquad [F_\Lambda] = 1$$

Now that we have defined how left-chiral superfields transform, we can build supersymmetric and gauge-invariant interactions. Consider a superpotential $\mathcal{W}$ that depends on a set of left-chiral superfields Φ_i, Φ_j, ... To have gauge invariance, we may only consider products of superfields whose charges add up to zero. Note that if the superpotential is invariant when the gauge parameter is a constant, it will remain invariant even after we make the gauge invariance local since it does not contain derivatives acting on the superfields.

Now consider the "kinetic" term for a left-chiral superfield $\Phi^\dagger\Phi$. When we perform a gauge transformation of this, we get

$$\Phi^\dagger\Phi \to e^{2iq(\Lambda-\Lambda^\dagger)}\Phi^\dagger\Phi \tag{13.1}$$

where we have used the fact that left-chiral superfields commute to move an exponential around. We also have taken the charge q to be a real number. We see that this
term is *not* gauge-invariant! It might seem that we could make it gauge-invariant
by requiring $\Lambda^\dagger = \Lambda$, but Λ is a left-chiral superfield, which cannot be real (the
hermitian conjugate of a left-chiral superfield does not have the form of a left-chiral
superfield because it will contain $\bar\theta$). So how are we to make $\Phi^\dagger \Phi$ gauge-invariant?
We can't simply drop it because it contains the kinetic terms of the component
fields. The way out is the same as in conventional field theories: We must introduce
a gauge superfield $\mathcal{V}$ whose transformation will compensate the variation appearing
in Eq. (13.1). We therefore replace $\Phi^\dagger \Phi$ by

$$\Phi^\dagger e^{2q\mathcal{V}} \Phi \tag{13.2}$$

with the gauge superfield defined to have a "super gauge transformation" given by

$$\mathcal{V} \rightarrow \mathcal{V} - i(\Lambda - \Lambda^\dagger) \tag{13.3}$$

It is conventional to write the exponential in the middle, as in Eq. (13.2), but there
is no special reason for this here because the superfields commute anyway. When
we consider nonabelian gauge theories, the superfields actually will be matrices,
and then the order will be important, obviously.

The reason for including a factor i in the gauge transformation in Eq. (13.3) is
to make the combination $i(\Lambda - \Lambda^\dagger)$ real. This then implies that we may take the
gauge superfield $\mathcal{V}$ to be a real superfield. Recall that the most general superfield
contains nine terms, as we saw in Exercise 11.2, which we may write as

$$\mathcal{V}(x) = A(x) + \theta \cdot \alpha(x) + \bar\theta \cdot \bar\beta(x) + \theta \cdot \theta B(x) + \bar\theta \cdot \bar\theta H(x) + \theta \sigma^\mu \bar\theta \, V_\mu(x)$$
$$+ \theta \cdot \theta \bar\theta \cdot \bar\gamma(x) + \bar\theta \cdot \bar\theta \theta \cdot \eta(x) + \theta \cdot \theta \bar\theta \cdot \bar\theta P(x) \tag{13.4}$$

where α, β, γ, and η are Weyl spinors, V_μ is a complex vector field, and A, B, H,
and P are all complex scalar fields. Let's now impose

$$\mathcal{V}^\dagger = \mathcal{V}$$

It turns out that this is consistent with supersymmetry (SUSY) transformation;
i.e., this reality condition is preserved by SUSY transformations. Imposing this
condition on Eq. (13.4) leads to

$$A^\dagger = A \qquad V_\mu^\dagger = V_\mu \qquad P^\dagger = P$$
$$B^\dagger = H \qquad \alpha = \beta \qquad \gamma = \eta$$

which reduces the field content to one real vector field, two Weyl spinors, two real and one complex scalar fields. This is still much more than the field content we needed in Chapter 10 to build the vector supermultiplet!

For reasons that will become clear very soon, a different representation is usually chosen for the real superfield $\mathcal{V}$:

$$
\mathcal{V} = C(x) + \frac{i}{\sqrt{2}} \theta \cdot \rho(x) - \frac{i}{\sqrt{2}} \bar{\theta} \cdot \bar{\rho}(x) + \frac{i}{4} \theta \cdot \theta \, (M(x) + i N(x))
$$

$$
- \frac{i}{4} \bar{\theta} \cdot \bar{\theta} \, (M^\dagger(x) - i N^\dagger(x)) + \frac{1}{2} \theta \sigma^\mu \bar{\theta} \, A_\mu(x)
$$

$$
+ \frac{1}{2\sqrt{2}} \theta \cdot \theta \left(\bar{\theta} \cdot \bar{\lambda} + \frac{1}{2} \bar{\theta} \bar{\sigma}^\mu \partial_\mu \rho \right) + \frac{1}{2\sqrt{2}} \bar{\theta} \cdot \bar{\theta} \left(\theta \cdot \lambda - \frac{1}{2} \theta \sigma^\mu \partial_\mu \bar{\rho} \right)
$$

$$
- \frac{1}{8} \theta \cdot \theta \, \bar{\theta} \cdot \bar{\theta} \left(D(x) + \frac{1}{2} \Box C(x) \right) \tag{13.5}
$$

where M, N, C, and D are all real scalar fields, A_μ is a real vector field, and λ and ρ are Weyl spinors.

This gauge superfield is often called a *vector superfield* because one of its *components* is the vector field A_μ. However, keep in mind that the vector superfield $\mathcal{V}$ itself is a *scalar* under Lorentz transformations, like the left-chiral superfields Φ_i.

The vector superfield is dimensionless, which can be seen by looking at the term containing A_μ. A vector field has a dimension equal to 1, but θ and $\bar{\theta}$ have dimension $-1/2$, which makes the term containing A_μ, and consequently the whole superfield, dimensionless.

Now let's see how these fields change under a gauge transformation [Eq. (13.3)]. We have (we will not write explicitly the x dependence of the component fields from now on)

$$
\mathcal{V} \rightarrow \mathcal{V} - i \, (\Lambda - \Lambda^\dagger)
$$

$$
\rightarrow \mathcal{V} - i \, (\phi_\Lambda - \phi_\Lambda^*) - i \, (\theta \cdot \chi_\Lambda - \bar{\theta} \cdot \bar{\chi}_\Lambda) - \frac{i}{2} \theta \cdot \theta \, F_\Lambda + \frac{i}{2} \bar{\theta} \cdot \bar{\theta} \, F_\Lambda^*
$$

$$
- \frac{1}{2} \theta \sigma^\mu \bar{\theta} \, \partial_\mu (\phi_\Lambda + \phi_\Lambda^*) - \frac{1}{4} \left(\theta \cdot \theta \bar{\theta} \bar{\sigma}^\mu \partial_\mu \chi_\Lambda - \bar{\theta} \cdot \bar{\theta} \theta \sigma^\mu \partial_\mu \bar{\chi}_\Lambda \right)
$$

$$
+ \frac{i}{16} \theta \cdot \theta \bar{\theta} \cdot \bar{\theta} \Box (\phi_\Lambda - \phi_\Lambda^*) \tag{13.6}
$$

where we have used Eqs. (12.47) and (A.50). We use a complex conjugation on ϕ_Λ and F_Λ instead of a dagger because these are complex functions, not quantum fields. Using Eq. (13.5), we can easily pick up how each of the component fields of $\mathcal{V}$ transform. Focus first on the terms with no Grassmann variables, as well as on those proportional to $\theta \cdot \theta \, \bar{\theta} \cdot \bar{\theta}$. These terms combined transform as

$$C - \frac{1}{8} \theta \cdot \theta \, \bar{\theta} \cdot \bar{\theta} \left(D + \frac{1}{2} \Box C \right) \rightarrow C - i \left(\phi_\Lambda - \phi_\Lambda^* \right) - \frac{1}{8} \theta \cdot \theta \, \bar{\theta} \cdot \bar{\theta} \left(D + \frac{1}{2} \Box C \right)$$

$$+ \frac{i}{16} \theta \cdot \theta \, \bar{\theta} \cdot \bar{\theta} \Box (\phi_\Lambda - \phi_\Lambda^*) \qquad (13.7)$$

from which we can read off the gauge transformations of the two fields to be

$$C \rightarrow C - i \left(\phi_\Lambda - \phi_\Lambda^* \right)$$
$$D \rightarrow D \qquad (13.8)$$

Now it should be clear why the last term of $\mathcal{V}$ was chosen to contain the apparently strange combination $D - \frac{1}{2} \Box C$. This choice allowed us to put all the gauge dependence into the field C, leaving a gauge-independent field D.

Working out the gauge transformations of the other component fields, one finds

$$\rho \rightarrow \rho - \sqrt{2} \, \chi_\Lambda$$
$$M + iN \rightarrow M + iN - 2 F_\Lambda$$
$$A_\mu \rightarrow A_\mu - \partial_\mu \left(\phi_\Lambda + \phi_\Lambda^* \right)$$
$$\lambda \rightarrow \lambda \qquad (13.9)$$

Again, the reason for the peculiar combinations of the fields ρ and λ in Eq. (13.5) was to have all the gauge dependence in one field, leaving the other one gauge-invariant.

We see that we have a lot of "gauge freedom" in Eqs. (13.8) and (13.9). In particular, by an appropriate choice of χ_Λ and F_Λ, we can use a gauge transformation to set the fields ρ, M, and N to zero. We also can choose the imaginary part of ϕ_Λ so that the field C also will vanish. This gauge, in which all these fields are set to zero, is called the *Wess-Zumino gauge*. In this gauge, the vector superfield is simply

$$\mathcal{V} = \frac{1}{2} \theta \sigma^\mu \bar{\theta} \, A_\mu + \frac{1}{2\sqrt{2}} \theta \cdot \theta \, \bar{\theta} \cdot \bar{\lambda} + \frac{1}{2\sqrt{2}} \bar{\theta} \cdot \bar{\theta} \, \theta \cdot \lambda - \frac{1}{8} \theta \cdot \theta \, \bar{\theta} \cdot \bar{\theta} D \qquad (13.10)$$

where the fields λ and D are gauge-invariant. From now on, we will only work with the gauge superfield in the Wess-Zumino gauge.

The gauge is not completely fixed, however. We still have one last gauge parameter, the real part of ϕ_Λ. Under this residual gauge transformation, the vector field changes by

$$A_\mu \rightarrow A_\mu - 2\,\partial_\mu\big[\text{Re}(\phi_\Lambda)\big]$$

which is again the standard form for the $U(1)$ gauge transformation of the gauge field A_μ apart from the factor of 2.

To summarize, we can write down gauge-invariant combinations of superfields at two conditions:

- Recall that the superpotential must be a holomorphic function of the left-chiral superfields; i.e., it may not contain their hermitian conjugates $\Phi_i^\dagger$. Therefore, in any product of left-chiral superfields appearing in the superpotential $\mathcal{W}(\Phi_i)$, the charges of the superfields must add up to zero in order to have gauge invariance.
- The kinetic term of left-chiral superfields $\Phi_i^\dagger \Phi_i$ must be replaced by $\Phi_i^\dagger e^{2q_i \mathcal{V}} \Phi_i$ (where the label i indicates the different superfields).

To obtain SUSY-invariant lagrangians (up to total derivatives, as always) from these terms, all we have to do is to extract the F term of the superpotential and the D term of the gauge-invariant kinetic term:

$$\mathcal{L} = \left(\mathcal{W}(\Phi_i)\Big|_F + \text{h.c.}\right) + \sum_i \Phi_i^\dagger e^{2q_i \mathcal{V}} \Phi_i\Big|_D \qquad (13.11)$$

Note that if there is only one charged left-chiral superfield, *the superpotential is automatically zero* because it is impossible to write down a holomorphic function of Φ that is gauge-invariant.

There is actually a third gauge-invariant and supersymmetric term that can be added to Eq. (13.11). This is a sneaky one. The trick is to notice that even though $\mathcal{V}$ is not gauge-invariant, its D term is! Indeed, the D term of $\mathcal{V}$ is simply [see Eq. (13.10)]

$$\mathcal{V}\Big|_D = -\frac{1}{2} D \qquad (13.12)$$

which is gauge-invariant [see Eq. (13.8)]. We therefore can add to Eq. (13.11) the term

$$-2\,\xi \mathcal{V}\Big|_D = \xi D$$

where the parameter ξ has the dimension of a mass squared. This is, of course, the Fayet-Illiopoulos term that we introduced in Section 10.2.

Incidentally, we now see why extracting the coefficient of the combination $\theta \cdot \theta \, \bar{\theta} \cdot \bar{\theta}/4$ is referred to as finding the D term: When this is done to a gauge superfield, it indeed pulls out the D field.

13.2 Explicit Interactions Between a Left-Chiral Multiplet and the Abelian Gauge Multiplet

Let us now work out explicitly the second term of Eq. (13.11), i.e.,

$$\Phi^{\dagger} e^{2qV} \Phi \Big|_{D}$$

in terms of component fields (we have already worked out the expansion of the superpotential, so we won't repeat that here).

Let us start by Taylor expanding e^{2qV}. We get

$$\Phi^{\dagger} e^{2qV} \Phi = \Phi^{\dagger} \Phi (1 + 2q\,V + 2q^2\,V^2) \tag{13.13}$$

The series terminates there because V^3 necessarily contains a Grassmann variable squared. We get a large number of terms, but of course, we need only retain those that have two θ and two $\bar{\theta}$ because we are interested only in the D term:

$$\Phi^{\dagger} e^{2qV} \Phi \Big|_{D} = \Phi^{\dagger} \Phi \Big|_{D} + 2q \, \Phi^{\dagger} \Phi V \Big|_{D} + 2q^2 \, \Phi^{\dagger} \Phi V^2 \Big|_{D} \tag{13.14}$$

We have already worked out the D term of $\Phi^{\dagger}\Phi$; it's simply the free Wess-Zumino lagrangian.

The last term of Eq. (13.14) is actually very simple because the only nonvanishing term in V^2 is

$$V^2 = \frac{1}{4}\theta\sigma^{\mu}\bar{\theta}\,\theta\sigma^{\nu}\bar{\theta}A_{\mu}A_{\nu}$$

$$= \frac{1}{8}\theta\cdot\theta\,\bar{\theta}\cdot\bar{\theta}A^{\mu}A_{\mu} \tag{13.15}$$

where we have used Eq. (A.56) for the last step. The only contribution to the D term of $\Phi^{\dagger}\Phi V^2$ therefore comes from the term with no Grassmann variable in $\Phi^{\dagger}\Phi$, which is $\phi^\dagger \phi$. Thus

$$2q^2 \, \Phi^\dagger \, \Phi \mathcal{V}^2|_D = q^2 \, \phi^\dagger \phi A^\mu A_\mu \tag{13.16}$$

Now let us consider the D term of $\Phi^\dagger \Phi \mathcal{V}$. Since $\mathcal{V}$ contains terms with at least one θ and one $\bar{\theta}$, we need only consider in the product $\Phi^\dagger \Phi$ terms with *at most* one θ and one $\bar{\theta}$. Keeping this in mind, we get

$$\Phi^\dagger \Phi = \phi^\dagger \phi + \phi^\dagger \theta \cdot \chi + \phi \bar{\theta} \cdot \bar{\chi} - \frac{i}{2} \phi^\dagger \theta \sigma^\mu \bar{\theta} \partial_\mu \phi + \frac{i}{2} \phi \theta \sigma^\mu \bar{\theta} \partial_\mu \phi^\dagger$$
$$+ \theta \cdot \chi \bar{\theta} \cdot \bar{\chi} + \cdots$$

where we have only written the terms that will contribute to the D term of $\Phi^\dagger \Phi \mathcal{V}$. Writing again only the relevant terms, we have

$$\Phi^\dagger \Phi \mathcal{V} = -\frac{1}{8} \theta \cdot \theta \, \bar{\theta} \cdot \bar{\theta} \phi^\dagger \phi D + \left(\frac{1}{2\sqrt{2}} \bar{\theta} \cdot \bar{\theta} \, \theta \cdot \chi \, \theta \cdot \lambda \phi^\dagger + \text{h.c.} \right)$$
$$- \left(\frac{i}{4} \theta \sigma^\mu \bar{\theta} \, \theta \sigma^\nu \bar{\theta} \phi^\dagger \, A_\mu \partial_\nu \phi + \text{h.c.} \right)$$
$$+ \frac{1}{2} \theta \sigma^\mu \bar{\theta} \, \theta \cdot \chi \bar{\theta} \cdot \bar{\chi} A_\mu + \cdots \tag{13.17}$$

We need to rewrite all the terms so that they are in the form

$$\text{Product of fields} \times \theta \cdot \theta \, \bar{\theta} \cdot \bar{\theta}$$

so that we can extract the D term. We can achieve this using identities we already have. For the second term of Eq. (13.17), we use Eq. (A.54); for the third term, we use Eq. (A.56), whereas for the last term, we use Eq. (12.47). This finally gives

$$\Phi^\dagger \Phi \mathcal{V} = \frac{1}{4} \theta \cdot \theta \, \bar{\theta} \cdot \bar{\theta} \left(-\frac{1}{2} \phi^\dagger \phi \, D + \frac{1}{2} \chi \sigma^\mu \bar{\chi} A_\mu \right.$$
$$\left. - \left(\frac{1}{\sqrt{2}} \chi \cdot \lambda \phi^\dagger + \frac{i}{2} A^\mu \phi^\dagger \partial_\mu \phi + \text{h.c.} \right) + \cdots \right)$$

from which we can read off the D term directly. Recall that what we are really interested in is the quantity

$$2q\Phi^\dagger\Phi\mathcal{V}\Big|_D = -q\,\phi^\dagger\phi\,D + q\,\chi\sigma^\mu\bar\chi A_\mu - \left(\sqrt{2}\,q\,\chi\cdot\lambda\,\phi^\dagger + i\,q\,A^\mu\,\phi^\dagger\,\partial_\mu\phi + \text{h.c.}\right) \tag{13.18}$$

Finally, the D term of $\Phi^\dagger e^{2q\mathcal{V}}\Phi$ is given by the sum of the free Wess-Zumino lagrangians, of Eqs. (13.16) and (13.18):

$$\Phi^\dagger e^{2q\mathcal{V}}\Phi\Big|_D = \partial_\mu\phi^\dagger\partial^\mu\phi + i\,\bar\chi\bar\sigma^\mu\partial_\mu\chi + FF^\dagger + q^2\,\phi^\dagger\phi\,A^\mu A_\mu - q\,\phi^\dagger\phi\,D$$

$$- \left(\sqrt{2}\,q\,\chi\cdot\lambda\,\phi^\dagger + i\,q\,A^\mu\,\phi^\dagger\,\partial_\mu\phi + \text{h.c.}\right) + q\,\chi\sigma^\mu\bar\chi A_\mu \tag{13.19}$$

This must be compared with the supersymmetric lagrangian for the interaction of a left-chiral Weyl spinor with an abelian gauge field that we worked out in Chapter 10 [see Eq. (10.54)]. To make the relation more explicit, consider the following terms appearing in Eq. (13.19):

$$\partial_\mu\phi^\dagger\partial^\mu\phi + q^2\,\phi^\dagger\phi\,A^\mu A_\mu - \left(i\,q\,A^\mu\,\phi^\dagger\,\partial_\mu\phi + \text{h.c.}\right)$$

$$= \partial_\mu\phi^\dagger\partial^\mu\phi + q^2\,\phi^\dagger\phi\,A^\mu A_\mu - i\,q\,A^\mu\,\phi^\dagger\,\partial_\mu\phi + i\,q\,A^\mu\,\phi\,\partial_\mu\phi^\dagger \tag{13.20}$$

These terms can be combined into

$$(D_\mu\phi)^\dagger\,(D^\mu\phi)$$

if we define

$$D_\mu \equiv \partial_\mu + i\,q\,A_\mu \tag{13.21}$$

the usual covariant derivative. It would be nice to also write the kinetic energy of the spinor in terms of a covariant derivative by combining the following two terms:

$$i\,\bar\chi\bar\sigma^\mu\partial_\mu\chi + q\,\chi\sigma^\mu\bar\chi A_\mu \tag{13.22}$$

but we see that we can't because the σ matrices in the two terms are of different types. Luckily, we are saved again by an identity, this time Eq. (A.50), which allows us to write the second term as

$$q \, \chi \sigma^\mu \bar{\chi} A_\mu = -q \, \bar{\chi} \bar{\sigma}^\mu \chi A_\mu$$

so that the two terms of Eq. (13.22) can be combined into

$$i \, \bar{\chi} \bar{\sigma}^\mu \partial_\mu \chi - q \, \bar{\chi} \bar{\sigma}^\mu \chi A_\mu = i \, \bar{\chi} \bar{\sigma}^\mu D_\mu \chi$$

with D_μ again defined by Eq. (13.21).

Using covariant derivatives, Eq. (13.19) therefore can be simplified to

$$
\begin{aligned}
\Phi^\dagger e^{2qV} \Phi \Big|_D &= (D_\mu \phi)^\dagger (D^\mu \phi) + i \, \bar{\chi} \bar{\sigma}^\mu D_\mu \chi + F F^\dagger - q \, \phi^\dagger \phi \, D \\
&\quad - \left(\sqrt{2} q \, \chi \cdot \lambda \, \phi^\dagger + \text{h.c.} \right)
\end{aligned}
\tag{13.23}
$$

which, apart from three missing terms, is exactly the lagrangian for the interaction of a left-chiral Weyl spinor with an abelian gauge field we obtained in Eq. (10.54)! It's in order to get this simple equality, with no overall factor, that a factor of 2 is included in the exponential containing V.

The missing terms are those that contain only the fields of the gauge superfield, namely, $A_\mu(x)$, $\lambda(x)$, and $D(x)$. We now turn our attention to figuring out how these terms can be recovered in the superfield approach.

13.3 Lagrangian of a Free Supersymmetric Abelian Gauge Theory in Superfield Notation

What we would like to introduce is a *field-strength superfield* $\mathcal{F}$ that would contain the gauge-invariant field strength $F_{\mu\nu}$ [recall that in the case of a $U(1)$ gauge theory, the field strength is gauge invariant; this is no longer true for nonabelian gauge theories, but we will discuss those later]. As a SUSY partner of this field strength, we would like to still use the same fields that appear in V, i.e., the photino field λ and the auxiliary field D, which are gauge invariants, as we have just seen. In this way, all the components of the field-strength superfield would be gauge invariant, and it would be trivial to get a gauge-invariant lagrangian.

Now, to build this field-strength superfield, we could do it the hard way: using brute force to find some combination of λ, $F_{\mu\nu}$, and D whose SUSY transformation will reproduce the correct transformations of the component fields. This is extremely long and extremely painful. Or we can do it the "easy" way: starting from the superfield $\mathcal{V}$ and applying the "SUSY derivatives" D^a and $\bar{D}^{\dot{a}}$ to generate the field-strength superfield. This is merely long and painful. We will choose the "long and painful" approach over the "extremely long and extremely painful" one. Actually, the calculation itself is not too bad; it's just painful to type!

We need to apply at least three SUSY derivatives to $\mathcal{V}$ because if we apply fewer, the photon field A_μ (with no derivative acting on it) will remain present, and our result therefore will not be gauge-invariant. This is obvious if you recall that D^a contains a derivative with respect to θ_a and that $\bar{D}^{\dot{a}}$ contains a derivative with respect to $\bar{\theta}_{\dot{a}}$ [see Eqs. (11.57), (11.58), and (11.61)]. The standard choice is to define the field-strength superfield $\mathcal{F}_a$ as

$$\mathcal{F}_a \equiv \bar{D} \cdot \bar{D} D_a \mathcal{V} \qquad (13.24)$$

The reason for choosing this particular combination of derivatives will soon become clear.

The superfield $\mathcal{F}$ carries a lower undotted index, which is just a fancy way of saying that it transforms like a left-chiral Weyl spinor under Lorentz transformations (unlike all the other superfields we have introduced so far, which are scalars).

Note that $\mathcal{F}_a$ is a left-chiral superfield! Indeed, because of Eq. (11.62), $\mathcal{F}_a$ satisfies

$$\bar{D}_{\dot{b}} \mathcal{F}_a = 0$$

which defines a left-chiral superfield. Unlike the Φ we worked with in Chapter 12, $\mathcal{F}_a$ is a *spinor* left-chiral superfield, not a scalar left-chiral superfield.

An obvious thing to check is whether this field-strength superfield is invariant under the super gauge transformation in Eq. (13.3) like its nonsupersymmetric counterpart $F_{\mu\nu}$. Under that transformation, we have

$$\mathcal{F}_a \to \mathcal{F}_a - i\,\bar{D} \cdot \bar{D} D_a \Lambda + i\,\bar{D} \cdot \bar{D} D_a \Lambda^\dagger \qquad (13.25)$$

where we have used the fact that the SUSY derivatives are linear operators to distribute them. Now recall that Λ is a left-chiral superfield and that applying D_a to the hermitian conjugate of a left-chiral superfield gives zero [see Eq. (12.4)]. Therefore, the last term of Eq. (13.25) is identically zero. We still have to show that the second term on the right-hand side also vanishes. We will use an anticommutator to move a $\bar{D}^{\dot{a}}$ so that it will act directly on the left-chiral superfield and then use the fact that $\bar{D}^{\dot{a}} \Lambda = 0$ because Λ is left-chiral. We have (recall that $\bar{D}_{\dot{a}}$ and $\bar{D}^{\dot{a}}$ anticommute)

$$
\begin{aligned}
\bar{D}_{\dot{a}} \bar{D}^{\dot{a}} D_b \Lambda &= -\bar{D}^{\dot{a}} \bar{D}_{\dot{a}} D_b \Lambda \\
&= -\bar{D}^{\dot{a}} \{\bar{D}_{\dot{a}}, D_b\} \Lambda + \bar{D}^{\dot{a}} D_b \bar{D}_{\dot{a}} \Lambda \\
&= -\bar{D}^{\dot{a}} \{D_b, \bar{D}_{\dot{a}}\} \Lambda \\
&= -i \, (\sigma^{\mu})_{b\dot{a}} \, \bar{D}^{\dot{a}} \partial_{\mu} \Lambda
\end{aligned}
\tag{13.26}
$$

where in the last step we have used Eq. (11.63). We obviously may move the operator $\bar{D}_{\dot{a}}$ through the partial derivative ∂_{μ} (they commute), so we end up with

$$
\begin{aligned}
\bar{D}_{\dot{a}} \bar{D}^{\dot{a}} D_b \Lambda &= -i \, (\sigma^{\mu})_{b\dot{a}} \partial_{\mu} \bar{D}^{\dot{a}} \Lambda \\
&= 0
\end{aligned}
$$

This completes the proof that the last two terms of Eq. (13.25) vanish and that $\mathcal{F}_a$ is gauge-invariant.

The next step is to calculate explicitly $\mathcal{F}_a$ in terms of component fields. This requires working out explicitly the right-hand side of Eq. (13.24). This calculation is nontrivial and requires a few special tricks, so we will devote a section to it.

13.4 The Abelian Field-Strength Superfield in Terms of Component Fields

We will now explicitly calculate the right-hande side of Eq. (13.24), where the vector superfield $\mathcal{V}$ is given in Eq. (13.10) (in the Wess-Zumino gauge). It turns out that the calculation is much simpler if we trade the variables x^{μ}, θ, and $\bar{\theta}$ for the variables y^{μ}, θ', and $\bar{\theta}'$ defined by

$$
y^{\mu} = x^{\mu} - \frac{i}{2} \theta \sigma^{\mu} \bar{\theta}
$$
$$
\theta' = \theta
$$
$$
\bar{\theta}' = \bar{\theta}
\tag{13.27}
$$

The relation between y^{μ} and x^{μ} is the same as before [see Eq. (12.3)]. Recall that σ^{μ} carries lower indices, so we may write the relation between y^{μ} and x^{μ} more

explicitly as

$$y^\mu = x^\mu - \frac{i}{2}\theta^a (\sigma^\mu)_{a\dot{b}} \bar{\theta}^{\dot{b}}$$

Now we will express the partial derivatives with respect to the first set of variables in terms of partial derivatives with respect to the second set. We have

$$\frac{\partial}{\partial x^\mu} = \frac{\partial y^\nu}{\partial x^\mu}\frac{\partial}{\partial y^\nu} + \frac{\partial \theta'^a}{\partial x^\mu}\frac{\partial}{\partial \theta'^a} + \frac{\partial \bar{\theta}'^a}{\partial x^\mu}\frac{\partial}{\partial \bar{\theta}'^a}$$

$$= \delta^\nu_\mu \frac{\partial}{\partial y^\nu}$$

$$= \frac{\partial}{\partial y^\mu} \qquad (13.28)$$

The derivatives with respect to the Grassmann variables are more interesting. For example,

$$\frac{\partial}{\partial \theta^a} = \frac{\partial \theta'^b}{\partial \theta^a}\frac{\partial}{\partial \theta'^b} + \frac{\partial \bar{\theta}'^b}{\partial \theta^a}\frac{\partial}{\partial \bar{\theta}'^b} + \frac{\partial y^\mu}{\partial \theta^a}\frac{\partial}{\partial y^\mu}$$

$$= \delta^b_a \frac{\partial}{\partial \theta'^b} - \frac{i}{2}(\sigma^\mu)_{a\dot{b}}\bar{\theta}^{\dot{b}}\frac{\partial}{\partial y^\mu}$$

$$= \frac{\partial}{\partial \theta'^a} - \frac{i}{2}(\sigma_\mu)_{a\dot{b}}\bar{\theta}'^{\dot{b}}\frac{\partial}{\partial y^\mu} \qquad (13.29)$$

where we have used the fact that $\bar{\theta} = \bar{\theta}'$ in the last step. A similar calculation leads to

$$\frac{\partial}{\partial \bar{\theta}^{\dot{a}}} = \frac{\partial}{\partial \bar{\theta}'^{\dot{a}}} + \frac{i}{2}\theta'^b (\sigma_\mu)_{b\dot{a}}\frac{\partial}{\partial y^\mu}$$

Let us now go back to the SUSY derivatives. We find [see Eqs. (11.57) and (11.58)]

$$D_a = \frac{\partial}{\partial \theta^a} - \frac{i}{2}(\sigma^\mu)_{a\dot{b}}\bar{\theta}^{\dot{b}}\frac{\partial}{\partial x^\mu}$$

$$= \frac{\partial}{\partial \theta'^a} - i(\sigma_\mu)_{a\dot{b}}\bar{\theta}'^{\dot{b}}\frac{\partial}{\partial y^\mu} \qquad (13.30)$$

whereas

$$\bar{D}_{\dot{a}} = \frac{\partial}{\partial \bar{\theta}'^{\dot{a}}}$$

It is the simplicity of this second expression that makes it worthwhile to change variables.

Before calculating the field-strength superfield, we need to express the vector superfield [Eq. (13.10)] in terms of the new variables. We simply replace x^{μ} by $y^{\mu} + i\,\theta'\sigma^{\mu}\bar{\theta}'/2$ and Taylor expand. Keeping in mind that any expression with three θ or three $\bar{\theta}$ vanishes identically, we get a simple result:

$$\mathcal{V} = \frac{1}{2}\theta\sigma^{\mu}\bar{\theta}\,A_{\mu}(x) + \frac{1}{2\sqrt{2}}\theta\cdot\theta\bar{\theta}\cdot\bar{\lambda}(x) + \frac{1}{2\sqrt{2}}\bar{\theta}\cdot\bar{\theta}\theta\cdot\lambda(x) - \frac{1}{8}\theta\cdot\theta\bar{\theta}\cdot\bar{\theta}D(x)$$

$$= \frac{1}{2}\theta'\sigma^{\mu}\bar{\theta}'\,A_{\mu}(y) + \frac{i}{4}\theta'\sigma^{\mu}\bar{\theta}'\theta'\sigma^{\nu}\bar{\theta}'\frac{\partial}{\partial y^{\nu}}A_{\mu}(y) + \frac{1}{2\sqrt{2}}\theta'\cdot\theta'\bar{\theta}'\cdot\bar{\lambda}(y)$$

$$+ \frac{1}{2\sqrt{2}}\bar{\theta}'\cdot\bar{\theta}'\theta'\cdot\lambda(y) - \frac{1}{8}\theta'\cdot\theta'\bar{\theta}'\cdot\bar{\theta}'D(y)$$

To simplify the notation, we will drop the prime on the Grassmann variables and not write explicitly the y dependence of the fields. We also will use ∂_{μ} for the derivative with respect to y^{μ}. The second term on the right-hand side can be simplified by using Eq. (A.56). We then get

$$\mathcal{V} = \frac{1}{2}\theta\sigma^{\mu}\bar{\theta}\,A_{\mu} + \frac{i}{8}\theta\cdot\theta\,\bar{\theta}\cdot\bar{\theta}\partial_{\mu}A^{\mu} + \frac{1}{2\sqrt{2}}\theta\cdot\theta\bar{\theta}\cdot\bar{\lambda}$$

$$+ \frac{1}{2\sqrt{2}}\bar{\theta}\cdot\bar{\theta}\theta\cdot\lambda - \frac{1}{8}\theta\cdot\theta\bar{\theta}\cdot\bar{\theta}\,D \tag{13.31}$$

We are finally ready to compute

$$\mathcal{F}_{a} = \bar{D}\cdot\bar{D}D_{a}\mathcal{V}$$

We start by applying to Eq. (13.31) the SUSY derivative

$$D_{a} = \frac{\partial}{\partial\theta'^{a}} - i\,(\sigma^{\mu})_{a\dot{b}}\,\bar{\theta}'^{\dot{b}}\frac{\partial}{\partial y^{\mu}}$$

$$= \frac{\partial}{\partial\theta^{a}} - i\,(\sigma^{\mu})_{a\dot{b}}\,\bar{\theta}^{\dot{b}}\partial_{\mu} \tag{13.32}$$

where we again have dropped the prime on the Grassmann variables and used the notation $\partial_\mu = \partial/\partial y^\mu$.

We get

$$
D_a \mathcal{V} = \frac{1}{2} (\sigma^\mu)_{a\dot{b}} \bar{\theta}^{\dot{b}} A_\mu + \frac{i}{4} \theta_a \bar{\theta} \cdot \bar{\theta} \partial_\mu A^\mu + \frac{1}{\sqrt{2}} \theta_a \bar{\theta} \cdot \bar{\lambda} + \frac{1}{2\sqrt{2}} \bar{\theta} \cdot \bar{\theta} \lambda_a
$$

$$
- \frac{1}{4} \theta_a \bar{\theta} \cdot \bar{\theta} D - \frac{i}{2} (\sigma^\mu)_{a\dot{b}} \bar{\theta}^{\dot{b}} \theta \sigma^\nu \bar{\theta} \partial_\mu A_\nu - \frac{i}{2\sqrt{2}} \theta \cdot \theta (\sigma^\mu)_{a\dot{b}} \bar{\theta}^{\dot{b}} \bar{\theta} \cdot \partial_\mu \bar{\lambda}
$$

$$
\tag{13.33}
$$

We now need to apply $\bar{D} \cdot \bar{D}$ to Eq. (13.33). Luckily for us, in terms of the new variables, this is simply

$$
\bar{D} \cdot \bar{D} = \bar{\partial}_{\dot{a}} \, \bar{\partial}^{\dot{a}}
$$

$$
= \frac{\partial}{\partial \bar{\theta}^{\dot{a}}} \frac{\partial}{\partial \bar{\theta}_{\dot{a}}}
\tag{13.34}
$$

We see that we may drop the two terms in Eq. (13.33) containing only one $\bar{\theta}$. As for the other terms, it will make our life easier if we can get all the $\bar{\theta}$ to be dotted together, i.e., in the form $\bar{\theta} \cdot \bar{\theta}$. For example, consider the combination

$$
(\sigma^\mu)_{a\dot{b}} \bar{\theta}^{\dot{b}} \theta \sigma^\nu \bar{\theta} = (\sigma^\mu)_{a\dot{b}} \bar{\theta}^{\dot{b}} \theta^c (\sigma^\nu)_{c\dot{d}} \bar{\theta}^{\dot{d}}
$$

This can be rewritten in a more convenient form following the steps

$$
(\sigma^\mu)_{a\dot{b}} \bar{\theta}^{\dot{b}} \theta^c (\sigma^\nu)_{c\dot{d}} \bar{\theta}^{\dot{d}} = -(\sigma^\mu)_{a\dot{b}} \theta^c (\sigma^\nu)_{c\dot{d}} \bar{\theta}^{\dot{b}} \bar{\theta}^{\dot{d}}
$$

$$
= -\frac{1}{2} \epsilon^{\dot{b}\dot{d}} (\sigma^\mu)_{a\dot{b}} \theta^c (\sigma^\nu)_{c\dot{d}} \bar{\theta} \cdot \bar{\theta} \qquad \text{[using Eq. (A.48)]}
$$

$$
= -\frac{1}{2} \epsilon^{\dot{b}\dot{d}} \epsilon^{ce} (\sigma^\mu)_{a\dot{b}} \theta_e (\sigma^\nu)_{c\dot{d}} \bar{\theta} \cdot \bar{\theta} \qquad \text{(writing } \theta^c = \epsilon^{ce} \theta_e)
$$

$$
= \frac{1}{2} (\sigma^\mu)_{a\dot{b}} (\bar{\sigma}^\nu)^{\dot{b}e} \theta_e \bar{\theta} \cdot \bar{\theta} \qquad \begin{array}{l} \text{[using } \epsilon^{ce} = -\epsilon^{ec} \text{ and} \\ \text{then Eq. (A.26)]} \end{array}
$$

$$
= \frac{1}{2} (\sigma^\mu \bar{\sigma}^\nu)_a^b \theta_b \bar{\theta} \cdot \bar{\theta}
\tag{13.35}
$$

where we have relabeled $e \to b$ in the last step. Using again Eq. (A.48), we can also prove

$$(\sigma^\mu)_{a\dot{b}}\,\bar{\theta}^{\dot{b}}(\partial_\mu\bar{\lambda}) \cdot \bar{\theta} = (\sigma^\mu)_{a\dot{b}}\,\bar{\theta}^{\dot{b}}(\partial_\mu\bar{\lambda})_{\dot{c}}\,\bar{\theta}^{\dot{c}}$$

$$= -(\sigma^\mu)_{a\dot{b}}(\partial_\mu\bar{\lambda})_{\dot{c}}\,\bar{\theta}^{\dot{b}}\,\bar{\theta}^{\dot{c}}$$

$$= -\frac{1}{2}\,\epsilon^{\dot{b}\dot{c}}\,(\sigma^\mu)_{a\dot{b}}(\partial_\mu\bar{\lambda})_{\dot{c}}\bar{\theta} \cdot \bar{\theta}$$

$$= -\frac{1}{2}\,(\sigma^\mu)_{a\dot{b}}(\partial_\mu\bar{\lambda})^{\dot{b}}\bar{\theta} \cdot \bar{\theta} \qquad (13.36)$$

Substituting Eqs. (13.35) and (13.36) into Eq. (13.33) and dropping the two terms that contain only one $\bar{\theta}$ (which won't survive the application of $\bar{D} \cdot \bar{D}$), we get

$$D_a\mathcal{V} = \frac{i}{4}\,\theta_a\,\bar{\theta} \cdot \bar{\theta}\partial_\mu A^\mu + \frac{1}{2\sqrt{2}}\,\bar{\theta} \cdot \bar{\theta}\,\lambda_a - \frac{1}{4}\,\theta_a\,\bar{\theta} \cdot \bar{\theta}\,D$$

$$- \frac{i}{4}\,(\sigma^\mu\bar{\sigma}^\nu)_a{}^b\,\theta_b\bar{\theta} \cdot \bar{\theta}\partial_\mu A_\nu + \frac{i}{4\sqrt{2}}\,\theta \cdot \theta\,(\sigma^\mu)_{a\dot{b}}\bar{\theta} \cdot \bar{\theta}\,(\partial_\mu\bar{\lambda}^{\dot{b}}) \qquad (13.37)$$

This can be simplified even further. Recall that

$$\sigma^{\mu\nu} \equiv \frac{i}{4}\,(\sigma^\mu\bar{\sigma}^\nu - \sigma^\nu\bar{\sigma}^\mu)$$

Using Eq. (A.9), this also may be written as

$$\sigma^{\mu\nu} = \frac{i}{2}\,(\sigma^\mu\bar{\sigma}^\nu - \eta^{\mu\nu}\,\mathbf{1})$$

which implies

$$(\sigma^\mu\bar{\sigma}^\nu)_a{}^b = \delta_a^b\,\eta^{\mu\nu} - 2i\,(\sigma^{\mu\nu})_a{}^b$$

Using this in Eq. (13.37), we obtain a simple result:

$$D_a\mathcal{V} = \left(\frac{1}{2\sqrt{2}}\,\lambda_a - \frac{1}{4}\,\theta_a\,D - \frac{1}{2}\,(\sigma^{\mu\nu})_a{}^b\,\theta_b\partial_\mu A_\nu + \frac{i}{4\sqrt{2}}\,\theta \cdot \theta\,(\sigma^\mu)_{a\dot{b}}\,(\partial_\mu\bar{\lambda}^{\dot{b}})\right)\bar{\theta} \cdot \bar{\theta}$$

All we now need is (see Exercise 13.1)

$$\bar{D} \cdot \bar{D}\,\bar{\theta} \cdot \bar{\theta} = 4 \tag{13.38}$$

which gives us

$$\mathcal{F}_a = \bar{D} \cdot \bar{D}D_a \mathcal{V} = \sqrt{2}\,\lambda_a - \theta_a\,D - 2\,(\sigma^{\mu\nu})_a{}^b\,\theta_b \partial_\mu A_\nu + \frac{i}{\sqrt{2}}\,\theta \cdot \theta\,(\sigma^\mu)_{a\dot{b}}\,(\partial_\mu \bar{\lambda}^{\dot{b}})$$

Using the antisymmetry of $\sigma^{\mu\nu}$ in its Lorentz indices, we may write

$$\sigma^{\mu\nu}\,\partial_\mu A_\nu = \frac{1}{2}\,(\sigma^{\mu\nu} - \sigma^{\nu\mu})\,\partial_\mu A_\nu$$

$$= \frac{1}{2}\,\sigma^{\mu\nu}\,(\partial_\mu A_\nu - \partial_\nu A_\mu)$$

$$= \frac{1}{2}\,\sigma^{\mu\nu}\,F_{\mu\nu} \tag{13.39}$$

Using this, and showing explicitly the y dependence of the fields, we obtain our final result:

$$\mathcal{F}_a(y) = \sqrt{2}\,\lambda_a(y) - D(y)\,\theta_a - F_{\mu\nu}(y)\,(\sigma^{\mu\nu})_a{}^b\,\theta_b + \frac{i}{\sqrt{2}}\,\theta \cdot \theta\,(\sigma^\mu)_{a\dot{b}}\,\partial_\mu \bar{\lambda}^{\dot{b}}(y) \tag{13.40}$$

At this point, we could rewrite the fields in terms of the spacetime coordinate x using Eq. (12.3), but this won't be necessary because we are ultimately interested in the F term of $\mathcal{F}^a \mathcal{F}_a$, in which we may simply set y equal to x [see Eq. (12.27)].

EXERCISE 13.1
Prove Eq. (13.38).

13.5 The Free Abelian Supersymmetric Lagrangian from the Superfield Approach

We are finally ready to construct the lagrangian for the free abelian gauge theory in the superfield approach. In sharp contrast with the usual field strength $F_{\mu\nu}$, which is a second-rank antisymmetric tensor (in other words, the coefficient of a two-form),

its supersymmetric extension is a spinor superfield. A lagrangian must be a Lorentz scalar, so the obvious thing to consider is $\mathcal{F}^a \mathcal{F}_a$. Recall that $\mathcal{F}_a$ is a left-chiral superfield, so to obtain a supersymmetric lagrangian, we will need to extract its F term. But let's first start with the calculation of

$$\mathcal{F}^a \mathcal{F}_a = \epsilon^{ab} \mathcal{F}_b \mathcal{F}_a$$

To alleviate the notation, we will not write explicitly that the fields depend on y, but this must be kept in mind.

Since we are ultimately interested in taking the F term of this expression, we need only retain the terms that contain two θ. These are

$$\mathcal{F}^a \, \mathcal{F}_a = 2 i \, \theta \cdot \theta \, \lambda \sigma^\mu \partial_\mu \bar\lambda + D^2 \theta \cdot \theta + \epsilon^{ab} \, F_{\mu\nu} \, (\sigma^{\mu\nu})_b^{\ c} \, \theta_c \, F_{\alpha\beta} \, (\sigma^{\alpha\beta})_a^{\ d} \, \theta_d + \cdots \tag{13.41}$$

where the dots indicate terms that will not contribute to the F term (to eliminate the term containing $D \partial_\mu A_\nu$, we have used the fact that $\epsilon^{ab} \theta^a \theta_c (\sigma^{\mu\nu})_a^{\ c}$ may be written as $\theta \cdot \theta \, Tr(\sigma^{\mu\nu})/2$ which is identically zero). In order to read off the F term, it is preferable to have all the θ combined in the form $\theta \cdot \theta$. We can do this for the the third term by using the identity (A.45):

$$\epsilon^{ab} \, F_{\mu\nu} \, (\sigma^{\mu\nu})_b^{\ c} \, \theta_c \, F_{\alpha\beta} \, (\sigma^{\alpha\beta})_a^{\ d} \, \theta_d = \frac{1}{2} \theta \cdot \theta \, \epsilon^{ab} \epsilon_{cd} \, F_{\mu\nu} \, (\sigma^{\mu\nu})_b^{\ c} \, F_{\alpha\beta} \, (\sigma^{\alpha\beta})_a^{\ d} \tag{13.42}$$

Following the same steps as in part b of Exercise 6.5, it is straightforward to prove

$$\epsilon^{ab} \, \epsilon_{cd} \, (\sigma^{\mu\nu})_b^{\ c} = -(\sigma^{\mu\nu})_d^{\ a}$$

so that Eq. (13.42) becomes

$$\frac{1}{2} \epsilon^{ab} \, \epsilon_{cd} \, \theta \cdot \theta \, (\sigma^{\mu\nu})_b^{\ c} \, F_{\mu\nu} \, (\sigma^{\alpha\beta})_a^{\ d} \, F_{\alpha\beta} = -\frac{1}{2} \theta \cdot \theta \, (\sigma^{\mu\nu})_d^{\ a} \, F_{\mu\nu} \, (\sigma^{\alpha\beta})_a^{\ d} \, F_{\alpha\beta}$$

$$= -\frac{1}{2} \theta \cdot \theta \, Tr\!\left(\sigma^{\mu\nu} \sigma^{\alpha\beta}\right) F_{\mu\nu} F_{\alpha\beta} \tag{13.43}$$

To make progress on this term, the following identity is required

$$Tr\!\left(\sigma^\mu \, \bar\sigma^\nu \, \sigma^\alpha \, \bar\sigma^\beta\right) = 2 \left(\eta^{\mu\nu} \eta^{\alpha\beta} + \eta^{\nu\alpha} \eta^{\mu\beta} - \eta^{\mu\alpha} \eta^{\nu\beta} - i \, \epsilon^{\mu\nu\alpha\beta}\right) \tag{13.44}$$

Using the definition

$$\sigma^{\mu\nu} = \frac{i}{4}\left(\sigma^\mu\,\bar\sigma^\nu - \sigma^\nu\,\bar\sigma^\mu\right)$$

and the corresponding expression for $\sigma^{\alpha\beta}$, and applying Eq. (13.44), a short calculation reveals that

$$\mathrm{Tr}\!\left(\sigma^{\mu\nu}\,\sigma^{\alpha\beta}\right) = -\frac{1}{16}\,\mathrm{Tr}\!\left(\sigma^\mu\,\bar\sigma^\nu\,\sigma^\alpha\,\bar\sigma^\beta - \sigma^\mu\,\bar\sigma^\nu\,\sigma^\beta\,\bar\sigma^\alpha - \sigma^\nu\,\bar\sigma^\mu\,\sigma^\alpha\,\bar\sigma^\beta\right.$$
$$\left. + \sigma^\nu\,\bar\sigma^\mu\,\sigma^\beta\,\bar\sigma^\alpha\right)$$
$$= -\frac{1}{2}\left(\eta^{\nu\alpha}\,\eta^{\mu\beta} - \eta^{\mu\alpha}\,\eta^{\nu\beta} - i\,\epsilon^{\mu\nu\alpha\beta}\right) \qquad (13.45)$$

Plugging this into Eq. (13.43), we get

$$-\frac{1}{2}\,\theta\cdot\theta\,\mathrm{Tr}\!\left(\sigma^{\mu\nu}\,\sigma^{\alpha\beta}\right)F_{\mu\nu}F_{\alpha\beta} = \frac{1}{4}\,\theta\cdot\theta\left(F_{\mu\nu}F^{\nu\mu} - F_{\mu\nu}F^{\mu\nu} - i\,\epsilon^{\mu\nu\alpha\beta}\,F_{\mu\nu}F_{\alpha\beta}\right)$$
$$= -\frac{1}{4}\,\theta\cdot\theta\left(2F_{\mu\nu}F^{\mu\nu} + i\,\epsilon^{\mu\nu\alpha\beta}\,F_{\mu\nu}F_{\alpha\beta}\right) \quad (13.46)$$

where we have used the antisymmetry of the field strength, i.e., $F^{\nu\mu} = -F^{\mu\nu}$. Using Eq. (13.46) for the third term of Eq. (13.41), we finally get

$$\mathcal{F}^a\mathcal{F}_a = 2i\,\theta\cdot\theta\,\lambda\sigma^\mu\partial_\mu\bar\lambda + D^2\theta\cdot\theta - \frac{1}{4}\,\theta\cdot\theta\left(2F_{\mu\nu}F^{\mu\nu} + i\,\epsilon^{\mu\nu\alpha\beta}\,F_{\mu\nu}F_{\alpha\beta}\right)$$

Recall that all the fields are functions of the variable y. However, if we extract the F term, i.e., the coefficient of $\theta\cdot\theta/2$, we may replace y by x, as you have shown in Exercise 12.1. We then get

$$\mathcal{F}_a\mathcal{F}^a\Big|_F = 4i\,\lambda(x)\sigma^\mu\partial_\mu\bar\lambda(x) + 2\,D^2(x) - F_{\mu\nu}(x)\,F^{\mu\nu}(x)$$
$$- \frac{i}{2}\,\epsilon^{\mu\nu\alpha\beta}\,F_{\mu\nu}(x)\,F_{\alpha\beta}(x)$$

This is starting to resemble the lagrangian of the supersymmetric free abelian gauge theory, [Eq. (10.8)]. The kinetic term for the spinor looks different, but looks can be deceiving. Indeed, using Eq. (A.50) followed by an integration by parts, we have

$$\lambda \sigma^\mu \partial_\mu \bar{\lambda} = -(\partial_\mu \bar{\lambda}) \bar{\sigma}^\mu \lambda$$

$$= \bar{\lambda} \bar{\sigma}^\mu \partial_\mu \lambda \tag{13.47}$$

which is of the same form as the the spinor kinetic term appearing in Eq. (10.8). Thus we may write

$$\mathcal{F}^a \mathcal{F}_a \Big|_F = 4i\,\bar{\lambda}(x)\,\bar{\sigma}^\mu \partial_\mu \lambda(x) + 2\,D^2(x) - F_{\mu\nu}(x)\,F^{\mu\nu}(x)$$

$$- \frac{i}{2}\,\epsilon^{\mu\nu\alpha\beta}\,F_{\mu\nu}(x)\,F_{\alpha\beta}(x) \tag{13.48}$$

We see that the first three terms are exactly four times the free abelian gauge theory shown in Eq. (10.8) (where we used the notation $\lambda^\dagger$ instead of $\bar{\lambda}$). However, we have an extra term, which is usually written as

$$i F_{\mu\nu} F^{*\mu\nu}$$

where

$$F^{*\mu\nu} \equiv \frac{1}{2}\,\epsilon^{\mu\nu\alpha\beta}\,F_{\alpha\beta}$$

is called the *dual field strength*. This term is actually a total derivative and is not considered unless the gauge fields have a nontrivial topology, which is why we did not include it in Chapter 10. We will ignore it from now on.

Our final expression therefore is

$$\boxed{\frac{1}{4}\,(\mathcal{F}^a \mathcal{F}_a)\Big|_F = \frac{1}{2}D^2 + i\,\bar{\lambda}\bar{\sigma}^\mu \partial_\mu \lambda - \frac{1}{4}\,F_{\mu\nu} F^{\mu\nu}} \tag{13.49}$$

which, apart from the Fayet-Iliopoulos term, reproduces exactly our supersymmetric lagrangian [Eq. (10.8)]! To include the Fayet-Iliopoulos term, we simply have to add the D term of $-2\xi V$ (see the discussion after Eq. (13.12)). Of course, we could simply write $\mathcal{F}^a \mathcal{F}_a$ as $\mathcal{F} \cdot \mathcal{F}$, using our notation for the spin dot products.

The obvious next step would be to apply a SUSY transformation to $\mathcal{F}_a$ using the supercharges in order to confirm the transformations of the component fields we established in Chapter 10, but we won't pursue this rather lengthy calculation here. Rest assured that it does work!

Our final SUSY and gauge-invariant lagrangian that couples a set of i left-chiral superfields to a $U(1)$ gauge field is the sum of Eqs. (13.11) and (13.49):

$$\mathcal{L} = \left[\mathcal{W}(\Phi_i)\big|_F + \text{h.c.}\right] + \sum_i \Phi_i^\dagger e^{2q_i \mathcal{V}} \Phi_i\Big|_D + \frac{1}{4}\mathcal{F}^a \mathcal{F}_a\Big|_F \qquad (13.50)$$

As noted at the end of Section 10.6, the superpotential is absent if there is only one left-chiral superfield that is charged under the gauge group because it is impossible to write down a gauge-invariant holomorphic function of Φ.

13.6 Supersymmetric QED

We will now apply the results of this chapter to build a supersymmetric version of quantum electrodynamics (QED). This will be useful as a warm-up exercise before we construct the supersymmetric extension of the standard model, which we will pursue in Chapter 15.

The first step is to determine what superfields we need. QED involves a photon field and the electron field, which comes in right-chiral and left-chiral states, together with the corresponding positron states. Clearly, we will introduce an abelian gauge superfield $\mathcal{V}$ that will contain the photon, its spin 1/2 left-chiral partner, the photino λ, and a corresponding auxiliary field D. In the Wess-Zumino gauge, the vector superfield is just what we wrote in Eq. (13.10). Of course, we also will use the field-strength superfield $\mathcal{F}$ given in Eq. (13.40).

The more tricky issue is the electron. There are four physical degrees of freedom associated with the electron field: the left-chiral and right-chiral electron states and the left-chiral and right-chiral positron states. However, it is conventional to build supersymmetric theories out of left-chiral superfields only. This can be done using a trick that we will describe below.

But first, let's introduce a left-chiral superfield $\mathcal{E}_e$ containing the left-chiral electron state:

$$\mathcal{E}_e = \tilde{\phi}_e + \theta \cdot \chi_e + \frac{1}{2}\theta \cdot \theta \, F_e$$

It's important to emphasize that $\mathcal{E}_e$ is a scalar under Lorentz transformations. As we have seen before, it is called a left-chiral superfield only because one of its components is the left-chiral Weyl spinor χ_e.

Here we have started using the notation that we will follow in Chapter 15 when we cover the minimal supersymmetric standard model (MSSM). First, we will denote all superfields by calligraphic capital letters. Also, it will always be understood that a spinor χ represents a left-chiral Weyl spinor, so we won't include a label L to avoid cluttering the notation. Finally, we will include a tilde (˜) over the fields that are superpartners of the known particles (the Higgs field will require a separate discussion).

To respect *CPT* invariance, our lagrangian must involve both $\mathcal{E}_e$ and $\mathcal{E}_e^\dagger$, so the physical spectrum contains both the left-chiral electron state and its hermitian conjugate, the right-chiral positron state. But there are two states still missing: the right-chiral electron state and its hermitian conjugate, the left-chiral positron state. To account for those, we have to include a *second* chiral superfield. We could make it a right-chiral superfield containing for spinor component the right-chiral electron state, but, as mentioned earlier, the convention is to express everything in terms of left-chiral spinors.

The trick is to introduce a left-chiral superfield whose spinor component is the left-chiral *positron* state! As we saw in Chapter 4, one always may trade off a *right-chiral particle* Weyl spinor for the corresponding *left-chiral antiparticle* spinor using [see Eq. (4.5)]

$$\eta_e = i\,\sigma^2\,\chi_{\bar{e}}^{\dagger T}$$

where we used the notation of Chapter 2: η is a right-chiral spinor, χ is a left-chiral spinor, and $\bar{e}$ represents the positron. We use $\bar{e}$ to represent the positron instead of e^+ because we want a notation that can be extended to denote all antiparticles, including those with no electric charge. Obviously, this bar has nothing to do with the bar notation introduced for dot products between Weyl spinors or the bar used to denote Dirac conjugation!

We will write the left-chiral superfield containing the left-chiral positron state as

$$\mathcal{E}_{\bar{e}} = \tilde{\phi}_{\bar{e}} + \theta \cdot \chi_{\bar{e}} + \frac{1}{2}\theta \cdot \theta F_{\bar{e}}$$

We take, of course, $\mathcal{E}_e$ and $\mathcal{E}_{\bar{e}}$ to have $U(1)$ charges $q = -e$ and $q = e$, respectively. In other words, under a gauge transformation, we have

$$\mathcal{E}_e \to e^{-2ie\Lambda}\mathcal{E}_e$$
$$\mathcal{E}_{\bar{e}} \to e^{2ie\Lambda}\mathcal{E}_{\bar{e}} \tag{13.51}$$

We are now ready to build our supersymmetric extension of QED. The gauge- and Lorentz-invariant quantities that we have at our disposal are

$$\mathcal{E}_e\,\mathcal{E}_{\bar{e}} \qquad \mathcal{F}^a\mathcal{F}_a \qquad \mathcal{E}_e^\dagger e^{-2e\mathcal{V}}\mathcal{E}_e \qquad \mathcal{E}_{\bar{e}}^\dagger e^{2e\mathcal{V}}\mathcal{E}_{\bar{e}} \tag{13.52}$$

as well as products of any combination of these terms and their hermitian conjugates (more about this on next page).

The first two combinations of Eq. (13.52) are left-chiral superfields because $\mathcal{E}_e$, $\mathcal{E}_{\bar{e}}$, and $\mathcal{F}$ are all left-chiral superfields (recall that a product of left-chiral superfields is also a left-chiral superfield). To obtain a supersymmetric lagrangian, we therefore must take the F term of these expressions. The last two are not chiral superfields, so we must take their D term to obtain supersymmetric lagrangians.

Now, let's look at the dimensions. The field-strength superfield $\mathcal{F}_a$ has dimension 3/2, whereas $\mathcal{V}$ is dimensionless, and both $\mathcal{E}_e$ and $\mathcal{E}_{\bar{e}}$ have dimension 1. Recall that projecting out the F term increases the dimension by 1 because $d^2\theta$ has dimension 1, whereas projecting out the D term increases the dimension by 2 because $d^2\theta\, d^2\bar{\theta}$ has dimension 2. Thus we must consider combinations of superfields of at most dimension 3 if we use the F term and at most dimension 2 if we use the D term. Therefore, the following terms are all dimension 4 supersymmetric lagrangians:

$$\mathcal{F}^a \mathcal{F}_a \Big|_F , \qquad \mathcal{E}_e^\dagger e^{-2eV} \mathcal{E}_e \Big|_D , \qquad \mathcal{E}_{\bar{e}}^\dagger e^{2eV} \mathcal{E}_{\bar{e}} \Big|_D \qquad (13.53)$$

These are, of course, the kinetic terms we obtained previously for these superfields, so no surprise here. We also have an interaction term of dimension 3:

$$\mathcal{E}_e \mathcal{E}_{\bar{e}} \Big|_F$$

which must be multiplied by a constant with dimensions of mass. It turns out that this coefficient corresponds precisely to the mass of a Dirac particle, as we will demonstrate below, so let's write this term as

$$m \mathcal{E}_e \mathcal{E}_{\bar{e}} \Big|_F \qquad (13.54)$$

It'e easy to see that we have exhausted all the possible terms that can contribute to a renormalizable, supersymmetric, and gauge-invariant lagrangian. For example, the F term of $\mathcal{E}_e \mathcal{E}_{\bar{e}} \mathcal{E}_e \mathcal{E}_{\bar{e}}$ has dimension 5.

We must worry about one last thing before writing down the supersymmetric version of QED: We must make sure that the lagrangian is real. All the terms in Eq. (13.53) are real, but Eq. (13.54) is not, so we must add to it its hermitian conjugate. Our supersymmetric QED lagrangian therefore is

$$\boxed{\mathcal{L}_{\text{super QED}} = \frac{1}{4} \mathcal{F}^a \mathcal{F}_a \Big|_F + \mathcal{E}_e^\dagger e^{-2eV} \mathcal{E}_e \Big|_D + \mathcal{E}_{\bar{e}}^\dagger e^{2eV} \mathcal{E}_{\bar{e}} \Big|_D + m\, \mathcal{E}_e \mathcal{E}_{\bar{e}} \Big|_F + m\, \mathcal{E}_e^\dagger \mathcal{E}_{\bar{e}}^\dagger \Big|_F}$$

$$(13.55)$$

where we have taken m to be a real quantity.

Note that the last two terms of Eq. (13.55) represent the superpotential of the theory (plus its hermitian conjugate), so we could write the lagrangian as

$$\mathcal{L}_{\text{super QED}} = \frac{1}{4} \mathcal{F}^a \mathcal{F}_a \Big|_F + \mathcal{E}_e^\dagger e^{-2eV} \mathcal{E}_e \Big|_D + \mathcal{E}_{\bar{e}}^\dagger e^{2eV} \mathcal{E}_{\bar{e}} \Big|_D + \left(\mathcal{W} \Big|_F + \text{h.c.} \right) \quad (13.56)$$

with

$$\mathcal{W} \equiv m\, \mathcal{E}_e \mathcal{E}_{\bar{e}}$$

Let's write this out in terms of the component fields. Actually, let's write the lagrangian with the auxiliary fields eliminated, as an illustration of how to use Eq. (10.56), which we repeat here for convenience (we will not include a Fayet-Iliopoulos term):

$$\mathcal{L} = \mathcal{L}_{F_i = D = 0} - \left| \frac{\partial \mathcal{W}}{\partial \phi_i} \right|^2 - \frac{1}{2} \left(q_i\, \phi_i^\dagger \phi_i \right)^2 \quad (13.57)$$

To see how this is used, we first write the superpotential $m\mathcal{E}_e\mathcal{E}_{\bar{e}}$ as a function of the corresponding scalar fields

$$\mathcal{W} = m\, \tilde{\phi}_e\, \tilde{\phi}_{\bar{e}}$$

This superpotential depends on two scalar fields, so we may take the ϕ_i appearing in Eq. (13.57) to be given by, say, $\phi_1 = \tilde{\phi}_e$ and $\phi_2 = \tilde{\phi}_{\bar{e}}$. We then find

$$
\begin{aligned}
-\left| \frac{\partial \mathcal{W}}{\partial \phi_i} \right|^2 &= -\frac{\partial \mathcal{W}}{\partial \phi_i} \left(\frac{\partial \mathcal{W}}{\partial \phi_i} \right)^\dagger \\
&= -\frac{\partial \mathcal{W}}{\partial \tilde{\phi}_e} \left(\frac{\partial \mathcal{W}}{\partial \tilde{\phi}_e} \right)^\dagger - \frac{\partial \mathcal{W}}{\partial \tilde{\phi}_{\bar{e}}} \left(\frac{\partial \mathcal{W}}{\partial \tilde{\phi}_{\bar{e}}} \right)^\dagger \\
&= -m^2\, \tilde{\phi}_{\bar{e}} \tilde{\phi}_{\bar{e}}^\dagger - m^2\, \tilde{\phi}_e \tilde{\phi}_e^\dagger \\
&= -m^2\, |\tilde{\phi}_{\bar{e}}|^2 - m^2\, |\tilde{\phi}_e|^2
\end{aligned} \quad (13.58)
$$

On the other hand, the last term of Eq. (13.57) is simply

$$
\begin{aligned}
-\frac{1}{2} \left(q_i\, \phi_i^\dagger \phi_i \right)^2 &= -\frac{1}{2} \left(-e\, |\tilde{\phi}_e|^2 + e\, |\tilde{\phi}_{\bar{e}}|^2 \right)^2 \\
&= -\frac{e^2}{2} \left(|\tilde{\phi}_e|^2 - |\tilde{\phi}_{\bar{e}}|^2 \right)^2
\end{aligned} \quad (13.59)
$$

All we have to do now is to write down Eq. (13.56) in terms of component fields with the auxiliary fields set to zero and add Eqs. (13.58) and (13.59).

We have actually worked out all the expressions appearing in Eq. (13.56) in terms of component fields before. The first term is simply Eq. (13.49), the second and third terms can be obtained from Eq. (13.23), and the last two terms can be read off from Eq. (12.28). We finally get

$$
\begin{aligned}
\mathcal{L}_{\text{super QED}} = {} & i\bar{\lambda}\bar{\sigma}^{\mu}\partial_{\mu}\lambda - \frac{1}{4}F_{\mu\nu}F^{\mu\nu} + (D_{\mu}\tilde{\phi}_{e})^{\dagger}(D^{\mu}\tilde{\phi}_{e}) + i\,\bar{\chi}_{e}\bar{\sigma}^{\mu}D_{\mu}\chi_{e} \\
& + \left(\sqrt{2}\,e\,\chi_{e}\cdot\lambda\tilde{\phi}_{e}^{\dagger} + \text{h.c.}\right) + (D_{\mu}\tilde{\phi}_{\bar{e}})^{\dagger}(D^{\mu}\tilde{\phi}_{\bar{e}}) + i\,\bar{\chi}_{\bar{e}}\bar{\sigma}^{\mu}D_{\mu}\chi_{\bar{e}} \\
& - \left(\sqrt{2}\,e\,\chi_{\bar{e}}\cdot\lambda\tilde{\phi}_{\bar{e}}^{\dagger} + \text{h.c.}\right) - m\,\chi_{e}\cdot\chi_{\bar{e}} - m\,\bar{\chi}_{e}\cdot\bar{\chi}_{\bar{e}} \\
& - m^{2}\,|\tilde{\phi}_{\bar{e}}|^{2} - m^{2}\,|\tilde{\phi}_{e}|^{2} - \frac{e^{2}}{2}\left(|\tilde{\phi}_{e}|^{2} - |\tilde{\phi}_{\bar{e}}|^{2}\right)_{2} \qquad (13.60)
\end{aligned}
$$

Here, it is understood that when applied to fields belong to the electron multiplet ($\tilde{\phi}_{e}$ and χ_{e}), the covariant derivative is (with the convention $e > 0$)

$$
D_{\mu} = \partial_{\mu} - ieA_{\mu}
$$

whereas, when acting on the positron multiplet, the covariant derivative is

$$
D_{\mu} = \partial_{\mu} + ieA_{\mu}
$$

Using Eq. (4.19), we see that the QED lagrangian is indeed contained in Eq. (13.60):

$$
\begin{aligned}
\mathcal{L}_{\text{QED}} = {} & \overline{\Psi}_{D}\gamma^{\mu}\left(\partial_{\mu} - ieA_{\mu}\right)\Psi_{D} - \frac{1}{4}F_{\mu\nu}F^{\mu\nu} \\
= {} & i\,\bar{\chi}_{e}\bar{\sigma}^{\mu}D_{\mu}\chi_{e} + i\,\bar{\chi}_{\bar{e}}\bar{\sigma}^{\mu}D_{\mu}\chi_{\bar{e}} - m\,\chi_{e}\cdot\chi_{\bar{e}} - m\,\bar{\chi}_{e}\cdot\bar{\chi}_{\bar{e}} - \frac{1}{4}F_{\mu\nu}F^{\mu\nu}
\end{aligned}
$$

$$(13.61)$$

A few points deserve comments. First, let's emphasize once more that we needed to introduce *two* complex scalar fields: $\tilde{\phi}_{e}$ and $\tilde{\phi}_{\bar{e}}$. The reason is clear: We have to introduce a superpartner for each chiral state of the electron, which we have taken to be the left-chiral particle and antiparticle states. So there are two supersymmetric partners to a massive Dirac fermion. The next key observation is that those superpartners have the same mass as the electron, a consequence of unbroken SUSY. This immediately rules out this lagrangian as a viable model for the real world because no scalar particle with the mass of an electron has ever been observed. What we will need to do is to find ways to break SUSY spontaneously, but we will keep this topic for next chapter.

13.7 Supersymmetric Nonabelian Gauge Theories

We have seen how superfields may be used to obtain supersymmetric lagrangians for a free abelian gauge field theory and for a chiral multiplet coupled to an abelian gauge field. Here, we will generalize these results to nonabelian gauge theories. The calculations are almost identical to what we did in previous sections, so we will simply quote the main results without showing any explicit derivation.

The gauge superfield $\mathcal{V}$ is now taken to be a matrix in the fundamental representation of the Lie algebra [which we take to be $su(n)$], i.e.,

$$\mathcal{V} \equiv \mathcal{V}^i T^i_F$$

where the sum is over the $N^2 - 1$ generators. This means that each component field of $\mathcal{V}$ appears in $N^2 - 1$ copies, given by [see Eq. (13.10)]

$$\mathcal{V}^i = \frac{1}{2}\theta\sigma^\mu\bar\theta\, A^i_\mu + \frac{1}{2\sqrt{2}}\,\theta\cdot\theta\,\bar\theta\cdot\bar\lambda^i + \frac{1}{2\sqrt{2}}\,\bar\theta\cdot\bar\theta\,\theta\cdot\lambda^i - \frac{1}{8}\,\theta\cdot\theta\,\bar\theta\cdot\bar\theta\, D^i$$

Consider a left-chiral superfield Φ belonging to the fundamental representation. This means that Φ is actually an N-component column vector that transforms as

$$\Phi \to \Phi' = \exp(2ig\Lambda)\Phi$$

with Λ being the matrix-valued quantity

$$\Lambda \equiv \Lambda^i T^i_F$$

We take the vector superfield to transform as

$$\exp(2g\mathcal{V}) \to \exp(2g\mathcal{V}') = \exp(2ig\Lambda^\dagger)\exp(2g\mathcal{V})\exp(-2ig\Lambda)$$

The quantity

$$\Phi^\dagger \exp(2g\mathcal{V})\,\Phi \tag{13.62}$$

is clearly gauge-invariant. This is, of course, exactly what we had for the abelian case except that here the superfields do not commute because they are matrix valued.

The proper generalization of the abelian field-strength superfield $\mathcal{F}_a$ given in Eq. (13.24) is

$$\mathcal{F}_a = \bar{D} \cdot \bar{D} \, \exp(-2g\mathcal{V}) \, D_a \exp(2g\mathcal{V}) \qquad (13.63)$$

where, as before, *the label a is a spinor index, not a gauge group index*. The field-strength superfield is a matrix-valued quantity, like the other superfields. Showing explicitly the matrix dependence, Eq. (13.63) reads

$$\mathcal{F}_a^i \, T_F^i = \bar{D} \cdot \bar{D} \, \exp(-2g\mathcal{V}^i T_F^i) D_a \exp\left(2g\mathcal{V}_i T_F^i\right)$$

The choice of Eq. (13.63) is motivated by the fact that the field-strength superfield can be shown to transform in the usual way, i.e.,

$$\mathcal{F}_a \rightarrow \exp(2ig\Lambda) \, \mathcal{F}_a \, \exp(-2ig\Lambda)$$

Therefore, the trace

$$\text{Tr}\left(\mathcal{F}^a \mathcal{F}_a\right)$$

is gauge-invariant.

We now have all the pieces needed to write down a supersymmetric lagrangian containing a nonabelian gauge superfield coupled to a left-chiral superfield. If there is only one left-chiral superfield (which we assume is not invariant under the gauge group), there is no superpotential because we can't write down any gauge-invariant holomorphic function of that superfield. There are then only two terms in the lagrangian: the kinetic energy terms for the gauge superfield and for the left-chiral superfield:

$$\mathcal{L} = \frac{1}{2} \, \text{Tr}(\mathcal{F}^a \mathcal{F}_a)\Big|_F + \Phi^\dagger e^{2g\mathcal{V}} \Phi\Big|_D$$

If there are several left-chiral superfields, we obviously include a kinetic energy term for each superfield, but in addition, we can now include a superpotential, i.e.,

$$\mathcal{W}(\Phi_i) + \text{h.c.}$$

assuming that the quantum numbers of the left-chiral superfields Φ_i make it possible to write some gauge-invariant combination. Here the subscript i labels the different left-chiral superfields, each of which is a gauge multiplet transforming in some representation of the group.

13.8 Quiz

1. How does the field-strength superfield transform under Lorentz transformations?

2. What is the dimension of the gauge superfield V?

3. In the Wess-Zumino gauge, what are the component fields of the gauge superfield?

4. In a nonabelian supersymmetric $SU(N)$ gauge theory, how many gauginos are there?

5. How many complex scalar fields are needed to build a supersymmetric version of QED? Why is that number of scalar fields required?

CHAPTER 14

SUSY Breaking

A key aspect of the theories we have considered so far is that all the particles of a supersymmetric multiplet necessarily have the same mass. This follows from the fact that the supercharges commute with the momentum operator, so applying supercharges on a state does not change the eigenvalue of $P^\mu P_\mu = m^2$. This is obviously a disaster from the point of view of phenomenology because this degeneracy is clearly not observed in nature (e.g., there is no scalar particle of the same mass and charge as the electron).

One can break supersymmetry (SUSY) explicitly by adding terms in the lagrangian that are not invariant under SUSY. Of course, if we allow for any type of SUSY-breaking interaction, we will loose all the benefits of SUSY, in particular the cancellation of power-law divergences. The tricky question then becomes how to identify the various SUSY-breaking interactions that do not spoil this cancellation—in which case SUSY is said to be it *softly broken*—and how to justify why only those should be included. We will come back to the issue of soft SUSY breaking later in this chapter.

A second, and more desirable, method to break a symmetry is through spontaneous symmetry breaking (SSB). In this case, the lagrangian possesses the symmetry, but the ground state does not, which then is reflected through the absence of symmetry in the spectrum of the states.

We will first discuss various means to spontaneously break SUSY and then come back to the issue of explicit SUSY breaking in Section 14.9.

14.1 Spontaneous Supersymmetry Breaking

When there is SSB, the charges generating the symmetry do not annihilate the vacuum, i.e.,

$$Q|0\rangle \neq 0$$

Let us see what consequence this has in the case of SUSY. Recall that in Section 7.5 we showed that if the charges Q_1 and Q_2 do not annihilate the vacuum, we get a strict inequality for the vacuum expectation value (vev) of the hamiltonian:

$$\langle 0|H|0\rangle > 0 \tag{14.1}$$

The hamiltonian can be broken up into two parts: the kinetic energies terms of the fields and the potential (in which we lump all the interactions among the fields as well as the mass terms). In the vacuum, we can take the vev of the kinetic energies terms to be zero (we are not considering field configurations with nontrivial topology), leaving us with the vev of the potential. The only fields that may have a nonzero vacuum expectation values without breaking Lorentz invariance are scalar fields, so the condition (14.1) reduces to

$$\langle 0|V(\phi_i)|0\rangle > 0$$

As we have seen in Section 10.7, *the entire scalar field potential comes from the auxiliary fields*, after they are eliminated using their equations of motion. Schematically, we may write the potential as

$$\boxed{V(\phi_i) = F_i F_i^\dagger + \frac{1}{2} D^2} \tag{14.2}$$

where it is understood that by F_i, $F_i^\dagger$, and D we mean the solutions to the equations of motion, namely,

$$\boxed{F_i^\dagger = -\frac{\partial \mathcal{W}}{\partial \phi_i}}$$

and, for an abelian gauge group,

$$D = q_i \, \phi_i^\dagger \phi_i - \xi \qquad (14.3)$$

We have included a Fayet-Illiopoulos term ξD in the lagrangian. We will see that this term plays a crucial role in some types of SUSY breaking.

In the case of a nonabelian gauge group, the D^2 in Eq. (14.2) stands for $D^a D^a$, and the solution of the equations of motion is

$$D^a = g \, \phi_i^\dagger \, T^a \, \phi_i$$

As we have mentioned before, gauge invariance precludes a Fayet-Illiopoulos term for nonabelian gauge theories.

The condition for SSB of SUSY can be restated in the following way: *There will be SSB of SUSY if, and only if, there is no possible scalar field configuration for which* $\langle 0|V|0 \rangle = 0$.

Consider the case of a single scalar field ϕ, which is too simplistic but will allow us to draw simple figures. The expectation values of four different potentials as a function of the scalar field are shown in Figures 14.1 through 14.4.

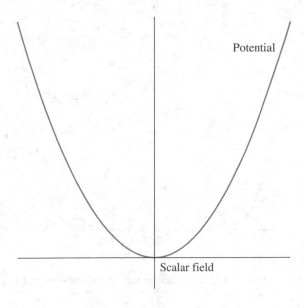

Figure 14.1   Potential that does not break any symmetry.

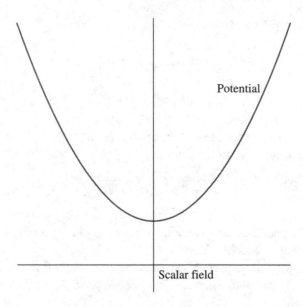

Figure 14.2 Potential that breaks SUSY but no internal symmetry.

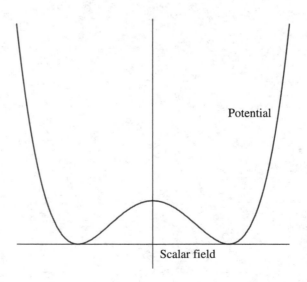

Figure 14.3 Potential that breaks an internal symmetry but not SUSY.

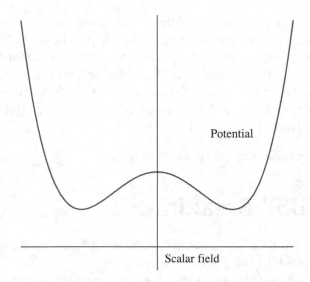

Figure 14.4 Potential that breaks both SUSY and an internal symmetry.

In Figure 14.1, no symmetry is broken. In Figure 14.2, SUSY is broken because there are no field configurations for which the expectation value of the potential is zero. In Figure 14.3, SUSY is not broken, but since the potential is not minimized for $\langle \phi \rangle = 0$, an internal (gauge) symmetry is broken. In Figure 14.4, both SUSY and gauge symmetry are broken.

Since the potential (14.2) is a sum of squares, the condition for SSB of SUSY can be rephrased in yet another, and more convenient form: *There will be SSB of SUSY if the potential is such that there is no possible scalar field configuration for which* $\langle 0|F_i|0\rangle$ *and* $\langle 0|D^a|0\rangle$ *are both simultaneously equal to zero.*

There are two ways to achieve this:

> ***First case:*** The superpotential is such that there are no scalar field conditions for which $\langle F_i \rangle = \langle F_i^\dagger \rangle = 0$ irrespective of the value of $\langle D \rangle$. In other words, the superpotential must be chosen such that[*]

$$ F_i^\dagger = -\frac{\partial \mathcal{W}}{\partial \phi_i} = 0 \tag{14.4}$$

> has no solutions. This is known as *F-type SUSY breaking.*

---

[*] From now on we will not write explicitly the angled brackets $\langle\ \rangle$ to denote the vevs of the scalar fields or of the potential.

Second case: The second possibility arises only if there is at least one Fayet-Illiopoulos term [and therefore at least one $U(1)$ gauge field]. Then two scenarios are possible. In one scenario, $\langle D \rangle = 0$ cannot be satisfied irrespective of the potential in F_i. In the second scenario, $\langle F_i \rangle = 0$ and $\langle D \rangle = 0$ cannot be satisfied simultaneously, so the breaking arises out of an interplay between the two types of auxiliary fields. Both scenarios are referred to as *D-type SUSY breaking*.

Let's consider these various situations in turn.

14.2 F-Type SUSY Breaking

Here we need to pick the superpotential such that there are no values of the scalar fields ϕ_i for which Eq. (14.4) has a solution.

Consider the simple case of a single left-chiral superfield, with the superpotential of Eq. (12.31). We will not include a term $C \, \Phi$ because in the minimal supersymmetric standard model (MSSM) there is no left-chiral superfield that is invariant under the gauge symmetry $SU(3)_c \times SU(2)_L \times U(1)_Y$, so a term linear in a chiral superfield would break gauge invariance. Consider, then,

$$ \mathcal{W} = \frac{1}{2} m \, \Phi^2 + \frac{1}{6} y \, \Phi^3 $$

To calculate $F^\dagger$, we first replace the superfield Φ by its scalar field component ϕ and then differentiate with respect to ϕ:

$$ F^\dagger = -\frac{\partial \mathcal{W}}{\partial \phi} = -m \, \phi - \frac{1}{2} y \, \phi^2 \qquad (14.5) $$

Clearly, SUSY is not spontaneously broken because we can set $F^\dagger$ to zero simply by choosing $\phi = 0$.

As a second example, consider supersymmetric quantum electrodynamics (QED). We saw in Section 13.6 that the superpotential is $\mathcal{W} = m \, \mathcal{E}_e \mathcal{E}_{\bar{e}}$. In terms of the scalar component fields, this corresponds to $\mathcal{W} = m \, \tilde{\phi}_e \tilde{\phi}_{\bar{e}}$. There are two auxiliary fields, and the two equations corresponding to Eq. (14.4) are

$$ m \, \tilde{\phi}_e = m \, \tilde{\phi}_{\bar{e}} = 0 $$

which clearly have the solution $\tilde{\phi}_e = \tilde{\phi}_{\bar{e}} = 0$.

It is not easy to construct a theory involving several left-chiral superfields that exhibit F-type SUSY breaking. The simplest such model was worked out by O'Raifeartaigh[34] and will be presented in the next section. The term *O'Raifeartaigh models* is sometimes used to refer to *all* models in which F-type SUSY breaking occurs.

14.3 The O'Raifeartaigh Model

This model requires three left-chiral superfields, whose superpotential $\mathcal{W}$ is taken to be

$$\mathcal{W} = m\,\Phi_1\Phi_3 + g\,\Phi_2\big(\Phi_3^2 - M^2\big) \tag{14.6}$$

All three parameters, m, g, and M^2, are taken to be real and positive. In this model, the conditions in Eq. (14.4) read

$$F_1^\dagger = -m\,\phi_3 = 0$$

$$F_2^\dagger = -g\,\phi_3^2 + g\,M^2 = 0$$

$$F_3^\dagger = -m\,\phi_1 - 2\,g\,\phi_2\,\phi_3 = 0 \tag{14.7}$$

which clearly have no solution because no value of ϕ_3 can fulfill the first two equations. Therefore, SUSY is spontaneously broken. To confirm this explicitly, we will find the masses of all the fields and show that the mass degeneracy between the scalars and the fermions is lifted, as expected when SUSY is broken. To do this, we must first find the vevs of the scalar fields, i.e., the values that minimize the potential.

The potential is

$$V = |F_1|^2 + |F_2|^2 + |F_3|^2$$

$$= \underbrace{m^2\,|\phi_3|^2}_{\equiv V_1} + \underbrace{g^2\,|M^2 - \phi_3^2|^2}_{\equiv V_2} + \underbrace{|m\,\phi_1 + 2\,g\,\phi_2\,\phi_3|^2}_{\equiv V_3} \tag{14.8}$$

Let us now find the values of the fields ϕ_1, ϕ_2, and ϕ_3 that minimize Eq. (14.8). For any value of ϕ_3, it is always possible to have V_3 equal to zero, so we only have to minimize the first two terms of the potential, $V_1 + V_2$. The analysis is simplified if we express the field ϕ_3 in terms of its real and imaginary parts, which we will write as

$$\phi_3 \equiv \frac{1}{\sqrt{2}}(R_3 + i\, I_3)$$

We then have

$$V_1 + V_2 = \frac{1}{2}\,(m^2 - 2\,g^2\,M^2)\,R_3^2 + \frac{1}{2}\,(m^2 + 2\,g^2\,M^2)\,I_3^2$$

$$+ \frac{g^2}{4}\,(R_3^2 + I_3^2)^2 + g^2\,M^4 \tag{14.9}$$

We now want to find the values of R_3 and I_3 that minimize the potential. All the terms except the first one are necessarily positive. The rest of the analysis depends on whether the combination $m^2 - 2g^2M^2$ is semipositive definite or negative. Let's consider these two possibilities in turn:

First case: $m^2 \geq 2\,g^2\,M^2$

If $m^2 > 2\,g^2\,M^2$, then all the terms in Eq. (14.9) are positive, and the minimum of $V_1 + V_2$ is obviously obtained by setting both components of ϕ_3 equal to zero; i.e., $R_3 = I_3 = 0$. If $m^2 = 2\,g^2\,M^2$, then the first term of Eq. (14.9) is zero and again the minimum is attained for $\phi_3 = 0$.

Now, let's go back to the third term of the potential,

$$V_3 = |m\,\phi_1 + 2\,g\,\phi_2\,\phi_3|^2$$

If we set $\phi_3 = 0$, the minimum is clearly obtained by setting $\phi_1 = 0$ as well, whereas ϕ_2 is left completely arbitrary. It is typical in supersymmetric theories to have the expectation value of some scalar field left arbitrary. The corresponding scalar field then is referred to as a *flat direction* of the potential because we can vary the value of the scalar field while staying at the minimum of the potential.

At the minimum of the potential, we therefore have the following vevs for the three auxiliary fields [see Eq. (14.7)]:

$$F_1 = F_3 = 0 \qquad F_2 = g\,M^2$$

and the minimum value of the potential is

$$V_{\min} = g^2\,M^4$$

We see that SUSY is broken because the minimum of the potential is not zero. Moreover, $V_{\min}$ is larger than zero, in agreement with the property $\langle 0|H|0\rangle > 0$.

Second case: $m^2 < 2g^2M^2$

In this case, all the terms in Eq. (14.9) containing the imaginary part I_3 are still positive, the potential is still minimized for $I_3 = 0$, in which case the first two terms of the potential simplify to

$$V_1 + V_2 = \frac{1}{2}(m^2 - 2g^2 M^2)R_3^2 + \frac{1}{4}g^2 R_3^4 + g^2 M^4$$

which is minimized for

$$R_3^2 = 2 M^2 - \frac{m^2}{g^2} \tag{14.10}$$

Note that this is necessarily positive because $m^2 < 2 g^2 M^2$. Let's look at the third term of the potential when $I_3 = 0$:

$$V_3 = |m\,\phi_1 + 2\,g\,\phi_2\,\phi_3|^2$$
$$= |m\,\phi_1 + 2\,g\,\phi_2\,R_3|^2$$

with R_3 being given by Eq. (14.10). This is minimized, and equal to zero, at the condition that the fields ϕ_1 and ϕ_2 are related by

$$\phi_1 = -2\frac{g}{m}\sqrt{2 M^2 - \frac{m^2}{g^2}}\,\phi_2 \tag{14.11}$$

Again, we see that the expectation value of one of the fields (which we may choose once more to be ϕ_2) at the minimum of potential is left completely arbitrary.

Using the values in Eqs. (14.10) and (14.11) and $I_3 = 0$, we find that the minimum value of the potential is now

$$V_{\min} = m^2\left(M^2 - \frac{m^2}{4g^2}\right)$$

Let's now look at the mass spectrum of this theory. When scalar fields have nonzero vacuum expectation values, figuring out the masses and the mass eigenstates is usually complicated by the presence of terms bilinear in the fields, i.e., by

mixing. Let us pause the discussion on SUSY breaking to review how to compute the masses of the scalar fields in such a situation.

14.4 Mass Spectrum: General Considerations

Consider two real scalar fields A and B with lagrangian

$$\mathcal{L} = \frac{1}{2} \partial_\mu A \partial^\mu A + \frac{1}{2} \partial_\mu B \partial^\mu B - V(A, B)$$

Let's say that the potential has a minimum at the vevs $A = v_A$ and $B = v_B$. Let us define the new fields $\tilde{A} \equiv A - v_A$ and $\tilde{B} \equiv B - v_B$. The minimum of the potential is then at $\tilde{A} = \tilde{B} = 0$. If we Taylor expand the potential around this minimum, we find

$$V(A, B) = V(\tilde{A} + v_A, \tilde{B} + v_B) \approx V(v_A, v_B) + \tilde{A} \frac{\partial V}{\partial \tilde{A}} + \tilde{B} \frac{\partial V}{\partial \tilde{B}}$$

$$+ \frac{1}{2} \tilde{B}^2 \frac{\partial^2 V}{\partial \tilde{B}^2} + \frac{1}{2} \tilde{A}^2 \frac{\partial^2 V}{\partial \tilde{A}^2} + \tilde{A}\tilde{B} \frac{\partial^2 V}{\partial \tilde{A} \partial \tilde{B}} + \cdots$$

where all the derivatives of the potential are evaluated at the minimum $\tilde{A} = \tilde{B} = 0$. If we are just interested in computing the mass spectrum, we can discard the first term because it is a constant. The first derivatives are zero because they are evaluated at a minimum. We therefore see that there is a term mixing the two fields if the mixed derivative of the potential is not zero. Putting all this in the lagrangian, and using the fact that $\partial_\mu A = \partial_\mu \tilde{A}$ and $\partial_\mu B = \partial_\mu \tilde{B}$, we get

$$\mathcal{L} = \frac{1}{2} \partial_\mu \tilde{A} \partial^\mu \tilde{A} + \frac{1}{2} \partial_\mu \tilde{B} \partial^\mu \tilde{B} - \frac{1}{2} V_{BB} \tilde{B}^2 - \frac{1}{2} V_{AA} \tilde{A}^2 - V_{AB} \tilde{A}\tilde{B} + \cdots$$

where

$$V_{AB} \equiv \frac{\partial^2 V}{\partial \tilde{A} \partial \tilde{B}} \bigg|_{\tilde{A} = \tilde{B} = 0}$$

and so on.

We now see the problem: There is a mixing of the two fields. Because of this, we cannot read off their masses directly from the lagrangian. What we need to do is to find a linear combination of the two fields that will decouple the quadratic terms.

Using the fact that the partial derivatives commute to write $V_{AB} = (V_{AB} + V_{BA})/2$, we see that the quadratic terms can be written in matrix form:

$$\mathcal{L}_{\text{quad}} = -\frac{1}{2}(\tilde{A}\ \tilde{B}) \begin{pmatrix} V_{AA} & V_{AB} \\ V_{BA} & V_{BB} \end{pmatrix} \begin{pmatrix} \tilde{A} \\ \tilde{B} \end{pmatrix}$$

The matrix V is usually called the *squared-mass matrix* $\mathbf{M}_{\text{sq}}$. What we need to do is to diagonalize this matrix. The eigenstates will be the actual mass eigenstates, and the corresponding eigenvalues are the physical (tree-level) mass squared. In this section we are interested only in the masses, not the eigenstates, so all we really want are the eigenvalues, which are simply

$$m^2 = \frac{V_{AA} + V_{BB} \pm \sqrt{\left(V_{AA} - V_{BB}\right)^2 + 4\, V_{AB}^2}}{2} \tag{14.12}$$

where we have used $V_{AB} = V_{BA}$.

This is the approach to follow to find the masses of the scalar fields. For the fermion masses, one does not read off the squared-mass matrix from the lagrangian, but the mass matrix $\mathbf{M}$. The squared-mass matrix is given by $V = \mathbf{M}_{\text{sq}} = \mathbf{M}^{\dagger}\mathbf{M}$, and the square of the fermion masses can be found from Eq. (14.12).

This is all quite abstract. We will put the theory in practice in the next section.

14.5 Mass Spectrum of the O'Raifeartaigh Model for $m^2 \geq 2\, g^2\, M^2$

Let us get back to the O'Raifeartaigh model. We will consider only the first case, i.e., $m^2 \geq 2\, g^2\, M^2$. In this case, both ϕ_1 and ϕ_3 have zero vevs, so they do not get shifted. We need only shift ϕ_2:

$$\phi_2 \rightarrow \phi_2 + \langle \phi_2 \rangle$$

where, as we have seen, the vev $\langle \phi_2 \rangle$ is completely arbitrary. We will take it to be real. If we substitute this back into the potential, we get

$$V = m^2\, |\phi_3|^2 + g^2\, |M^2 - \phi_3^2|^2 + |m\, \phi_1 + 2\, g\, \phi_2\, \phi_3 + 2\, g\, \langle \phi_2 \rangle \phi_3|^2$$

We now write all three fields in terms of their real and imaginary parts:

$$\phi_i \equiv \frac{1}{\sqrt{2}} (R_i + i \, I_i)$$

which leads to

$$V = \frac{1}{2}(m^2 - 2g^2 M^2) R_3^2 + \frac{1}{2}(m^2 + 2g^2 M^2) I_3^2 + \frac{g^2}{4}\left(R_3^2 + I_3^2\right)^2 + g^2 M^4$$

$$+ \left(\frac{m}{\sqrt{2}} R_1 + g \, R_2 \, R_3 - g \, I_2 \, I_3 + \sqrt{2} \, g \, \langle \phi_2 \rangle \, R_3 \right)^2$$

$$+ \left(\frac{m}{\sqrt{2}} I_1 + g \, R_2 \, I_3 + g \, I_2 \, R_3 + \sqrt{2} \, g \, \langle \phi_2 \rangle \, I_3 \right)^2$$

From this, one can find the masses of the scalar fields using the approach described in the preceding section. If we choose the expectation value $\langle \phi_2 \rangle$ equal to zero, the result is particularly simple. The masses of all the fields are (see Exercise 14.1)

$$m_{R_1} = \sqrt{m^2 - g^2 M^2} \qquad m_{I_1} = \sqrt{m^2 + g^2 M^2}$$

$$m_{R_2} = m_{I_2} = 0 \qquad m_{R_3} = m_{I_3} = m \tag{14.13}$$

The last two results can be summarized by saying that the complex scalar fields ϕ_2 and ϕ_3 have masses

$$m_{\phi_2} = 0 \quad \text{and} \quad m_{\phi_3} = m$$

whereas ϕ_1 sees its real and imaginary parts acquiring different masses through SSB.

EXERCISE 14.1
Find the masses of the scalar fields for $\langle \phi_2 \rangle \neq 0$, and then show that by setting $\langle \phi_2 \rangle = 0$, you recover the result in Eq. (14.13).

To find the mass of the fermions, we go back to the general expression for the lagrangian of left-chiral superfields, Eq. (8.62), which shows that the fermion mass terms arise necessarily from the terms in

$$\mathcal{L}_{\text{fermion}} = -\frac{1}{2} \sum_{i \neq j} \frac{\partial^2 \mathcal{W}}{\partial \phi_i \partial \phi_j} \chi_i \cdot \chi_j + \text{h.c.} \qquad (14.14)$$

that do not contain any scalar fields. In other words, the fermion mass terms all come from the terms in the superpotential, which are either quadratic or bilinear in the scalar fields. To use this equation, we must express the superpotential in terms of the shifted fields, which, in our case, just means that we replace ϕ_2 by $\phi_2 + \langle \phi_2 \rangle$ in Eq. (14.6):

$$\mathcal{W} = m \, \phi_1 \, \phi_3 + g \, \phi_2 \left(\phi_3^2 - M^2 \right) + g \, \langle \phi_2 \rangle \left(\phi_3^2 - M^2 \right)$$

The lagrangian (14.14) then is given by

$$\mathcal{L}_{\text{fermion}} = -\frac{1}{2} m \, \chi_1 \cdot \chi_3 - g \, \langle \phi_2 \rangle \, \chi_3 \cdot \chi_3 + \text{h.c.} + \cdots$$

where we wrote only the terms that will contribute to the mass matrix of the fermions. We immediately see that the Weyl spinor χ_2 is massless. The mass matrix in the basis χ_1, χ_3 is

$$\mathbf{M} = \begin{pmatrix} 0 & m \\ m & 2 \, g \langle \phi_2 \rangle \end{pmatrix} \qquad (14.15)$$

so that the squared mass matrix is

$$\mathbf{M}_{\text{sq}} = \mathbf{M}^\dagger \mathbf{M} = \begin{pmatrix} m^2 & 2 \, gm \langle \phi_2 \rangle \\ 2 \, gm \langle \phi_2 \rangle & m^2 + 4 \, g^2 \langle \phi_2 \rangle^2 \end{pmatrix}$$

whose eigenvalues are

$$m_f^2 = m^2 + 2 \, g^2 \, \langle \phi_2 \rangle^2 \pm 2 \, g \, \langle \phi_2 \rangle \sqrt{g^2 \, \langle \phi_2 \rangle^2 + m^2}$$
$$= \left(\sqrt{g^2 \, \langle \phi_2 \rangle^2 + m^2} \pm g \langle \phi_2 \rangle \right)^2 \qquad (14.16)$$

so the masses of the two fermions are

$$m_f = \sqrt{g^2 \, \langle \phi_2 \rangle^2 + m^2} \pm g \langle \phi_2 \rangle \qquad (14.17)$$

With $\langle \phi_2 \rangle = 0$, we see that both fermions have a mass equal to m. It is easy to find the spinors that are mass eigenstates when $\langle \phi_2 \rangle = 0$ because then the mass matrix in Eq. (14.15) is simply

$$\mathbf{M} = \begin{pmatrix} 0 & m \\ m & 0 \end{pmatrix}$$

whose two eigenstates are clearly

$$\chi = \frac{\chi_1 \pm \chi_3}{\sqrt{2}} \tag{14.18}$$

Now, if we compare what we found for the boson masses, Eq. (14.13), with the fermion masses we just calculated, we see that the degeneracy between one of the complex scalar fields and one of the fermions is lifted.

To summarize the situation when $\langle \phi_2 \rangle = 0$, there is one complex scalar field and one fermion that are both massless (ϕ_2 and χ_2), and there is a second pair of boson and fermion with a mass m [ϕ_3 and either combination appearing in Eq. (14.18)], but the third boson/fermion pair is not degenerate in mass because the remaining fermion has a mass m, whereas we have two real scalar fields with masses $\sqrt{m^2 \pm g^2 M^2}$. Therefore, SUSY is spontaneously broken: the lagrangian is supersymmetric, but the vacuum breaks the symmetry.

Spontaneous symmetry breaking of SUSY or, to be more precise, of global SUSY (which is what we consider in this book) always introduces fermion massless modes as opposed to the usual Goldstone bosons. These massless fermion modes therefore are referred to as *goldstinos*. In the O'Raifeartaigh model, there is one goldstino, the spinor χ_2. Note that it is the fermion that belongs to the same multiplet as the auxiliary field whose vev is nonzero, i.e., F_2. This is a general feature of F-type SSB.

EXERCISE 14.2

Find the mass spectrum of the bosons and fermions both in the limit $g \to 0$, $M \neq 0$ and in the limit $M \to 0$, $g \neq 0$. You will observe that the mass degeneracy between the fermionic and bosonic degrees of freedom reappears. Note that this holds no matter what $\langle \phi_2 \rangle$ is equal to.

There is unfortunately an aspect of F-type SUSY breaking that makes it unattractive from a phenomenological point of view. Before addressing this issue, we need to introduce yet another "superquantity," the *supertrace*.

14.6 The Supertrace

The *supertrace* is defined as

$$\mathrm{STr}(\mathcal{M}^2) \equiv \sum_{\mathrm{particles}} (-1)^{2s}\,(2s+1)m^2_{\mathrm{particle}}$$

where it is understood that the sum is performed over all the physical particles appearing in the theory (not over the auxiliary fields), and s denotes the spin of the particle. It is also understood that there is one term for each *real* scalar field and one term for each Weyl spinor.

In the case of theories with left-chiral superfields only, there are only scalar fields and spinor fields, so the sum reduces to two terms:

$$\mathrm{STr}(\mathcal{M}^2) = \sum_{\mathrm{scalars}} m^2 - 2 \sum_{\mathrm{spinors}} m^2 \tag{14.19}$$

This sum obviously vanishes in supersymmetric theories because for each Weyl spinor, there are two real scalar fields (one complex scalar field) with the same mass. What is astonishing is that *the supertrace still vanishes when SUSY is spontaneously broken with F-type breaking*. We can check that this is indeed the case with the masses obtained in last section. When we set $\langle \phi_2 \rangle = 0$, we obtained the following squared masses for the bosons:

Squared masses of the bosons $= 0, 0, m^2, m^2, m^2 - g^2 M^2, m^2 + g^2 M^2$

so that

$$\sum_{\mathrm{scalars}} m^2 = 4\,m^2$$

By the way, the reason it is called a *supertrace* is that the sum of the squared masses is simply given by the trace of the squared-mass matrix $\mathbf{M}_{\mathrm{sq}}$.

Now, let's look at the fermions. We have three fermions with masses 0, m, and m, so

$$2 \sum_{\mathrm{spinors}} m^2 = 4\,m^2$$

Again, we could have simply taken the trace of the squared fermion mass matrix and multiplied it by 2.

The end result is that the supertrace (14.19) vanishes indeed. We obtain the same result if we keep $\langle \phi_2 \rangle$ arbitrary (see Exercise 14.3).

EXERCISE 14.3
Verify that the supertrace still vanishes if we don't set $\langle \phi_2 \rangle = 0$.

The key point to note here is that the reason the supertrace still vanishes even if we have SSB is that one of the scalar squared masses *increased* by an amount $g^2 M^2$ relative to its "unbroken" value of m^2, whereas another scalar squared mass *decreased by the same amount*.

This is an important result and turns out to be a general feature of F-type SUSY breaking: The heaviest boson state is heavier than all the fermions, and as a consequence of the vanishing of the supertrace, the lightest boson state also must be lighter than all fermion states. This makes F-type SUSY breaking phenomenologically unappealing because it implies that each known fermion would have a lighter scalar partner. For example, this would imply the existence of a charged scalar particle lighter than the electron, which is ruled out experimentally.

So let us turn our attention to D-type SUSY breaking.

14.7 D-Type SUSY Breaking

For this type of symmetry breaking to occur, a Fayet-Illiopoulos term is necessary. This means that at least one abelian gauge field must be present. We will first illustrate a simple situation where $\langle D \rangle = 0$ cannot hold irrespective of the expectation values of the auxiliary fields F_i, so we are free to ignore their contribution to the potential.

The simplest model in which this type of spontaneous symmetry breaking may occur is a $U(1)$ gauge theory with a single charged left-chiral superfield. If there is only one charged chiral superfield, there is no superpotential because it is impossible to write down a function of Φ alone that is gauge-invariant. As we saw in Chapter 13, the lagrangian then is (with ξ taken to be real)

$$\Phi^\dagger e^{-2q\mathcal{V}} \Phi \Big|_D + \mathcal{F}\mathcal{F} \Big|_F + \xi D \Big|_F \tag{14.20}$$

From Eqs. (14.2) and (14.3), we have

$$D = q \left(\phi^\dagger \phi - \frac{\xi}{q} \right) \tag{14.21}$$

and the potential may be written as

$$V = \frac{q^2}{2} \left(|\phi|^2 - \frac{\xi}{q} \right)^2 \qquad (14.22)$$

The reason for factoring out the charge is that it makes clear that the effect of the Fayet-Iliopoulos term depends crucially on whether ξ/q is positive or negative.

Case $\xi/q < 0$

In this case, it is clear from Eq. (14.21) that there is no field ϕ for which $\langle D \rangle = 0$ (because $\phi^\dagger \phi$ is semipositive definite), and thus SUSY is spontaneously broken. From Eq. (14.22), we find that the potential is minimum when $\phi = 0$, in which case

$$V_{\mathrm{min}} = \frac{1}{2} \xi^2$$

Let us now have a look at the mass spectrum to confirm explicitly that SUSY is spontaneously broken. The fields of the gauge potential superfield A_μ and the photino λ are massless. Since there is no superpotential, there is no mass term for the left-chiral spinor component of the superfield Φ either. If we write the complex scalar field in terms of two real scalar fields as usual, i.e.,

$$\phi = \frac{1}{\sqrt{2}} (R + i\,I)$$

and retain only the terms that will contribute to the squared-mass matrix, the potential (14.22) is equal to

$$V(\phi) = \frac{|q\xi|}{2} (R^2 + I^2) + \cdots$$

We see that both real scalar fields acquire a mass equal to $\sqrt{|q\xi|}$. Thus SUSY is indeed spontaneously broken: The spinor field χ is massless, but its scalar partners are massive.

Notice how different the situation is from F-type breaking: here, all the bosons have seen their mass *increase* relative to the "unbroken" value of $m = 0$. This is more encouraging from a phenomenological point of view because this would provide an explanation for why the SUSY partners of the known fermions would have escaped detection so far. However, the present theory is just a toy model, and the situation turns out to be not so encouraging when we try to apply this approach to the MSSM, as we will see in Chapter 16.

Let's find out what happens to the supertrace. Obviously, it does not vanish anymore:

$$
\mathrm{STr}(\mathcal{M}^2) = \sum_{\text{scalars}} m^2 - 2 \sum_{\text{spinors}} m^2
$$

$$
= 2\,|q\xi| \tag{14.23}
$$

There is, however, a generalization of the rule $\mathrm{STr}(\mathcal{M}^2) = 0$ that is valid for D-type SSB. The proof is not difficult, but we will simply quote the result for abelian gauge theories with only one auxiliary field acquiring a vev (see Section 7.7 of Ref. 5 for a derivation):

$$
\mathrm{STr}(\mathcal{M}^2) = 2\,\langle D \rangle \sum_i q_i \tag{14.24}
$$

where the sum is over all the charges of the left-chiral superfields present in the theory. In the case of F-type breaking, there is no D field with a vev, so we recover the result that the supertrace then is zero. In our example, $\langle D \rangle = |\xi|$, and there is only one left-chiral superfield, of charge q, so the formula does indeed give

$$
\mathrm{STr}(\mathcal{M}^2) = 2\,|q\xi|
$$

as we found in Eq. (14.23).

Note that even though SUSY is spontaneously broken, the gauge symmetry remains unbroken because at the minimum of the potential, $\phi = 0$. We therefore have a potential that corresponds to Figure 14.2.

Case $\xi/q > 0$

In this case, the potential is minimized for

$$
|\phi| = \sqrt{\frac{\xi}{q}}
$$

and the potential at the minimum vanishes, i.e.,

$$
V_{\min} = 0
$$

SUSY is therefore unbroken, but since the minimum of the potential occurs for a nonzero expectation value of the field, the gauge symmetry is broken spontaneously (we are in the situation depicted in Figure 14.3). This is clear because the kinetic energy of the scalar field is $D^\mu \phi \, D_\mu \phi^\dagger$, which contains the term $q^2 \, A^\mu A_\mu \, \phi^\dagger \phi$. Therefore, at the minimum of the potential, the gauge field acquires a mass equal to $2\,q\,\xi$. A more detailed analysis reveals that the imaginary part of the scalar field has been eaten by the gauge field, whereas the real part acquires a mass equal to the mass of the gauge boson.

14.8 Second Example of D-Type SUSY Breaking

Now we will consider an example where $\langle D \rangle = 0$ and $\langle F \rangle = 0$ cannot hold simultaneously. The simplest model in which this occurs is supersymmetric QED [see Eq. (13.55)] with a Fayet-Illiopoulos term. The superpotential is $\mathcal{W} = m\,\mathcal{E}_e\,\mathcal{E}_{\bar{e}}$. Using Eq. (14.3), we find

$$D = e\,|\tilde{\phi}_{\bar{e}}|^2 - e\,|\tilde{\phi}_e|^2 - \xi \tag{14.25}$$

It is clear that we can find (nonzero) values of the scalar fields to make this vanish, no matter what the sign of ξ is.

Now consider the expressions for the $F_e^\dagger$ and $F_{\bar{e}}^\dagger$:

$$F_e^\dagger = -\frac{\partial \mathcal{W}}{\partial \tilde{\phi}_e} = -m\,\tilde{\phi}_{\bar{e}}$$

$$F_{\bar{e}} = -m\,\tilde{\phi}_e \tag{14.26}$$

These two can be set equal to zero if we simply take $\tilde{\phi}_e = \tilde{\phi}_{\bar{e}} = 0$, but *then $\langle D \rangle$ is not equal to zero*! Conversely, if we choose values of the scalar fields such that $\langle D \rangle = 0$, the expectation values of the $F_i^\dagger$ are not zero. Therefore, SUSY is spontaneously broken.

The potential is given by

$$V = m^2\big(|\tilde{\phi}_e|^2 + |\tilde{\phi}_{\bar{e}}|^2\big) + \frac{1}{2}\,\big(e\,|\tilde{\phi}_{\bar{e}}|^2 - e\,|\tilde{\phi}_e|^2 - \xi\big)^2$$

$$= (m^2 - e\xi)|\tilde{\phi}_e|^2 + (m^2 + e\xi)|\tilde{\phi}_{\bar{e}}|^2 + \frac{1}{2}\,\big(e\,|\tilde{\phi}_{\bar{e}}|^2 - e\,|\tilde{\phi}_e|^2\big)^2 + \frac{1}{2}\,\xi^2 \tag{14.27}$$

We will take ξ to be positive. The minimization of the potential depends on the sign of the quantity $m^2 - e\xi$. We will only consider $m^2 - e\xi > 0$, in which case,

all the terms in Eq. (14.27) are positive, and the potential is obviously minimized for $\tilde{\phi}_e = \tilde{\phi}_{\bar{e}} = 0$, so the gauge symmetry is unbroken. The value of the potential at its minimum is $V_{\min} = \frac{1}{2}\xi^2$.

The masses of the scalar can be read directly from Eq. (14.27) because there are no bilinear terms in the two scalar fields. The squared masses then are simply $m^2 \pm e\xi$. It is also easy to check that the electron mass is unaffected by the spontaneous breakdown of SUSY and therefore remains equal to m.

We can now check that Eq. (14.24) is indeed satisfied. The left-hand side is equal to

$$\text{STr}(\mathcal{M}^2) = m^2 + e\xi + m^2 - e\xi - 2m^2 = 0$$

This indeed agrees with the right-hand side because the sum of the charges is equal to zero!

Now comes the kicker. The MSSM contains the hypercharge abelian gauge symmetry, so one could have hoped to use a Fayet-Illiopoulos term to generate spontaneous SUSY breaking, but, as is the case in the standard model, *the sum of the hypercharges of all the particles is zero* (which is required to ensure cancellation of anomalies).

Therefore, in the context of the MSSM, D-type SUSY breaking suffers from the same problem that we had with F-type breaking: since the supertrace is zero, light sleptons and squarks are predicted, in conflict with experiments.

We therefore must look somewhere else for an explanation of SUSY breaking.

14.9 Explicit SUSY Breaking

Since there is no (yet) known phenomenologically viable mechanism to spontaneously break SUSY in the MSSM, the only option left is to break it explicitly, i.e., by adding by hand terms that violate the symmetry. Of course, if we do this haphazardly, we will lose the feature that got us excited in SUSY in the first place: the cancellation of quadratic divergences. We have to be more subtle. We must look for SUSY-breaking terms that will not spoil this cancellation. Such terms are said to break SUSY *softly*. To see how this can be done (if it is possible at all), it is instructive to get back to the Wess-Zumino model and to have a closer look at the cancellation of divergences we worked out in Chapter 9.

The Wess-Zumino model was given, in Majorana form, in Eqs. (8.69) through (8.71). There is a lot of symmetry in this lagrangian: All the particles have the same mass, and that mass also appears as a coupling constant in the cubic interactions. In addition, the same coupling constant g appears in all interactions. What we want to understand is which of these features are absolutely necessary for the quadratic divergences to cancel. To do that, let us rewrite the lagrangian with distinct masses and coupling constants in all the terms. We then write the free lagrangian as

$$\frac{1}{2} \partial_\mu A \partial^\mu A - \frac{1}{2} m_A^2 A^2 + \frac{1}{2} \partial_\mu B \partial^\mu B - \frac{1}{2} m_B^2 B^2 + \frac{1}{2} \overline{\Psi}_M (i\,\gamma^\mu \partial_\mu - m_F)\,\Psi_M$$

while we represent the interactions by

$$L_{I1} = -\frac{1}{2} g_1^2 A^4 - g_2^2 A^2 B^2 - \frac{1}{2} g_3^2 B^4$$

$$L_{I2} = -m_F\, g_4\, A^3 - m_F\, g_5\, A\, B^2$$

$$L_{I3} = -g_6\, A\, \overline{\Psi}_M \Psi_M$$

$$L_{I4} = -i\, g_7\, B\, \overline{\Psi}_M \gamma_5 \Psi_M$$

Note that we have chosen to set the parameter m that appears in L_{I2} equal to the mass of the fermion, which we can obviously do in all generality because we have introduced the arbitrary coupling constants g_4 and g_5.

Unbroken SUSY corresponds to

$$m_A = m_B = m_F \qquad g_1, \dots, g_7 \equiv g$$

Of course, what really matters is the relationship between the coupling constants, not their absolute values. So we shall fix one of them to be equal to g and then parametrize the breaking of SUSY by how much we can allow the other coupling constants to deviate from their supersymmetric value, which is also equal to g.

It is a simple matter to go back to the calculation of the B propagator of Chapter 9 and keep track of the masses and coupling constants appearing in each diagram. In that chapter, all quadratically divergent integrals were proportional to the integral I_d. Since we are now keeping track of the masses, let us have a closer look at this integral. It is given by

$$I_d = \int \frac{d^4 p}{(2\pi)^4} e^{-ip\cdot(x-y)} \frac{i}{p^2 - m_B^2 + i\epsilon} \frac{i}{p^2 - m_B^2 + i\epsilon} \int \frac{d^4 q}{(2\pi)^4} \frac{1}{q^2 - m^2 - i\epsilon}$$

where m may either be m_A, m_B, or m_F. However, the *quadratic divergence* arising from I_d is *independent of the mass m*. The logarithmically divergent contribution and the finite piece do depend on m, but since we are concerned only with maintaining the cancellation of the quadratic divergences, we may safely set to zero the mass m in I_d.

Let us look at the tadpole diagrams that contributed to the one-loop correction to the B propagator. Going through the calculation we performed in Chapter 9 while keeping track of the masses and coupling constants, it's easy to find

$$A \text{ tadpole (Figure 9.9)} = 6i \, g_4 \, g_5 \, \frac{M^2}{m_A^2} \, I_d^{\text{quad}}$$

$$B \text{ tadpole (Figure 9.8)} = 2i \, g_5^2 \, \frac{m_F^2}{m_A^2} \, I_d^{\text{quad}}$$

$$\text{Fermion tadpole (Figure 9.10)} = -8i \, g_5 \, g_6 \, \frac{m_F^2}{m_A^2} \, I_d^{\text{quad}}$$

where I_d^{quad} denotes the quadratically divergent term in I_d, which is independent of the mass of the particle circulating in the loop as we discussed earlier.

For these quadratic divergences to cancel out, we must impose

$$3 \, g_4 + g_5 = 4 \, g_6 \tag{14.28}$$

Imposing the cancellation of the quadratic divergences in the one-loop tadpole corrections to the A propagator leads to the same result.

The first thing to observe, and probably the most surprising, is that *there is no condition on the masses of the scalar particles*! This is surprising because the degeneracy in the masses in a multiplet is a trademark of unbroken SUSY, so one might have expected it to be necessary for the cancellation of quadratic divergences. What we have shown, at least in one simple case, is that this is not so. It turns out that this remains true for all processes and to all orders of perturbation theory (see Ref. 47 for more details).

The key point is that we could add arbitrary corrections to the scalar masses, i.e.,

$$\mathcal{L}_{\text{SB}} = \delta m_A \, A^2 + \delta m_B \, B^2$$

where SB stands for "soft breaking." These terms obviously break SUSY because they lift the degeneracy of the masses in the multiplet, but they break it *softly* because they do not ruin the cancellation of quadratic divergences.

The fact that we may change the mass of the scalars is excellent news for phenomenology because we can make the scalar partners of known fermions as massive as we wish, thereby avoiding the problem of light scalar particles encountered in our models of spontaneous symmetry breaking. Of course, this is not so nice from a theoretical point of view because it introduces new parameters and makes the theory less predictive.

We also see that we have much more freedom with the coupling constants g_i than unbroken SUSY allows. Let's set g_6 equal to g. Condition (14.28) then is

$$3\, g_4 + g_5 = 4\, g \tag{14.29}$$

whereas unbroken SUSY implies that $g_4 = g_5 = g$. The general solution to Eq. (14.29) is obviously $g_5 = 4\, g - 3\, g_4$ with g_4 arbitrary. Let us write g_4 as

$$g_4 = g + \delta g_4$$

where δg_4 represents the difference of g_4 relative to the unbroken SUSY value, $g_4 = g$. Then $g_5 = g - 3\, \delta g_4$ and the sum of interaction lagrangians L_{12} and L_{13} is

$$-g\, A\, \overline{\Psi}_M \Psi_M - m_F\, g\, (A^3 + A\, B^2) - m_F\, \delta g_4\, (A^3 - 3\, A\, B^2)$$

The first two terms are interactions that were present in the Wess-Zumino model. What we have shown is that we may add the combination of fields $A^3 - 3AB^2$ times an arbitrary constant with dimension of mass without spoiling the cancellation of quadratic divergences, at least in this particular calculation. This interaction thus is a soft SUSY-breaking term.

Actually, it is now obvious why this does not generate any quadratic divergences in the tadpole diagrams. Consider the contribution of this soft SUSY-breaking interaction to the one-point function of the A field, to one loop:

$$\langle 0|\, T\left(A(x)\, A^3(z) - 3A(x)\, A(z)\, B^2(z)\right)|0\rangle \tag{14.30}$$

Both terms lead to a quadratic divergence, but there are three ways to contract the fields of the first term, whereas there is only one way to contract the fields of the first term, so the two cancel out.

The expression $A^3 - 3\, A\, B^2$ does not look very pleasing, until one realizes that it takes a strikingly simple form when expressed in terms of the original complex scalar field $\phi = (A + i\, B)/\sqrt{2}$. Indeed, one finds that this combination is simply proportional to

$$A^3 - 3\, A\, B^2 \propto (\phi^3 + \text{h.c.})$$

In other words, this soft SUSY-breaking term is simply a cubic interaction of the complex scalar field (plus its hermitian conjugate, to get a real lagrangian).

Let's introduce a new constant δm through

$$-m_F\, \delta g_4\, (A^3 - 3\, A\, B^2) \equiv \delta m_F\, (\phi^3 + \text{h.c.}) \tag{14.31}$$

so that our soft SUSY-breaking lagrangian now contains three terms:

$$\mathcal{L}_{\text{SB}} = \delta m_A \, A^2 + \delta m_B \, B^2 - \delta m_F \, (\phi^3 + \text{h.c.})$$

It is also possible to rewrite the mass corrections in terms of the complex scalar field ϕ as follows:

$$\delta m_A^2 \, A^2 + \delta m_B^2 \, B^2 = \left(\delta m_A^2 + \delta m_B^2\right) \phi^\dagger \phi + \left(\delta m_A^2 - \delta m_B^2\right)(\phi\phi + \text{h.c.})$$

$$\equiv \delta m_+^2 \, \phi^\dagger \phi + \delta m_-^2 \, (\phi\phi + \text{h.c.})$$

To summarize, we have found that if we modify the Wess-Zumino lagrangian by adding the following SUSY-breaking terms:

$$\mathcal{L}_{\text{WZ}} \Longrightarrow \mathcal{L}_{\text{WZ}} + \delta m_+^2 \, \phi^\dagger \phi + \delta m_-^2 \, (\phi\phi + \text{h.c.}) + \delta m_F (\phi^3 + \text{h.c.}) \qquad (14.32)$$

the cancellation of the quadratic divergences is not spoiled, at least in the tadpole diagrams and to one loop. Of course, this is a very special case, but it can be shown that it remains valid *to all orders of perturbation theory and in all processes*!

Are there other soft SUSY-breaking terms? If we look at the nontadpole diagrams of either the A or the B propagator, we obtain other combinations of fields that do not spoil the cancellation of the quadratic divergences, but it turns out that those results are not valid to all orders of perturbation theory and in all processes. In other words, although those terms break SUSY softly in the one-loop corrections to the scalar field propagators, in general, they spoil the cancellation of quadratic divergences. For example, the quadratic divergences of the nontadpole corrections to the A propagator still cancel if we introduce an interaction proportional to the combination

$$A^4 - 6 \, A^2 \, B^2$$

However, this introduces a quadratic divergence in the one-loop B propagator: The A^4 term does not contribute (it gives a disconnected diagram), but the term $-6A^2B^2$ gives a contribution, i.e.,

$$-6 \int d^4z \, D_B(x - z) \, D_B(y - z) \, D_A(z - z)$$

that is proportional to the quadratically divergent integral I_d [see Eq. (9.16)].

So what are the general conditions for an interaction to break SUSY softly? A necessary condition for a term to break SUSY softly is that it must be *superrenormalizable*; i.e., it must be of dimension 3 at most (ignoring any overall constant). The interaction $\delta M\,\phi^3$ is superrenormalizable, but $A^4 - 6\,A^2\,B^2$ is not. However, it turns out that this is not a sufficient condition. Consider the interaction

$$m\,\phi^2\,\phi^\dagger + \text{h.c.} = \frac{m}{\sqrt{2}}\,(A^3 + A\,B^2) \tag{14.33}$$

This is superrenormalizable, yet it introduces quadratic divergences [e.g., the contribution of Eq. (14.33) to the one-point function of A is quadratically divergent, which is obvious after the discussion following Eq. (14.30)].

So what are the possible terms that break SUSY softly? It turns out that for a theory containing only left-chiral multiplets, there are two possible types of contributions:

- Mass terms for the complex scalar fields ϕ_i.
- Interactions among the ϕ_i that are *both superrenormalizable and holomorphic in* ϕ, i.e., dependent on ϕ_i only, not on $\phi_i^\dagger$. This last criterion is the reason why ϕ^3 breaks SUSY softly but $\phi^2\phi^\dagger$ does not.

Apart from a term linear in ϕ, we see that Eq. (14.32) contains all the possible soft SUSY-breaking terms for a theory with a single chiral multiplet. Note that the allowed terms are exactly of the same form as those that are permitted to build a supersymmetric theory in terms of the superfield Φ! Indeed, the most general lagrangian we can build out of left-chiral superfields Φ_i is the sum of $\Phi_i^\dagger\Phi_i$ and a holomorphic function of the Φ_i with a maximum of three superfields (plus the hermitian conjugates). This is, of course, no coincidence. The proof is very similar to the proof of the renormalization theorems mentioned in Chapter 9, which themselves rely heavily on special properties of holomorphic interactions. The proof is beyond the scope of this book, but details can be found in Refs. [5], [43], and [47].

For theories with gauge fields, there is one more type of soft-breaking term available:

- Mass terms for the gauginos, the fermion superpartners of the gauge bosons.

As usual, we take λ to represent the gauginos. In the case of an abelian gauge theory, we would add the photino mass term

$$-\frac{1}{2}\,\delta m_\lambda\,(\lambda \cdot \lambda + \bar{\lambda} \cdot \bar{\lambda}) \tag{14.34}$$

For a nonbelian gauge group, there is a gauge group index on the gauginos, so the expression in parentheses must be replaced by $\lambda^a \cdot \lambda^a + $ h.c. The obvious question is: If we start adding terms that break SUSY softly, what is the justification for not including *all* types of SUSY-breaking terms, losing all the benefits of SUSY? The conventional wisdom is to assume that there is a "hidden (or shadow) supersymmetric sector" composed of particles that are almost completely decoupled from the particles of the MSSM (to be presented in Chapter 15), interacting with them only through some "messenger interaction." In some models, this messenger interaction is gravity, whereas in other models, it is a gauge interaction.

Furthermore, it is assumed that SUSY is spontaneously broken in this hidden sector. One then can show that with appropriate conditions on the lagrangians, the messenger interaction can "communicate" the SSB of SUSY in the hidden sector to the particles' MSSM in such a way that only soft SUSY breaking is generated in the "visible sector" of the MSSM. When the messenger interaction is gravity, this mechanism of SUSY breaking is known as *gravity-mediated SUSY breaking* (also called *SUGRA-mediated SUSY breaking*, where *SUGRA* stands for "supergravity"). When the messenger interaction is a gauge interaction, the breaking of supersymmetry is also called *gauge-mediated SUSY breaking*. Another popular mechanism, in which there is a particular coupling of the hidden and visible sectors is referred to as *anomaly-mediated SUSY breaking*. The study of these mechanisms is beyond the scope of this book. The interested reader is invited to consult Refs. [8], [43], and [47] and the references therein for more details.

Let us conclude this chapter by mentioning another completely different mechanism of SUSY breaking that we haven't touched on at all in this chapter. We have only discussed SUSY breaking through the effect of the fundamental fields present in the theory. It is possible, however, to construct scenarios in which SUSY is broken by *condensates* of the fields (in the same spirit as the technicolor models). The references already cited discuss such scenarios of *dynamical symmetry breaking*.

14.10 Quiz

1. Why is it not possible to have D-type breaking for a nonabelian gauge theory?
2. What makes F-type breaking unappealing phenomenologically?
3. What is the condition on the potential to have spontaneous symmetry breaking of a gauge symmetry but no SSB of SUSY?
4. When is the supertrace equal to zero?
5. What is a goldstino?

CHAPTER 15

Introduction to the Minimal Supersymmetric Standard Model

15.1 Lightning Review of the Standard Model

It is assumed that you, the gentle reader, are already familiar with the basic structure of the standard model at the level of an introductory course on particle physics. We will nevertheless review it very briefly in case your class was at 8:30 in the morning and you slept through part of the semester. This review also will serve the purpose of introducing some notation that will carry over to the minimal supersymmetric standard model (MSSM).

The standard model is based on the gauge group $SU(3)_c \otimes SU(2)_L \otimes U(1)_Y$. This means that all the terms of the lagrangian must be invariant under this symmetry. The symmetry is broken spontaneously via the Higgs mechanism to the subgroup $U(1)_{em} \otimes SU(3)_c$. The fermions of the theory, as well as the scalar Higgs field, transform in the fundamental representations of the unbroken group (more about this below), whereas the gauge bosons transform in the adjoint representations.

The terms in the lagrangian can be grouped into the following five categories:

- The kinetic terms of the fermions
- The kinetic terms of the gauge bosons
- The kinetic term of the Higgs boson
- The Higgs potential
- The Yukawa terms coupling the Higgs field to the fermions

Here, by *kinetic terms*, we mean the expressions made gauge-invariant by replacing the partial derivatives with the proper gauge-covariant derivatives (we will be more specific soon).

This is it. There are no mass terms; all masses are generated through the Higgs mechanism. The Higgs mechanism is a complex doublet, so it has four degrees of freedom. On spontaneous symmetry breaking, three of these degrees of freedom become the longitudinal modes of the $W^\pm$ and Z, which become massive, leaving a single Higgs scalar particle. The masses of the fermions are also generated by spontaneous symmetry breaking (SSB) via the Yukawa terms. The standard model does not include neutrino masses, so we already know that it must be modified to accommodate such terms, but this can be done in a straightforward manner (although it depends on whether the neutrinos are Majorana or Dirac particles). The MSSM is also written without neutrino masses, and again, these can be added easily. We will not include neutrino masses in our presentation of the MSSM.

Before writing down the standard model (SM) lagrangian, one must first specify the quantum numbers of the fermions and Higgs bosons. This then determines how these fields can be coupled together and what their kinetic terms will look like. The standard model is a chiral theory in the sense that part of the theory, the weak interaction to be more precise, treats left- and right-chiral states of the fermions differently. In other words, the $SU(2)_L$ quantum numbers of the left- and right-chiral states of the fermions are different.

The Fermions: Quantum Numbers and Kinetic Energy Terms

Let's be more specific. The rest of this section is not only useful as a review of the standard model, but it is also a good preparation for construction of the MSSM because in both cases the key difficulty is in assembling terms that are invariant under the unbroken gauge group.

The kinetic term of the fermions is always of the form

$$\overline{\Psi} i \gamma^{\mu} D_{\mu} \Psi$$

with the explicit form of the gauge-covariant derivative D_{μ} depending on the quantum numbers of the fermion field Ψ.

Consider the right-chiral electron state η_e (it is understood that we will be using η for right-chiral states and χ for left-chiral states). The quantum numbers of this state under $SU(3)_c \otimes SU(2)_L \otimes U(1)_Y$ are $\mathbf{1}, \mathbf{1}, -2$. A particle with quantum number $\mathbf{1}$ under $SU(3)_c$ is said to be a *singlet* under the color group, and this implies that it does not transform at all under that gauge group. It is therefore "blind" to the color force. We see that it is also a singlet under $SU(2)_L$ and therefore also blind to the weak force. On the other hand, it interacts via the $U(1)_Y$ force because its hypercharge [the quantum number of association with the $U(1)_Y$ group] is -2. For particles that are singlet under $SU(2)_L$, the hypercharge is given simply by twice the electric charge of the particle, i.e., $Y = 2Q$.

Note that interpretation of the quantum numbers for the abelian group $U(1)_Y$ and the nonabelian groups $SU(3)_c$ and $SU(2)_L$ is different in one important aspect: A particle that would be blind to the $U(1)_Y$ force would have an hypercharge of 0, not 1. For this reason, we will distinguish the quantum numbers of the two nonabelian groups from the hypercharge by writing the former in boldface. This is also convenient to remind us of the order in which the quantum numbers are given (i.e., hypercharge last).

Now let us see what this tells us about the kinetic term of the η_e. First, let's emphasize that all the terms in the SM lagrangian are written in terms of four-component Dirac spinors. In order to isolate the right-chiral state of the electron, we must apply a projection operator P_R to the Dirac spinor Ψ_e, i.e.,

$$\Psi_{e_R} = P_R \Psi_e$$

The kinetic energy term therefore is

$$\overline{\Psi}_{e_R} i \gamma^{\mu} D_{\mu} \Psi_{e_R}$$

For a field singlet under $SU(2)_L$ and $SU(3)_c$ and with a hypercharge Y, we take the covariant derivative to be

$$D_{\mu} = \partial_{\mu} + i g' \frac{Y}{2} B_{\mu}(x)$$

where $B_\mu(x)$ is the gauge field associated with $U(1)_Y$, and g' is its coupling constant. Be aware that another popular convention is to use Y instead of $Y/2$, in which case all the hypercharge assignments are half the values we have with our present convention. Also, many authors use a negative sign for all the gauge field terms appearing in covariant derivatives. To translate equations written with one convention into the other convention, one simply has to switch the signs of all the coupling constants.

For the field Ψ_{e_R}, which has a hypercharge of -2, this gives

$$D_\mu \Psi_{e_R} = (\partial_\mu - i g' B_\mu) \Psi_{e_R}$$

Consider now the left-chiral fields Ψ_{e_L} and $\Psi_{v_{e_L}}$. These form a doublet under $SU(2)_L$ that we will denote by Ψ_L (where the subscript L means both lepton and left-handed). To be explicit,

$$\Psi_L = \begin{pmatrix} \Psi_{v_{eL}} \\ \Psi_{eL} \end{pmatrix}$$

where $\Psi_{v_{eL}} = P_L \Psi_{v_e}$ and $\Psi_{eL} = P_L \Psi_e$. The choice of the order (neutrino up and electron down) comes from the requirement that the difference $Q - T_3$ must give the same value for both components. The quantum number T_3 of a state is defined as the eigenvalue of the matrix $\tau_3/2$ when applied to that state (τ_3 is simply the third Pauli matrix; the reason for this new notation will be explained below). Thus the particle at the top has $T_3 = +1/2$, and the one at the bottom has $T_3 = -1/2$. We see that $Q - T_3$ is equal to $-1/2$ for both of them.

The hypercharge of an $SU(2)$ doublet is given by $Y = 2(Q - T_3)$, so it is equal to -1 for the L_1 doublet. This is obviously a singlet under color $SU(3)$ because leptons do not interact via the strong force. The complete list of quantum numbers of Ψ_{L_1} therefore is $\mathbf{1}, \mathbf{2}, -1$.

The gauge-covariant derivative of this field then is

$$D_\mu \Psi_L = \left(\partial_\mu + i g \frac{\tau^i}{2} W_\mu^i - i \frac{g'}{2} B_\mu \right) \Psi_L$$

where the three W_μ^i are the $SU(2)_L$ gauge fields and we have used $Y = -1$.

The matrices τ_i are simply the Pauli matrices σ_i. The reason for using a different notation here is to emphasize that the τ_i are acting on $SU(2)_L$ doublets, not on two-component Weyl spinors. If we write this covariant derivative explicitly, we get

$$D_\mu \Psi_L = \left[\partial_\mu + \frac{ig}{2} \begin{pmatrix} W_3 & W_{1\mu} - i W_{2\mu} \\ W_{1\mu} + i W_{2\mu} & -W_{3\mu} \end{pmatrix} - \frac{ig'}{2} B_\mu \begin{pmatrix} 1 & 0 \\ 0 & 1 \end{pmatrix} \right] \Psi_L$$

$$(15.1)$$

Next, let's look at the kinetic energy of the up and down quarks. The left-chiral components of these two quarks are combined into an $SU(2)_L$ doublet with quantum numbers $\mathbf{3}, \mathbf{2}, 1/3$. The first quantum number, $\mathbf{3}$, tells us that they transform under the triplet representation of $SU(3)_c$, which is the fundamental representation for that group. We may represent this doublet by the shorthand notation

$$\Psi_{Q_L} \equiv \begin{pmatrix} \Psi_{u_L} \\ \Psi_{d_L} \end{pmatrix}$$

where the index 1 is simply a reminder that these are the quarks of the first generation. The color indices are not shown explicitly.

The gauge-covariant derivative of fields with quantum numbers $\mathbf{3}, \mathbf{2}, 1/3$ is

$$D_\mu \Psi_{Q_L} = \left(\partial_\mu + i g_s \frac{\lambda^a}{2} G_\mu^a + i g \frac{\tau^i}{2} W_\mu^i + i \frac{g'}{6} B_\mu \right) \Psi_{Q_L}$$

where the G_μ^a are the gluon fields, the λ^a are the Gell-Mann matrices, and the index a runs from 1 to 8. The gluon fields are often denoted by A_μ^a in the literature, but we will reserve the symbol A_μ for the photon field. Of course, if the $SU(3)_c$ indices are shown explicitly on the gluon fields, there is no danger of confusion, but it is convenient to sometimes not show these indices explicitly, as we will do in the next section.

To complete the list of fermions of the first generation, we still need to include the right-chiral states of the up and down quarks Ψ_{u_R} and Ψ_{d_R}. These are $SU(2)_L$ singlets and color triplets, and the hypercharge assignments are $Y = 4/3$ for u_R and $Y = -2/3$ for d_R, which leads to the following covariant derivatives:

$$D_\mu \Psi_{u_R} = \left(\partial_\mu + i g_s \frac{\lambda^a}{2} G_\mu^a + i \frac{2g'}{3} B_\mu \right) \Psi_{u_R}$$

$$D_\mu \Psi_{d_R} = \left(\partial_\mu + i g_s \frac{\lambda^a}{2} G_\mu^a - i \frac{g'}{3} B_\mu \right) \Psi_{d_R}$$

Kinetic Terms of the Gauge Bosons

The gauge-invariant kinetic energy terms of the $U(1)_Y$, $SU(2)_L$, and $SU(3)_c$ gauge fields are, respectively (see Section 10.3 for a review of nonabelian gauge theories),

$$\mathcal{L}_{GB} = -\frac{1}{4} F^{\mu\nu} F_{\mu\nu} - \frac{1}{2} \text{Tr}\big(W_{\mu\nu} W^{\mu\nu}\big) - \frac{1}{2} \text{Tr}\big(G_{\mu\nu} G^{\mu\nu}\big)$$

where

$$F_{\mu\nu} \equiv \partial_\mu B_\nu - \partial_\nu B_\mu$$
$$W_{\mu\nu} \equiv \partial_\mu W_\nu - \partial_\nu W_\mu - i\, g\, [W_\mu, W_\nu]$$
$$G_{\mu\nu} \equiv \partial_\mu G_\nu - \partial_\nu G_\mu - i\, g_s\, [G_\mu, G_\nu]$$

The weak bosons and gluon fields appearing in those equations are, of course, the matrix-valued quantities:

$$W_\mu \equiv W_\mu^i \frac{\tau^i}{2} \qquad G_\mu \equiv G_\mu^a \frac{\lambda^a}{2}$$

Unfortunately, we are using the same symbol, λ^a, for the Gell-Mann matrices and for the spin components of the fermion superpartners of the gauge fields, the gauginos λ^a, but the context is usually clear enough to avoid any confusion.

Kinetic Term of the Higgs Field

Consider now the Higgs field H, which has quantum numbers **1**, **2**, 1. It is a singlet with respect to the color group, so it does not interact directly with the gluons (there are, of course, indirect interactions, i.e., interactions through loop diagrams), and it is an $SU(2)_L$ doublet that we will write as

$$H = \begin{pmatrix} H^+ \\ H^0 \end{pmatrix}$$

where each of the fields H^+ and H^0 are complex scalar fields, and the labels indicate their electric charge. The kinetic term is given by

$$(D^\mu H)^\dagger D_\mu H \tag{15.2}$$

where D_μ is the gauge-covariant derivative corresponding to the quantum numbers given above, namely,

$$D_\mu H = \left(\partial_\mu + i g \frac{\tau^i}{2} W^i_\mu + i \frac{g'}{2} B_\mu \right) H \tag{15.3}$$

Interaction Terms

Let us now turn our attention to the interaction terms. The interactions between the gauge bosons and the other fields (fermions and Higgs) all arise from the gauge-covariant derivatives listed earlier. The self-interactions of the nonabelian gauge bosons are all contained in their kinetic terms.

We'll turn our attention now to the interactions between the fermions and the Higgs field, which are all given by Yukawa terms. Let's first restrict our attention to the fermions of the first generation; the generalization to all three generations will be straightforward.

The Higgs field is an $SU(2)$ doublet, so it must be combined with another $SU(2)$ doublet to give a quantity invariant under $SU(2)$ transformations. On the other hand, it is a color singlet, so it must be combined with a color singlet (i.e., a color-invariant). Finally, it has a hypercharge of 1, so it must be combined with a quantity with hypercharge -1. To summarize, the Higgs field must be coupled with a quantity that has quantum numbers $\mathbf{1}, \bar{\mathbf{2}}, -1$. We therefore need to build a quantity out of two fermion fields that have these quantum numbers.

Let's first consider only the leptons. One possibility is the combination

$$\overline{\Psi}_L \Psi_{e_R} \tag{15.4}$$

This is an $SU(2)$ antidoublet (because of the hermitian conjugation) and a color singlet. Taking into account the fact the Dirac conjugate

$$\overline{\Psi}_L = \Psi^\dagger_L \gamma^0 \tag{15.5}$$

changes the sign of the hypercharge (because of the hermitian conjugate), we see that Eq. (15.4) has hypercharge $1 - 2 = -1$ and therefore has all the right quantum numbers to be coupled with the Higgs field in a gauge-invariant manner. A gauge invariant interaction is therefore

$$\overline{\Psi}_L H \Psi_{e_R} \tag{15.6}$$

We must, of course, multiply it by an arbitrary constant that we will call y_e, the Yukawa coupling for the electron field (do not confuse Yukawa couplings, represented by lower case y, with the hypercharge assignments, denoted by capital Y). This constant is dimensionless because Eq. (15.6) has dimension $3/2 + 1 + 3/2 = 4$. We also must add to Eq. (15.6) its hermitian conjugate to get a real contribution to the lagrangian of the standard model. The final result is that we must include the following contribution to the standard model:

$$y_e \overline{\Psi}_L H \Psi_{e_R} + \text{h.c.}$$

On SSB, this term generates the electron mass.

There is also a Yukawa coupling between the Higgs field and the quarks. A quark combination that has quantum numbers $\mathbf{1}, \mathbf{2}, -1$ is $\overline{\Psi}_{Q_1} \Psi_{d_R}$, so we have a second gauge-invariant Yukawa term to write:

$$y_d \overline{\Psi}_{Q_L} H \Psi_{d_R} + \text{ h.c.} \tag{15.7}$$

After SSB, this term will generate a mass for the down quark.

In order to obtain a mass for the up quark as well, we must find a way to couple the Higgs field with Ψ_{Q_L} and Ψ_{u_R}. This works okay with respect to the $SU(2)$ and $SU(3)$ quantum numbers, but we run into trouble with the hypercharge. A term $\overline{\Psi}_{Q_L} H \Psi_{u_R}$ has a total hypercharge of $-1/3 + 1 + 4/3 = 2$, so it is not invariant under $U(1)_Y$. The solution is to use the fact that the quantity $i\tau_2 H^{\dagger T}$ transforms the same way as H itself under $SU(2)_L$, i.e. as a doublet (see for example Section 12.1.3 of Ref. 2). This quantity therefore has quantum numbers $\mathbf{1}, \mathbf{2}, -1$ so the interaction $\overline{\Psi}_{Q_L} i\tau_2 H^{\dagger T} \Psi_{u_R}$ is gauge invariant.

We will have to combine $SU(2)$ doublets using the matrix $i\tau_2$ very often in the SM and in the MSSM, so it would be nice to have a shorthand notation for this product. Some authors write $\overline{\Psi}_{Q_L} i\tau_2 H^{\dagger T}$ simply as $\overline{\Psi}_{Q_L} \cdot H^{\dagger T}$ (or, rather, as $\overline{\Psi}_{Q_L} \cdot H^*$) but this might be mistaken for the spinor dot product notation we have introduced in chapter 2. Others actually leave out the $i\tau_2$ altogether, which is even more confusing.

To make the notation as clear as possible, we will introduce the following explicit (but nonstandard) definition

$$\boxed{\overline{\Psi}_{Q_L} i\tau_2 H^{\dagger T} \equiv \overline{\Psi}_{Q_L} \circ H^{\dagger T}}$$

From now on, whenever you encounter the symbol $\circ$, it will appear between two $SU(2)$ doublets and will mean that a matrix $i\tau_2$ must be inserted there.

Using this notation, our third gauge invariant Yukawa term is

$$y_u \, \overline{\Psi}_{Q_L} \circ H^{\dagger T} \, \Psi_{u_R} + \text{h.c.} \tag{15.8}$$

Recall that by $H^{\dagger T}$, we mean the column vector whose components are the Hermitian conjugates of the fields, i.e.,

$$H^{\dagger T} = \begin{pmatrix} (H^+)^\dagger \\ (H^0)^\dagger \end{pmatrix}$$

$$= \begin{pmatrix} H^- \\ H^{0\dagger} \end{pmatrix}$$

The fact that we absolutely need $H^\dagger$ as well as H to obtain masses for the quarks (after SSB) in the standard model will have profound consequences when we consider its supersymmetric extension.

The next step is to generalize the Yukawa interactions to three generations:

$$\mathcal{L}_{\text{Yuk}} = y_e^{ij} \, (\overline{\Psi}_L)_i \, H (\Psi_{e_R})_j + y_d^{ij} \, (\overline{\Psi}_{Q_L})_i \, H (\Psi_{Q_R})_j$$
$$+ \, y_u^{ij} \, (\overline{\Psi}_{Q_L})_i \circ H^{\dagger T} (\Psi_{Q_R})_j + \text{h.c.} \tag{15.9}$$

where it is implicit that the subscripts i and j label the generations.

The Higgs Potential

The only thing missing is the Higgs potential. It is conventionally written as

$$V(H) = -\mu^2 \, H^\dagger H + \lambda (H^\dagger H)^2 \tag{15.10}$$

and the constants μ^2 and λ are chosen to be positive in order to generate SSB. We now turn our attention to SSB in the standard model. This is a very useful warm up in preparation for the study of symmetry breaking in the MSSM, which is much more involved than in the standard model.

15.2 Spontaneous Symmetry Breaking in the Standard Model

Spontaneous symmetry breaking in the standard model is very simple; one can see by simple inspection that Eq. (15.10) has a minimum at the condition that both λ and μ^2 are positive real numbers and that the minimum occurs when the fields have a vacuum expectation value (vev) given by

$$\langle H^\dagger H \rangle \equiv v^2 \qquad \text{with } v = \frac{\mu}{\sqrt{2\lambda}}$$

We may always use an $SU(2)$ transformation to bring the vev of H in the standard form, i.e.,

$$\langle H \rangle_{\text{min}} = \begin{pmatrix} 0 \\ v \end{pmatrix}$$

In another popular normalization, the vev is instead defined as

$$\langle H^\dagger H \rangle \equiv \frac{\tilde{v}^2}{2} \qquad \text{with } \tilde{v} = \frac{\mu}{\sqrt{\lambda}} \tag{15.11}$$

The next step is to define a shifted field H_0', i.e.,

$$H_0' \equiv H_0 - v$$

that has a zero vev, i.e., $\langle H_0' \rangle = 0$. We now replace H_0 by $H_0' + v$ in the kinetic energy of the Higgs field, [Eq. (15.2)]. The gauge-covariant derivative $D_\mu H$ becomes

$$D_\mu H = D_\mu \begin{pmatrix} H^+ \\ H_0' \end{pmatrix} + D_\mu \begin{pmatrix} 0 \\ v \end{pmatrix} \tag{15.12}$$

with D_μ given in Eq. (15.3). To be explicit, the second term of Eq. (15.12) is

$$D_\mu \begin{pmatrix} 0 \\ v \end{pmatrix} = \frac{i\,v}{2} \begin{pmatrix} g\,W_{1\mu} - i\,g\,W_{2\mu} \\ -g\,W_{3\mu} + g'\,B_\mu \end{pmatrix}$$

Inserting this into the kinetic energy term of the Higgs [Eq. (15.2)] and writing down only the terms bilinear or quadratic in the gauge fields, which will contribute to the masses of the gauge bosons, we get

$$\mathcal{L}_{\mathrm{MGB}} = \frac{v^2}{4}\,g^2\left(W_1^2 + W_2^2\right) + \frac{v^2}{4}\,(g\,W_3 - g'\,B)^2 \qquad (15.13)$$

We finally introduce the $W^{\pm}$ and Z gauge fields by defining

$$\mathcal{L}_{\mathrm{MGB}} \equiv m_W^2\,W^{+\mu}W_{\mu}^{-} + \frac{1}{2}m_Z^2 Z_{\mu}Z^{\mu}$$

with

$$m_W^2\,W^{+\mu}W_{\mu}^{+\dagger} \equiv \frac{v^2 g^2}{4}\left(W_1^2 + W_2^2\right)$$

$$= \frac{v^2\,g^2}{4}\left(W_{1\mu} + i\,W_{2\mu}\right)\left(W_1^{\mu} - i\,W_2^{\mu}\right)$$

We define

$$W_{\mu}^{\pm} \equiv \frac{1}{\sqrt{2}}\left(W_{1\mu} \pm i\,W_{2\mu}\right)$$

where the factor of $1/\sqrt{2}$ is simply a normalizing factor. This identification then implies the important relation

$$\boxed{m_W^2 = \frac{v^2 g^2}{2}} \qquad (15.14)$$

In addition, by considering weak interactions involving the W bosons at small energies, one can relate the ratio g^2/m_W^2 to Fermi's constant G_F. The explicit relation is

$$\frac{g^2}{8m_W^2} = \frac{G_F}{\sqrt{2}} \qquad (15.15)$$

Let us turn our attention to the Z boson. We define

$$\frac{1}{2}m_Z^2 Z_{\mu}Z^{\mu} \equiv \frac{v^2}{4}(g\,W_3 - g'B)^2 \qquad (15.16)$$

so that Z is proportional to the combination $g W_{3\mu} - g' B_\mu$. Normalizing, we therefore have

$$Z_\mu = \frac{1}{\sqrt{g^2 + g'}} \left(g W_{3\mu} - g' B_\mu \right)$$

Using this in Eq. (15.16), we get a relation for the mass of the Z:

$$\boxed{m_Z^2 = \frac{v^2}{2} (g^2 + g')}$$ (15.17)

Obviously, the linear combination orthogonal to Z_μ is massless. This corresponds to the photon field:

$$A_\mu \equiv \frac{1}{\sqrt{g^2 + g^2}} \left(g' W_{3\mu} + g B_\mu \right)$$

It is conventional to introduce an angle relating the two coupling constants called the *Weinberg* (or *weak*) *mixing angle* θ_W through the relation

$$\boxed{\tan \theta_W \equiv \frac{g'}{g}}$$

This implies

$$\cos \theta_W = \frac{g}{\sqrt{g^2 + g'}}$$

$$\sin \theta_W = \frac{g'}{\sqrt{g^2 + g'}}$$ (15.18)

in which case the Z and photon fields take a nicer form:

$$Z_\mu = \cos \theta_W W_{3\mu} - \sin \theta_W B_\mu$$
$$A_\mu = \sin \theta_W W_{3\mu} + \cos \theta_W B_\mu$$

Inverting those relations, we find

$$W_{3\mu} = \cos \theta_W Z_\mu + \sin \theta_W A_\mu$$
$$B_\mu = \cos \theta_W A_\mu - \sin \theta_W Z_\mu$$

Particles couple to the B_μ with a strength g'; therefore, they will couple to the photon field A_μ with a strength $g' \cos \theta_w$, which, by definition, is the electromagnetic coupling constant e, i.e.,

$$\boxed{e = g' \cos \theta_W} \qquad (15.19)$$

Applying the same reasoning to the $W_{3\mu}$ we also get

$$\boxed{e = g \sin \theta_W} \qquad (15.20)$$

We also could have obtained Eq. (15.20) directly from Eq. (15.19) using the definition of the Weinberg angle.

Note that from Eqs. (15.17), (15.13), and (15.18), we get

$$m_W = \cos \theta_W m_Z$$

so these three parameters are not independent. This relation is often expressed in terms of the so-called ρ parameter, defined as

$$\rho \equiv \frac{m_W^2}{m_Z^2 \cos^2 \theta_W} \qquad (15.21)$$

which is therefore equal to 1 at tree level. This prediction of the standard model is confirmed experimentally and is a key test of the gauge structure and pattern of symmetry breaking of the theory. One of the challenges of theories "beyond the standard model" is to reproduce this result.

Let us summarize the situation regarding SSB in the gauge sector of the standard model. The lagrangian contains four parameters: g, g', μ and λ. Physical results depend only on g, g', and the combination $\mu/\sqrt{2\lambda} \equiv v$. These three parameters can be expressed in terms of any three measurements among G_F, e, M_Z, M_W, and θ_W. In using these relations, one must keep in mind that coupling constants run with energy, so one must be careful to use all values evaluated at a common scale. We will discuss the running of coupling constants and the measured values of the standard model parameters more in Section 16.5. Here, let us simply mention that experimental results can be used to calculate

$$v \approx 174 \, \text{GeV}$$

$$g \approx 0.651$$

$$g' \approx 0.357$$

where all values are evaluated at the Z mass. With the normalization (15.11), we have $\tilde{v} \approx 246$ GeV, which you may have seen quoted in the literature.

Now we turn our attention to the MSSM. But before doing this, a short comment about notation is in order.

15.3 Aside on Notation

In writing down the MSSM, we are going to use a notation that will be the obvious generalization of the one introduced to describe supersymmetric quantum electrodynamics (QED) in Section 13.6. If you haven't covered that section yet, or if it has been quite a while, it would be a good idea to read (or reread) it before tackling the MSSM. To summarize,

- We will work only with left-chiral Weyl spinors. The right-chiral particle states are traded off for the corresponding left-chiral antiparticle spinors.

- We will use calligraphic capital letters for all the superfields ($\mathcal{E}, \mathcal{F}, \mathcal{U}$, etc). This will make it easier to distinguish superfields from "ordinary" fields (i.e., fields that are functions of spacetime only).

- All the scalar fields appearing in left-chiral multiplets with one exception, will be represented by ϕ (with labels indicating which multiplet it belongs to). The exception will be for the Higgs scalar fields, which will be denoted by H (with some labels). All the left-chiral Weyl spinors will be denoted by χ, and all the auxiliary fields will be written as F, with appropriate labels.

- The gauge fields will be denoted by the same symbols as in the standard model: B_μ for the hypercharge gauge field, A_μ^a for the color fields, and W_μ^i for the weak $SU(2)$ fields. Their supersymmetric left-chiral Weyl partners will be denoted by λ with appropriate labels (see next section), and the corresponding auxiliary fields will be written as D.

- A bar over the name of a particle denotes the corresponding antiparticle. For example, $\chi_{\bar{u}}$ is the left-chiral state of the anti-up quark.

- We will use a tilde over the symbols of supersymmetric partners of known particles, which are all particles that haven't been observed yet. For example, $\tilde{\phi}_{\bar{u}}$ is the scalar SUSY partner of the left-chiral anti-up quark, and $\tilde{\lambda}_c^a$ will be the eight spinor partners of the color gauge fields.

There is no universal standard for the notation used in presentation of the MSSM. The notation introduced here has been chosen to maximize clarity, hopefully.

15.4 The Left-Chiral Superfields of the MSSM

We are now ready to construct the superfields of the MSSM. An important point is that as we saw in Chapter 10, supersymmetry (SUSY) transformations commute with gauge transformations. This implies that *all the fields making up a SUSY multiplet must have the same quantum numbers under* $SU(3)_c \times SU(2)_L \times U(1)_Y$. This is a key observation.

As mentioned several times, we trade the right-chiral particle states for the left-chiral antiparticle states. Let's start with the left-chiral positron state $\chi_{\bar{e}}$, which has quantum numbers $\mathbf{1}, \mathbf{1}, \mathbf{2}$. Note that the sign of the hypercharge is opposite to the one of the left-chiral electron spinor χ_e. The corresponding superfield is identical to the one we used in the supersymmetric extension of QED (see Section 13.6), with the difference that we will use a numeral to refer to the generation to which the corresponding component fields belong, i.e.,

$$\mathcal{E}_1 = \tilde{\phi}_{\bar{e}} + \theta \cdot \chi_{\bar{e}} + \frac{1}{2}\theta \cdot \theta\, F_{\bar{e}} \tag{15.22}$$

The corresponding superfields for the second and third generations will be written as

$$\mathcal{E}_2 \equiv \tilde{\phi}_{\bar{\mu}} + \theta \cdot \chi_{\bar{\mu}} + \frac{1}{2}\theta \cdot \theta\, F_{\bar{\mu}}$$

$$\mathcal{E}_3 \equiv \tilde{\phi}_{\bar{\tau}} + \theta \cdot \chi_{\bar{\tau}} + \frac{1}{2}\theta \cdot \theta\, F_{\bar{\tau}}$$

In the MSSM, as in the standard model, there is no right-chiral neutrino field.

Consider now the left-chiral state of the electron. This state is paired up with the electron neutrino left-chiral state to form the $SU(2)_L$ doublet of the first-generation leptons familiar from the standard model:

$$\begin{pmatrix} \chi_{\nu_e} \\ \chi_e \end{pmatrix}$$

with quantum numbers $\mathbf{1}, \mathbf{2}, -1$. We will write the corresponding superfield as

$$\mathcal{L}_1 \equiv \begin{pmatrix} \tilde{\phi}_{\nu_e} \\ \tilde{\phi}_e \end{pmatrix} + \theta \cdot \begin{pmatrix} \chi_{\nu_e} \\ \chi_e \end{pmatrix} + \frac{1}{2}\theta \cdot \theta \begin{pmatrix} F_{\nu_e} \\ F_e \end{pmatrix}$$

where the index 1 on the lepton superfield refers to the generation. Note that in this chapter and the next one, a calligraphic $\mathcal{L}$ with a numerical subscript will represent a lepton superfield, not a lagrangian!

Now, let us turn our attention to the quark superfields. The superfield containing the left-chiral anti-up quark will be denoted as

$$\mathcal{U}_1 \equiv \tilde{\phi}_{\bar{u}} + \theta \cdot \chi_{\bar{u}} + \frac{1}{2}\theta \cdot \theta\, F_{\bar{u}}$$

with quantum numbers $\bar{\mathbf{3}}, \mathbf{1}, -4/3$ (again, we are suppressing the color indices). The superfield containing the left-chiral anti-down quark is

$$\mathcal{D}_1 \equiv \tilde{\phi}_{\bar{d}} + \theta \cdot \chi_{\bar{d}} + \frac{1}{2}\theta \cdot \theta\, F_{\bar{d}}$$

with quantum numbers $\bar{\mathbf{3}}, \mathbf{1}, 2/3$.

We also need a superfield containing the $SU(2)_L$ left-chiral quark, which we write as

$$\mathcal{Q}_1 \equiv \begin{pmatrix} \tilde{\phi}_u \\ \tilde{\phi}_d \end{pmatrix} + \theta \cdot \begin{pmatrix} \chi_u \\ \chi_d \end{pmatrix} + \frac{1}{2}\theta \cdot \theta \begin{pmatrix} F_u \\ F_d \end{pmatrix} \tag{15.23}$$

with quantum numbers $\mathbf{3}, \mathbf{2}, 1/3$. For the second-generation quarks, the strange and charm, we of course use $\mathcal{D}_2, \mathcal{U}_2$, and $\mathcal{L}_2$, and so on.

Note that the superfields $\mathcal{E}_i, \mathcal{U}_i$, and $\mathcal{D}_i$ [which are all $SU(2)_L$ singlets] contain the *antiparticle* left-chiral spinors, whereas the $SU(2)_L$ doublets $\mathcal{L}_i$ and $\mathcal{Q}_i$ contain the *particle* left-chiral states.

Consider now the Higgs sector of the MSSM. Here, there is a twist. Recall that in the standard model we needed the conjugate Higgs field to generate a mass for the up quark. On the other hand, we have seen in previous chapters that SUSY-invariant interactions between left-chiral superfields are obtained from the F term of the superpotential, *which must be holomorphic* in the chiral superfields! Therefore, we will not be able to use $\mathcal{H}^\dagger$ to construct SUSY-invariant Yukawa-type interactions, and some of the standard model fermions will remain massless even after SSB. This is very bad, so we need to figure out a way to take care of this.

The solution is obviously to introduce a *second* Higgs superfield, with quantum numbers opposite to those of the first. This means that the MSSM *contains twice as many Higgs fields* as in the standard model. In the standard model, there is a single $SU(2)$ Higgs doublet of complex fields corresponding to four degrees of freedom. Three of those are the Goldstone modes that get "eaten" by gauge bosons to make the $W^\pm$ and Z_0 massive, leaving a single Higgs particle. In the MSSM, there are eight Higgs degrees of freedom, three of which also will get eaten, leaving this time five observable Higgs particles. We will study the properties of these particles in Chapter 16.

We therefore introduce two Higgs superfields. The first one will be taken as having the same quantum numbers as the standard model Higgs field, namely, **1, 2**, 1. We will call it $\mathcal{H}_u$ and write its components as

$$\mathcal{H}_u = \begin{pmatrix} H_u^+ \\ H_u^0 \end{pmatrix} + \theta \cdot \begin{pmatrix} \tilde{\chi}_u^+ \\ \tilde{\chi}_u^0 \end{pmatrix} + \frac{1}{2}\theta \cdot \theta \begin{pmatrix} F_u^+ \\ F_u^0 \end{pmatrix} \qquad (15.24)$$

As in the standard model, the subscripts indicate the electric charge of the particle, which can be obtained using, as usual, $Q = T_3 + Y/2$ (with the states in the upper position having $T_3 = 1/2$ and the ones below having $T_2 = -1/2$). The reason for the subscript u is that this Higgs generates a mass for the up quark.

Since the second Higgs doublet superfield will have to play the role that $H^\dagger$ played in the standard model, we must assign to it the opposite hypercharge to $\mathcal{H}_u$. We obviously take it to be a color singlet. We therefore assign the quantum numbers **1, 2**, −1 to this second Higgs doublet and write it as

$$\mathcal{H}_d = \begin{pmatrix} H_d^0 \\ H_d^- \end{pmatrix} + \theta \cdot \begin{pmatrix} \tilde{\chi}_d^0 \\ \tilde{\chi}_d^- \end{pmatrix} + \frac{1}{2}\theta \cdot \theta \begin{pmatrix} F_d^0 \\ F_d^- \end{pmatrix} \qquad (15.25)$$

The spin 1/2 superpartners of the Higgs are referred to as *Higgsinos*. The quantum numbers of the MSSM left-chiral superfields are summarized in Table 15.1.

Table 15.1   Quantum Numbers of the MSSM Left-Chiral Superfields

Superfield	$SU(3)_c$	$SU(2)_L$	$U(1)_Y$
$\mathcal{Q}_i$	**3**	**2**	1/3
$\mathcal{H}_u$	**1**	**2**	1
$\mathcal{H}_d$	**1**	**2**	−1
$\mathcal{U}_i$	$\bar{\mathbf{3}}$	**1**	−4/3
$\mathcal{D}_i$	$\bar{\mathbf{3}}$	**1**	2/3
$\mathcal{L}_i$	**1**	**2**	−1
$\mathcal{E}_i$	**1**	**1**	2

15.5 The Gauge Vector Superfields

We will denote the hypercharge gauge superfield by $\mathcal{B}$. In the Wess-Zumino gauge, it can be read off [see Eq. (13.10)] as

$$\mathcal{B} \equiv \frac{1}{2} \theta \sigma^\mu \bar{\theta} \, B_\mu + \frac{1}{2\sqrt{2}} \theta \cdot \theta \bar{\theta} \cdot \bar{\lambda}_Y + \frac{1}{2\sqrt{2}} \bar{\theta} \cdot \bar{\theta} \theta \cdot \lambda_Y - \frac{1}{8} \theta \cdot \theta \bar{\theta} \cdot \bar{\theta} \, D_Y \quad (15.26)$$

For color, we introduce an octet of gauge fields G_μ^a, as in the standard model, which means that we must introduce an octet of left-chiral fermions λ^a as well as an octet of auxiliary fields D^a. We use $\mathcal{G}^a$ to represent the corresponding superfield, which has the same expansion as Eq. (15.26) with λ_Y replaced by λ_c^a and D_Y replaced by D_c^a (where the label c stands for color).

It is natural to use $\mathcal{W}^i$ to represent the $SU(2)_L$ gauge superfield, which should not be confused with the superpotential $\mathcal{W}$ (with no index). The component fields are W_μ^i, λ_L^i, and D_L^i.

To be honest, we will not do much with any of the component fields just given and have defined them for the sake of completeness only. So don't sweat it if all the notation is confusing; we will be concerned mainly with the superfields.

15.6 The MSSM Lagrangian

We can organize the contributions to the MSSM lagrangian into five categories:

- The kinetic energy terms of the leptons and quarks
- The kinetic energy terms of the Higgs fields
- The kinetic energy terms of the gauge bosons
- The superpotential
- The SUSY-breaking terms

Let us introduce each type of term in turn.

Kinetic Energy Terms of the Leptons, Quarks, and Higgs Fields

It is straightforward to write down these terms after our work in Chapter 13. For example, consider the kinetic energy of the first-generation quark doublet $\mathcal{Q}_1$. It is given by $\Phi^\dagger \exp(\partial q \mathcal{V})\Phi$, which here takes the form

$$\mathcal{L}_{\text{kin}} = \mathcal{Q}_1^\dagger \, \exp\left(\frac{1}{3}\, g' \mathcal{B} + g_s \, \mathcal{G}^a \lambda^a + g \, \mathcal{W}^i \tau^i\right) \mathcal{Q}_1 \bigg|_D \qquad (15.27)$$

where we have used $Y = 1/3$.

We will write the lagrangian explicitly in terms of the component fields, just to illustrate what kind of interactions are generated. Before doing so, it proves convenient to assign names to the $SU(2)_L$ doublets containing the component fields in Eq. (15.23). To keep things as simple as possible, let's just use

$$\mathcal{Q}_1 \equiv \tilde{\phi} + \theta \cdot \chi + \frac{1}{2}\, \theta \cdot \theta F \qquad (15.28)$$

where

$$\tilde{\phi} \equiv \begin{pmatrix} \tilde{\phi}_u \\ \tilde{\phi}_d \end{pmatrix}$$

and so on.

In terms of these doublets, Eq. (15.27) gives the following lagrangian:

$$\begin{aligned}
\mathcal{L}_{\text{kin}} =\, & D_\mu \tilde{\phi}^\dagger \, D^\mu \tilde{\phi} + i\, \chi^\dagger \, \bar{\sigma}^\mu \, D_\mu \chi + F^\dagger F \\
& - \frac{\sqrt{2}}{6}\, g' \, \tilde{\phi}^\dagger \, \lambda_Y \cdot \chi - \frac{\sqrt{2}}{6}\, g' \, \tilde{\phi} \, \bar{\lambda}_Y \cdot \bar{\chi} - \frac{1}{6}\, g' \, \tilde{\phi}^\dagger \, D_Y \, \tilde{\phi} \\
& - \frac{g}{\sqrt{2}}\, \tilde{\phi}^\dagger \, \tau^i \chi \cdot \lambda_L^i - \frac{g}{\sqrt{2}}\, \tilde{\phi} \, \tau^i \bar{\chi} \cdot \bar{\lambda}_L^i - \frac{g}{2}\, \tilde{\phi}^\dagger \, \tau^i \, \tilde{\phi} \, D_L^i \\
& - \frac{g_s}{\sqrt{2}}\, \tilde{\phi}^\dagger \, \lambda_G^a \chi \cdot \lambda_C^a - \frac{g_s}{\sqrt{2}}\, \tilde{\phi} \, \lambda_G^a \bar{\chi} \cdot \bar{\lambda}_C^a - \frac{g_s}{2}\, \tilde{\phi}^\dagger \, \lambda_G^a \, \tilde{\phi} \, D_C^a
\end{aligned} \qquad (15.29)$$

where Λ_G^a stands for the Gell-Mann matrices and D_μ is the gauge-covariant derivative corresponding to the quantum numbers of $\mathcal{Q}_1$, i.e., **3**, **2**, $1/3$.

It is now that all our work on superspace truly pays off. We know that this complicated lagrangian is supersymmetric by construction. Imagine how difficult it would be to construct this using the "brute force" approach we followed in Chapters 5, 8, and 10!

Consider now the superfield $\mathcal{U}_1$ with quantum numbers $\bar{\mathbf{3}}, \mathbf{1}, -4/3$. The kinetic term of this superfield is

$$
\begin{aligned}
\mathcal{L}_{\text{kin}} &= \mathcal{U}_1^\dagger \, \exp\left(-\frac{4}{3} g' \, \mathcal{B} + g_s \, \mathcal{G}^a (\lambda^a)^*\right) \mathcal{U}_1 \bigg|_D \\
&= D_\mu \tilde{\phi}_{\bar{u}}^\dagger \, D^\mu \tilde{\phi}_{\bar{u}} + i \, \chi_{\bar{u}}^\dagger \, \bar{\sigma}^\mu \, D_\mu \chi_{\bar{u}} + F_{\bar{u}}^\dagger \, F_{\bar{u}} \\
&\quad + \frac{2\sqrt{2}}{3} g' \, \tilde{\phi}^\dagger \lambda_Y \cdot \chi + \frac{2\sqrt{2}}{3} g' \, \tilde{\phi} \, \bar{\lambda}_Y \cdot \bar{\chi} + \frac{2}{3} g' \, \tilde{\phi}^\dagger D_Y \, \tilde{\phi} \\
&\quad - \frac{g_s}{\sqrt{2}} \tilde{\phi}^\dagger (\lambda_G^a)^* \chi \cdot \lambda_C^a - \frac{g_s}{\sqrt{2}} \tilde{\phi} (\lambda_G^a)^* \bar{\chi} \cdot \bar{\lambda}_C^a - \frac{g_s}{2} \tilde{\phi}^\dagger (\lambda_G^a)^* \tilde{\phi} D_C^* \quad (15.30)
\end{aligned}
$$

The only interesting thing to mention here is that these terms contain the standard model kinetic terms written in terms of the left-chiral particle and antiparticle Weyl spinors. For example, if we focus on the $SU(3)_c$ kinetic terms for the spinors in Eqs. (15.29) and (15.28), we have

$$
i \chi_u^\dagger \bar{\sigma}^\mu \left(\partial_\mu + \frac{i}{2} g_s \, \lambda^a G_\mu^a\right) \chi_u + i \, \chi_{\bar{u}}^\dagger \bar{\sigma}^\mu \left(\partial_\mu - i \, g_s \, (\lambda^a)^* G_\mu^a\right) \chi_{\bar{u}}
$$

$$
+ i \, \chi_d^\dagger \bar{\sigma}^\mu \left(\partial_\mu + \frac{i}{2} g_s \, \lambda^a G_\mu^a\right) \chi_d \tag{15.31}
$$

We showed in Section 10.4 that the first two terms are equivalent to the quantum chromodynamics (QCD) part of kinetic term for the up quark written in terms of a Dirac spinor, i.e.,

$$
i \, \bar{\Psi} \gamma^\mu \left(\partial_\mu + \frac{i}{2} \, g_s \, G_\mu^a \lambda^a\right) \Psi \tag{15.32}
$$

As for the down quark, the kinetic energy for the superfield $\mathcal{D}$ will provide

$$
i \, \chi_d^\dagger \bar{\sigma}^\mu \left(\partial_\mu - i \, g_s \, (\lambda^a)^* G_\mu^a\right) \chi_{\bar{d}} \tag{15.33}
$$

which, when combined with the last term of Eq. (15.31), corresponds to the QCD part of the standard model kinetic term for the down quark. It's clear that the standard model kinetic terms are all contained in our supersymmetric expressions.

We won't write down the kinetic terms for the leptons and the Higgs superfields because they are obvious to obtain after our examples above (note that $\mathcal{Q}$ is the most complex case because it is not a singlet under any of the three gauge groups).

The Kinetic Energy Terms for the Gauge Bosons

These are simply (see Sections 13.5 and 13.7)

$$\mathcal{L}_{\text{GB}} = \frac{1}{4}\mathcal{F}_Y \cdot \mathcal{F}_Y\Big|_F + \frac{1}{2}\text{Tr}(\mathcal{F}_L \cdot \mathcal{F}_L)\Big|_F + \frac{1}{2}\text{Tr}(\mathcal{F}_C \cdot \mathcal{F}_C)\Big|_F \qquad (15.34)$$

where $\mathcal{F}_Y$, $\mathcal{F}_L$, and $\mathcal{F}_C$ are the field-strength superfields of the $U(1)_Y$, $SU(2)_L$, and $SU(3)_C$ gauge fields, respectively. Here, the dot product denotes the usual spinor dot product (recall that the field-strength superfields are Weyl spinors).

The corresponding expansions in terms of component fields can be found in Eqs. (10.8) and (10.29).

15.7 The Superpotential of the MSSM

Now we are ready to look for the superpotential of the MSSM, which is a function of these superfields of the theory, i.e., the lepton, quark, and Higgs left-chiral superfields. This will contain the Yukawa couplings of the fermion with the Higgs field, as a subset.

What we need is to build something that is an $SU(3)_C \times SU(2)_L \times U(1)_Y$ invariant out of the scalar left-chiral superfields. This will automatically be Lorentz-invariant because these superfields are invariant under Lorentz transformations. And we will get a SUSY-invariant lagrangian by projecting out the F term, as usual. So the only thing we have to worry about is to make sure that we respect gauge invariance.

Again, note how the superfield approach makes it easy to construct supersymmetric theories! Imagine the difficulty we would be facing if we were working with the component fields and trying to ensure invariance under SUSY following the brute-force approach of Chapters 5, 8, and 10! The calculations were already long when we were dealing with only one left-chiral and one gauge supermultiplet, so you can imagine the nightmare if we were trying this approach with a theory as complex as the MSSM.

Let us consider a single family for now. It will be trivial to generalize the superpotential to three families later on.

Let's start with the superfield $\mathcal{Q}_1$. It transforms as **3** under the color group, so we must pair it up with something that transforms as $\bar{\mathbf{3}}$ in order to get a color singlet, i.e., a color invariant. We have two possibilities: $\mathcal{U}_1$ and $\mathcal{D}_1$.

Consider first the combination $\mathcal{U}_1\mathcal{Q}_1$. This has a hypercharge of $1/3 - 4/3 = -1$ and is a color singlet and a doublet under $SU(2)_L$. We must pair it up with a field that has a hypercharge of 1 and is a weak doublet and a color singlet. Therefore, we are looking for a field with quantum numbers **1**, **2**, 1. The field $\mathcal{H}_u$ fits the bill. A technical detail we need to know is that the $SU(2)$-invariant way to couple two $SU(2)$ doublets is to sandwich between them the matrix $i\tau_2$. We have thus found our first candidate for the superpotential:

$$(\mathcal{U}_1 \mathcal{Q}_1) \circ \mathcal{H}_u$$

which, of course, will have to be multiplied by an arbitrary constant. We are suppressing the color indices but will put in parentheses the superfields which have color indices that are contracted.

If we now consider $\mathcal{D}_1 \mathcal{Q}_1$, with quantum numbers **1**, **2**, 1, we need to couple this with a field with quantum numbers **1**, **2**, -1. This time we have two possibilities: $\mathcal{H}_d$ and $\mathcal{L}_1$, so we can construct the two gauge invariants

$$(\mathcal{D}_1 \mathcal{Q}_1) \circ \mathcal{H}_d \qquad (\mathcal{D}_1 \mathcal{Q}_1) \circ \mathcal{L}_1$$

However, we will see a bit below that the second term will have to be rejected for phenomenological reasons, and this will force us to impose a new type of symmetry. But let's keep going with building gauge invariants. We have exhausted all the possible terms containing $\mathcal{Q}_1$.

Consider now terms containing $\mathcal{H}_u$ but with no $\mathcal{Q}_1$ (to avoid listing again some of the terms already found). In other words, we consider only pairing $\mathcal{H}_u$ with itself or with other fields appearing below it in the table.

Consider pairing $\mathcal{H}_u$ with itself. $\mathcal{H}_u \circ \mathcal{H}_u$ has quantum numbers **1**, **1**, 2. We have one possibility: We can pair it with $\mathcal{L}_1 \circ \mathcal{L}_1$:

$$\mathcal{H}_u \circ \mathcal{H}_u \mathcal{L}_1 \circ \mathcal{L}_1$$

but the F term of this expression is of dimension 5 (recall that extracting the F term increases the dimension by 1), so we do not consider it (and is actually zero! See below).

We also can construct invariants containing only $\mathcal{H}_u$ and a single other field. To get an invariant, we need to pair up $\mathcal{H}_u$ with a field with quantum numbers **1**, **2**, -1. We have two candidates! So it is possible to build two invariants containing only two fields:

$$\mathcal{H}_u \circ \mathcal{H}_d \qquad \mathcal{H}_u \circ \mathcal{L}_1$$

There are no other invariants that we can build out of $\mathcal{H}_u$ (and, again, that do not contain $\mathcal{Q}_1$).

Let us turn to $\mathcal{H}_d$. We can pair it up with itself, in which case we need to multiply by the $\mathcal{E}_1$ superfield:

$$\mathcal{H}_d \circ \mathcal{H}_d \mathcal{E}_1$$

but this vanishes identically because for any two-component vector A, we have $A \circ A = A^T i \tau_2 A = 0$.

$\mathcal{H}_d$ can be paired up with $\mathcal{L}_1$ to form an $SU(2)$-invariant. But then, to get the hypercharge equal to zero, we need to pair them up with $\mathcal{E}_1$, so we get

$$\mathcal{E}_1 \mathcal{L}_1 \circ \mathcal{H}_d$$

We cannot build any new invariants containing only two of the antiquark super-fields $\mathcal{U}_1$ and $\mathcal{D}_1$ because there is no way to make this invariant under $SU(3)_c$ (one can indeed show using Young tableaux that $\bar{\mathbf{3}} \otimes \bar{\mathbf{3}} = \mathbf{3} \oplus \bar{\mathbf{6}}$). However, it is possible to build a color singlet out of three $\bar{\mathbf{3}}$ representations. Explicitly, the color singlet is given by

$$f_C^{abc} \mathcal{U}_1^a \mathcal{D}_1^b \circ \mathcal{D}_1^c$$

where we showed explicitly the color indices and f_c^{abc} are the structure constants of QCD.

Consider now $\mathcal{L}_1$. Our last possible invariant is

$$(\mathcal{L}_1 \circ \mathcal{L}_1)\,\mathcal{E}_1$$

Thus we have uncovered eight gauge-invariant terms built out of the chiral super-fields. But we have a problem. Some of these terms violate either baryon or lepton number conservation, and this has nasty implications in terms of phenomenology. Lepton and baryon number–violating interactions might lead to rapid decay of the proton (which is not observed) or to large flavor-changing neutral current (such as unobserved muon decay to an electron and a photon $\mu \to e\gamma$). The undesirable interactions contain unequal numbers of lepton and antilepton fields or unequal numbers of quark and antiquark fields:

$$(\mathcal{D}_1 \mathcal{Q}_1) \circ \mathcal{L}_1 \qquad \mathcal{H}_u \circ \mathcal{L}_1 \qquad \mathcal{L}_1 \circ \mathcal{L}_1 \mathcal{E}_1 \qquad f_c^{abc}\, \mathcal{U}_1^a \mathcal{D}_1^b \circ \mathcal{D}_1^c \qquad (15.35)$$

The first three violate lepton number conservation, whereas the last one violates baryon number conservation. The "good" interactions are

$$(\mathcal{U}_1 \mathcal{Q}_1) \circ \mathcal{H}_u \qquad (\mathcal{D}_1 \mathcal{Q}_1) \circ \mathcal{H}_d \qquad \mathcal{E}_1 \mathcal{L}_1 \circ \mathcal{H}_d \qquad \mathcal{H}_u \circ \mathcal{H}_d \qquad (15.36)$$

Why don't we have such a problem in the standard model? Well, the standard model is actually quite special in that respect. The particle content is such that there are no gauge-invariant interactions that violate lepton or baryon number conservation. Therefore, gauge invariance is what saves the day for the standard model (although nonperurbative effects are known to violate both symmetries, but only by extremely small amounts at the energies we can reach in particle accelerators).

Unfortunately, the situation is not as rosy for the MSSM as we just saw. Why not simply discard the unpleasant terms, then? The problem is that in quantum field theory, if we set to zero some interaction even though it does not violate any symmetry of the theory, loop corrections will necessarily reintroduce the offending term in the lagrangian. One always could fine-tune the bare coupling constants to set the problematic interactions to zero order by order in the loop expansion, but this just reintroduces a naturalness problem (actually, several naturalness problems!), which is what we wanted to avoid in the first place!

So what is the way out? We have used gauge invariance and SUSY to restrict the types of interactions, but that's not enough. Clearly, what we need is an extra symmetry, one that will kill all the undesirable terms while keeping the good ones! This symmetry preferably should be a global one so as to not require extra gauge bosons and all the complications that would follow. It turns out that we can indeed define a global $U(1)$ symmetry that does exactly what we need! Admittedly, it may sound a bit unsatisfactory to invent an extra symmetry to get rid of things we don't like, but the fact that it is possible at all is admirable in itself and suggests that there must be some underlying but not yet understood principle at work. This should not be a disappointment; after all, nobody expects a supersymmetric standard model to be a fundamental theory.

To understand this new symmetry, let's have a closer look at what the difference is between the good and the bad terms. A generic left-chiral superfield is given by

$$\Phi = \phi + \theta \cdot \chi + \frac{1}{2}\theta \cdot \theta F \tag{15.37}$$

The F term of a product of three left-chiral superfields $\mathcal{X}\mathcal{Y}\mathcal{Z}$ can be written schematically as

$$\mathcal{X}\,\mathcal{Y}\,\mathcal{Z} \simeq \phi_x\,\phi_y\,F_z + \phi_x\,\chi_y \cdot \chi_z + \text{permutations} \tag{15.38}$$

After eliminating the auxiliary fields, the terms proportional to F will become part of the scalar potential. Let's focus on the spinor terms.

The key observation is the following: The standard model fermions correspond to the spinor component fields of the lepton and quark superfields. The Higgs scalars, correspond to the scalar field components of the Higgs superfields (by *scalar field components*, we obviously mean the scalar components that are not multiplied by any Grassmann variables, not the auxiliary fields). We will refer to these particles as *standard particles* because they are present in the standard model, with the only exception that there are more Higgs scalars.

On the other hand, the superpartners of these standard particles are the scalar components of the lepton and quark superfields and the spinor component of the Higgs superfields. If we use a label ST for the standard particles and SP for their superpartners, then the F terms of the good interactions (15.36) containing two spinors are of the form

$$\chi_{ST} \cdot \chi_{ST} \phi_{ST} \qquad \chi_{SP} \cdot \chi_{ST} \phi_{SP} \qquad \chi_{SP} \chi_{SP} \qquad (15.39)$$

whereas the bad interactions are of the form

$$\chi_{ST} \cdot \chi_{ST} \phi_{SP} \quad \text{or} \quad \chi_{ST} \cdot \chi_{SP} \qquad (15.40)$$

Do you see the pattern? All the unwanted terms contain a single superpartner, whereas the good terms contain either no superpartners or two of them. We can summarize this simply by saying that terms with even numbers of superpartners are allowed, whereas we want to eliminate the terms with an odd number of superpartners.

Clearly, we need a symmetry that acts differently on the component fields of the superfields in order to distinguish the standard particles from their superpartners. Since the different component fields multiply different powers of the Grassmann variable θ, it is natural to introduce a $U(1)$ symmetry acting the following way:

$$\theta \to \theta' = e^{i\varphi}\theta \qquad \bar{\theta} \to \bar{\theta}' = e^{-i\varphi}\bar{\theta} \qquad (15.41)$$

where φ is a $U(1)$ global phase. This transformation is often written as

$$\begin{pmatrix} \theta \\ \bar{\theta} \end{pmatrix} \to e^{i\gamma_5\varphi} \begin{pmatrix} \theta \\ \bar{\theta} \end{pmatrix} \qquad (15.42)$$

when the Grassmann variables θ and $\bar{\theta}$ are assembled into a four-component spinor. This symmetry is known as *R-symmetry*.

We now impose that left-chiral superfields transform with an overall phase under *R*-symmetry. If a superfield Φ transforms with a phase $\Phi \to \exp(ik\varphi)$, where k is a real constant, then Eqs. (15.37) and (15.41) show that the component fields transform as

$$\phi \to e^{ik\varphi}\phi \qquad \chi \to e^{i(k-1)\varphi}\chi \qquad F \to e^{i(k-2)\varphi}F \qquad (15.43)$$

It is more useful to restrict this symmetry to the value $\varphi = \pi$, in which case the R-symmetry is simply referred to as *R-parity* (or sometimes, *matter parity*) because the phase is equal to either -1 or $+1$.

Now let's impose that the MSSM lagrangian be invariant under R-parity. We take the gauge superfields to be invariant under this symmetry. It is then clear that all the kinetic energy terms of the left-chiral superfields are invariant because they contain the product of the hermitian conjugate of each superfield with the superfield itself. On the other hand, the superpotential is not automatically invariant.

Recall that extracting the F term is equivalent to computing the following integral:

$$\mathcal{W}\big|_F = \int d^2\theta \, \mathcal{W}$$

We see that in order for this term to be invariant under R-parity, the superpotential must be invariant (because $d^2\theta$ is invariant under R-parity).

Now we need to assign the R-parity of the MSSM left-chiral superfields. We want to keep terms with two fermion superfields (quark or lepton) and one Higgs or with two Higgs superfields, but we need to eliminate terms with three fermion superfields or one Higgs and one fermion superfield. Obviously, we should take the two Higgs superfields to be even under R-parity, whereas we take all quark and lepton superfields to be odd.

As an aside, there is a simple formula that gives the R-parity of the component fields:

$$R = (-1)^{3B+L+2s} \tag{15.44}$$

where B and L are the baryon and lepton quantum numbers, and s is the spin. Because of the factor of $(-1)^{2s}$, particles and their superpartners have opposite R-parities. This formula is sometimes written as

$$R = (-1)^{3(B-L)+2s} \tag{15.45}$$

which is obviously equivalent to Eq. (15.44) because the difference in the exponent is $6L$, which is always even.

Consider, for example, the spinor component $\chi_{\tilde{e}}$ of the superfield $\mathcal{E}$. In this case, $B = 0$, $L = -1$, and $s = 1/2$, so this field has even parity. Since it multiplies a factor θ which is odd under R-parity, the superfield $\mathcal{E}$ has odd parity, in agreement with our previous assignment. It is easy to check that all the standard model particles (and the Higgs scalars) have $R = 1$, whereas all their superpartners have $R = -1$.

We see that imposing invariance under R-parity eliminates all interactions violating conservation of the baryon and lepton quantum numbers. This turns out to be true only because we are working with renormalizable interactions. When nonrenormalizable interactions are included, in the spirit of effective field theories, this correspondence is no longer valid, and therefore one may write baryon and lepton number violating terms that respect R-parity (but, of course, they are suppressed by factors of some high energy scale).

One striking implication of invariance under R-parity is that there are no interactions coupling a single superpartner to two standard particles. Therefore, *the lightest supersymmetric particle (LSP) is absolutely stable*! It cannot decay to anything. It means that the universe must be filled with these particles. Experiments carried on the charge-to-mass ratio of matter rules out the possibility that such particles may be charged. The LSP therefore must be electrically neutral. There is much leeway in the masses of the superpartners, as we will discuss briefly in Chapter 16, so different models have different particles playing the role of the LSP, but it is usually taken to be the neutralino, the scalar superpartner to the electron neutrino. This opens up the exciting possibility that the LSP could account for dark matter (in supergravity theories, the gravitino is also a candidate of choice). More details can be found in Refs. 8 and 47.

Other consequences of R-parity that are relevant for their search at particle accelerators is that superpartners can be produced only in pairs and that a superpartner may decay only into standard particles plus an odd number of superpartners. More detailed discussion of the phenomenology of the MSSM can be found in the two books cited above, as well as in Refs. 1 and 5.

15.8 The General MSSM Superpotential

Now let's generalize to include all three generations. We get

$$\mathcal{W} = y_u^{ij} \mathcal{U}_i \left(\mathcal{Q}_j \circ \mathcal{H}_u \right) - y_d^{ij} \mathcal{D}_i \left(\mathcal{Q}_j \circ \mathcal{H}_d \right) - y_e^{ij} \mathcal{E}_i \left(\mathcal{L}_j \circ \mathcal{H}_d \right) + \mu \, \mathcal{H}_u \circ \mathcal{H}_d \tag{15.46}$$

where the indices i and j label the generations. The signs in the various terms have been chosen in order for the mass terms for the quark and leptons come out with the right sign if we take the Yukawa couplings to be positive, as we will briefly discuss in the last chapter.

This is the superpotential. Of course, to obtain the lagrangian, we must extract the F term. The use of the symbol μ for the coefficient of the term with two Higgs superfields (sometimes referred to as the μ-term) is conventional. This parameter may be complex and should not be confused with the real parameter μ of the standard model. Note that we haven't included a Fayet-Illiopoulos term for the $U(1)_y$ symmetry. The reason is that D-type supersymmetry breaking is not possible in the MSSM because it conflicts with the absence of light squarks and sleptons, as we will discuss in Section 16.1 (and already mentioned in Chapter 14). Of course, this does not give any theoretical justification for dropping that term. The real reason is that we will have to include supersymmetry soft breaking terms which contain all the possible interactions that would have been generated by a Fayet-Illiopoulos term anyway, so there is no need to introduce one here.

The parameters y_u^{ij}, y_d^{ij}, and y_e^{ij} correspond precisely to the Yukawa couplings present in the standard model (more about this in Chapter 16). We therefore see that, quite remarkably, the minimal supersymmetric extension of the standard model contains the *same number of parameters as the standard model* since we have traded the real parameters μ and λ (or, if you prefer, v and λ) for the MSSM complex parameter μ. This is amazing since one could have expected that enlarging the standard model to make it supersymmetric would have entailed the introduction of a plethora of new constants. However, this is for the theory with unbroken supersymmetry; when we will break SUSY, the number of new parameters will unfortunately skyrocket.

We have succeeded in obtaining a supersymmetric version of the standard model, but unfortunately, it is not phenomenologically viable. The problem is that SUSY is unbroken (and it turns out, the $SU(2)_L \otimes U(1)_Y$ gauge symmetry is unbroken as well). Indeed, it is easy to see that that if we write our superpotential in terms of the scalar component fields, it is minimized—and equal to zero—when all the scalar fields are equal to zero. Therefore, SUSY is unbroken, and each particle of the standard model should have a superpartner of equal mass, which is ruled out experimentally.

The question then becomes: How should we break SUSY? It turns out that the simple mechanisms of SSB described in Chapter 14 are not phenomenologically viable, and we are forced to include terms that explicitly break SUSY. Of course, we shall include only terms that break SUSY softly because we don't want to ruin the most attractive aspect of SUSY as far as phenomenology is concerned: the cancellation of quadratic divergences. We will discuss these issues in more details in Chapter 16.

15.9 Quiz

1. Why are two Higgs superfields required in the MSSM?

2. How many new parameters (relative to the standard model) did we need to introduce to build its supersymmetric extension?

3. What are the R-parity of the left-chiral superfields of the MSSM? Write down a dimension 5 interaction that would respect R-parity.

4. What are the quantum numbers of the two Higgs superfields?

5. How do the Grassmann variables θ and $\bar{\theta}$ transform under R-parity?

CHAPTER 16

Some Phenomenological Implications of the MSSM

In this chapter we will briefly discussed some phenomenological implications of the minimal supersymmetric standard model (MSSM). This is a vast subject that could easily fill an entire book just by itself. A choice therefore had to be made between providing an overview of a large number of results and giving an indepth coverage of only one or two aspects of particular interest. We have opted for the second approach because it is more appropriate for a self-teaching book whose purpose is to demystify a topic. Moreover, there are already many good reviews available, several of which are mentioned in Section 16.7, whose purpose is precisely to offer summaries of the MSSM phenomenology.

16.1 Supersymmetry Breaking in the MSSM

We learned about two mechanisms to spontaneously break supersymmetry (SUSY) in Chapter 14. But, as we also discussed, both lead to the prediction that some superpartners would be lighter than the known leptons, a prediction that is in conflict with experimental observations. The prediction of the existence of light superpartners was a consequence of the vanishing of the supertrace. In the case of F-type breaking, the vanishing of the supertrace is automatic, whereas in the case of D-type breaking, it follows from the fact that the hypercharges of the MSSM particles add up to zero, making the right-hand side of Eq. (14.24) vanish. The fact that the hypercharges add up to zero is bad news for spontaneous supersymmetry breaking (SSB), but it is necessary for the cancellation of certain anomalies.

As discussed in Section 14.9, one way out is to introduce explicit SUSY-breaking terms that we take to break SUSY softly, in order to retain the nice ultraviolet behavior that cures the hierarchy problem. As also mentioned in that section, the presence of these terms may be justified in the context of models where there is a hidden sector interacting with the MSSM particles through some messenger interaction. This topic is beyond our scope (see Refs. 43 and 47 for more details); we will simply introduce the soft SUSY-breaking terms by hand without justifying their existence any further.

Recall the three types of possible soft SUSY-breaking terms, as described in Section 14.9. These are mass terms for the scalar particles (by *mass terms*, we include terms that are bilinear in two scalar fields), interaction terms between scalar fields that are holomorphic and superrenormalizable (in other words, terms of the form $\phi_i \phi_j \phi_k$, where the indices don't have to be different), and finally, mass terms for the spin 1/2 superpartners of the gauge fields, the gauginos. Because the mass terms and the cubic interactions may, in principle, mix different families, this amounts to a very large number of new parameters, 105 to be precise! More specifically, there are

- 5 real parameters and 3 phases that violate charge conjugation C and parity P (*CP* violation) in the fermion sector of the gauge and Higgs supermultiplets (i.e., in the higgsinos and gauginos sector)
- 21 masses, 36 mixing angles, and 40 CP-violating phases in the slepton and squark sector

These parameters are, of course, constrained by experimental results on *CP* violation, flavor changing neutral currents, etc. But there is still quite a bit of leeway. A popular model is *minimal supergravity* or *minimal SUGRA*, in which there are only *five* additional parameters relative to the MSSM. More detail can be found in Section 9.2 of Ref. 1 and in Section 8.1.1 of Ref. 5, as well as in Ref. 10.

Now that SUSY breaking is "taken care of," we need to make sure that the electroweak gauge symmetry is also broken and that the predicted masses for the gauge bosons and Higgs fields do not conflict with experiments. We will therefore focus on this issue in the next section.

16.2 The Scalar Potential, Electroweak Symmetry Breaking, and All That

We have many fields that contribute to the scalar potential: the sleptons, the squarks, and the Higgs fields, which means that the scalar potential is fairly long. However, we are interested only in the value of the potential when the fields are set equal to their vacuum expectation values (vevs) to see if the electroweak gauge symmetry is broken spontaneously.

One thing is clear: We do not want to spontaneously break down charge or color invariance. Therefore, we must take all the scalar fields that have a nonzero charge or that carry color to have vanishing vevs. This includes all sleptons (except the sneutrino) as well as the squarks and the charged Higgs. In addition, we do not want R-parity to be broken spontaneously, to avoid the reappearance of baryon and lepton number violating terms, so we take the vev of the sneutrino, which is odd under R-parity, to be zero as well. We therefore only need to include the Higgs field terms when we write down the scalar potential. Note that the only fields that may have a nonzero vev are the neutral Higgs scalar fields H_u^0 and H_d^0 with hypercharges 1 and -1, respectively.

Four types of terms contribute to the scalar potential after elimination of the auxiliary fields:

$$
V = \sum_i \left| \frac{\partial W}{\partial \phi_i} \right|^2 + \frac{1}{2} D^2 + \frac{1}{2} D^a D^a + V_{\text{SSB}}
$$

$$
= \sum_i \left| \frac{\partial W}{\partial \phi_i} \right|^2 + \frac{g'^2}{2} \left(\sum_i \frac{Y_i}{2} H_i^\dagger H_i \right)^2
$$

$$
+ \frac{g^2}{2} \left(H_u^\dagger \frac{\tau^i}{2} H_u + H_d^\dagger \frac{\tau^i}{2} H_d \right) \left(H_u^\dagger \frac{\tau^i}{2} H_u + H_d^\dagger \frac{\tau^i}{2} H_d \right) + V_{\text{SSB}} \quad (16.1)
$$

where V_{SSB} is the soft SUSY-breaking potential, to which we will come back shortly.

The first D term is for the $U(1)_Y$ gauge symmetry, and the second D term is for the $SU(2)_L$ gauge symmetry (we will be more explicit shortly). There is no corresponding $SU(3)_C$ term because the Higgs fields are color singlets. In this equation, ϕ_i stands for the four Higgs fields H_u^+, H_u^0, H_d^-, H_d^0 and the $SU(2)$ doublets H_u and H_d are defined as

$$ H_u \equiv \begin{pmatrix} H_u^+ \\ H_u^0 \end{pmatrix} \quad \text{and} \quad H_d \equiv \begin{pmatrix} H_d^0 \\ H_d^- \end{pmatrix} \tag{16.2} $$

Now let's consider each term in Eq. (16.2) in turn. To calculate the first term, we simply express the part of the superpotential that contains the Higgs superfields in terms of the corresponding scalar fields. The term in the superpotential that contains only the Higgs superfields is

$$ \mu\, \mathcal{H}_u \circ \mathcal{H}_d \big|_F + \text{h.c.} = \sum_i \mu\, \mathcal{H}_u^T\, i\tau_2\, \mathcal{H}_d + \text{h.c.} \tag{16.3} $$

Now we replace the superfield doublets by their scalar field components H_u and H_d given in Eq. (16.2) and expand:

$$ \mathcal{W} = \sum_i \mu \big(H_u^+ H_d^- - H_u^0 H_d^0 \big) + \text{h.c.} \tag{16.4} $$

The first term of Eq. (16.2) is therefore quite simple:

$$ \left| \frac{\partial \mathcal{W}}{\partial \phi_i} \right|^2 = |\mu|^2 \big(|H_u^0|^2 + |H_u^+|^2 + |H_d^0|^2 + |H_d^-|^2 \big) $$

The $U(1)_Y$ contribution in Eq. (16.1) is

$$ \frac{g'^2}{8} \big(|H_u^+|^2 + |H_u^0|^2 - |H_d^0|^2 - |H_d^-|^2 \big)^2 $$

where we used the fact that the hypercharge assignments of H_u and H_d are 1 and -1, respectively.

The $SU(2)$ contribution is

$$ \frac{g^2}{2} \left(H_u^\dagger \frac{\tau^i}{2} H_u + H_d^\dagger \frac{\tau^i}{2} H_d \right) \left(H_u^\dagger \frac{\tau^i}{2} H_u + H_d^\dagger \frac{\tau^i}{2} H_d \right) $$

Expanding this out, we get

$$\frac{g^2}{8}\left(|H_u^+|^2 + |H_u^0|^2 - |H_d^0|^2 - |H_d^-|^2\right)^2$$

$$+ \frac{g^2}{2}\left(H_u^+ H_d^{0\dagger} + H_u^0 H_d^{-\dagger}\right)\left(H_u^{+\dagger} H_d^0 + H_u^{0\dagger} H_d^-\right)$$

The SUSY-breaking terms are

$$V_{\text{SSB}} = \delta m_{H_u}^2 H_u^\dagger H_u + \delta m_{H_d}^2 H_d^\dagger H_d + (bH_u \circ H_d + \text{h.c.}) \qquad (16.5)$$

Watch out here! The invariant $SU(2)$ dot product between two $SU(2)$ doublets involved the usual $i\tau_2$, but the $SU(2)$-invariant dot product between a doublet and an antidoublet involves the *identity* 2×2 matrix, which is why there is no dot-product symbol or matrices between the doublets in the first two terms of Eq. (16.5). Expanding, we have

$$V_{\text{SSB}} = \delta m_{H_u}^2 \left(H_u^{+\dagger} \quad H_u^{0\dagger}\right)\mathbf{1} \begin{pmatrix} H_u^+ \\ H_u^0 \end{pmatrix} + \delta m_{H_d}^2 \left(H_d^{0\dagger} \quad H_d^{-\dagger}\right)\mathbf{1} \begin{pmatrix} H_d^0 \\ H_d^- \end{pmatrix}$$

$$+ b\left(H_u^+ \quad H_u^0\right)i\tau^2 \begin{pmatrix} H_d^0 \\ H_d^- \end{pmatrix} - b^*\left(H_d^{0\dagger} \quad H_d^{-\dagger}\right)i\tau^2 \begin{pmatrix} H_u^{+\dagger} \\ H_u^{0\dagger} \end{pmatrix} \qquad (16.6)$$

where the extra minus sign in the last term comes from the fact that the dagger of $i\tau_2$ is $-i\tau_2$. So we finally get

$$V_{\text{SSB}} = \delta m_{H_u}^2 |H_u^+|^2 + \delta m_{H_u}^2 |H_u^0|^2$$

$$+ \delta m_{H_d}^2 |H_d^0|^2 + \delta m_{H_d}^2 |H_d^-|^2 + b(H_u^+ H_d^- - H_u^0 H_d^0 + \text{h.c.})$$

Combining all terms together, we find that the full scalar potential is

$$V = \left(|\mu|^2 + \delta m_{H_u}^2\right)\left(|H_u^+|^2 + |H_u^0|^2\right) + \left(|\mu|^2 + \delta m_{H_d}^2\right)\left(|H_d^0|^2 + |H_d^-|^2\right)$$

$$+ \left(b\left(H_u^+ H_d^- - H_u^0 H_d^0\right) + \text{h.c.}\right)$$

$$+ \frac{g^2 + g'^2}{8}\left(|H_u^+|^2 + |H_u^0|^2 - |H_d^0|^2 - |H_d^-|^2\right)^2 + \frac{g^2}{2}|H_d^{0\dagger} H_u^+ + H_d^{-\dagger} H_u^0|^2$$

Let us define

$$a_1 \equiv |\mu|^2 + \delta m_{H_u}^2$$

$$a_2 \equiv |\mu|^2 + \delta m_{H_d}^2$$

$$c \equiv \frac{g^2 + g'^2}{8} \tag{16.7}$$

Using this notation, the full scalar potential is

$$V = a_1 \left(|H_u^+|^2 + |H_u^0|^2 \right) + a_2 \left(|H_d^0|^2 + |H_d^-|^2 \right)$$

$$+ b \left(H_u^+ H_d^- - H_u^0 H_d^0 + \text{h.c.} \right) + c \left(|H_u^+|^2 + |H_u^0|^2 - |H_d^0|^2 - |H_d^-|^2 \right)^2$$

$$+ \frac{g^2}{2} |H_d^{0\dagger} H_u^+ + H_d^{-\dagger} H_u^0|^2 \tag{16.8}$$

Note that the full scalar potential contains five parameters. Of those, c and g have known values; $c \approx 0.069$ and $g \approx 0.651$. The other three are free parameters and could be either positive or negative.

16.3 Finding a Minimum

The situation is obviously quite a bit more complex than in the standard model because we have twice as many Higgs fields.

As in the standard model, the equations are simplified if we use the $SU(2)$ invariance to set a component of *either* H_u or H_d equal to zero at the minimum (in other words, we can set the vev of one component to be zero without any loss of generality). Let's choose to set $H_u^+ = 0$, which, of course, also implies $H_u^{+\dagger} = 0$ at the minimum.

In addition, we will impose that H_d^- has a zero vev because otherwise electromagnetism would be broken spontaneously, which is not observed experimentally. Of course, if we were to find out that it is impossible to have spontaneous symmetry breaking of $SU(2)_L \times U(1)_Y$ to $U(1)_{em}$ if we impose $\langle H_d^- \rangle = 0$, the MSSM would be ruled out experimentally from the very start. As you might expect, this won't be the case, and we will indeed find that a zero value of the vev of H_d^- does not clash with SSB of the gauge symmetry.

A first consistency check is to verify that the full potential has indeed an extremum at $H_u^+ = H_d^- = 0$ (from now on, we drop the expectation value symbols around the vevs). In other words, we must verify that the partial derivatives

$$\frac{\partial V}{\partial H_u^+} \qquad \frac{\partial V}{\partial H_d^-} \qquad \frac{\partial V}{\partial H_u^{+\dagger}} \qquad \text{and} \qquad \frac{\partial V}{\partial H_d^{-\dagger}}$$

vanish when we set $H_u^+ = H_d^- = 0$ (and, of course, $H_u^{+\dagger} = H_d^{-\dagger} = 0$ as well). That this is actually the case is easy to see just by looking at the potential because there are no terms containing only one of these four fields.

Therefore, at the minimum, the Higgs potential reduces to

$$V_n = a_1 |H_u^0|^2 + a_2 |H_d^0|^2 - b\big(H_u^0 H_d^0 + \text{h.c.}\big) + c\big(|H_u^0|^2 - |H_d^0|^2\big)^2$$

where the label n is to indicate that this potential contains only the neutral fields H_u^0 and H_d^0.

Note that the only term that depends on the phase of the fields is the term proportional to b. Without loss of generality, we then can take b to be real and positive because its phase always can be absorbed into the phase of the fields. Actually, $b = 0$ is also possible (nothing says that this term *must* be present), so we take

$$b \geq 0$$

We will make one last simplification before finally working out the pattern of SSB: We will assume that CP (the symmetry under the combined operations of parity and charge conjugation) is not broken spontaneously, as is the case in the standard model. The consequence is that any nonzero vev must be real, in which case the potential at the minimum is simply

$$V_n = a_1\big(H_u^0\big)^2 + a_2\big(H_d^0\big)^2 - 2b H_u^0 H_d^0 + c\big[\big(H_u^0\big)^2 - \big(H_d^0\big)^2\big]^2 \qquad (16.9)$$

Now we want to investigate if there are stable minima for nonzero values of H_u^0 and H_d^0. To simplify the notation, let's write the potential as

$$V_n = a_1 x^2 + a_2 y^2 - 2bxy + c(x^2 - y^2)^2 \qquad (16.10)$$

Note that b and c are positive, whereas a_1 and a_2 could be either positive or negative [see Eq. (16.7)].

One thing we must ensure is that the potential goes to *plus* infinity as we move away from the origin in any direction. If we fix, say, x to any value and take y to infinity (or vice versa), the last term, $c(x^2 - y^2)^2$, will always dominate all the other terms, and since c is positive, we are safe. However, if we move away from the origin along the line $x = y$, this term vanishes, and we must be more careful. In this case, the potential reduces to

$$V_n(x = y) = (a_1 + a_2 - 2b)x^2$$

This is a parabola. If we want this to have a minimum (as opposed to a maximum), the overall coefficient must be positive. This implies that $a_1 + a_2 > 2b$. And since b is itself larger or equal to zero, we have

$$\boxed{a_1 + a_2 > 2b \geq 0} \tag{16.11}$$

Recalling the definitions in Eq. (16.7), this translates into

$$2|\mu|^2 + \delta m_{H_u}^2 + \delta m_{H_d}^2 > 2b > 0$$

There are three possibilities:

- Both a_1 and a_2 are positive with $a_1 + a_2 > 2b$
- a_1 is positive and a_2 negative with $a_1 > 2b + |a_2|$
- a_1 is negative and a_2 positive with $a_2 > 2b + |a_1|$

Now that we have ensured that the potential goes to infinity in any direction, we also want to ensure that the origin, $x = y = 0$, is *not* a minimum because then the vev of all the fields would be zero, and there would be no spontaneous symmetry breaking of the gauge symmetry. We already know that along the line $y = x$, we have a minimum at the origin. The only way for the origin not to be a minimum of the potential (along any direction) is to impose it to be a saddle point at the origin, which leads to the condition

$$\left. \begin{vmatrix} \dfrac{\partial^2 V_n}{\partial x^2} & \dfrac{\partial^2 V_n}{\partial x \partial y} \\[2ex] \dfrac{\partial^2 V_n}{\partial x \partial y} & \dfrac{\partial^2 V_n}{\partial y^2} \end{vmatrix} \right|_{x=y=0} < 0$$

which leads to the condition

$$\boxed{a_1 a_2 < b^2} \tag{16.12}$$

or, equivalently,

$$\left(|\mu|^2 + \delta m_{H_u}^2\right)\left(|\mu|^2 + \delta m_{H_d}^2\right) < b^2$$

To summarize, conditions (16.11) and (16.12) are required in order for our potential to have a minimum and for that minimum to *not* be at the origin.

Let us now assume that both conditions are satisfied and look for a minimum with nonzero values of x and y. Recall that x denotes H_u^0 and y represents H_d^0, so let us call the coordinates of the minimum $x_{\min} \equiv v_u$ and $y_{\min} \equiv v_d$. So we impose

$$\left.\frac{\partial V_n}{\partial x}\right|_{x=v_u, y=v_d} = \left.\frac{\partial V_n}{\partial y}\right|_{x=v_u, y=v_d} = 0$$

This gives the two conditions

$$a_1 v_u - b v_d + 2c v_u \left(v_u^2 - v_d^2\right) = 0$$
$$a_2 v_d - b v_u - 2c v_d \left(v_u^2 - v_d^2\right) = 0 \tag{16.13}$$

Thus, at this point, we have seven independent parameters: the five parameters of the scalar potential [see Eq. (16.8)], which are $a_1, a_2, b, c,$ and g, and the two vevs v_u and v_d. As constraints, we have the two equalities (16.13) as well as the two inequalities, (16.11) and (16.12). We will soon come back to these constraints, but first, let us find what the masses of the gauge bosons can tell us about our parameters.

16.4 The Masses of the Gauge Bosons

Calculation of the gauge boson masses proceeds along exactly the same lines as in the standard model (see Section 15.2). The MSSM lagrangian contains the following kinetic terms for the Higgs scalars:

$$(D_\mu H_u)^\dagger (D^\mu H_u) + (D_\mu H_d)^\dagger (D^\mu H_d) \tag{16.14}$$

where

$$D_\mu = \partial_\mu + ig\frac{\vec{\tau} \cdot \vec{W}_\mu}{2} + i\frac{g'}{2}Y B_\mu$$

$$= \partial_\mu + \frac{ig}{2}\begin{pmatrix} W_3 & W_{1\mu} - i W_{2\mu} \\ W_{1\mu} + i W_{2\mu} & -W_{3\mu} \end{pmatrix} + \frac{ig'Y}{2}B_\mu\begin{pmatrix} 1 & 0 \\ 0 & 1 \end{pmatrix} \tag{16.15}$$

Supersymmetry Demystified

Recall that

$$H_u = \begin{pmatrix} H_u^+ \\ H_u^0 \end{pmatrix}$$

At the minimum of the potential, this becomes

$$H_u = \begin{pmatrix} 0 \\ v_u \end{pmatrix}$$

Similarly, H_d at the minimum of the potential is

$$H_d = \begin{pmatrix} v_d \\ 0 \end{pmatrix}$$

Using the fact that the hypercharge of H_u is equal to 1, we get

$$D_\mu \begin{pmatrix} 0 \\ v_u \end{pmatrix} = \frac{ig}{2} v_u \begin{pmatrix} W_{1\mu} - iW_{2\mu} \\ -W_{3\mu} \end{pmatrix} + \frac{ig'}{2} v_u B_\mu \begin{pmatrix} 0 \\ 1 \end{pmatrix}$$

and using the fact that H_d has a hypercharge of -1,

$$D_\mu \begin{pmatrix} v_d \\ 0 \end{pmatrix} = \frac{ig}{2} v_d \begin{pmatrix} W_{3\mu} \\ W_{1\mu} + iW_{2\mu} \end{pmatrix} - \frac{ig'}{2} v_d B_\mu \begin{pmatrix} 1 \\ 0 \end{pmatrix}$$

Inserting these expressions in Eq. (16.14), we get the following contribution to the masses of the gauge bosons:

$$\mathcal{L}_{\mathrm{MGB}} = \frac{g^2}{4} (v_u^2 + v_d^2)(W_{1\mu} W_1^\mu + W_{2\mu} W_2^\mu + W_{3\mu} W_3^\mu)$$

$$- \frac{gg'}{2} (v_u^2 + v_d^2) W_{3\mu} B^\mu + \frac{g'^2}{4} (v_u^2 + v_d^2) B^\mu B_\mu$$

We expect to be able to rewrite this in a form similar to Eq. (15.13), and indeed, we can reorganize the terms to get

$$\mathcal{L}_{\mathrm{MGB}} = \frac{g^2}{4} (v_u^2 + v_d^2)(W_1^2 + W_2^2) + \frac{(v_u^2 + v_d^2)}{4} (g^2 W_3^2 - 2gg' W_{3\mu} B^\mu + g'^2 B^2)$$

$$= \frac{g^2}{4} (v_u^2 + v_d^2)(W_1^2 + W_2^2) + \frac{(v_u^2 + v_d^2)}{4} (g W_3 - g' B)^2 \qquad (16.16)$$

Following exactly the same steps as in Section 15.2, we get the same expressions for the $W^{\pm}$, Z, and A^{μ} (photon fields) with masses now given by

$$m_W^2 = \frac{g^2}{2}\left(v_u^2 + v_d^2\right)$$

$$m_Z^2 = (g^2 + g'^2)\frac{\left(v_u^2 + v_d^2\right)}{2}$$

and, of course, the mass of the photon being zero. We see that these relations *are exactly the same as in the standard model with* v^2 *replaced by* $v_u^2 + v_d^2$. Let us *define*

$$v_{susy}^2 \equiv v_u^2 + v_d^2$$

so that the relations for the gauge boson masses have exactly the same form as in the standard model with v^2 replaced by v_{susy}^2. Note that

$$c = \frac{m_Z^2}{4v_{susy}^2} \tag{16.17}$$

In addition, studying the weak interaction at low energies yields of course, the same relation with Fermi's constant as in the standard model [Eq. (15.15)]. Using those results, we see that the parameters g, g', and the combination v_{susy} are all fixed by experiments and have the same values as in the standard model. It is customary to trade the parameters g and g' (and therefore c) for v_{susy} and the gauge boson masses m_W and m_Z, so we will take our seven parameters to be

$$m_Z, m_W, v_{susy}, a_1, a_2, b$$

plus any combination of v_u and v_d that is linearly independent of v_{susy}^2. As we will see below, the ratio v_u/v_d proves convenient.

We now turn our attention to the two conditions (16.13) and the two inequalities (16.11) and (16.12). Let us first go back to Eq. (16.13). The first one is

$$a_1 v_u - b v_d + 2c v_u\left(v_u^2 - v_d^2\right) = 0$$

Isolating a_1, we get [using Eq. (16.17)]

$$a_1 = b\frac{v_d}{v_u} - 2c\left(v_u^2 - v_d^2\right) = b\frac{v_d}{v_u} - \frac{\left(v_u^2 - v_d^2\right)}{\left(v_u^2 + v_d^2\right)}\frac{m_Z^2}{2}$$

Note that it might seem at first sight that a_1 depends on four parameters, but actually, it depends only on three because v_u and v_d only appear in the ratio v_u/v_d. Thus we can rewrite a_1 in terms of only three parameters: b, m_Z, and the dimensionless ratio v_u/v_d. Let us define

$$\tan \beta \equiv \frac{v_u}{v_d}$$

Since v_u and v_d are real and positive quantities, $0 \leq \beta \leq \pi/2$. All expressions that depend only on the ratio v_u/v_d can be expressed in terms of the angle β. A few useful relations are

$$\frac{v_d}{v_u} = \cot \beta \qquad \frac{v_u^2}{v_u^2 + v_d^2} = \sin^2 \beta \qquad \frac{v_d^2}{v_u^2 + v_d^2} = \cos^2 \beta$$

$$\frac{\left(v_u^2 - v_d^2\right)}{\left(v_u^2 + v_d^2\right)} = \sin^2 \beta - \cos^2 \beta = -\cos 2\beta$$

$$\frac{v_d^2 - v_u^2}{v_u v_d} = \cot \beta - \tan \beta = 2 \cot 2\beta$$

$$\frac{v_u^2 + v_d^2}{v_u v_d} = \tan \beta + \cot \beta = 2 \operatorname{cosec} 2\beta \tag{16.18}$$

We therefore finally get

$$a_1 = b \cot \beta + \frac{m_Z^2}{2} \cos 2\beta \tag{16.19}$$

Similarly, we find [still using Eqs. (16.13), (16.17) and (16.18)]

$$a_2 = b \tan \beta - \frac{m_Z^2}{2} \cos 2\beta \tag{16.20}$$

Let's now turn our attention to the inequalities (16.11) and (16.12). Using the preceding expressions for a_1 and a_2, Eq. (16.11) translates into

$$\frac{1}{\sin 2\beta} > 1$$

This is satisfied at the condition that $\beta \neq \pi/4$, or in other words, the two vevs v_u and v_d *must not be equal*:

$$\boxed{v_u \neq v_d}$$
(16.21)

The inequality (16.12) leads to the condition

$$b^2 - m_Z^2 b \cos(2\beta) \cot 2\beta - \frac{m_Z^4}{4} \cos^2(2\beta) < b^2$$

which implies

$$m_Z^2 b \cos(2\beta) \cot 2\beta + \frac{m_Z^2}{4} \cos^2(2\beta) > 0$$

Since $b \geq 0$ and $\beta \leq \pi/2$, this is always satisfied except at $\beta = \pi/4$, once more. Thus we get again the condition (16.21).

To summarize, we started this section with seven parameters: a_1, a_2, b, m_Z, m_W, and v_{susy}^2 and any combination of v_u and v_d independent of v_{susy}^2. We have seen that it proves convenient to choose for this combination the ratio v_u/v_d, which can be expressed in terms of the angle β. We then used the constraints on the vevs to eliminate a_1 and a_2 in terms of the other parameters. We therefore have five parameters left:

$$m_W^2, \ m_Z^2, \ v_{\text{susy}}^2, \ b, \ \beta$$

Of these, the first three are determined experimentally, leaving us with two free parameters with only the constraint $\beta \neq \pi/4$.

It turns out that when we determine the masses of the Higgs bosons, we will be able to express everything in terms of only four of these parameters: m_Z, m_W, β, and b; the parameter v_{susy}^2 will play no role.

Note that all our results are only valid at tree level. Therefore the restriction $\beta \neq \pi/4$ (and therefore $v_u \neq v_d$) should not be seen as exact. In any case, it makes more sense to ignore it given that an incredible fine-tuning would be required for the vevs to correspond to the only point of parameter space that needs to be excluded! We won't discuss loop corrections except for a small comment at the end of Section 16.5.

16.5 Masses of the Higgs

It is important to keep in mind that the four scalar fields $(H_u^0, H_u^+, H_d^0, H_d^0)$ are all complex, so there are eight degrees of freedom. To make this explicit, it is convenient to express all the fields in terms of their real and imaginary parts. To simplify our calculations, we will use the notation

$$H_u^0 = \frac{1}{\sqrt{2}}\left(R_u^0 + i I_u^0\right)$$

where

$$R_u^0 \equiv \operatorname{Re}\left(H_u^0\right) \qquad I_u^0 \equiv \operatorname{Im}\left(H_u^0\right)$$

and so on. The complete Higgs potential (not evaluated at the minimum) given in Eq. (16.8) then takes the form

$$
\begin{aligned}
V = & \frac{a_1}{2}\left(\left(R_u^+\right)^2 + \left(I_u^+\right)^2 + \left(R_u^0\right)^2 + \left(I_u^0\right)^2\right) + \frac{a_2}{2}\left(\left(R_d^0\right)^2 + \left(I_d^0\right)^2 + \left(R_d^-\right)^2 + \left(I_d^-\right)^2\right) \\
& + b\left(R_u^+ R_d^- - I_u^+ I_d^- - R_u^0 R_d^0 + I_u^0 I_d^0\right) \\
& + \frac{c}{4}\left(\left(R_u^+\right)^2 + \left(I_u^+\right)^2 + \left(R_u^0\right)^2 + \left(I_u^0\right)^2 - \left(R_d^0\right)^2 - \left(I_d^0\right)^2 - \left(R_d^-\right)^2 - \left(I_d^-\right)^2\right)^2 \\
& + \frac{g^2}{8}\left(R_d^0 R_u^+ + I_d^0 I_u^+ + R_d^- R_u^0 + I_d^- I_u^0\right)^2 \\
& + \frac{g^2}{8}\left(R_d^0 I_u^+ - I_d^0 R_u^+ - I_d^- R_u^0 + I_u^0 R_d^-\right)^2
\end{aligned}
\tag{16.22}
$$

We have not expressed a_1 and a_2 and g and c in terms of $v_{\text{susy}}, m_W, m_Z, \beta$, and b because this makes the intermediate calculations cumbersome. We will change parameters when we express our final results for the Higgs masses.

The only fields with a nonzero vev are R_u^0 and R_d^0:

$$\left\langle R_u^0\right\rangle = \sqrt{2} v_u$$

$$\left\langle R_d^0\right\rangle = \sqrt{2} v_d$$

To work out the masses of all the Higgs fields, we proceed as in Section 14.4: We take the second derivative of the potential with respect to all the Higgs fields and evaluate the resulting expression at the minimum, $R_u^0 = \sqrt{2} v_u, R_d^0 = \sqrt{2} v_d$ and all the other fields set to zero. Only a fraction of the terms appearing in Eq. (16.22) survive when we do that. They can be grouped into three categories:

- The terms quadratic in a field. These are

$$\frac{a_1}{2}\left((R_u^+)^2 + (I_u^+)^2 + (R_u^0)^2 + (I_u^0)^2\right) + \frac{a_2}{2}\left((R_d^0)^2 + (I_d^0)^2 + (R_d^-)^2 + (I_d^-)^2\right)$$

- The terms containing the product of two different fields. These are

$$b\left(R_u^+ R_d^- + I_u^0 I_d^0 - R_u^0 R_d^0 - I_u^+ I_d^-\right)$$

- The quadratic terms that have at least two fields that are either R_u^0 or R_d^0. These are

$$\frac{c}{2}\left((R_u^0)^2 - (R_d^0)^2\right)\left((R_u^+)^2 + I_u^{+2} + I_u^{02} - I_d^{02} - R_d^{-2} - I_d^{-2}\right)$$

$$+\frac{c}{4}\left((R_u^0)^2 - (R_d^0)^2\right)^2 + \frac{g^2}{8}\left((R_u^+)^2(R_d^0)^2 + (R_u^0)^2(R_d^-)^2\right.$$

$$\left.+ 2R_u^+ R_d^0 R_u^0 R_d^- + (I_u^+)^2(R_d^0)^2 + (R_u^0)^2(I_d^-)^2 - 2I_u^+ I_d^- R_u^0 R_d^0\right)$$

All the other terms will not contribute once we take the second derivatives with respect to the fields and set them equal to their vevs.

We see that there is a mixing of masses between the pairs (I_u^0, I_d^0), (R_u^0, R_d^0), $(R_u^+ R_d^-)$, and (I_u^+, I_d^-), so we will consider each pair separately.

Consider first the pair (I_u^0, I_d^0). The relevant terms are

$$\frac{a_1}{2}(I_u^0)^2 + \frac{a_2}{2}(I_d^0)^2 + bI_u^0 I_d^0 + \frac{c}{2}\left((R_u^0)^2 - (R_d^0)^2\right)\left((I_u^0)^2 - (I_d^0)^2\right)$$

So we get [using Eqs. (16.17) to (16.20)]

$$\frac{\partial^2 V}{\partial (I_u^0)^2}\bigg|_{\text{vev}} = a_1 + 2c(v_u^2 - v_d^2) = b \cot \beta$$

$$\frac{\partial^2 V}{\partial (I_d^0)^2}\bigg|_{\text{vev}} = a_2 - 2c(v_u^2 - v_d^2) = b \tan \beta$$

$$\frac{\partial^2 V}{\partial I_u^0 \partial I_d^0}\bigg|_{\text{vev}} = b$$

where the subscript vev means that all the fields are set equal to their vacuum expectation value. The squared mass matrix is then

$$\mathbf{V} = b \begin{pmatrix} \cot \beta & 1 \\ 1 & \tan \beta \end{pmatrix}$$

The squared masses are given by the solutions of

$$\lambda^2 - \lambda b \big(\tan(\beta) + \cot(\beta)\big) = 0 \tag{16.23}$$

which are

$$0 \quad \text{and} \quad 2 b \cosec(2\beta)$$

The massless state provides the longitudinal mode of the Z_0, following the same mechanism as in the standard model. Following a standard practice, we will denote the field corresponding to the massive state by A_0, which has, therefore, a mass squared:

$$m_{A_0}^2 \equiv 2 b \cosec(2\beta) \tag{16.24}$$

We can use this relation to trade the parameter b for the mass m_{A_0}, so that our five parameters are now

$$m_W, m_Z, v_{\text{SUSY}}, \beta, \quad \text{and} \quad m_{A_0} \tag{16.25}$$

We won't work out the explicit mass eigenstates, i.e., the linear combinations of I_u^0 and I_d^0, that correspond to these masses.

Since in principle β can be very small, there is no upper limit on this mass. Next, consider the fields R_u^0 and R_d^0. The relevant terms are

$$\frac{a_1}{2} \left(R_u^0\right)^2 + \frac{a_2}{2} \left(R_d^0\right)^2 - b \, R_u^0 R_d^0 + \frac{c}{4} \left(\left(R_u^0\right)^2 - \left(R_d^0\right)^2\right)^2$$

This time, we find the squared mass matrix to be more complicated. We get (on the basis of R_u^0, R_d^0)

$$V_{11} = a_1 + \left(6 v_u^2 - 2 v_d^2\right) c = b \cot \beta + \sin^2(\beta) \, m_Z^2$$

$$V_{22} = a_2 + \left(6 v_d^2 - 2 v_u^2\right) c = b \tan \beta + \cos^2(\beta) \, m_Z^2$$

$$V_{12} = V_{21} = -4 v_u v_d \, c - b = -\frac{\sin 2\beta}{2} m_Z^2 - b \tag{16.26}$$

We must find the eigenvalues of the matrix

$$
\mathbf{V} = \begin{pmatrix} b\cot\beta + \sin^2(\beta)m_Z^2 & -\dfrac{\sin(2\beta)}{2}m_Z^2 - b \\[2mm] -\dfrac{\sin(2\beta)}{2}m_Z^2 - b & b\tan\beta + \cos^2(\beta)m_Z^2 \end{pmatrix}
$$

which can be shown to be

$$
b\sec 2\beta + \frac{m_Z^2}{2} \pm \frac{1}{2}\sqrt{(2b\cot 2\beta + m_Z^2\cos(2\beta))^2 + 4(\sin(2\beta)\,m_Z^2/2 + b)^2}
$$

$$(16.27)$$

The Higgs fields corresponding to those two masses are referred to as the H_0 [when the plus sign is used in Eq. (16.27)] and the h_0 (when the minus sign is used). The expression for the masses simplifies tremendously if we express b in terms of m_{A_0} using Eq. (16.24), which leads to

$$
\frac{m_{A_0}^2 + m_Z^2}{2} \pm \frac{1}{2}\sqrt{\left(m_{A_0}^2 - m_Z^2\right)^2 \cos^2 2\beta + \left(m_{A_0}^2 + m_Z^2\right)^2 \sin^2 2\beta}
$$

Writing $\sin^2 2\beta$ as $1 - \cos^2 2\beta$, we finally get the two masses to be

$$
m_{h^0}^2 = \frac{m_{A_0}^2 + m_Z^2}{2} - \frac{1}{2}\sqrt{\left(m_{A_0}^2 + m_Z^2\right)^2 - 4\,m_{A_0}^2\,m_Z^2\,\cos^2 2\beta}
$$

and

$$
m_{H^0}^2 = \frac{m_{A_0}^2 + m_Z^2}{2} + \frac{1}{2}\sqrt{\left(m_{A_0}^2 + m_Z^2\right)^2 - 4\,m_{A_0}^2 m_Z^2\,\cos^2 2\beta}
\qquad (16.28)
$$

Consider now the pair $(R_u^+ R_d^-)$. The relevant terms in the lagrangian are

$$
\frac{a_1}{2}\left(R_u^+\right)^2 + \frac{a_2}{2}\left(R_d^-\right)^2 + bR_u^+ R_d^- + \frac{c}{2}\left(\left(R_u^0\right)^2 - \left(R_d^0\right)^2\right)\left(\left(R_u^+\right)^2 - \left(R_d^-\right)^2\right)
$$

$$
+ \frac{g^2}{8}\left(\left(R_u^+\right)^2\left(R_d^0\right)^2 + \left(R_u^0\right)^2\left(R_d^-\right)^2 + 2R_u^+ R_d^0 R_u^0 R_d^-\right)
$$

so that

$$\left.\frac{\partial^2 V}{\partial \left(R_u^+\right)^2}\right|_{vev} = a_1 + 2c\left(v_u^2 - v_d^2\right) + \frac{g^2}{2}v_d^2 = b\cot\beta + m_W^2\cos^2\beta$$

$$\left.\frac{\partial^2 V}{\partial \left(R_d^-\right)^2}\right|_{vev} = a_2 - 2c\left(v_u^2 - v_d^2\right) + \frac{g^2}{2}v_u^2 = b\tan\beta + m_W^2\sin^2\beta$$

$$\left.\frac{\partial^2 V}{\partial R_u^+ \partial R_d^-}\right|_{vev} = b + \frac{g^2}{2}v_u v_d = b + \frac{m_W^2}{2}\sin(2\beta) \tag{16.29}$$

This gives the following mass matrix:

$$\mathbf{V} = \begin{pmatrix} b\cot\beta + m_W^2\cos^2\beta & b + m_W^2\sin(2\beta)/2 \\ b + m_W^2\sin(2\beta)/2 & b\tan\beta + m_W^2\sin^2\beta \end{pmatrix}$$

The eigenvalues are

$$b\,\sec(2\beta) + \frac{m_W^2}{2} \pm \frac{1}{2}\sqrt{\left[2\cot(2\beta)b + m_W^2\cos(2\beta)\right]^2 + 4\left[b + m_W^2\sin(2\beta)/2\right]^2}$$

Again, using $b = m_{A^0}^2 \sin(2\beta)/2$ simplifies greatly the result, which turns out to be

$$0 \quad \text{and} \quad m_{A^0}^2 + m_W^2$$

The state with zero mass gets eaten through the Higgs mechanism. The massive state is a new scalar field. Because its mass contains m_{A^0}, it does not have any upper limit.

Finally, we have the pair I_u^+, I_d^-. It's easy to see that we get all the same terms as for the pair R_u^+, R_d^- except that the cross-terms in $I_u^+ I_d^-$ have the opposite signs as the cross-terms in $R_u^+ R_d^-$. This means that the off-diagonal elements of the mass matrix will have the opposite sign to what we previously had, but this does not change the eigenvalues, so we again get a massless state and a massive state with mass $m_{A^0}^2 + m_W^2$.

The two massless states that we just found provide the longitudinal modes of the $W^\pm$. The two charged massive states are referred to as H^+ and H^- with degenerate masses:

$$m_{H^\pm}^2 = m_W^2 + m_{A^0}^2$$

To summarize, we have found that the spectrum of the Higgs sector in the MSSM consists into three massless and five massive states. The three massless states play the same role as the massless states in the standard model: They become the longitudinal modes of the massive $W^{\pm}$ and Z. One massive state, the A^0, has a mass squared

$$m_{A_0}^2 = \frac{2b}{\sin(2\beta)}$$

Since there is no lower limit on β, there is no upper limit on how massive this state could be. This is not very exciting from the point of view of phenomenology because there is no way to "kill" the model: If the A^0 is not observed, it will simply put an upper limit on β. The masses of the other four states all can be expressed in terms of m_W, m_Z, m_{A^0}, and β:

$$m_{H^{\pm}}^2 = m_W^2 + m_{A_0}^2$$

$$m_{H^0} = \frac{m_{A_0}^2 + m_Z^2}{2} + \frac{1}{2}\sqrt{\left(m_{A_0}^2 + m_Z^2\right)^2 - 4m_{A_0}^2 m_Z^2 \cos^2 2\beta}$$

$$m_{h^0}^2 = \frac{m_{A_0}^2 + m_Z^2}{2} - \frac{1}{2}\sqrt{\left(m_{A_0}^2 + m_Z^2\right)^2 - 4m_{A_0}^2 m_Z^2 \cos^2 2\beta} \quad (16.30)$$

If we consider the limit $\beta \to 0$ (and $b \neq 0$), we get the following limits for the Higgs masses:

$$m_{A^0} \to \infty$$

$$m_{H^{\pm}} \to \infty$$

$$m_{H^0} \to \infty$$

$$m_{h^0} \to m_Z|\cos(2\beta)|$$

On the other hand, in the limit $b \to 0$ (and $\beta \neq 0$) we get

$$m_{A^0} \to 0$$

$$m_{H^{\pm}} \to m_W$$

$$m_{H^0} \to m_Z$$

$$m_{h^0} \to 0$$

The important point to notice is that the mass of the h^0 does *not* tend to infinity in any limit. Even better, it turns out that the value corresponding to the first limit,

$m_Z|\cos(2\beta)|$, is the *upper limit* to the mass of the h^0! This can be seen more clearly if we invert the last equation of Eq. (16.30) to get

$$m_{A^0}^2 = m_{h^0}^2 \frac{m_Z^2 - m_{h^0}^2}{m_Z^2 \cos^2(2\beta) - m_{h^0}^2}$$

which shows clearly that

$$m_{h^0} < m_Z|\cos(2\beta)| \qquad (16.31)$$

This means that the MSSM predicts a Higgs particle lighter than the Z! But this is catastrophic because it is already experimentally excluded[4]:

$$m_H \geq 114.4 \ \text{GeV (confidence level of 95\%)}$$

However, it is important to keep in mind that all the results we have obtained in this section are only tree-level relations. They are modified by loop corrections, and sometimes the corrections are sizable. Calculations of loop diagrams in the MSSM are beyond the scope of this book. Let's just mention that the main correction to the bound (16.31) arises from loop diagrams containing the top quark and its supersymmetric partner, the *stop* (or *top squark*), and is given by

$$m_{h^0} < \sqrt{m_Z^2 + \frac{3g^2}{8\pi^2} \frac{m_t^2}{M_W^2} \ \ln\left(\frac{m_{\tilde{t}}^2}{m_t^2}\right)}$$

Note that if SUSY were unbroken, the stop mass would be equal to the top mass, and the logarithmic correction would be identically zero. Using a mass of 1 TeV as an estimate of the stop mass, this brings up the bound on the h^0 to approximately

$$m_{h^0} < 130 \ \text{GeV}$$

which is not yet excluded experimentally.

16.6 Masses of the Leptons, Quarks, and Their Superpartners

Let's first have a look at the masses of the leptons and quarks in the MSSM. Consider the term in the superpotential equal to

$$y_u^{ij} \mathcal{U}_i \left(\mathcal{Q}_j \circ \mathcal{H}_u \right)$$

We would like to extract from this the coupling of the standard model particles to the Higgs bosons. The Higgs bosons are the scalar part of the superfield $\mathcal{H}_u$. The standard model particles are contained in $\mathcal{U}_i$ and $\mathcal{Q}_j$. The F terms containing a scalar field and two fermions when we expand the product of the three chiral superfields, is $-\phi \chi_A \cdot \chi_B / 2$. Therefore, we finally get

$$-\frac{y_u^{ij}}{2} \left(\bar{u}_{Li} \cdot u_j H_u^0 + \bar{u}_{Lj} \cdot u_i H_u^0 - \bar{u}_{Li} \cdot d_j H_u^+ - \bar{u}_{Lj} \cdot d_i H_u^+ \right)$$

The first two terms will generate masses for the fermions. Setting $H_u^0 = v_u$, this gives

$$-\frac{y_u^{ij}}{2} v_u \left(\bar{u}_{Li} \cdot u_j + \bar{u}_{Lj} \cdot u_i \right) = -y_u^{ij} v_u \bar{u}_{Li} \cdot u_j$$

using the symmetry of the Yukawa couplings.

If we add the hermitian conjugate, we get a Dirac mass term. After going to a diagonal basis, this will lead to masses for the up, charm, and top quarks of the form

$$m_{u,c,t} = v_u y_{u,c,t}$$

where the $y_{u,c,t}$ are linear combinations of the y_u^{ij}.

Now consider another term in the superpotential that leads to a fermion mass:

$$-y_d^{ij} \mathcal{D}_i \mathcal{Q}_j \circ \mathcal{H}_d$$

Keeping only the terms that yield fermion masses, we get (we pick up an extra minus sign due to the fact that the vev of $i\tau^2 \mathcal{H}_d$ is a doublet with components

$(0, -v_d)$ in contrast with the vev of $i\tau^2 \mathcal{H}_u$ which had a vev equal to $(v_u, 0)$)

$$-\frac{y_d^{ij}}{2} H_d^0 (\bar{d}_i \cdot d_j + \bar{d}_j \cdot d_i) = -y_d^{ij} v_d \bar{d}_i \cdot d_j$$

If we add the hermitian conjugate, this generates Dirac mass terms. After diagonalization, this yields masses

$$m_{d,s,b} = v_d \, y_{d,s,b}$$

The third and final term will give the masses to the leptons:

$$-y_e^{ij} \mathcal{E}_i \left(\mathcal{L}_j \circ \mathcal{H}_d \right)$$

Again, keeping only the relevant terms (that generate fermion masses), we get

$$-\frac{y_e^{ij}}{2} H_d^0 (\bar{e}_{Li} \cdot e_{Lj} + \bar{e}_{Li} \cdot e_{Li}) = -y_e^{ij} H_d^0 \bar{e}_i \cdot e_{Lj}$$

Again, after setting the Higgs field to its vev, this generates Dirac masses of the form $v_d^0 y_e^{ij}$. After diagonalizing, we get lepton masses

$$m_{e,\mu,\tau} = v_d \, y_{e,\mu,\tau}$$

The end result is that we get *exactly* the same type of fermion mass terms as in the standard model, so in this respect, the MSSM can't be differentiated from the standard model.

On the other hand, once we introduce soft supersymmetry breaking terms, which include mass terms for scalar fields, we lose all predictive power for the masses of the sleptons and squarks.

16.7 Some Other Consequences of the MSSM

We have seen that a prediction of the minimal supersymmetric standard model (MSSM) is a relatively light Higgs scalar field. The phenomenology of the MSSM is extremely rich, and as mentioned earlier, this topic alone could fill an entire book.

Let us only mention again that if *R*-parity is conserved, the lightest supersymmetric particle (LSP) is absolutely stable, and because of this, it should fill the universe and represents therefore a promising dark matter candidate. The exact nature of the LSP is model-dependent, with the two most likely candidates being the neutralino and the gravitino (which appears only in theories of supergravity, i.e., when SUSY is gauged).

Another consequence of R-parity is that the superpartners of the known particles can only be created in pairs. If they are weakly interacting, their existence thus could show up in the form of missing energy in scattering events, missing energy that would be equal to at least twice the mass of the LSP.

Unfortunately, we are running out of space (and I am running out of energy!), so we won't be able to discuss any of these fascinating topics more. To learn more about the phenomenology of the MSSM, you are invited to consult References 1, 5, 8, 10, 12, 29, 32, 35 and 47.

16.8 Coupling Unification in Supersymmetric GUT

We will cover one last topic. If you have read about SUSY in particle physics before, you have probably heard about a very exciting feature of the theory: a near-perfect unification of the coupling constants at the so-called unification scale. To be precise, this is not a feature of the MSSM itself but of supersymmetric *grand unified theories* (GUT). Let's see how this works.

First, let's review the situation in nonsupersymmetric theories and why things don't look so good for them. We need to consider the "running" of coupling constants with energy. This is given by the renormalization group (RG) equation

$$\frac{1}{\alpha(E)} = \frac{1}{\alpha(E_0)} + \frac{\beta}{2\pi} \ln\left(\frac{E}{E_0}\right)$$

where E_0 and E are two energy scales, and β is the "beta function" of the theory calculated in perturbation theory (an explicit expression will be provided shortly). α is generically defined as the square of the coupling constant over 4π. We therefore define

$$\alpha_{g'} \equiv \frac{g'^2}{4\pi} \qquad \alpha_g \equiv \frac{g^2}{4\pi} \qquad \alpha_s \equiv \frac{g_s^2}{4\pi}$$

with g', g, and g_s being the coupling constants of the $U(1)_Y$, $SU(2)_L$, and $SU(3)_C$ groups, as introduced in Section 15.2. The term *coupling constants* is misleading because these parameters vary with energy. Note that the α's themselves are also often referred to as the *coupling constants of the three forces*.

The coupling constants g', g, and g_s are not the parameters extracted directly from experiments. Instead, the following quantities are usually quoted (see Ref. 4):

$$\alpha_{em}^{-1}(M_Z) \approx 128$$

$$\alpha_s^{-1}(M_Z) \approx 8.47$$

$$\sin^2 \theta_W(M_Z) \approx 0.2312$$

where, as indicated, all the values are obtained at the energy scale $E = M_Z$. Our goal is not to offer a detailed numerical analysis of the data, so we will not bother with the uncertainties. The errors are in the last digit of each result quoted, α_s having the largest percentage uncertainty at $\pm 1.7\%$.

The fine structure constant α_{em} is, of course, defined as

$$\alpha_{em} = \frac{e^2}{4\pi}$$

$$= \frac{g^2}{4\pi} \sin^2 \theta_W$$

where we have used $e = g \sin \theta_W$. Therefore, we have

$$\frac{1}{\alpha_g(M_Z)} = \frac{\sin^2 \theta_W}{\alpha_{em}(M_Z)}$$

$$\approx 29.6 \tag{16.32}$$

On the other hand, $g' = g \tan \theta_W = e/\cos \theta_W$, so

$$\frac{1}{\alpha_{g'}(M_Z)} = \frac{\cos^2 \theta_W}{\alpha_{em}(M_Z)}$$

$$\approx 98.4 \tag{16.33}$$

Now that we have the values of all the inverse constants at the Z mass, the next step is to calculate the beta functions β_1, β_2, and β_3 corresponding to each group and to "run" the coupling constants to see if they converge to a single value at some energy scale. This scale then would be interpreted as a *unification scale*, or *GUT scale*, at which the three forces would merge into a single, unified gauge theory.

However, it does not make sense to do this directly with the coupling constants of the standard model. To understand why, consider the gauge-covariant derivative of a weak doublet of quarks, for example. It is of the form

$$D_\mu = \partial_\mu + i g_s \frac{\lambda^a}{2} A_\mu^a + i g \frac{\tau^i}{2} W_\mu^i + i g' \frac{Y}{2} B_\mu$$

We chose the example of a weak doublet of quarks simply to have all three gauge fields present in the covariant derivative.

We see right away that the value of g' is totally arbitrary because any change in it can be compensated by a change in the hypercharge assignments of all the particles. We could divide the value of g' by an arbitrary factor c', and all physical results would remain identical if we would in addition multiply by the same factor, the hypercharge of all particles. In other words, we may make the replacement

$$g'Y \rightarrow \frac{g'}{c'} c'Y$$

and define g'/c' as a new coupling constant without affecting any physical result! It would simply rescale the hypercharge assignments of all the particles without changing the physics.

The same freedom is present in the nonabelian gauge groups. We may rescale the $SU(2)_L$ and $SU(3)_c$ generators by factors of c and c_s, respectively, if we also rescale the corresponding coupling constants by the same factors. In other words, we could make the replacements

$$g\tau^i \rightarrow \left(\frac{g}{c}\right) c\tau^i$$

$$g_s\lambda^a \rightarrow \left(\frac{g_s}{c_s}\right) c_s\lambda^a$$

without affecting any physical result.

To put it another way, for a nonabelian gauge group,[*] we impose a normalization condition on the generators of the form

$$\text{Tr}(G_i G_j) = N\delta_{ij} \tag{16.34}$$

where the G_i are the generators. The normalization constant is arbitrary; it is usually chosen to be $1/2$, but it could be changed at the condition of rescaling appropriately the corresponding coupling constant (here, the trace is understood as being

---

[*] More specifically, for a *simple* group.

calculated with respect to the fermions present in the theory, as we will demonstrate explicitly shortly).

With this in mind, we can rewrite the covariant derivative as

$$D_\mu = \partial_\mu + i \left(\frac{g_s}{c_s} \right) \frac{c_s \lambda^a}{2} A_\mu^a + i \left(\frac{g}{c} \right) \frac{c \tau^i}{2} W_\mu^i + i \left(\frac{g'}{c'} \right) \frac{c' Y}{2} B_\mu$$

The problem is now clear. Since the standard model is based on a direct product of three groups, all three coupling constants can be *independently* rescaled. We can use the beta functions of the three gauge groups to run the coupling constants, but if we can choose arbitrarily which value to take for each coupling constant at the Z mass, it is meaningless to ask if they meet at some well-defined GUT scale.

There is, however, a way to set the *relative scale* of the coupling constants. The key observation is that within each nonabelian group, there is only a single coupling constant even though there are several generators. Consider the $SU(2)_L$ interaction, for example. We could not write the covariant derivative as

$$D_\mu = \partial_\mu + i g_1 \frac{\tau^1}{2} W_\mu^i + i g_2 \frac{\tau^2}{2} W_\mu^i + i g_3 \frac{\tau^3}{2} W_\mu^i$$

with the g_i being unequal. The reason is that the group structure of the interaction, through Eq. (16.34), imposes a *common* normalization to all generators, implying that $g_1 = g_2 = g_3$.

Going back to the issue of the three independent coupling constants of the standard model, the solution is now obvious: If the three forces of the standard model are unified into a single theory at some high energy scale, their gauge groups must be subgroups of a larger, grand unified group. This implies that there is a *common* normalization condition that must be imposed on *all* the generators of the standard model, including the hypercharge!

The overall normalization chosen does not matter, only the fact that we must use the same one for all the generators. For the $SU(2)_L$ and $SU(3)_C$, we will pick τ^3 and λ^3 as the representative generators because they are diagonal in the basis used for the particles of the standard model, which will make our life easier (any generator can be used in principle, but if they are not diagonal, more work is needed to find their eigenvalues; note that λ^8 is also diagonal and could be used as easily as λ^3). Imposing that all the generators of the standard model have the same normalization then leads to the relation

$$\text{Tr} \left(c' \frac{Y}{2} \right)^2 = \text{Tr} \left(c \frac{\tau^3}{2} \right)^2 = \text{Tr} \left(c_s \frac{\lambda^3}{2} \right)^2 \tag{16.35}$$

Recall that T^3 is defined as the operator $\tau^3/2$, so the second term of Eq. (16.35) may be written as

$$\text{Tr}\left(c\frac{\tau^3}{2}\right)^2 = \text{Tr}(cT^3)^2$$

Since the overall scale is irrelevant, let's set $c_s = 1$ [in other words, we choose to set the normalization for the generators of the unified group equal to the normalization of the $SU(3)_c$ group of the standard model]. With this choice, the normalization of the other generators is uniquely fixed:

$$c^2 = \frac{1}{4}\frac{\text{Tr}(\lambda^3)^2}{\text{Tr}(T^3)^2}$$

$$c'^2 = \frac{\text{Tr}(\lambda^3)^2}{\text{Tr}(Y)^2} \tag{16.36}$$

The values of these coefficients determine the coupling constants we must evolve with the renormalization group. These are g_s, g/c, and g'/c'. In other words, we will evolve α_s, α_g/c^2 and $\alpha_{g'}/c'^2$.

By *taking the trace*, we mean summing over all the eigenvalues corresponding to the particles of the standard model. The hypercharge assignments of the standard model particles are given in Chapter 12. The eigenvalues of T^3 and the hypercharge are related by $Y = 2(Q - T^3)$, where, by a slight abuse of the language, we use the symbol T^3 to represent the eigenvalue corresponding to the operator with the same name, and where Q is the electric charge. The eigenvalues of λ^3 are, of course, zero for the leptons. Each generation contributes equally to the traces in Eq. (16.35), so we can simply impose the relation for the first-generation. The quantum numbers of the first-generation particles are listed in Table 16.1.

Keeping in mind that each quark comes in three colors, we get, for the first generation,

$$\text{Tr}(Y^2) = 3\left(\frac{1}{3}\right)^2 + 3\left(\frac{4}{3}\right)^2 + 3\left(\frac{1}{3}\right)^2 + 3\left(-\frac{2}{3}\right)^2 + (-1)^2 + (-2)^2 + (-1)^2$$

$$= \frac{40}{3} \tag{16.37}$$

It is also straightforward to calculate

$$\text{Tr}(T^3)^2 = 3\left(\frac{1}{2}\right)^2 + 3\left(-\frac{1}{2}\right)^2 + \left(-\frac{1}{2}\right)^2 + \left(\frac{1}{2}\right)^2$$

$$= 2 \tag{16.38}$$

Table 16.1 Quantum Numbers (Except Color) of the First-Generation
Fermions of the Standard Model

Particle	Y	T^3
u_L	1/3	1/2
u_R	4/3	0
d_L	1/3	−1/2
d_R	−2/3	0
e_L	−1	−1/2
e_R	−2	0
ν_L	−1	1/2

For the trace over the quarks, we must first remind ourselves of the definition of λ^3:

$$\lambda_3 = \begin{pmatrix} 1 & 0 & 0 \\ 0 & -1 & 0 \\ 0 & 0 & 0 \end{pmatrix}$$

We see that the eigenvalues of λ_3 of each color state are respectively 1, −1, and 0. The first generation contains four quarks (the left- and right-chiral up and down quarks), so we get

$$\text{Tr}(\lambda_3)^2 = 4(1^2 + (-1)^2 + 0^2) = 8 \tag{16.39}$$

Using Eqs. (16.37), (16.38), and (16.39) in Eq. (16.36), we finally get

$$c = 1 \quad c' = \sqrt{\frac{3}{5}}$$

What this is telling us is that the generators of the grand unified theory must be taken to be

$$\frac{\lambda^3}{2} \quad \frac{\tau^3}{2} \quad \text{and} \quad \sqrt{\frac{3}{5}}\frac{Y}{2}$$

Thus the coupling constants that we have to evolve with the RG equations are

$$g_s \qquad g \qquad \text{and} \qquad \sqrt{\frac{5}{3}}\, g'$$

It's important to emphasize that the angle θ_W remains defined in terms of the original, unscaled constants g and g' through $\tan\theta_W = g'/g$. In addition, the relations

$$\alpha_g = \frac{\alpha_{em}}{\sin^2\theta_W} \qquad \alpha_{g'} = \frac{\alpha_{em}}{\cos^2\theta_W}$$

remain valid, so the numerical values obtained in Eqs. (16.32) and (16.33) are unaffected by our rescaling of the hypercharge operator. The only change is that instead of running $\alpha_{g'}$, we must run $5\alpha_{g'}/3$. Let's define

$$\tilde{\alpha} \equiv \frac{5}{3}\alpha_{g'}$$

of which the inverse at the Z mass is equal to

$$\frac{1}{\tilde{\alpha}(M_Z)} = \frac{3}{5}\frac{1}{\alpha_{g'}(M_Z)}$$

$$= 59.0$$

The three RG equations then are

$$\frac{1}{\alpha_s(E)} = 8.47 + \frac{\beta_S}{2\pi}\ln\left(\frac{E}{M_Z}\right)$$

$$\frac{1}{\alpha_g(E)} = 29.6 + \frac{\beta_g}{2\pi}\ln\left(\frac{E}{M_Z}\right)$$

$$\frac{1}{\tilde{\alpha}(E)} = 59.0 + \frac{\tilde{\beta}}{2\pi}\ln\left(\frac{E}{M_Z}\right) \tag{16.40}$$

We now need to evaluate the beta functions of the three groups. For $SU(N)$ gauge theories, the beta function to one loop is given by

$$\beta[SU(N)] = \frac{11}{3}N - \frac{1}{3}n_f - \frac{1}{6}n_s \tag{16.41}$$

where n_f and n_s are the number of Weyl fermions and complex scalar fields in the fundamental representation of the gauge group, respectively (see Ref. 37 for a derivation of this result). In Eq. (16.41), the effects of the standard model gauge bosons are already included in the first term.

For $SU(3)_c$, we need only include the quark fields. Because we count the number of quarks in the fundamental representation, we do *not* count the three color states of a given quark as independent (the three color states of a given quark simply represent the three states of a single fundamental representation). In the standard model, the quark states are the left- and right-chiral up, down, charm, strange, top, and bottom quarks, for a total of 12 fermion states. The only scalar particle, the Higgs, is a color singlet, so $n_s = 0$. Setting $N = 3$, $n_f = 12$, and $n_s = 0$ in Eq. (16.41), we find

$$\beta_s = 7$$

For $SU(2)_L$, we included only the left-chiral fermion states because the right-chiral states are singlet under that group. The left-chiral electron and neutrino states belong to a common fundamental representation of $SU(2)_L$, so they count for only 1 in the calculation of n_f. The left-chiral up and down quark doublets count for 3 because now states of different colors are different fields as far as $SU(2)_L$ is concerned. So the fermions of the first generation contribute a total of 4 to n_f, and therefore, the three generations combined yield a total $n_f = 12$. The Higs field is a complex doublet, so it counts only as 1 for n_s. We finally get, with $N = 2$, $n_f = 12$, and $n_s = 1$,

$$\beta_g = \frac{19}{6}$$

For a $U(1)$ group, the formula for the beta function to one loop is

$$\beta[U(1)] = -\frac{1}{20} \sum \left(2Y_f^2 + Y_s^2\right) \tag{16.42}$$

where the sum is over all the fermions and scalar fields with hypercharges Y_f and Y_s, respectively, Again, the effects of the gauge bosons are already included.

Using the hypercharge assignments of the fermions given in Table 16.1, we get (the overall factor of 3 accounts for the three generations, and there is an additional factor of 3 for the three color states of each quark):

$$\sum Y_f^2 = 3 \left[3\left(\frac{1}{3}\right)^2 + 3\left(\frac{4}{3}\right)^2 + 3\left(\frac{1}{3}\right)^2 + 3\left(-\frac{2}{3}\right)^2 \right.$$
$$\left. + (-1)^2 + (-2)^2 + (-1)^2 \right]$$
$$= 40 \tag{16.43}$$

Recall that the Higgs has $Y = 1$. The Higgs field is an $SU(2)_L$ doublet of complex fields. As far as the $U(1)_Y$ force is concerned, these two complex fields are independent, which gives us

$$\sum_s Y_s^2 = 2 \qquad (16.44)$$

Substituting Eqs. (16.43) and (16.44) in Eq. (16.42), we finally get

$$\tilde{\beta} = -\frac{41}{10}$$

To summarize, the three RG equations (16.40) are

$$\frac{1}{\alpha_s(E)} = 8.47 + 7\frac{\ln\left(\frac{E}{M_Z}\right)}{2\pi}$$

$$\frac{1}{\alpha_g(E)} = 29.6 + \frac{19}{6}\frac{\ln\left(\frac{E}{M_Z}\right)}{2\pi}$$

$$\frac{1}{\tilde{\alpha}(E)} = 59.0 - \frac{41}{10}\frac{\ln\left(\frac{E}{M_Z}\right)}{2\pi}$$

To see if these three lines meet at a common point, we could plot them but a simple analytical test can be made. If we have three straight lines with slopes m_1, m_2, and m_3 and with y intercepts b_1, b_2, and b_3, it is easy to show that the condition for those three lines to meet at the same point is

$$\left(\frac{b_1 - b_2}{b_3 - b_2}\right)\left(\frac{m_3 - m_2}{m_1 - m_2}\right) = 1 \qquad (16.45)$$

Using the three slopes and y intercepts of our RG equations, the expression on the left of Eq. (16.45) is found to be equal to 1.36, which clearly shows that the lines do not meet at a single point. It is important to emphasize that our whole derivation is only valid to one loop because the expressions for the beta functions and the RG equations are only valid to that order. However, the three lines miss one another by so much that there is no hope that higher-order corrections could cure the problem.

Let us now repeat the calculation for the MSSM. The values of the coupling constants at the Z mass remain the same (of course, because they are measured values!), but now the beta functions will be different because of the different particle content. To start with, Eq. (16.41) must be modified because of the superpartners of the gauge bosons that, it turns out, contribute $-2N/3$, so the $SU(N)$ beta function now reads

$$\beta_{SU(N)} = \frac{9}{3}N - \frac{1}{3}n_f - \frac{1}{6}n_s$$

For $SU(3)_c$, we have, as before, 12 quark states, but now in addition, we have the 12 gluinos, the scalar superpartners of the gluons. Using $n_f = n_s = 12$, we find for the MSSM,

$$\beta_S = 3$$

For $SU(2)_L$, we have the 12 fermion states of the standard model, but in addition, we have the fermionic partners of the Higgs fields, the higgsinos. There are two Higgs doublets in the MSSM, so the total n_f is equal to 14. All these fermions have scalar partners, so $n_s = 14$ as well. Putting all this together, we find

$$\beta_g = -1$$

Note that β_g has changed sign relative to its value in the standard model!

The formula for the $U(1)$ beta function remains Eq. (16.42). To each standard model fermion with a certain hypercharge assignment corresponds a scalar sfermion with the same hypercharge. Therefore, excluding the Higgs fields and their superpartners, we have

$$\sum \left(2Y_f^2 + Y_s^2\right) = 80 + 40$$
$$= 120 \tag{16.46}$$

The two Higgs doublets comprise four complex scalar fields with hypercharges ± 1. The corresponding four higgsinos have the same hypercharges, so for those fields, we find

$$\sum \left(2Y_f^2 + Y_s^2\right) = 12 \tag{16.47}$$

Substituting Eqs. (16.46) and (16.47) into Eq. (16.42), we finally get

$$\tilde{\beta} = -\frac{33}{5}$$

Thus, for the MSSM, Eq. (16.45) is replaced by

$$\frac{1}{\alpha_s(E)} = 8.47 + 3\frac{\ln\left(\frac{E}{M_Z}\right)}{2\pi}$$

$$\frac{1}{\alpha_g(E)} = 29.6 - \frac{\ln\left(\frac{E}{M_Z}\right)}{2\pi}$$

$$\frac{1}{\tilde{\alpha}(E)} = 59.0 - \frac{33}{5}\frac{\ln\left(\frac{E}{M_Z}\right)}{2\pi}$$

which leads to

$$\left(\frac{b_1 - b_2}{b_3 - b_2}\right)\left(\frac{m_3 - m_2}{m_1 - m_2}\right) = 1.006$$

This is an amazing result: The three coupling constants now *do* meet up at a certain energy (taking into account the uncertainty in our experimental values)! We can, of course, estimate this energy: We simply set equal two of the coupling constants and solve for E. This yields

$$E \approx e^{33} M_Z$$
$$= 2 \times 10^{16} \text{ GeV}$$

which is the unification scale in a grand unified extension of the MSSM.

A generic feature of grand unified theories is the decay of protons. You may have heard that the $SU(5)$ GUT is ruled out experimentally because it predicts too rapid a decay. The MSSM gut fares better because its GUT scale is about twenty times larger than the $SU(5)$ grand unification scale, which raises the proton lifetime. A factor of twenty does not seem like such a big deal, until one notices that the proton lifetime is extremely sensitive to the GUT scale, being proportional to E_{GUT}^4. An increase by a factor of twenty in the GUT scale therefore translates into an increase by over five orders of magnitude in the proton lifetime. This is sufficient to keep the MSSM grand unified theory in the game.

At this energy, all three values of α^{-1} are approximately 24, which means that the coupling constant α itself is about

$$\alpha(\text{GUT scale}) \approx 0.041$$

Note that at this scale,

$$g = \sqrt{\frac{5}{3}} g'$$

which implies that

$$\tan \theta_W \equiv \frac{g'}{g}$$

$$= \sqrt{\frac{3}{5}}$$

or

$$\sin^2 \theta_W = \frac{3}{8}$$

This is, of course, not an exact result. First, as pointed out earlier, our calculation is only valid at one loop order. More crucial is the fact that the equations that we used for the beta functions assume that no mass terms explicitly break the symmetries. As we have seen, however, in the MSSM we have to include such explicit SUSY-breaking masses. There is much more to say about supersymmetry and the MSSM in particular but alas, it's now time to end our introduction to the topic. We hope you enjoyed the ride!

16.9 Quiz

1. How many parameters are needed to describe the most general soft SUSY-breaking terms? How many parameters are needed in the minimal supergravity model?

2. How many observable Higgs fields (i.e., not including the states that get absorbed by the gauge bosons) are there in the MSSM?

3. In a supersymmetric GUT theory, which inverse coupling constants increase with energy and which ones decrease? Is this the same behavior as in a GUT extension of the standard model?

4. One of the Higgs fields of the MSSM is of particular interest from a phenomenological point of view. Explain why.

5. How many parameters were necessary to describe the soft SUSY-breaking terms of the Higgs sector?

Final Exam

1. What are the dimensions of the fields ϕ, χ, and F appearing in the Wess-Zumnino model, as well as the dimension of the infinitesimal spinor ζ that parametrizes SUSY transformations?

2. What is the result of the "integral" $\int d\theta_1 d\theta_2\, \theta_1 \theta_2$?

3. Using the antisymmetric ϵ symbol, prove that $\bar{\chi} \cdot \bar{\lambda} = \bar{\lambda} \cdot \bar{\chi}$.

4. What is the action of the supercharges $Q_1, Q_2, Q_1^\dagger$ and $Q_2^\dagger$ on a one-particle state?

5. Write down the kinetic energy term for a left-chiral Weyl spinor and show that it is a real quantity.

6. What was the motivation for introducing an auxiliary scalar field in the Wess-Zumino model?

7. Why is it not possible to use a Fayet-Iliopoulos D term to break SUSY spontaneously in the MSSM?

8. Consider a supersymmetric theory with the potential for the scalar fields depicted in Figure 1 on the next page. Is SUSY broken spontaneously in this theory?

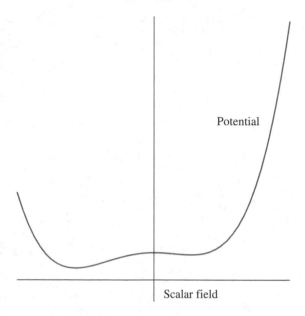

Figure 1 Potential for Question 8.

9. In a supersymmetric abelian gauge theory, write down the most general transformation for the auxiliary field D that has the right dimension and correct Lorentz transformation properties (without looking at Chapter 10!). It must of course contain the SUSY parameter ζ and one of the other fields present in the supermultiplet.

10. Consider the interactions and mass terms of the Wess-Zumino model. If we do not impose invariance under supersymmetry and do not assume any relation between the coupling constants and masses, how many parameters are there? How many parameters are there when invariance under SUSY is imposed?

11. What constraint is obeyed by left-chiral superfields?

12. Write down the mass terms of a Dirac fermion and of a Majorana fermion in terms of Weyl spinors.

13. Why can't we take a superfield to depend only on the variables x^μ and θ (and not on $\bar{\theta}$)?

14. Write the components of a left-chiral spinor with upper dotted indices $\bar{\chi}^{\dot{1}}$ and $\bar{\chi}^{\dot{2}}$ in terms of the components with lower undotted indices χ_1 and χ_2.

15. What is a gaugino?

16. What do we mean by the *D term* of a product of superfields?

17. What are *helicity* and *chirality*?

18. Write down the supersymmetric lagrangian of a free left-chiral superfield in terms of the corresponding component fields.

19. What is the LSP?

20. If R-parity is a valid symmetry, what implication does this have for the LSP?

21. Could we add a term of the form $c\Phi$ to the lagrangian of the MSSM, where Φ is any of the left-chiral superfield in the theory?

22. Why does the component field of a superfield that has the highest dimension necessarily transform with a total derivative under SUSY?

23. Why did we need to introduce R-parity in the MSSM?

24. What is the value of the Weinberg (or weak) angle predicted at the unification scale in grand unified theories?

25. A renormalizable lagrangian is of dimension 4, and left-chiral scalar superfields Φ_i have dimension 1, so why do we include only terms with up to three left-chiral superfields in the superpotential?

26. Why is it crucial to include loop effects when considering the mass of the h^0?

27. Why do you impose that the vacuum expectation values of the sleptons and squarks be zero?

28. What is the mass term for a Majorana fermion?

29. How is the relation between the W and Z masses different from what we have in the standard model?

30. There are three types of terms that break SUSY softly. What are they?

31. List the component fields of a left-chiral superfield Φ (which is a scalar under Lorentz transformation) and show that the bosonic and fermionic degrees of freedom match both on-shell and off-shell.

32. What is the superpotential of the Wess-Zumino model in terms of the left-chiral superfields Φ_i?

33. What is the most general supersymmetric lagrangian for an abelian gauge theory with a set of left-chiral superfield Φ_i of charges q_i? Write the answer in terms of $\mathcal{V}$, $\mathcal{F}_a$ and Φ_i.

34. How does the field-strength superfield $\mathcal{F}_a$ transform under Lorentz transformations?

35. Write $\chi \cdot \lambda$, $\bar{\chi} \cdot \bar{\lambda}$ and $\bar{\chi}\bar{\sigma}^\mu\lambda$ with the indices explicitly shown (using dotted and undotted indices in the appropriate positions).

36. Why do the coupling constants evolve differently in the standard model and in the MSSM (give a physical explanation as opposed to a purely mathematical one)?

37. What are the F and D terms of the following expression:

$$\alpha(x) \cdot \theta + \bar{\beta}(x) \cdot \bar{\theta} + A(x)\theta \cdot \theta + B(x)\bar{\theta} \cdot \bar{\theta} + \chi(x) \cdot \theta\bar{\theta} \cdot \bar{\theta} + C(x)\theta \cdot \theta\bar{\theta} \cdot \bar{\theta}?$$

38. Write down the potential for the neutral Higgs fields in the MSSM, in the absence of any soft SUSY breaking terms (you may consult Section 16.3). Show that for this potential, both SUSY and the gauge symmetry are unbroken.

39. What is the R-parity of all the particles of the MSSM? You do not need to list them all. Simply give the R-parity for the quarks, leptons, Higgs scalars and gauge bosons, as well as for their supersymmetric partners, the squarks, sleptons, Higgsinos and gauginos.

40. Consider the left-chiral superfields $\mathcal{H}_u$ (quantum numbers $\mathbf{1}, \mathbf{2}, 1$), $\mathcal{H}_d(\mathbf{1}, \mathbf{2}, -1)$, and $\mathcal{E}_i(\mathbf{1}, \mathbf{1}, 2)$. Write down the most general gauge invariant lagrangian density containing up to four of these superfields.

41. What is a Fayet-Iliopoulos term? Why isn't there such a term in the MSSM?

42. Calculate

$$\frac{\partial}{\partial\theta^1} \frac{\partial}{\partial\theta_1} (\theta_2\theta_1)$$

43. What are the chirality and helicity operators?

44. In Section 7.5 we had to prove that

$$\text{Tr}\big((-1)^{N_f}\{Q_a, Q_b^\dagger\}\big) = 0$$

where the trace is over all the fields of a SUSY supermultiplet and N_f is equal to 1 for a fermionic state and zero for a bosonic state. Prove this again, without looking at Chapter 7!

45. Why does a small Higgs mass (in comparison with a GUT unification scale) represent a naturalness problem in the standard model?

46. When evaluating Feynman diagrams with fermion loops, what is the difference between diagrams containing Dirac fermion loops and diagrams containing Majorana fermion loops?

47. Using the notation of dotted and undotted indices, prove $(\chi \cdot \lambda)^\dagger = \bar{\lambda} \cdot \bar{\chi}$.

48. What is the spin of the gauginos? In what representation of the gauge group do they transform?

49. What is the definition of the supertrace? Why is it equal to zero when SUSY is unbroken?

50. Consider a supersymmetric abelian gauge theory with a single charged left-chiral superfield Φ. What is the most general superpotential for such a theory?

References

1. Aitchison, I. J. R.: *Supersymmetry in Particle Physics*, Cambridge University Press, New York, 2007.
2. Aitchison, I. J. R., and Hey, A. J. G.: *Gauge Theories in Particle Physics*, Vol. I: *From Relativistic Quantum Mechanics to QED*, and Vol. II: *QCD and the Electroweak Theory*, Taylor and Francis, London, 2004.
3. Alvarez-Gaumé, A., and Hassan, S. F.: "Introduction to S-Duality and $N = 2$ Supersymmetric Gauge Theories (A Pedagogical Review of the Work of Seiberg and Witten)," arXiv:hep-th/9701069v1; accessed January 1997.
4. Amsler, C., et al.: *Phys. Lett.* **B667**:1, 2008.
5. Baer, H., and Tata, X.: *Weak Scale Supersymmetry*, Cambridge University Press, New York, 2006.
6. Bailin, D., and Love, A.: *Supersymmetric Gauge Field Theory and String Theory*, Institute of Physics Publishing, Bristol, 1994.
7. Berezin, F. A.: *The Method of Second Quantization*, Academic Press, New York, 1966.
8. Binétruy, P.: *Supersymmetry: Theory, Experiment, and Cosmology*, Oxford University Press, New York, 2006.
9. Burgess, C.: "Introduction to Effective Field Theory," arXiv:hep-th/0701053v2; accessed January 2007.

10. Chung, D. J. H., Everett, L. L., Kane, G. L., et al.: "The Soft Supersymmetry-Breaking Lagrangian: Theory and Applications," arXiv:hep-ph/0312378v1; accessed December 2003.

11. Coleman, S., and Mandula, J.: *Phys. Rev.* **159**:1251, 1967.

12. Dine, M.: *Supersymmetry and String Theory*, Cambridge University Press, New York, 2007.

13. Dreiner, H. K., Haber, H. E., and Martin, S. P.: "Two-Component Spinor Techniques and Feynman Rules for Quantum Field Theory and Supersymmetry," arXiv:hep-ph/0812.1594v1; accessed December 2008.

14. Fayet, P., and Iliopoulos, J.: *Phys. Lett.* **51B**:461, 1974.

15. Ferrara, S. (ed.): *Supersymmetry*, North Holland/World Scientific, Amsterdam, The Netherlands, 1987.

16. Gervais, J.-L., and Sakita, B.: *Nucl. Phys.* **B34**: 632, 1971. Reprinted in Ref. 40.

17. Gof'land, Y. A., and Likhtman, E. P.: *JETPS Lett.* **13**:323, 1971. Reprinted in Ref. 15.

18. Girardello, L., and Grisaru, M.: *Nucl. Phys.* **B194**:65, 1982.

19. Greiner, W., and Reinhardt, J.: *Quantum Electrodynamics*, Springer, New York, 2008; Greiner, W., Schramm, S., Stein, E., and Bromley, D. A.: *Quantum Chromodynamics*, Springer, New York, 2007; Greiner, W., and Müller, B., *Gauge Theory of Weak Interactions*, Springer, New York, 2009.

20. Greiner, W., and Reinhardt, J.: *Field Quantization*, Springer-Verlag, Berlin, 1996.

21. Griffiths, D.: *Introduction to Elementary Particles*, Wiley-VCH, New York, 2008.

22. Grisaru, M., Siegel, W., and Rocek, M.: *Nucl. Phys.* **B159**:429, 1979.

23. Haag, R., Lopuszanski, J., and Sohnius, M.: *Nucl. Phys.* **B88**:257, 1975.

24. Hatfield, B.: *Quantum Field Theory of Point Particles and Strings*, Addison-Wesley, Reading, Mass., 1992.

25. Jacob, M. (ed.): *Supersymmetry and Supergravity: A Reprint Volume of Physics Reports*, North Holland/World Scientific, 1986.

26. Lepage, G. P.: "What Is Renormalization?" arXiv:hep-ph/0506330v1; accessed June 2005.

27. Lykken, J. D.: "Introduction to Supersymmetry," arXiv:hep-th/9612114v1; accessed December 1996.

28. Maggiore, M.: *A Modern Introduction to Quantum Field Theory*, Oxford University Press, New York, 2005.

29. Martin, S. P.: "A Supersymmetry Primer," arXiv:hep-ph/9709356v5; accessed December 2008.

30. McMahon, D.: *Quantum Field Theory Demystified*, McGraw-Hill, New York, 2008.

31. Müller-Kirsten, H. J. W., and Wiedemann, A.: *Supersymmetry: An Introduction with Conceptual and Calculational Details*, World Scientific, Singapore, 1987.

32. Murayama, H.: "Supersymmetry Phenomenology," arXiv:hep-ph/0002232v2; accessed March 2000.

33. Neveu, A., and Schwarz, J. H.: *Nucl. Phys.* **B31**:86, 1971, and *Phys. Rev.* **D4**:1109, 1971. Both reprinted in Ref. 40.

34. O'Raifeartaigh, L., *Nucl. Phys.* **B96**:331, 1975.

35. Paige, F. E.: "Supersymmetry Signatures at the CERN LHC," arXiv:hep-ph/9801254v1; accessed January 1998.

36. Peskin, M. E.: "Supersymmetry in Elementary Particle Physics," arXiv:hep-ph/0801.1928v1; accessed January 2008.

37. Peskin, M. E., and Schroeder, D. V.: *An Introduction to Quantum Field Theory*, Addison-Wesley, Reading, Mass., 1995.

38. Polonsky, N.: "Supersymmetry: Structure and Phenomena," arXiv:hep-ph/0108236v1; accessed August 2001.

39. Ramond, P.: *Phys. Rev.* **D3**:2415, 1971. Reprinted in Ref. 40.

40. Schwarz, J. H. (ed.): *Superstrings: The First 15 Years of Superstring Theory*, World Scientific, 1985.

41. Salam, A., and Strathdee, J.: *Nucl. Phys.* **B76**:477, 1974.

42. Seiberg, N.: *Phys. Lett.* **B318**:469, 1993 (arXiv:hep-ph/9309335).

43. Shirman, Y.: "TASI 2008 Lectures: Introduction to Supersymmetry and Supersymmetry Breaking," arXiv:hep-ph/0907.0039v1; July 2009.

44. Srednicki, M.: *Quantum Field Theory*, Cambridge University Press, New York, 2007.

45. t' Hooft, G.: "Naturalness, Chiral Symmetry, and Spontaneous Chiral Symmetry Breaking," in *Recent Developments in Gauge Theories (NATO ASI Series B: Physics, Vol. 59)*, Plenum Press, New York, 1979.

46. Volkov, D. V., and Akulov, V. P.: *Phys. Lett.* **46B**:109, 1973. Reprinted in Ref. 15.

47. Weinberg, S.: *The Quantum Theory of Fields*, Vol. III: *Supersymmetry*, Cambridge University Press, New York, 2000.

48. Weinberg, S.: *The Quantum Theory of Fields*, Vol. I: *Foundations*, and Vol. II: *Modern Applications*, Cambridge University Press, New York, 2005.

49. Wess, J., and Bagger, J.: *Supersymmetry and Supergravity*, Princeton University Press, Princeton, N.J., 1992.

50. Wess, J., and Zumino, B.: *Nucl. Phys.* **B70**:39, 1974, and *Phys. Lett.* **49B**:52, 1974. Both articles are reprinted in Ref. 15.

51. West, P. C.: "Introduction to Rigid Supersymmetric Theories," arXiv:hep-th/9805055v1; accessed May 1998.

52. Wigner, E. P.: *Annals of Mathematics* **40**:149, 1939.

APPENDIX A

Useful Identities

Gamma Matrices and Trace Identities

In the following, a and b stand for arbitrary four-vectors. The following list is far from being exhaustive; we are only providing the few identities needed in this book. For a more complete set of rules, please consult one of the books on quantum field theory listed in the References section.

$$(\gamma^5)^2 = 1$$
$$\gamma^5\gamma^\mu = -\gamma^\mu\gamma^5$$
$$\mathrm{Tr}\gamma^\mu = \mathrm{Tr}\slashed{a} = 0$$
$$\mathrm{Tr}(\slashed{a}\slashed{b}) = 4a \cdot b$$
$$\mathrm{Tr}(\gamma^5\slashed{a}\gamma^5\slashed{b}) = -4a \cdot b$$
$$\mathrm{Tr}(\gamma^5) = \mathrm{Tr}(\gamma^5\slashed{a}) = 0$$

We also need the charge conjugation matrix

$$C \equiv -i\gamma^2\gamma^0 = \begin{pmatrix} i\sigma^2 & 0 \\ 0 & -i\sigma^2 \end{pmatrix} \tag{A.1}$$

which obeys the following identities:

$$C^2 = (C^T)^2 = -1$$
$$\gamma^5 C^T = C^T \gamma^5$$

Identities Involving Pauli Matrices Only

$$\sigma^\mu \equiv (\mathbf{1}, \vec{\sigma}) \tag{A.2}$$

$$\bar{\sigma}^\mu \equiv (\mathbf{1}, -\vec{\sigma}) \tag{A.3}$$

$$\sigma^i \sigma^j = \delta^{ij}\mathbf{1} + i\epsilon^{ijk}\sigma^k \tag{A.4}$$

$$\sigma^2 \sigma^\mu \sigma^2 = \bar{\sigma}^{\mu T} \tag{A.5}$$

$$\sigma^2 \bar{\sigma}^\mu \sigma^2 = \sigma^{\mu T} \tag{A.6}$$

$$\sigma^2 \sigma^{\mu T} \sigma^2 = \bar{\sigma}^\mu \tag{A.7}$$

$$\sigma^2 \bar{\sigma}^{\mu T} \sigma^2 = \sigma^\mu \tag{A.8}$$

Here are a few identities involving products of σ^μ and $\bar{\sigma}^\mu$ matrices:

$$\sigma^\mu \bar{\sigma}^\nu + \sigma^\nu \bar{\sigma}^\mu = 2\eta^{\mu\nu}\mathbf{1} \tag{A.9}$$

$$\bar{\sigma}^\mu \sigma^\nu + \bar{\sigma}^\nu \sigma^\mu = 2\eta^{\mu\nu}\mathbf{1} \tag{A.10}$$

$$\bar{\sigma}^\nu \sigma^\mu \bar{\sigma}^\rho = \eta^{\mu\nu}\bar{\sigma}^\rho - \eta^{\nu\rho}\bar{\sigma}^\mu + \eta^{\mu\rho}\bar{\sigma}^\nu - i\epsilon^{\nu\mu\rho\delta}\bar{\sigma}_\delta \tag{A.11}$$

$$\text{Tr}(\sigma^\mu \bar{\sigma}^\nu) = 2\eta^{\mu\nu} \tag{A.12}$$

$$\text{Tr}(\sigma^\nu \bar{\sigma}^\mu \sigma^\lambda \bar{\sigma}^\rho) = 2(\eta^{\nu\mu}\eta^{\lambda\rho} + \eta^{\mu\lambda}\eta^{\nu\rho} - \eta^{\nu\lambda}\eta^{\mu\rho} - i\epsilon^{\nu\mu\lambda\rho}) \tag{A.13}$$

By definition, the matrices σ^μ and $\bar{\sigma}^\mu$ carry the following indices:

$$\bar{\sigma}^\mu = (\bar{\sigma}^\mu)^{\dot{a}b} \tag{A.14}$$

$$\sigma^\mu = (\sigma^\mu)_{a\dot{b}} \tag{A.15}$$

We also define

$$\sigma^{\mu\nu} \equiv \frac{i}{4}(\sigma^\mu\bar{\sigma}^\nu - \sigma^\nu\bar{\sigma}^\mu) \tag{A.16}$$

$$\bar{\sigma}^{\mu\nu} \equiv \frac{i}{4}(\bar{\sigma}^\mu\sigma^\nu - \bar{\sigma}^\nu\sigma^\mu) \tag{A.17}$$

which carry the following indices

$$\sigma^{\mu\nu} = (\sigma^{\mu\nu})_a^{\ b} \tag{A.18}$$

$$\bar{\sigma}^{\mu\nu} = (\bar{\sigma}^{\mu\nu})^{\dot{a}}_{\ \dot{b}} \tag{A.19}$$

Identities involving ϵ:

$$\epsilon^{ab} = (i\sigma^2)^{ab} \tag{A.20}$$

$$\epsilon^{\dot{a}\dot{b}} = \epsilon^{ab} \tag{A.21}$$

$$\epsilon_{ab} = (-i\sigma^2)_{ab} \tag{A.22}$$

$$\epsilon_{\dot{a}\dot{b}} = \epsilon_{ab} \tag{A.23}$$

$$\epsilon^{ab}\epsilon_{bc} = \epsilon_{cb}\epsilon^{ba} = \delta^a_c \tag{A.24}$$

$$\epsilon^{\dot{a}\dot{b}}\epsilon_{\dot{b}\dot{c}} = \epsilon_{\dot{c}\dot{b}}\epsilon^{\dot{b}\dot{a}} = \delta^{\dot{a}}_{\dot{c}} \tag{A.25}$$

$$\epsilon^{ca}\epsilon^{\dot{d}\dot{b}}(\sigma^\mu)_{a\dot{b}} = (\bar{\sigma}^\mu)^{\dot{d}c} \tag{A.26}$$

$$\epsilon_{cb}\epsilon_{\dot{c}\dot{a}}(\bar{\sigma}^\mu)^{\dot{a}b} = (\sigma^\mu)_{c\dot{d}} \tag{A.27}$$

$$\epsilon^{\dot{c}\dot{b}}\epsilon_{\dot{d}\dot{a}}(\bar{\sigma}^{\mu\nu})^{\dot{a}}_{\ \dot{b}} = (\bar{\sigma}^{\mu\nu})^{\dot{c}}_{\ \dot{d}} \tag{A.28}$$

Definitions of the Various Components of Weyl Spinors

In the following, the relations are written in terms of a left-chiral spinor χ, but they would apply as well to a right-chiral spinor. In all the following equations, the barred spinor components always may be replaced by the hermitian conjugate of the unbarred components using

$$\bar{\chi}^{\dot{a}} = \chi^{a\dagger} \tag{A.29}$$

$$\bar{\chi}_{\dot{a}} = \chi_a^\dagger \tag{A.30}$$

The matrix $i\sigma^2$ may be used to raise both dotted and undotted indices:

$$
\begin{aligned}
\chi^a &= (i\sigma^2)^{ab} \chi_b \\
&= \epsilon^{ab} \chi_b \\
\bar{\chi}^{\dot{a}} &= (i\sigma^2)^{\dot{a}\dot{b}} \bar{\chi}_{\dot{b}} \\
&= \epsilon^{\dot{a}\dot{b}} \bar{\chi}_{\dot{b}}
\end{aligned}
\tag{A.31}
$$

To be explicit,

$$\chi^1 = \chi_2 \qquad \chi^2 = -\chi_1 \tag{A.32}$$

$$\bar{\chi}^{\dot{1}} = \bar{\chi}_{\dot{2}} \qquad \bar{\chi}^{\dot{2}} = -\bar{\chi}_{\dot{1}} \tag{A.33}$$

The matrix $(-i\sigma^2)$ may be used to lower either type of index:

$$\bar{\chi}_{\dot{a}} = (-i\sigma^2)_{\dot{a}\dot{b}} \bar{\chi}^{\dot{b}} \tag{A.34}$$

$$= \epsilon_{\dot{a}\dot{b}} \bar{\chi}^{\dot{b}} \tag{A.35}$$

Identities Involving Two Spinors

A general relation is (see Section 2.8)

$$\alpha^T A \beta = -\beta^T A^T \alpha \tag{A.36}$$

where α and β are arbitrary Weyl spinors.
The spinor dot products are defined by

$$
\begin{aligned}
\chi \cdot \lambda &= \chi^T (-i\sigma^2)\lambda \\
\bar{\chi} \cdot \bar{\lambda} &= \chi^\dagger i\sigma^2 \lambda^{\dagger T}
\end{aligned}
\tag{A.37}
$$

Writing the indices explicitly, we have

$$\chi \cdot \lambda = \chi^a \lambda_a = \chi_a (-i\sigma^2)^{ab} \lambda_b = \chi_2 \lambda_1 - \chi_1 \lambda_2 \qquad (A.38)$$

$$\bar\chi \cdot \bar\lambda = \bar\chi_{\dot a} \bar\lambda^{\dot a} = \bar\chi_{\dot a}(i\sigma^2)^{\dot a \dot b} \bar\lambda_{\dot b} = -\bar\chi_{\dot 2}\bar\lambda_{\dot 1} + \bar\chi_{\dot 1}\bar\lambda_{\dot 2} \qquad (A.39)$$

Using the ϵ notation, we also may write the spinor dot products as

$$\chi \cdot \lambda = \chi_a \epsilon^{ab} \lambda_b = \chi^a \epsilon_{ab} \lambda^b \qquad (A.40)$$

$$\bar\chi \cdot \bar\lambda = \bar\chi_{\dot a} \epsilon^{\dot a \dot b} \bar\lambda_{\dot b} = \bar\chi^{\dot a} \epsilon_{\dot a \dot b} \bar\lambda^{\dot b} \qquad (A.41)$$

The order in the spinor dot products does not matter:

$$\chi \cdot \lambda = \lambda \cdot \chi$$
$$\bar\chi \cdot \bar\lambda = \bar\lambda \cdot \bar\chi \qquad (A.42)$$

Or, in terms of components,

$$\chi^a \lambda_a = \lambda^a \chi_a$$
$$\bar\chi_{\dot a} \bar\lambda^{\dot a} = \bar\lambda_{\dot a} \bar\chi^{\dot a} \qquad (A.43)$$

The effect of hermitian conjugation on the spinor dot products is

$$(\lambda \cdot \chi)^\dagger = \bar\lambda \cdot \bar\chi$$
$$(\bar\lambda \cdot \bar\chi)^\dagger = \lambda \cdot \chi \qquad (A.44)$$

The product of two components of a spinor may be rewritten in terms of the spinor dotted with itself using

$$\chi_a \chi_b = -\frac{1}{2}(i\sigma^2)_{ab} \chi \cdot \chi = \frac{1}{2}\epsilon_{ab} \chi \cdot \chi \qquad (A.45)$$

$$\bar\chi_{\dot a} \bar\chi_{\dot b} = \frac{1}{2}(i\sigma^2)_{\dot a \dot b} \bar\chi \cdot \bar\chi = -\frac{1}{2}\epsilon_{\dot a \dot b} \bar\chi \cdot \bar\chi \qquad (A.46)$$

$$\chi^a \chi^b = -\frac{1}{2}(i\sigma^2)^{ab} \chi \cdot \chi = -\frac{1}{2}\epsilon^{ab} \chi \cdot \chi \qquad (A.47)$$

$$\bar\chi^{\dot a} \bar\chi^{\dot b} = \frac{1}{2}(i\sigma^2)^{\dot a \dot b} \bar\chi \cdot \bar\chi = \frac{1}{2}\epsilon^{\dot a \dot b} \bar\chi \cdot \bar\chi \qquad (A.48)$$

Some identities involving σ^μ and $\bar{\sigma}^\mu$:

$$\bar{\chi}\bar{\sigma}^\mu\lambda = -\lambda\sigma^\mu\bar{\chi} \tag{A.49}$$

$$\chi\sigma^\mu\bar{\lambda} = -\bar{\lambda}\bar{\sigma}^\mu\chi \tag{A.50}$$

$$(\bar{\chi}\bar{\sigma}^\mu\lambda)^\dagger = \bar{\lambda}\bar{\sigma}^\mu\chi \tag{A.51}$$

$$(\chi\sigma^\mu\bar{\lambda})^\dagger = \lambda\sigma^\mu\bar{\chi} \tag{A.52}$$

Identities Involving Three Left-Chiral Spinors

$$\alpha_a(\beta \cdot \gamma) + \beta_a(\gamma \cdot \alpha) + \gamma_a(\alpha \cdot \beta) = 0 \tag{A.53}$$

Identities Involving Four Left-Chiral Spinors

Particularly useful are *Weyl spinor Fierz identities*:

$$\chi \cdot \alpha\chi \cdot \beta = -\frac{1}{2}\chi \cdot \chi\alpha \cdot \beta \tag{A.54}$$

$$\bar{\chi} \cdot \bar{\alpha}\bar{\chi} \cdot \bar{\beta} = -\frac{1}{2}\bar{\chi} \cdot \bar{\chi}\bar{\alpha} \cdot \bar{\beta} \tag{A.55}$$

$$(\lambda\sigma^\mu\bar{\chi})(\lambda\sigma^\nu\bar{\chi}) = \frac{1}{2}\eta^{\mu\nu}\lambda \cdot \lambda\bar{\chi} \cdot \bar{\chi} \tag{A.56}$$

$$(\bar{\chi}\bar{\sigma}^\mu\lambda)(\bar{\chi}\bar{\sigma}^\nu\lambda) = \frac{1}{2}\eta^{\mu\nu}\lambda \cdot \lambda\bar{\chi} \cdot \bar{\chi} \tag{A.57}$$

APPENDIX B

Solutions to Exercises

2.1 The case $\mu = \nu = 0$ is trivially true because $\sigma^0 = \bar{\sigma}^0 = 1$ (the 2×2 unit matrix) and $\eta^{00} = 1$.

Consider now $\mu = 0$ and $\nu = i$. We use $\bar{\sigma}^{\mu=i} = -\sigma^i$. The left-hand side then is

$$-\sigma^i + \sigma^i = 0$$

On the other hand, the right-hand side also vanishes because $\eta^{\mu\nu}$ is diagonal, so $\eta^{0i} = \eta^{i0} = 0$.

Obviously, if $\nu = 0$ and $\mu = i$, both sides of the equation are also equal to zero.

The case of two spatial indices is more interesting. Setting $\mu = i$ and $\nu = j$ (where i and j may or may not be equal), the left-hand side of Eq. (2.25) is

$$\sigma^i \bar{\sigma}^j + \sigma^j \bar{\sigma}^i = -\sigma^i \sigma^j - \sigma^j \sigma^i$$

where we have again used the definitions of σ^μ and $\bar{\sigma}^\mu$. Using now (2.9), we find

$$-\sigma^i\sigma^j - \sigma^j\sigma^i = -\{\sigma^i, \sigma^j\}$$
$$= -2\delta^{ij}\mathbf{1} \tag{B.1}$$

As for the right-hand side of Eq. (2.25), it is equal to η^{ij} times the unit matrix. This is zero when $i \neq j$ and -1 when $i = j$, which is nothing other than $-\delta^{ij}$! Therefore, the right-hand side is

$$2\eta^{ij}\mathbf{1} = -2\delta^{ij}\mathbf{1}$$

which again agrees with the left-hand side. We thus have shown the validity of Eq. (2.24) for all possible values of the indices.

2.2 We have

$$(\eta'_R)^\dagger\chi'_L = \left[\left(\mathbf{1} + \frac{i}{2}\vec{\epsilon}\cdot\vec{\sigma} - \frac{1}{2}\vec{\beta}\cdot\vec{\sigma}\right)\eta_R\right]^\dagger\left(\mathbf{1} + \frac{i}{2}\vec{\epsilon}\cdot\vec{\sigma} + \frac{1}{2}\vec{\beta}\cdot\vec{\sigma}\right)\chi_L$$

$$= \eta_R^\dagger\left(\mathbf{1} + \frac{i}{2}\vec{\epsilon}\cdot\vec{\sigma} - \frac{1}{2}\vec{\beta}\cdot\vec{\sigma}\right)^\dagger\left(\mathbf{1} + \frac{i}{2}\vec{\epsilon}\cdot\vec{\sigma} + \frac{1}{2}\vec{\beta}\cdot\vec{\sigma}\right)\chi_L$$

$$= \eta_R^\dagger\left(\mathbf{1} - \frac{i}{2}\vec{\epsilon}\cdot\vec{\sigma} - \frac{1}{2}\vec{\beta}\cdot\vec{\sigma}\right)\left(\mathbf{1} + \frac{i}{2}\vec{\epsilon}\cdot\vec{\sigma} + \frac{1}{2}\vec{\beta}\cdot\vec{\sigma}\right)\chi_L$$

$$= \eta_R^\dagger\left(\mathbf{1} - \frac{i}{2}\vec{\epsilon}\cdot\vec{\sigma} - \frac{1}{2}\vec{\beta}\cdot\vec{\sigma} + \frac{i}{2}\vec{\epsilon}\cdot\vec{\sigma} + \frac{1}{2}\vec{\beta}\cdot\vec{\sigma}\right)\chi_L$$

$$= \eta_R^\dagger\chi_L \tag{B.2}$$

where in the third step we used the fact that the Pauli matrices are hermitian, and in the fourth step we have only kept the terms linear in $\vec{\epsilon}$ and $\vec{\beta}$. The proof that $\chi_L^\dagger\eta_R$ is also invariant is similar.

2.3 Let us first consider the transformation of $\eta^\dagger\sigma^0\eta$, which is simply $\eta^\dagger\eta$ because σ^0 is the identity matrix (we will not show explicitly the label R on the right-chiral spinor η in this solution). We get (following steps that are similar to the previous exercise)

$$(\eta')^\dagger\eta' = \eta^\dagger\left(\mathbf{1} - \frac{i}{2}\vec{\epsilon}\cdot\vec{\sigma} - \frac{1}{2}\vec{\beta}\cdot\vec{\sigma} + \frac{i}{2}\vec{\epsilon}\cdot\vec{\sigma} - \frac{1}{2}\vec{\beta}\cdot\vec{\sigma}\right)\eta$$

$$= \eta^\dagger\eta - \eta^\dagger\vec{\beta}\cdot\vec{\sigma}\eta$$

$$= \eta^\dagger\eta - \vec{\beta}\cdot\eta^\dagger\vec{\sigma}\eta$$

$$= \eta^\dagger\sigma^0\eta - \vec{\beta}\cdot\eta^\dagger\vec{\sigma}\eta \tag{B.3}$$

This is of the same form as the transformation of the zeroth component of a four-vector [see the transformation of the energy E in Eq. (2.33)] with $\eta^\dagger \vec{\sigma} \eta$ playing the role of the three-vector part.

Now consider the transformation of $\eta^\dagger \sigma^j \eta$ [we used an index j instead of i only because it will make it less confusing to use Eq. (2.14)]. We get, to first order in $\vec{\beta}$ and $\vec{\epsilon}$,

$$(\eta')^\dagger \sigma^j \eta' = \eta^\dagger \left(1 - \frac{i}{2}\vec{\epsilon}\cdot\vec{\sigma} - \frac{1}{2}\vec{\beta}\cdot\vec{\sigma}\right)\sigma^j\left(1 + \frac{i}{2}\vec{\epsilon}\cdot\vec{\sigma} - \frac{1}{2}\vec{\beta}\cdot\vec{\sigma}\right)\eta$$

$$= \eta^\dagger \sigma^j \eta + \eta^\dagger\left(-\frac{i}{2}\vec{\epsilon}\cdot\vec{\sigma}\sigma^j - \frac{1}{2}\vec{\beta}\cdot\vec{\sigma}\sigma^j + \frac{i}{2}\sigma^j\vec{\epsilon}\cdot\vec{\sigma} - \frac{1}{2}\sigma^j\vec{\beta}\cdot\vec{\sigma}\right)\eta$$

$$\text{(B.4)}$$

Consider the sum of the second and fourth terms on the right-hand side. Using Eq. (2.14), we obtain

$$\eta^\dagger\left(-\frac{i}{2}\vec{\epsilon}\cdot\vec{\sigma}\sigma^j + \frac{i}{2}\sigma^j\vec{\epsilon}\cdot\vec{\sigma}\right)\eta = -\left(\vec{\epsilon}\times(\eta^\dagger\vec{\sigma}\eta)\right)_j$$

For the third and fifth terms on the right-hand side of Eq. (B.4), we use instead Eq. (2.9):

$$\eta^\dagger\left(-\frac{1}{2}\vec{\beta}\cdot\vec{\sigma}\sigma^j - \frac{1}{2}\sigma^j\vec{\beta}\cdot\vec{\sigma}\right)\eta = \eta^\dagger\left(-\frac{1}{2}\beta_i\left(\sigma_i\sigma_j + \sigma_j\sigma_i\right)\right)\eta$$

$$= -\beta_i\delta_{ij}\eta^\dagger\eta$$

$$= -\beta_j\eta^\dagger\sigma^0\eta \qquad \text{(B.5)}$$

We therefore get for Eq. (B.4)

$$(\eta')^\dagger\vec{\sigma}\eta' = \eta^\dagger\vec{\sigma}\eta - \vec{\epsilon}\times(\eta^\dagger\vec{\sigma}\eta) - \vec{\beta}\,\eta^\dagger\sigma^0\eta$$

This is of the same form as the transformation of the three-vector part of a four-vector [see the transformation of $\vec{p}$ in Eq. (2.33)] with, again, $\eta^\dagger\sigma^0\eta$ playing the role of the zeroth component and $\eta^\dagger\vec{\sigma}\eta$ playing the role of the three-vector.

We have therefore proved that $\eta^\dagger\sigma^\mu\eta$ indeed transforms as a four-vector. The proof that the quantity $\chi^\dagger\bar{\sigma}^\mu\chi$ is also a four-vector is similar.

Supersymmetry Demystified

2.4 Consider taking the hermitian conjugate of $i\sigma^2\chi_L^{\dagger T}$ and contracting this with $-i\sigma^2\eta_R^{\dagger T}$. We get

$$\left(i\sigma^2\chi_L^{\dagger T}\right)^\dagger\left(-i\sigma^2\eta_R^{\dagger T}\right) = -\left(\sigma^2\chi_L^{\dagger T}\right)^\dagger\sigma^2\eta_R^{\dagger T}$$

$$= -\chi_L^T(\sigma^2)^\dagger\sigma^2\eta_R^{\dagger T}$$

$$= -\chi_L^T\sigma^2\sigma^2\eta_R^\dagger$$

$$= -\chi_L^T\eta_R^{\dagger T}$$

$$= \eta_R^\dagger\chi_L \qquad \text{(B.6)}$$

where we have used the fact that σ^2 is hermitian and that $\sigma^2\sigma^2 = \mathbf{1}$. This invariant is already listed in Eq. (2.42).

Is is now obvious that

$$\left(-i\sigma^2\eta_R^{\dagger T}\right)^\dagger\left(i\sigma^2\chi_L^{\dagger T}\right) = -\eta_R^T\chi_L^{\dagger T} = \chi_L^\dagger\eta_R$$

which is also on the list of Eq. (2.42).

2.5 For a the vector quantity, we may simply take Eq. (2.45), which is a Lorentz invariant, and drop the derivative. The result,

$$\chi_L^\dagger\bar\sigma^\mu i\chi_L$$

is a contravariant vector.

To get a second rank tensor, we need a second sigma matrix. Let's start with

$$\bar\sigma^\mu\chi_L$$

From Eqs. (2.43) and (2.44), we see that if the index μ is contracted with something, this behaves like a right-chiral spinor. Equation (2.43) tells us that we must multiply a right-chiral spinor by the matrix σ^ν to get something with well-defined Lorentz properties, so let's consider

$$\sigma^\nu\bar\sigma^\mu\chi_L \qquad \text{(B.7)}$$

Imagining again for an instant that the Lorentz indices are contracted, we know from Eq. (2.43) that this quantity transforms like a left-chiral spinor. The first term in Eq. (2.42) shows that we must contract a left-chiral spinor with $-\chi_L^T i\sigma^2$ to get a Lorentz scalar. We therefore have that

$$-\chi_L^T i\sigma^2 \sigma^\nu \bar{\sigma}^\mu \chi_L$$

transforms like a second-rank contravariant tensor.

2.6 For example, we have

$$
\begin{aligned}
\lambda \cdot \chi &= \lambda^T(-i\sigma^2)\chi \\
&= -\chi^T(-i\sigma^2)^T\lambda \qquad \text{using Eq. (2.48)} \\
&= \chi^T(-i\sigma^2)\lambda \qquad \text{using the antisymmetry of } -i\sigma^2 \\
&= \chi \cdot \lambda
\end{aligned}
\tag{B.8}
$$

What happens therefore is that we pick up two minus signs: one from the anticommutativity of the spinors and one from the antisymmetry of the matrix $i\sigma^2$. Likewise, one can show that $\bar{\lambda} \cdot \bar{\chi} = \bar{\chi} \cdot \bar{\lambda}$.

2.7 We will only prove the first one because the two proofs are almost identical. In terms of explicit components, we have

$$\theta \cdot \chi = \theta^T(-i\sigma^2)\chi = \theta_2\chi_1 - \theta_1\chi_2$$

and a similar expression for $\theta \cdot \lambda$. Therefore,

$$
\begin{aligned}
\theta \cdot \chi\, \theta \cdot \lambda &= (\theta_2\chi_1 - \theta_1\chi_2)(\theta_2\lambda_1 - \theta_1\lambda_2) \\
&= -\theta_2\chi_1\theta_1\lambda_2 - \theta_1\chi_2\theta_2\lambda_1
\end{aligned}
\tag{B.9}
$$

(all the other terms are identically zero because all the parameters here are Grassmann quantities).

On the other hand, we have

$$
\begin{aligned}
-\frac{1}{2}\theta \cdot \theta\, \chi \cdot \lambda &= -\frac{1}{2}(\theta_2\theta_1 - \theta_1\theta_2)(\chi_2\lambda_1 - \chi_1\lambda_2) \\
&= -\theta_2\theta_1(\chi_2\lambda_1 - \chi_1\lambda_2) \\
&= -\theta_1\chi_2\theta_2\lambda_1 - \theta_2\chi_1\theta_1\lambda_2
\end{aligned}
$$

This is identical to Eq. (B.9), which completes the proof. An alternative derivation using Eq. (2.54) is

$$\theta \cdot \chi \, \theta \cdot \lambda = \theta_a \, (-i\sigma^2)_{ab} \, \chi_b \, \theta_c \, (-i\sigma^2)_{cd} \, \lambda_d$$

$$= -(-i\sigma^2)_{ab} \, (-i\sigma^2)_{cd} \, \theta_a \, \theta_c \, \chi_b \, \lambda_d$$

$$= -\frac{1}{4}(-i\sigma^2)_{ab} \, (-i\sigma^2)_{cd} \, (i\sigma^2)_{ac} \, (i\sigma^2)_{bd} \, \theta \cdot \theta \, \chi \cdot \lambda$$

$$= -\frac{1}{4}(-i\sigma^2)_{ab} \, (i\sigma^2)_{bd} \, (-i\sigma^2)_{dc} \, (i\sigma^2)_{ca} \, \theta \cdot \theta \, \chi \cdot \lambda$$

$$= -\frac{1}{4}\text{Tr}(\sigma^2\sigma^2\sigma^2\sigma^2) \, \theta \cdot \theta \, \chi \cdot \lambda$$

$$= -\frac{1}{4}\text{Tr}(\mathbf{1}) \, \theta \cdot \theta \, \chi \cdot \lambda$$

$$= -\frac{1}{2} \, \theta \cdot \theta \, \chi \cdot \lambda$$

3.1 By definition,

$$\eta \cdot \chi = \eta^T (-i\sigma^2)\chi$$

where on the right-hand side it is implicit that the two spinors have lower undotted components. To be more explicit, we have

$$\eta \cdot \chi = -\eta_1 \chi_2 + \eta_2 \chi_1 \tag{B.10}$$

On the other hand,

$$\chi \cdot \eta = -\chi_1 \eta_2 + \chi_2 \eta_1 \tag{B.11}$$

The expressions (B.10) and (B.11) are equal because $-\eta_1\chi_2 = \chi_2\eta_1$ and $\eta_2\chi_1 = -\chi_1\eta_2$.

3.2 We must express $\bar{\eta}_{\dot{a}}$ in terms of components with upper dotted indices because this is the type of index that is implicitly assigned to all the right-chiral spinors in Eq. (2.42). Using Eq. (3.17), we get

$$\bar{\eta}_{\dot{a}}\bar{\eta}^{\dot{a}} = (-i\sigma^2)_{\dot{a}\dot{b}}\bar{\eta}^{\dot{b}}\bar{\eta}^{\dot{a}}$$

$$= \bar{\eta}^{\dot{b}}(-i\sigma^2)_{\dot{a}\dot{b}}\bar{\eta}^{\dot{a}}$$

$$= \bar{\eta}^{\dot{b}}(i\sigma^2)_{\dot{b}\dot{a}}\bar{\eta}^{\dot{a}}$$

which corresponds to the expression $\eta^T i\sigma^2 \eta$ of Eq. (2.42).

3.3 We have

$$\chi \cdot \chi = \chi_2 \chi_1 - \chi_1 \chi_2$$
$$= 2\chi_2 \chi_1$$
$$= -2\chi_1 \chi_2 \qquad \text{(B.12)}$$

Using Eq. (3.16), we can rewrite this as

$$\chi \cdot \chi = -2\chi^1 \chi^2$$
$$= 2\chi^2 \chi^1 \qquad \text{(B.13)}$$

We could, of course, have instead used

$$\chi \cdot \chi = \chi^a \chi_a = \chi^1 \chi_1 + \chi^2 \chi_2$$

and then Eq. (3.16) to recover Eq. (B.13). Either way, the result is

$$\chi^2 \chi^1 = \frac{1}{2} \chi \cdot \chi \qquad \chi^1 \chi^2 = -\frac{1}{2} \chi \cdot \chi$$

And, of course, $\chi^1 \chi^1 = \chi^2 \chi^2 = 0$. On the other hand,

$$(i\sigma^2)^{12} = 1 \qquad (i\sigma^2)^{21} = -1$$

whereas the diagonal elements are zero. We therefore find

$$\chi^a \chi^b = -\frac{1}{2} (i\sigma^2)^{ab} \chi \cdot \chi$$

as was to be shown.
    On the other hand, using Eq. (3.13), we have

$$\bar{\chi} \cdot \bar{\chi} = \bar{\chi}_i \bar{\chi}^i + \bar{\chi}_2 \bar{\chi}^2 = -\bar{\chi}^2 \bar{\chi}^i + \bar{\chi}^i \bar{\chi}^2$$

so that

$$\bar{\chi}^2 \bar{\chi}^i = -\frac{1}{2} \bar{\chi} \cdot \bar{\chi} \qquad \bar{\chi}^i \bar{\chi}^2 = \frac{1}{2} \bar{\chi} \cdot \bar{\chi}$$

which gives us

$$\bar{\chi}^{\dot{a}} \bar{\chi}^{\dot{b}} = \frac{1}{2} (i\sigma^2)^{\dot{a}\dot{b}} \, \bar{\chi} \cdot \bar{\chi}$$

3.4 This is actually a trick question. One can tell that this expression is zero without doing any work. Indeed, $\chi \cdot \chi$ contains one factor of χ_1 and one factor of χ_2, whereas $\chi \cdot \eta$ has two terms that contain either χ_1 or χ_2. Therefore, all terms contain the square of either χ_1 or χ_2, and the whole expression is identically zero.

3.5 Following the hint, we take the trace of both sides of Eq. (2.24). This gives

$$\text{Tr}(\sigma^\mu \bar{\sigma}^\nu) + \text{Tr}(\sigma^\nu \bar{\sigma}^\mu) = 2\eta^{\mu\nu}\text{Tr}(\mathbf{1})$$

From the properties of the Pauli matrices, each trace on the left is nonzero if and only if $\mu = \nu$. Therefore the two traces are equal to one another and we have

$$2\text{Tr}(\sigma^\mu \bar{\sigma}^\nu) = 4\eta^{\mu\nu}$$

which, after dividing by 2, completes the proof.

4.1 We have

$$\left(\chi^\dagger \bar{\sigma}^\mu i \partial_\mu \chi\right)^\dagger = (i\partial_\mu \chi)^\dagger (\bar{\sigma}^\mu)^\dagger \chi$$

Note that we did not pick up a minus sign when we switched the order of the spinors, according to the rule (2.56). Applying the hermitian conjugate to the parentheses and using the fact that the Pauli matrices are hermitian, we get

$$(i\partial_\mu \chi)^\dagger (\bar{\sigma}^\mu)^\dagger \chi = (-i\partial_\mu \chi^\dagger)\bar{\sigma}^\mu \chi$$
$$= \chi^\dagger \bar{\sigma}^\mu i \partial_\mu \chi$$

using an integration by parts in the last step. This is the original kinetic term, as was to be shown.

4.2 We have

$$-m\left(\chi_{\bar{p}} \cdot \chi_p + \bar{\chi}_p \cdot \bar{\chi}_{\bar{p}}\right)^\dagger = -m\left(\chi_{\bar{p}} \cdot \chi_p\right)^\dagger - m\left(\bar{\chi}_p \cdot \bar{\chi}_{\bar{p}}\right)^\dagger$$
$$= -m\bar{\chi}_{\bar{p}} \cdot \bar{\chi}_p - m\chi_p \cdot \chi_{\bar{p}} \qquad \text{(B.14)}$$

using in the last step Eq. (A.44). This is indeed equal to Eq. (4.11) because the order in spinor dot products does not matter.

4.3 The second term of Eq. (4.13) agrees with the second term of Eq. (4.12). Let's focus on the first term. First, use Eq. (A.5) to get rid of the σ^2 matrices:

$$i\chi_{\bar{e}}^T \sigma^2 \sigma^\mu \partial_\mu \sigma^2 \chi_{\bar{e}}^{\dagger T} = i\chi_{\bar{e}}^T \bar{\sigma}^{\mu T} \partial_\mu \chi_{\bar{e}}^{\dagger T}$$

Next, we use Eq. (A.36) to get rid of all the annoying transpose operations:

$$i\chi_{\bar{e}}^T \bar{\sigma}^{\mu T} \partial_\mu \chi_{\bar{e}}^{\dagger T} = -i\left(\partial_\mu \chi_{\bar{e}}^\dagger\right)\bar{\sigma}^\mu \chi_{\bar{e}}$$

Finally, we do an integration by parts to get the derivative to act on the spinor on the right:

$$-i\left(\partial_\mu \chi_{\bar{e}}^\dagger\right)\bar{\sigma}^\mu \chi_{\bar{e}} = i\chi_{\bar{e}}^\dagger \bar{\sigma}^\mu \partial_\mu \chi_{\bar{e}}$$

which is the first term of Eq. (4.13).

4.4 In Eq. (4.14), an integration by parts has been performed on one of the derivatives, which introduced a minus sign. We must replace ∂_μ by $\partial_\mu - ieA_\mu$ *before* any integration by parts on ∂_μ has been made. After an integration by parts, we have to make the replacement $\partial_\mu \to \partial_\mu + ieA_\mu$, but this gets confusing because we must keep track of which partial derivative has been integrated by parts!

5.1 For the hermitian conjugate of the scalar field transformation, the proof is trivial if we use Eq. (A.44) and the fact that the order does not matter in spinor dot products:

$$(\delta\phi)^\dagger = (\zeta \cdot \chi)^\dagger = \bar{\xi} \cdot \bar{\chi} = \bar{\chi} \cdot \bar{\xi} = \chi^\dagger(i\sigma^2)\zeta^*$$

where the last equality follows from the definition of the dot product between barred spinors.

Now consider

$$(\delta\chi)^\dagger = \left[-C^*(\partial_\mu\phi)\,\sigma^\mu i\sigma^2\zeta^*\right]^\dagger$$

Using the fact that $(AB)^\dagger = B^\dagger A^\dagger$, we get

$$(\delta\chi)^\dagger = -C(\partial_\mu\phi^\dagger)\zeta^T(i\sigma^2)^\dagger(\sigma^\mu)^\dagger$$

Now we use the fact that Pauli matrices are hermitian to obtain

$$(\delta\chi)^\dagger = C(\partial_\mu\phi^\dagger)\zeta^T i\sigma^2\sigma^\mu$$

where the change of sign comes uniquely from the complex conjugation of the i factor.

5.2 We have

$$\sigma^\mu\bar\sigma^\nu\partial_\mu\partial_\nu\phi = (\sigma^0\partial_t + \sigma^i\partial_i)(\sigma^0\partial_t - \sigma^j\partial_j)\phi$$
$$= \partial_t^2\phi - \sigma^i\sigma^j\partial_i\partial_j\phi$$

Now we use the fact that $\sigma^i\sigma^j = \delta^{ij} + i\epsilon^{ijk}\sigma^k$. The second term is antisymmetric in the indices i and j, so it won't contribute when contracted with the symmetric expression $\partial_i\partial_j\phi$. We therefore end up with

$$\partial_t^2\phi - \partial_i\partial_i\phi = \partial_\mu\partial^\mu\phi$$

6.1 Using the antisymmetry of the coefficients $\omega^{\mu\nu}$, we write

$$\phi(x^\mu + \omega^{\mu\nu}x_\nu) \approx \varphi(x) + \omega^{\mu\nu}x_\nu\partial_\mu\phi(x)$$
$$= \phi(x) + \frac{1}{2}(\omega^{\mu\nu} - \omega^{\nu\mu})x_\nu\partial_\mu\phi(x)$$
$$= \phi(x) + \frac{1}{2}\omega^{\mu\nu}x_\nu\partial_\mu\phi(x) - \frac{1}{2}\omega^{\nu\mu}x_\nu\partial_\mu\phi(x)$$
$$= \phi(x) + \frac{1}{2}\omega^{\mu\nu}(x_\nu\partial_\mu - x_\mu\partial_\nu)\phi(x)$$

where in the last step we simply relabeled $\mu \leftrightarrow \nu$ in the last term.

Using Eq. (6.12), we have

$$-\left[\frac{1}{2}\omega^{\mu\nu}M_{\mu\nu}, \phi\right] = -\frac{i}{2}\omega^{\mu\nu}(x_\nu\partial_\mu - x_\mu\partial_\nu)\phi \qquad (B.15)$$

where the minus sign on the left is due to the fact that the argument of the exponential in our definition of U is negative. We immediately get the result we were seeking, i.e.,

$$[M_{\mu,\nu}, \phi] = i(x_\nu\partial_\mu - x_\mu\partial_\nu)\phi \qquad (B.16)$$

6.2 Using the derivation made in the solution of Exercise 6.1, we write

$$\phi\left(x^\mu + \omega^{\mu\nu}x_\nu\right) \approx \varphi + \frac{1}{2}\omega^{\mu\nu}(x_\nu\partial_\mu - x_\mu\partial_\nu)\phi$$

which we set equal to

$$e^{\frac{i}{2}\omega^{\mu\nu}\hat{M}_{\mu\nu}}\varphi \approx \phi + \frac{i}{2}\omega^{\mu\nu}\hat{M}_{\mu\nu}\phi$$

which immediately leads to

$$\hat{M}_{\mu\nu} = i(x_\mu\partial_\nu - x_\nu\partial_\mu)$$

6.3 Consider, for example,

$$[\hat{M}_{\mu\nu}, \hat{P}_\lambda] = i^2[x_\mu\partial_\nu - x_\nu\partial_\mu, \partial_\lambda]$$

The trick to handle this is familiar from quantum mechanics: We simply imagine that the whole expression is acting on a test function. With this in mind, we get

$$\begin{aligned}
\left[\hat{M}_{\mu\nu}, \hat{P}_\lambda\right] &= -x_\mu\partial_\nu\partial_\lambda + (\partial_\lambda x_\mu)\partial_\nu + x_\mu\partial_\nu\partial_\lambda + x_\nu\partial_\mu\partial_\lambda - (\partial_\lambda x_\nu)\partial_\mu - x_\nu\partial_\lambda\partial_\mu \\
&= (\partial_\lambda x_\mu)\partial_\nu - (\partial_\lambda x_\nu)\partial_\mu \\
&= \eta_{\lambda\mu}\partial_\nu - \eta_{\lambda\nu}\partial_\mu \qquad\qquad\qquad\qquad\qquad (B.17)
\end{aligned}$$

You might have expected the derivative of the coordinates to give a Kronecker delta instead of a metric. The reason we get the metric is clear if we explicitly write the variable with respect to which the derivative is taken. For example,

$$\partial_\lambda x_\mu = \frac{\partial x_\mu}{\partial x^\lambda}$$

If both indices are equal to zero, we get

$$\frac{\partial x_0}{\partial x^0} = 1$$

because $x_0 = x^0$ (with our choice of metric). On the other hand, when the indices are spatial, we have

$$\frac{\partial x_i}{\partial x^i} = -1$$

because $x_i = -x^i$, so $\partial_\lambda x_\mu = \eta_{\lambda\mu}$.

Coming back to Eq. (B.17), we may use $\partial_\nu = -i\hat{P}_\nu$ and $\partial_\mu = -i\hat{P}_\mu$ to finally obtain

$$[\hat{M}_{\mu\nu}, \hat{P}_\lambda] = i(\eta_{\lambda\nu}\hat{P}_\mu - \eta_{\lambda\mu}\hat{P}_\nu)$$

We thus have shown that the representation as differential operators obeys the desired algebra. But the algebra is a fundamental property of the group, which is independent of the representation used, so our derivation gives the general result.

It is straightforward to verify the commutator of two Lorentz generators using the same approach.

6.4 We consider for the first transformation a translation with parameter a^λ and for second transformation a Lorentz transformations with parameters $\omega^{\mu\nu}$:

$$\delta_\omega \delta_a \phi - \delta_a \delta_\omega \phi = \delta_\omega(a^\lambda \partial_\lambda \phi) - \frac{1}{2}\omega^{\mu\nu}\delta_a(x_\nu \partial_\mu \phi - x_\mu \partial_\nu \phi)$$

$$= \frac{1}{2}a^\lambda \omega^{\mu\nu}\partial_\lambda(x_\nu \partial_\mu \phi - x_\mu \partial_\nu \phi) - \frac{1}{2}a^\lambda \omega^{\mu\nu}(x_\nu \partial_\mu \partial_\lambda \phi - x_\mu \partial_\nu \partial_\lambda \phi)$$

$$= \frac{1}{2}a^\lambda \omega^{\mu\nu}(\eta_{\nu\lambda}\partial_\mu \phi - \eta_{\lambda\mu}\partial_\nu \phi)$$

$$= \frac{i}{2}a^\lambda \omega^{\mu\nu}(\eta_{\nu\lambda}[P_\mu, \phi] - \eta_{\lambda\mu}[P_\nu, \phi])$$

From Eq. (6.40), this is equal to

$$\left[\left[a^\lambda P_\lambda, -\frac{1}{2}\omega^{\mu\nu}M_{\mu\nu}\right], \phi\right]$$

where we used $\exp(ia^\lambda P_\lambda)$ for the operator generating translations and $\exp(-i\omega^{\mu\nu}M_{\mu\nu}/2)$ for the operator generating Lorentz transformations. We

thus find

$$i\eta_{\nu\lambda}[P_\mu, \phi] - i\eta_{\lambda\mu}[P_\nu, \phi] = [[P_\lambda, -M_{\mu\nu}], \phi]$$

or

$$i\eta_{\nu\lambda}P_\mu - i\eta_{\lambda\mu}P_\nu = [M_{\mu\nu}, P_\lambda]$$

as was to be shown.

6.5 a. Let's work out the right-hand side of Eq. (6.82). Consider the first term:

$$\delta_\omega \delta_\zeta \phi = \delta_\omega(-\zeta^T i\sigma^2 \chi) \tag{B.18}$$

From Eq. (6.81), we have

$$\delta_\omega \delta_\zeta \phi = -\zeta^T i\sigma^2 \delta_\omega \chi$$

$$= -\frac{i}{2}\omega^{\mu\nu}\left(i\zeta^T\sigma^2\sigma_{\mu\nu}\chi + x_\nu\zeta^T\sigma^2\partial_\mu\chi - x_\mu\zeta^T\sigma^2\partial_\nu\chi\right) \tag{B.19}$$

On the other hand,

$$\delta_\zeta \delta_\omega \phi \approx \delta_\zeta\left[\frac{1}{2}\omega^{\mu\nu}\left(x_\nu\partial_\mu\phi - x_\mu\partial_\nu\phi\right)\right]$$

$$= -\frac{i}{2}\omega^{\mu\nu}\left(x_\nu\zeta^T\sigma^2\partial_\mu\chi - x_\mu\zeta^T\sigma^2\partial_\nu\chi\right) \tag{B.20}$$

We see the reason the order of the two transformations matters is entirely due to the spin generators in the Lorentz transformations, which appear in Eq. (B.19) but not in Eq. (B.20) because in the first case the Lorentz transformation acts on a spinor, whereas in the second case it acts on a scalar field.

Substituting Eqs. (B.19) and (B.20) into Eq. (6.82), we get

$$-\frac{1}{2}\left[[\zeta \cdot Q, \omega^{\mu\nu}M_{\mu\nu}], \phi\right] = \frac{1}{2}\omega^{\mu\nu}\zeta^T\sigma^2\sigma_{\mu\nu}\chi \tag{B.21}$$

Using Eq. (6.85), we have

$$-\frac{1}{2}\left[[\zeta \cdot Q, \omega^{\mu\nu}M_{\mu\nu}], \phi\right] = \frac{i}{2}\omega^{\mu\nu}\zeta^T\sigma^2\sigma_{\mu\nu}[Q, \phi] \tag{B.22}$$

from which we conclude

$$[\zeta \cdot Q, M_{\mu\nu}] = -i\zeta^T \sigma^2 \sigma_{\mu\nu} Q \tag{B.23}$$

Let's write the indices explicitly. As usual, we assign lower indices to both ζ and Q and make sure that repeated indices appear once in an upper position and once in a lower position, which uniquely fixes the positions of all the indices. This gives us

$$\left[\zeta_a(-i\sigma^2)^{ab} Q_b, M_{\mu\nu}\right] = -i\zeta_a(\sigma^2)^{ab}(\sigma_{\mu\nu})_b^{\ c} Q_c \tag{B.24}$$

Note that the matrices $\sigma_{\mu\nu}$ are antisymmetric, so we need to keep track of which index comes first, which is why the c is shifted a bit, to clearly indicate that it is the second index. Since ζ_a is arbitrary, we finally obtain

$$[Q_b, M_{\mu\nu}] = (\sigma_{\mu\nu})_b^{\ c} Q_c$$

as was to be shown.

b. Let's first evaluate

$$\delta_\omega \delta_\zeta \phi^\dagger = \delta_\omega (\bar{\chi} \cdot \bar{\zeta})$$
$$= (\delta_\omega \bar{\chi}_{\dot{a}}) \bar{\zeta}^{\dot{a}} \tag{B.25}$$

We need the Lorentz transformation of $\bar{\chi}$, which we can obtain simply by taking the hermitian conjugate of Eq. (6.81):

$$\delta_\omega \bar{\chi} = \frac{1}{2}\omega^{\mu\nu}\left(-i\bar{\chi}\,\bar{\sigma}_{\mu\nu} + x_\nu \partial_\mu \bar{\chi} - x_\mu \partial_\nu \bar{\chi}\right) \tag{B.26}$$

where we have used $(\sigma_{\mu\nu})^\dagger = \bar{\sigma}_{\mu\nu}$, which can be confirmed easily using the definitions in Eqs. (6.78) and (6.79).

Now let's write Eq. (B.26) with the indices explicitly shown. In Eq. (6.81), the spinors have lower undotted indices. This means that in Eq. (B.26), the spinors also have lower indices, except that they are now dotted because we have taken the hermitian conjugate. Therefore, the transformation is

$$\delta_\omega \bar{\chi}_{\dot{a}} = \frac{1}{2}\omega^{\mu\nu}\left[-i\bar{\chi}_{\dot{b}}\,(\bar{\sigma}_{\mu\nu})^{\dot{b}}_{\ \dot{a}} + x_\nu \partial_\mu \bar{\chi}_{\dot{a}} - x_\mu \partial_\nu \bar{\chi}_{\dot{a}}\right] \tag{B.27}$$

The indices of $\bar{\sigma}_{\mu\nu}$ have been chosen to match the rest of the expression, but we can check that they are indeed correct by using Eqs. (3.33) and (3.34) in the definition of $\bar{\sigma}_{\mu\nu}$ [Eq. (6.79)]. Indeed, the right-hand side of Eq. (6.79) is

$$\frac{i}{4}\left[(\bar{\sigma}_\mu)^{\dot{a}b}(\sigma_\nu)_{b\dot{c}} - (\bar{\sigma}_\nu)^{\dot{a}b}(\sigma_\mu)_{b\dot{c}}\right]$$

which carries indeed indices $(\ )^{\dot{a}}_{\dot{c}}$.

Substituting Eq. (B.27) into Eq. (B.25), we finally obtain

$$\delta_\omega\delta_\zeta\phi^\dagger = \frac{1}{2}\omega^{\mu\nu}\left[-i\,\bar{\chi}_{\dot{b}}\,(\bar{\sigma}_{\mu\nu})^{\dot{b}}_{\ \dot{a}}\,\bar{\xi}^{\dot{a}} + x_\nu\,(\partial_\mu\bar{\chi}_{\dot{a}})\bar{\xi}^{\dot{a}} - x_\mu\,(\partial_\nu\bar{\chi}_{\dot{a}})\bar{\xi}^{\dot{a}}\right]$$

$$= \frac{1}{2}\omega^{\mu\nu}\left[-i\,\bar{\chi}\,(\bar{\sigma}_{\mu\nu}\bar{\xi}) + x_\nu\,(\partial_\mu\bar{\chi})\cdot\bar{\xi} - x_\mu\,(\partial_\nu\bar{\chi})\cdot\bar{\xi}\right] \quad \text{(B.28)}$$

where we are using the shorthand notation discussed in Chapter 3 to avoid having to show the indices explicitly.

On the other hand, we find [using Eq. (6.23)]

$$\delta_\zeta\delta_\omega\phi^\dagger = \frac{1}{2}\omega^{\mu\nu}\,\delta_\zeta\left(x_\nu\partial_\mu\phi^\dagger - x_\mu\partial_\nu\phi^\dagger\right)$$

$$= \frac{1}{2}\omega^{\mu\nu}\left[x_\nu(\partial_\mu\bar{\chi})\cdot\bar{\xi} - x_\mu(\partial_\nu\bar{\chi})\cdot\bar{\xi}\right]$$

so that, finally, Eq. (6.87) gives

$$-\frac{1}{2}\left[\left[\bar{Q}_{\dot{a}}\bar{\xi}^{\dot{a}},\,\omega^{\mu\nu}M_{\mu\nu}\right],\,\phi^\dagger\right] = -\frac{i}{2}\omega^{\mu\nu}\,\bar{\chi}_{\dot{b}}\,(\bar{\sigma}_{\mu\nu})^{\dot{b}}_{\ \dot{a}}\,\bar{\xi}^{\dot{a}} \quad \text{(B.29)}$$

which can be simplified to

$$\left[\left[\bar{Q}_{\dot{a}},\,M_{\mu\nu}\right],\,\phi^\dagger\right] = i\,\bar{\chi}_{\dot{b}}\,(\bar{\sigma}_{\mu\nu})^{\dot{b}}_{\ \dot{a}} \quad \text{(B.30)}$$

The only missing ingredient is the commutator of $\bar{Q}_{\dot{a}}$ with $\phi^\dagger$. Instead of Eq. (6.84), we now have

$$\left[\bar{Q}\cdot\bar{\xi},\,\psi^\dagger\right] = -i\,\bar{\chi}\cdot\bar{\xi} \quad \text{(B.31)}$$

which implies

$$\left[\bar{Q}_{\dot{b}}, \phi^{\dagger}\right] = -i\,\bar{\chi}_{\dot{b}} \tag{B.32}$$

Using this to replace $\bar{\chi}_{\dot{b}}$ on the right-hand side of Eq. (B.30), we obtain our final result, i.e.,

$$\left[\bar{Q}_{\dot{a}}, M_{\mu\nu}\right] = -\bar{Q}_{\dot{b}}\,(\bar{\sigma}^{\mu\nu})^{\dot{b}}_{\ \dot{a}}$$

which confirms the first relation in Eq. (6.86). To obtain the second relation, we express the supercharges in terms of their components with upper dotted indices:

$$\epsilon_{\dot{a}\dot{c}}\left[\bar{Q}^{\dot{c}}, M_{\mu\nu}\right] = -\epsilon_{\dot{b}\dot{d}}\,\bar{Q}^{\dot{d}}\,(\bar{\sigma}^{\mu\nu})^{\dot{b}}_{\ \dot{a}}$$

To get rid of the ϵ on the left side, let's contract both sides with $\epsilon^{\dot{e}\dot{a}}$. On the left, we may use Eq. (A.25), namely $\epsilon^{\dot{e}\dot{a}}\epsilon_{\dot{a}\dot{c}} = \delta^{\dot{e}}_{\dot{c}}$, so the equation becomes

$$\left[\bar{Q}^{\dot{e}}, M_{\mu\nu}\right] = -\epsilon^{\dot{e}\dot{a}}\,\epsilon_{\dot{b}\dot{d}}\,\bar{Q}^{\dot{d}}\,(\bar{\sigma}^{\mu\nu})^{\dot{b}}_{\ \dot{a}}$$

$$= \epsilon^{\dot{e}\dot{a}}\,\epsilon_{\dot{d}\dot{b}}\,\bar{Q}^{\dot{d}}\,(\bar{\sigma}^{\mu\nu})^{\dot{b}}_{\ \dot{a}}$$

where we have used the antisymmetry of the ϵ symbol. Using Eq. (6.88) (which we will prove in a second), we finally obtain the second commutator of Eq. (6.86), after relabeling some of the indices.

That leaves us the proof of Eq. (6.88). Let's do it! Using Eq. (6.79) with all the indices raised, we have

$$\epsilon^{\dot{c}\dot{b}}\epsilon_{\dot{d}\dot{a}}\,(\bar{\sigma}^{\mu\nu})^{\dot{a}}_{\ \dot{b}} = \frac{i}{4}\epsilon^{\dot{c}\dot{b}}\epsilon_{\dot{d}\dot{a}}\,(\bar{\sigma}^{\mu})^{\dot{a}e}(\sigma^{\nu})_{e\dot{b}} - \mu \leftrightarrow \nu$$

$$= \frac{i}{4}\epsilon^{\dot{c}\dot{b}}\epsilon_{\dot{d}\dot{a}}\,\delta^{e}_{f}(\bar{\sigma}^{\mu})^{\dot{a}f}(\sigma^{\nu})_{e\dot{b}} - \mu \leftrightarrow \nu$$

$$= \frac{i}{4}\epsilon^{\dot{c}\dot{b}}\epsilon_{\dot{d}\dot{a}}\,\epsilon_{fg}\epsilon^{ge}(\bar{\sigma}^{\mu})^{\dot{a}f}(\sigma^{\nu})_{e\dot{b}} - \mu \leftrightarrow \nu \quad \text{[using Eq. (A.24)]}$$

$$= -\frac{i}{4}\,(\sigma^{\mu})_{g\dot{d}}(\bar{\sigma}^{\nu})^{\dot{c}g} - \mu \leftrightarrow \nu \quad \text{[using Eqs. (A.26) and (A.27)]}$$

$$= (\bar{\sigma}^{\mu\nu})^{\dot{c}}_{\ \dot{d}} \tag{B.33}$$

as was to be shown.

6.6 When the spinor dot products are written fully, Eq. (6.92) is

$$\alpha_a[\beta^T(-i\sigma^2)\gamma] + \beta_a[\gamma^T(-i\sigma^2)\alpha] + \gamma_a[\alpha^T(-i\sigma^2)\beta] = 0$$

with

$$\beta^T(-i\sigma^2)\gamma = \beta_2\gamma_1 - \beta_1\gamma_2$$

and so on. The identity then is

$$\alpha_a\beta_2\gamma_1 - \alpha_a\beta_1\gamma_2 + \beta_a\gamma_2\alpha_1 - \beta_a\gamma_1\alpha_2 + \gamma_a\alpha_2\beta_1 - \gamma_a\alpha_1\beta_2 = 0$$

If we set $a = 1$, the left side is

$$\alpha_1\beta_2\gamma_1 - \alpha_1\beta_1\gamma_2 + \beta_1\gamma_2\alpha_1 - \beta_1\gamma_1\alpha_2 + \gamma_1\alpha_2\beta_1 - \gamma_1\alpha_1\beta_2$$

To see the cancellation explicitly, let's move all the components in the order $\alpha\beta\gamma$. Including a minus sign every time we move a component passed another one, we get

$$\alpha_1\beta_2\gamma_1 - \alpha_1\beta_1\gamma_2 + \alpha_1\beta_1\gamma_2 - \alpha_2\beta_1\gamma_1 + \alpha_2\beta_1\gamma_1 - \alpha_1\beta_2\gamma_1$$

which is clearly zero because all terms cancel pairwise. The proof with $a = 2$ is similar.

6.7 In the case of the scalar field, we want to show that the new term in the transformation of the spinor field does not ruin Eq. (6.61), which was of the desired form. We have

$$\delta_\zeta\phi = \zeta \cdot \chi$$

so

$$\delta_\beta\delta_\zeta\phi = \delta_\beta(\zeta \cdot \chi)$$
$$= \text{previous expression} + F\zeta \cdot \beta \qquad (B.34)$$

On the other hand,

$$\delta_\zeta\delta_\beta\phi = \text{previous expression} + F\beta \cdot \zeta \qquad (B.35)$$

so

$$\delta_\beta \delta_\zeta \phi - \delta_\zeta \delta_\beta \phi = \text{previous expression} + F\zeta \cdot \beta - F\beta \cdot \zeta$$

$$= -i\left(\zeta^\dagger \bar{\sigma}^\mu \beta - \beta^\dagger \bar{\sigma}^\mu \zeta\right)\partial_\mu \phi + F\zeta \cdot \beta - F\beta \cdot \zeta \qquad \text{(B.36)}$$

using our result (6.61). But clearly, the last two terms cancel out because $\zeta \cdot \beta = \beta \cdot \zeta$! So our result of the commutator of two SUSY transformations on the scalar field is, thankfully, unaffected.

Consider now the transformation of the spinor field. We have

$$\delta_\beta \delta_\zeta \chi_a = \delta_\beta\left[(-i\sigma^\mu i\sigma^2 \zeta^*)_a \partial_\mu \phi + F\zeta_a\right]$$

$$= \text{previous expression} - i(\beta^\dagger \bar{\sigma}^\mu \partial_\mu \chi)\zeta_a \qquad \text{(B.37)}$$

whereas

$$\delta_\zeta \delta_\beta \chi_a = \text{previous expression} - i(\zeta^\dagger \bar{\sigma}^\mu \partial_\mu \chi)\beta_a \qquad \text{(B.38)}$$

Therefore,

$$\delta_\beta \delta_\zeta \chi_a - \delta_\zeta \delta_\beta \chi_a = \text{previous expression} - i(\beta^\dagger \bar{\sigma}^\mu \partial_\mu \chi)\zeta_a + i(\zeta^\dagger \bar{\sigma}^\mu \partial_\mu \chi)\beta_a$$

$$\text{(B.39)}$$

with the previous expression being given in Eq. (6.93). Substituting Eq. (6.93) into Eq. (B.39), we get that the two offending terms in Eq. (6.93) that did not vanish off-shell are exactly canceled by the two extra terms in Eq. (B.39), leaving us with the result we were wishing for:

$$\delta_\beta \delta_\zeta \chi_a - \delta_\zeta \delta_\beta \chi_a = -i\left(\zeta^\dagger \bar{\sigma}^\mu \beta - \beta^\dagger \bar{\sigma}^\mu \zeta\right)\partial_\mu \chi_a$$

It remains to check the auxiliary field. This is done easily:

$$\delta_\beta \delta_\zeta F = \delta_\beta(-i\zeta^\dagger \bar{\sigma}^\mu \partial_\mu \chi)$$

$$= -i\zeta^\dagger \bar{\sigma}^\mu \sigma^\nu \sigma^2 \beta^* \partial_\mu \partial_\nu \phi - i(\zeta^\dagger \bar{\sigma}^\mu \beta)\partial_\mu F \qquad \text{(B.40)}$$

The first term on the right-hand side can be simplified using Eq. (5.13):

$$-i\zeta^\dagger \bar{\sigma}^\mu \sigma^\nu \sigma^2 \beta^* \partial_\mu \partial_\nu \phi = -i\zeta^\dagger \sigma^2 \beta^* \Box \phi$$

which is simply

$$-\bar{\xi} \cdot \bar{\beta} \Box \phi$$

by definition. Using this in Eq. (B.40), we have

$$\delta_\beta \delta_\zeta F = -\bar{\xi} \cdot \bar{\beta} \Box \phi - i(\zeta^\dagger \bar{\sigma}^\mu \beta) \partial_\mu F \qquad (B.41)$$

From this we readily obtain

$$\delta_\zeta \delta_\beta F = -\bar{\beta} \cdot \bar{\xi} \Box \phi - i(\beta^\dagger \bar{\sigma}^\mu \zeta) \partial_\mu F$$

so that

$$\delta_\beta \delta_\zeta F - \delta_\zeta \delta_\beta F = -\bar{\xi} \cdot \bar{\beta} \Box \phi + \bar{\beta} \cdot \bar{\xi} \Box \phi - i(\zeta^\dagger \bar{\sigma}^\mu \beta - \beta^\dagger \bar{\sigma}^\mu \zeta) \partial_\mu F$$

$$= -i(\zeta^\dagger \bar{\sigma}^\mu \beta - \beta^\dagger \bar{\sigma}^\mu \zeta) \partial_\mu F \qquad (B.42)$$

where we have used the fact that $\bar{\xi} \cdot \bar{\beta} = \bar{\beta} \cdot \bar{\xi}$ to cancel the two terms containing the scalar field. We finally see that, indeed, the auxiliary field also obeys Eq. (6.100).

7.1 We write

$$P_\mu W^\mu = W^\mu P_\mu + [P_\mu, W^\mu]$$

The first term on the right-hand side is zero, from Eq. (7.3). The commutator is equal to

$$[P_\mu, W^\mu] = \frac{1}{2} \epsilon^{\mu\nu\rho\sigma} [P_\mu, M_{\rho\sigma}] P_\nu$$

$$= -\frac{i}{2} \epsilon^{\mu\nu\rho\sigma} (\eta_{\mu\sigma} P_\rho - \eta_{\mu\rho} P_\sigma) \qquad (B.43)$$

using Eq. (6.34). This is equal to zero because the flat metric is symmetric, whereas the Levi-Civita symbol is totally antisymmetric.

7.2 We have to simplify

$$\frac{1}{2} \epsilon^{ijk} M_{jk} = \frac{1}{2} \epsilon^{ijk} (\sigma_{jk} + x_j p_k - x_k p_j)$$

Consider first the spin part:

$$\frac{1}{2}\epsilon^{ijk}\sigma_{jk} = \frac{i}{8}\,\epsilon^{ijk}(\sigma_j\,\bar{\sigma}_k - \sigma_k\,\bar{\sigma}_j)$$

$$= \frac{i}{8}\,\epsilon^{ijk}(-\sigma_j\,\sigma_k + \sigma_k\,\sigma_j)$$

$$= \frac{i}{8}\,\epsilon^{ijk}(-\sigma^j\,\sigma^k + \sigma^k\,\sigma^j)$$

$$= \frac{i}{8}\,\epsilon^{ijk}\,[\sigma^k,\sigma^j]$$

$$= -\frac{1}{4}\,\epsilon^{ijk}\,\epsilon^{kja}\,\sigma^a$$

$$= \frac{1}{4}\,\epsilon^{ijk}\,\epsilon^{ajk}\,\sigma^a$$

$$= \frac{1}{2}\,\delta^{ia}\,\sigma^a$$

$$= \frac{1}{2}\,\sigma^i$$

which is simply the spin S^i! Now consider the spatial part:

$$\frac{1}{2}\,\epsilon^{ijk}\,(x_j\,p_k - x_k\,p_j)$$

Since both the Levi-Civita symbol and the expression in parenthesis are anti-symmetric in the indices j and k, we may write this as

$$\epsilon^{ijk}\,x_j\,p_k$$

which is of course the ith component of the total orbital angular momentum, $L^i = (\vec{r} \times \vec{p})^i$. Our final result is therefore

$$\frac{1}{2}\,\epsilon^{ijk}\,M_{jk} = S^i + L^i$$

7.3 We have

$$W^0|p\rangle_0 = \frac{1}{2}\epsilon^{0ijk}M_{jk}P_i|p\rangle_0$$

But the only P_i that does not give zero when acting on this state is P_3, which is equal to $-P^3$. Therefore,

$$W^0|p\rangle_0 = -\frac{1}{2}\epsilon^{03jk}M_{jk}P^3|p\rangle_0$$

$$= -\frac{1}{2}\epsilon^{03jk}M_{jk}E|p\rangle_0$$

$$= \frac{1}{2}\epsilon^{3jk}M_{jk}E|p\rangle_0$$

$$= (S_z + L_z)E|p\rangle_0$$

$$= Es_z|p\rangle_0 \tag{B.44}$$

where we have used Eq. (7.8) (we use the notation $S^3 = S_z$ and $L^3 = L_z$). Substituting this into Eq. (7.22) leads again to

$$h = s_z = \vec{s}\cdot\hat{p}$$

7.4 Following the same steps as in Section 7.2 and using the second relation of Eq. (6.86), we have

$$[\bar{Q}^{\dot{a}}, W^0] = \frac{1}{2}\epsilon^{03ij}[\bar{Q}^{\dot{a}}, M_{ij}]P_3$$

$$= -\frac{1}{2}\epsilon^{03ij}\bar{\sigma}_{ij}\bar{Q}P_3 \tag{B.45}$$

The only differences with what we did earlier is therefore, $\bar{Q}^{\dot{a}}$ and that we simply get $\bar{\sigma}_{ij}$ instead of σ_{ij}. However, from the definition of $\sigma_{\mu\nu}$ and $\bar{\sigma}_{\mu\nu}$, it is clear that when both indices are spatial, the two expressions are equal, i.e.,

$$\bar{\sigma}_{ij} = \sigma_{ij}$$

Indeed, $\sigma_{\mu\nu}$ and $\bar{\sigma}_{\mu\nu}$ differ only when one index is equal to 0 and the second index is spatial.

Our final result is thus identical to Eqs. (7.28) and (7.29) except for the fact that we have supercharges with upper dotted indices:

$$[\bar{Q}^{\dot{1}}, W_0] = -\frac{1}{2}\bar{Q}^{\dot{1}}P_3$$

$$[\bar{Q}^{\dot{2}}, W_0] = \frac{1}{2}\bar{Q}^{\dot{2}}P_3$$

For the commutator with the charges $\bar{Q}_{\dot{a}}$, we could repeat the above steps using the first relation of Eq. (6.86) or, more simply, use $\bar{Q}^{\dot{1}} = \bar{Q}_{\dot{2}}$ and $\bar{Q}^{\dot{2}} = -\bar{Q}_{\dot{1}}$ to get

$$\left[\bar{Q}_{\dot{2}}, W_0\right] = -\frac{1}{2}\bar{Q}_{\dot{2}}P_3$$

$$\left[\bar{Q}_{\dot{1}}, W_0\right] = \frac{1}{2}\bar{Q}_{\dot{1}}P_3$$

We can finally express this in terms of components with lower undotted indices using $\bar{Q}_{\dot{1}} = Q_1^\dagger$ and $\bar{Q}_{\dot{2}} = Q_2^\dagger$, which confirms Eq. (7.32).

7.5 The Majorana charge spinor is, as already shown in Eq. (7.43),

$$Q_M = \begin{pmatrix} Q_2^\dagger \\ -Q_1^\dagger \\ Q_1 \\ Q_2 \end{pmatrix} \tag{B.46}$$

whereas

$$\bar{Q}_M = (Q_M)^\dagger \gamma^0 = \left(Q_1^\dagger, \ Q_2^\dagger, \ Q_2, \ -Q_1\right) \tag{B.47}$$

It is convenient to write γ^μ as

$$\gamma^\mu = \begin{pmatrix} 0 & \bar{\sigma}^\mu \\ \sigma^\mu & 0 \end{pmatrix} \tag{B.48}$$

The anticommutator (7.44) then gives 16 relations. For example,

$$\{Q_{M1}, \bar{Q}_{M1}\} = (\gamma^\mu)_{11} P_\mu$$

which, using Eqs. (B.46), (B.47), and (B.48), gives

$$\{Q_2^\dagger, Q_1^\dagger\} = 0$$

Of the 16 anticommutators, 8 are equal to zero and reproduce Eqs. (6.70) and (6.71). As an example of a nonzero anticommutator, set $a = 3$ and $b = 1$ to find

$$\{Q_{M3}, \bar{Q}_{M1}\} = \{Q_1, Q_1^\dagger\} = (\gamma^\mu)_{31} = (\sigma^\mu)_{11} P_\mu \tag{B.49}$$

On the other hand, with $a = 2$ and $b = 4$, we find

$$\{Q_{M2}, \bar{Q}_{M4}\} = \{-Q_1^\dagger, -Q_1\} = \{Q_1^\dagger, Q_1\} = (\bar{\sigma}^\mu)_{22} P_\mu \qquad (B.50)$$

If we recall that $\bar{\sigma}^0 = \sigma^0 = \mathbf{1}$ and that $\bar{\sigma}^i = -\sigma^i$, it is easy to convince ourselves that $(\bar{\sigma}^\mu)_{11} = (\sigma^\mu)_{22}$, so we indeed find that both Eq. (B.49) and Eq. (B.50) yield

$$\{Q_1, Q_1^\dagger\} = (\sigma^\mu)_{11} P_\mu$$

Two of the other nonzero anticommutators are

$$\{Q_{M3}, \bar{Q}_{M2}\} = \{Q_1, Q_2^\dagger\} = (\sigma^\mu)_{12} P_\mu$$
$$\{Q_{M1}, \bar{Q}_{M4}\} = \{Q_2^\dagger, -Q_1\} = (\bar{\sigma}^\mu)_{12} P_\mu \qquad (B.51)$$

This time we need to use $(\bar{\sigma}^\mu)_{12} = -(\sigma^\mu)_{12}$, so both equations in Eq. (B.51) are actually equivalent and in agreement with Eq. (6.72). There are four more nonzero anticommutators that, after using $(\bar{\sigma}^\mu)_{21} = -(\sigma^\mu)_{21}$ and $(\bar{\sigma}^\mu)_{22} = (\sigma^\mu)_{11}$, lead to

$$\{Q_2, Q_1^\dagger\} = (\sigma^\mu)_{21} P_\mu$$
$$\{Q_2, Q_2^\dagger\} = (\sigma^\mu)_{22} P_\mu \qquad (B.52)$$

in agreement with Eq. (6.72).

7.6 Let's start with

$$\left[\zeta \cdot Q, \phi(\vec{y}, t)\right] = \int d^3x \left[\partial_\nu \phi^\dagger(\vec{x}, t), \phi(\vec{y}, t)\right] \zeta^T (-i\sigma^2)\sigma^\nu \chi(\vec{x}, t)$$

where we have used the fact that the spinor field commutes with the scalar field. The commutator is nonzero only when the derivative on $\phi^\dagger(\vec{x}, t)$ is a time derivative, so we must set $\nu = 0$. We then get

$$\left[\zeta \cdot Q, \phi(\vec{y}, t)\right] = -i \int d^3x \, \delta^3(\vec{x} - \vec{y}) \zeta^\dagger (-i\sigma^2)\sigma^0 \chi(\vec{x}, t)$$
$$= -i\zeta \cdot \chi(\vec{y}, t)$$

as was to be shown. It is clear from this derivation that we trivially have $[\bar{\zeta} \cdot \bar{Q}, \phi] = 0$ because $\bar{Q}$ does not contain $\dot{\phi}^\dagger$.

Now let's turn to the commutator with the spinor field. To keep the notation simple, we will use x for the coordinates $(\vec{x}, t)$ and y for $(\vec{y}, t)$. Let's then look at

$$\left[\bar{\xi} \cdot \bar{Q}, \chi_a(y)\right] = \left[\bar{Q} \cdot \bar{\xi}, \chi_a(y)\right]$$

$$= \int d^3x \left(\partial_\nu \phi(x)\right) \left(\chi^\dagger(x) \sigma^\nu (i\sigma^2) \zeta^*, \chi_a(y)\right) \qquad \text{(B.53)}$$

Let's go through this one slowly. Focus on the commutator, which, with spinor indices shown explicitly, is equal to

$$\left[\chi_b^\dagger(x)(\sigma^\nu(i\sigma^2))^{bc} \zeta_c^*, \chi_a(y)\right] = i\chi_b^\dagger(x)(\sigma^\nu\sigma^2)^{bc} \zeta_c^* \chi_a(y)$$

$$- i\chi_a(y)\chi_b^\dagger(x)(\sigma^\nu\sigma^2)^{bc} \zeta_c^*$$

$$= i\left(-\chi_b^\dagger(x)\chi_a(y) - \chi_a(y)\chi_b^\dagger(x)\right)(\sigma^\nu\sigma^2)^{bc} \zeta_c^*$$

$$= -i\{\chi_b^\dagger(x), \chi_a(y)\}(\sigma^\nu\sigma^2)^{bc} \zeta_c^*$$

$$= -i\delta_{ab}\delta^3(\vec{x} - \vec{y})(\sigma^\nu\sigma^2)^{bc} \zeta_c^*$$

where we have used the fact that ζ_c^* and $\chi_a(y)$ anticommute. Inserting this into Eq. (B.53), we finally obtain

$$\left[\bar{\xi} \cdot \bar{Q}, \chi_a(y)\right] = -i(\partial_\nu \phi(y))\delta_{ab}(\sigma^\nu\sigma^2)^{bc} \zeta_c^*$$

which can be written without showing the indices explicitly as

$$\left[\bar{\xi} \cdot \bar{Q}, \chi(y)\right] = -i(\partial_\nu \phi(y))\sigma^\nu\sigma^2 \zeta^*$$

in agreement with Eq. (6.52).

7.7 The most general symmetric second-rank covariant tensor that we can build of the metric, $\eta_{\mu\nu}$ and the momenta is

$$S_{\mu\nu} = c_1 \eta_{\mu\nu} + c_2 \hat{P}_\mu \hat{P}_\nu$$

If we apply this to a two particle state, we simply get

$$S_{\mu\nu}|p\,k\rangle = (2c_1 \eta_{\mu\nu} + c_2 p_\mu p_\nu + c_2 k_\mu k_\nu)|p\,k\rangle$$

To simplify the notation, let's define $A \equiv p_i^a$, $B \equiv p_i^b$, $C \equiv p_f^a$, and $D \equiv p_f^b$. Then, conservation of $S\mu\nu$ leads to

$$2c_1 \eta_{\mu\nu} + c_2 A_\mu A_\nu + c_2 B_\mu B_\nu = 2c_1 \eta_{\mu\nu} + c_2 C_\mu C_\nu + c_2 D_\mu D_\nu$$

or, equivalently,

$$(A_\mu - B_\mu)(A_\nu - B_\nu) + (A_\mu + B_\mu)(A_\nu + B_\nu) = (C_\mu - D_\mu)(C_\nu - D_\nu)$$
$$+ (C_\mu + D_\mu)(C_\nu + D_\nu)$$

In addition, conservation of four-momentum implies

$$A_\mu + B_\mu = C_\mu + D_\mu$$

so that we can rewrite the conservation of $S_{\mu\nu}$ as

$$(A_\mu - B_\mu)(A_\nu - B_\nu) + (A_\mu + B_\mu)(C_\nu + D_\nu) = (C_\mu - D_\mu)(C_\nu - D_\nu)$$
$$+ (A_\mu + B_\mu)(C_\nu + D_\nu)$$

or

$$(A_\mu - B_\mu)(A_\nu - B_\nu) = (C_\mu - D_\mu)(C_\nu - D_\nu)$$

There are only two possibilities for the relation between the momenta before and after the collision. Either $A_\mu - B_\mu = C_\mu - D_\mu$ or $A_\mu - B_\mu = D_\mu - C_\mu$. Now, if we use again the conservation of four-momentum, these two possibilities immediately lead to either $A = C$ and $B = D$ (no scattering) or $A = D$ and $B = C$ (exchange of four-momenta), as was to be shown.

10.1 If we take the hermitian conjugate of the first term, we get

$$\left(\zeta^\dagger \bar{\sigma}^\mu \lambda\right)^\dagger = \lambda^\dagger (\bar{\sigma}^\mu)^\dagger \zeta$$

Note that when we apply a hermitian conjugation to a product of spinors, we exchange their order without introducing a minus sign [see Eq. (2.56) and the discussion preceding it]. Since the Pauli matrices are hermitian, we simply get

$$\left(\zeta^\dagger \bar{\sigma}^\mu \lambda\right)^\dagger = \lambda^\dagger \bar{\sigma}^\mu \zeta$$

which is precisely the second term appearing in Eq. (10.2). It is now obvious that the hermitian conjugate of the second term of Eq. (10.2) is equal to the first term. Therefore,

$$\left(\zeta^\dagger \bar{\sigma}^\mu \lambda + \lambda^\dagger \bar{\sigma}^\mu \zeta\right)^\dagger = \zeta^\dagger \bar{\sigma}^\mu \lambda + \lambda^\dagger \bar{\sigma}^\mu \zeta$$

as was to be shown.

10.2 a. We simply have

$$\frac{1}{4}\mathrm{Tr}(\sigma^i \sigma^j) = \frac{1}{4}\mathrm{Tr}(\delta^{ij}\mathbb{1} + i\epsilon^{ijk}\sigma^k) = \delta^{ij}\mathrm{Tr}(\mathbb{1}) = \delta^{ij}$$

b. By definition,

$$(T^i_{\text{AD}})^{jk} = -i\epsilon^{ijk}$$

or, to be explicit,

$$T^1_{\text{AD}} = i \begin{pmatrix} 0 & 0 & 0 \\ 0 & 0 & -1 \\ 0 & 1 & 0 \end{pmatrix} \quad T^2_{\text{AD}} = i \begin{pmatrix} 0 & 0 & 1 \\ 0 & 0 & 0 \\ -1 & 0 & 0 \end{pmatrix} \quad T^3_{\text{AD}} = i \begin{pmatrix} 0 & -1 & 0 \\ 1 & 0 & 0 \\ 0 & 0 & 0 \end{pmatrix}$$

$$\tag{B.54}$$

It is a simple matter to check that

$$\mathrm{Tr}(T^i_{\text{AD}})^2 = 2$$

and that they satisfy

$$[T^i_{\text{AD}}, T^j_{\text{AD}}] = i\epsilon^{ijk} T^k_{\text{AD}}$$

10.3 Obviously, the terms in Eq. (10.37) would be absent. On the other hand, we would have more choices than in Eq. (10.39). Writing down only the terms of dimension 5/2 or less, we find

$$\phi, \chi, F, \phi^2, \phi\chi, \phi^\dagger\phi, \phi^\dagger\chi$$

plus their hermitian conjugates. Now we pair these up with the fields $F_{\mu\nu}$, λ, and D to form interactions of dimension 4 or less. We get the following eight possible combinations (when a term is not real, we add to it its hermitian conjugate to obtain a real interaction and have assumed all the constants to be real):

$$\mathcal{L}_{\text{int}} = C_1(D\phi^2 + D\phi^{\dagger 2}) + C_2(D\phi + D\phi^\dagger) + C_3(DF + DF^\dagger) + C_4 D\phi\phi^\dagger$$
$$+ C_5(\phi\chi \cdot \lambda + \phi^\dagger\bar{\chi} \cdot \bar{\lambda}) + C_6(\phi^\dagger\chi \cdot \lambda + \phi\bar{\chi} \cdot \bar{\lambda})$$
$$+ C_7(\phi\lambda \cdot \lambda + \phi^\dagger\bar{\lambda} \cdot \bar{\lambda}) + C_8(\phi^\dagger\lambda \cdot \lambda + \phi\bar{\lambda} \cdot \bar{\lambda})$$

11.1 Using

$$\theta \cdot \chi = -\theta_1 \chi_2 + \theta_2 \chi_1 = \chi_2 \theta_1 - \chi_1 \theta_2$$

and $\theta \cdot \theta = -2\theta_1\theta_2$, we get

$$\phi = A \qquad \chi_2 = B \qquad \chi_1 = -C \qquad F = -D$$

11.2 The nine possible terms containing only one quantum field are (the fields $A, B, \ldots$ are scalar fields, U_μ, V_μ, and W_μ are obviously vector fields, and the Greek letters represent left-chiral spinor fields)

$$\mathcal{S}(x, \theta, \bar{\theta}) = A + \theta \cdot \chi + \bar{\theta} \cdot \bar{\lambda} + \theta \cdot \theta \, B + \bar{\theta} \cdot \bar{\theta} \, C + \theta \sigma^\mu \bar{\theta} \, U_\mu$$
$$+ (\theta \sigma^\mu \bar{\theta}) \, \theta \cdot \theta \, V_\mu + (\theta \sigma^\mu \bar{\theta}) \, \bar{\theta} \cdot \bar{\theta} \, W_\mu + \theta \cdot \theta \, \bar{\theta} \cdot \bar{\theta} \, D \qquad \text{(B.55)}$$

The parentheses have been included only to make the grouping of terms more explicit. Any terms with more θ or $\bar{\theta}$ will be identically zero. We do not consider terms with σ^μ replaced by $\bar{\sigma}^\mu$ because these are not independent terms owing to the identity (A.49).

11.3 Consider the first expression. We need to express all the Grassmann variables in terms of components with upper undotted indices because the derivative is with respect to this type of component. We get

$$\frac{\partial}{\partial \theta^a} \theta^b \theta_b = \frac{\partial}{\partial \theta^a} \epsilon_{bc} \theta^b \theta^c$$
$$= \epsilon_{bc} \, \delta_a^b \, \theta^c - \epsilon_{bc} \, \theta^b \, \delta_a^c$$
$$= \epsilon_{ac} \, \theta^c - \epsilon_{ba} \, \theta^b$$
$$= \epsilon_{ac} \, \theta^c + \epsilon_{ab} \, \theta^b$$
$$= 2\theta_a$$

The proof of the second identity is similar.

11.4 By definition,

$$\zeta^a \partial_a = \zeta^a \frac{\partial}{\partial \theta^a} = \zeta^1 \frac{\partial}{\partial \theta^1} + \zeta^2 \frac{\partial}{\partial \theta^2}$$

Using Eq. (11.26), we have

$$\frac{\partial}{\partial \theta^2} = -\frac{\partial}{\partial \theta_1} \qquad \frac{\partial}{\partial \theta^1} = \frac{\partial}{\partial \theta_2}$$

We may write this as

$$\partial_2 = -\partial^1 \qquad \partial_1 = \partial^2$$

Note that these relations have the *opposite* sign as the relations in Eq. (11.26). The reason why is obviously owing to the fact that a partial derivative with a covariant index denotes a differentiation with respect to a variable with a contravariant index.

Using $\zeta^1 = \zeta_2$ and $\zeta_1 = -\zeta^2$, we finally find

$$\zeta^1 \frac{\partial}{\partial \theta^1} + \zeta^2 \frac{\partial}{\partial \theta^2} = \zeta_1 \frac{\partial}{\partial \theta_1} + \zeta_2 \frac{\partial}{\partial \theta_2}$$

so that

$$\zeta^a \frac{\partial}{\partial \theta^a} = \zeta_a \frac{\partial}{\partial \theta_a}$$

which completes the proof. Since the relation between the $\bar{\theta}^{\dot{a}}$ and $\bar{\theta}_{\dot{b}}$ is the same as the relation between the θ^a and θ_b, this also implies that

$$\bar{\zeta}^{\dot{a}} \frac{\partial}{\partial \bar{\theta}^{\dot{a}}} = \bar{\zeta}_{\dot{a}} \frac{\partial}{\partial \bar{\theta}_{\dot{a}}}$$

11.5 Let's apply $\epsilon_{\dot{b}\dot{a}}$ to Eq. (11.37):

$$\epsilon_{\dot{b}\dot{a}} \bar{Q}^{\dot{a}} = i \epsilon_{\dot{b}\dot{a}} \bar{\partial}^{\dot{a}} - \frac{1}{2} \epsilon_{\dot{b}\dot{a}} (\bar{\sigma}^\mu)^{\dot{a}c} \theta_c \partial_\mu$$

We may lower the indices of $\bar{Q}^{\dot{a}}$ and $\bar{\partial}^{\dot{a}}$ [using Eq. (11.42)], i.e.,

$$\bar{Q}_{\dot{b}} = -i \bar{\partial}_{\dot{b}} - \frac{1}{2} \epsilon_{\dot{b}\dot{a}} (\bar{\sigma}^\mu)^{\dot{a}c} \theta_c \partial_\mu$$

Now, in order to use Eq. (A.27), we need to write θ_c as $\epsilon_{cd}\theta^d$ and then switch the order of the indices using $\epsilon_{cd} = -\epsilon_{dc}$ to finally get

$$
\bar{\mathcal{Q}}_{\dot{b}} = -i\bar{\partial}_{\dot{b}} + \frac{1}{2}\epsilon_{\dot{b}\dot{a}}\epsilon_{dc}(\bar{\sigma}^\mu)^{\dot{a}c}\theta^d\,\partial_\mu
$$

$$
= -i\bar{\partial}_{\dot{b}} + \frac{1}{2}(\sigma^\mu)_{d\dot{b}}\theta^d\partial_\mu \tag{B.56}
$$

which agrees with Eq. (11.43).

11.6 If we pull out from the anticommutators all we can, we get the condition

$$
iC_{\dot{a}c}\left\{\theta^c, \frac{\partial}{\partial\theta^b}\right\} + \frac{i}{2}(\sigma^\mu)_{b\dot{a}}\hat{\mathcal{P}}_\mu\left\{\frac{\partial}{\partial\bar{\theta}^{\dot{a}}}, \bar{\theta}^{\dot{d}}\right\} = 0
$$

Using Eqs. (11.46) and (11.47), we simply get

$$
iC_{\dot{a}c}\,\delta^c_b + \frac{i}{2}(\sigma^\mu)_{b\dot{d}}\,\hat{\mathcal{P}}_\mu\delta^{\dot{d}}_{\dot{a}} = iC_{\dot{a}b} + \frac{i}{2}\,(\sigma^\mu)_{b\dot{a}}\,\hat{\mathcal{P}}_\mu
$$

$$
= 0
$$

which immediately yields the result [Eq. (11.56)].

12.1 This is obvious if we consider a generic component field $A(y)$ (which could stand for ϕ, χ, F, or any other field) and express it in terms of x and the Grassmann variables [see Eq. (12.7)]:

$$
A(y) = A\left(x^\mu - \frac{i}{2}\theta\sigma^\mu\bar{\theta}\right)
$$

$$
= A(x) - \frac{i}{2}\theta\sigma^\mu\bar{\theta}\partial_\mu A(x) - \frac{1}{16}\theta\cdot\theta\bar{\theta}\cdot\bar{\theta}\Box A(x) \tag{B.57}
$$

The product of left-chiral superfields will contain the sum of products of such component fields. Clearly, only the first term in the expansion of each component field, the term $A(x)$, may contribute to the F term because the other terms in the expansion (B.57) contain at least one $\bar{\theta}$.

12.2 We will use Eq. (A.54), which says that, for any three Weyl spinors α, β, and γ, the following holds:

$$
\gamma\cdot\alpha\gamma\cdot\beta = -\frac{1}{2}\gamma\cdot\gamma\alpha\cdot\beta \tag{B.58}
$$

In terms of explicit components, this gives

$$\gamma^a \alpha_a \gamma^b \beta_b = -\frac{1}{2} \gamma^c \gamma_c \alpha^d \beta_d \tag{B.59}$$

Now consider Eq. (12.47):

$$(\theta \cdot \partial_\mu \chi)(\theta \sigma^\mu \bar{\lambda})$$

The parentheses were added only to show the grouping more clearly. Writing the indices explicitly on all quantities, this is

$$\theta^a \, \partial_\mu \chi_a \theta^b (\sigma^\mu)_{b\dot{c}} \bar{\lambda}^{\dot{c}}$$

This is of the form of the left side of Eq. (B.59) with the identification $\gamma = \theta$ and

$$\alpha_a = \partial_\mu \chi_a \qquad \beta_b = (\sigma_\mu)_{b\dot{c}} \bar{\lambda}^{\dot{c}}$$

so that Eq. (B.58) gives

$$\theta \sigma^\mu \bar{\lambda} \theta \cdot \partial_\mu \chi = -\frac{1}{2} \theta \cdot \theta \partial_\mu \chi^d (\sigma_\mu)_{d\dot{c}} \bar{\lambda}^{\dot{c}}$$

$$= -\frac{1}{2} \theta \cdot \theta (\partial_\mu \chi) \sigma_\mu \bar{\lambda}$$

as was to be shown.

Now consider Eq. (12.48), which may be written as

$$\bar{\theta} \cdot \bar{\lambda}(\partial_\mu \chi) (\sigma_\mu)\bar{\theta} = \bar{\lambda}_{\dot{a}} \bar{\theta}^{\dot{a}} (\partial_\mu \chi^d) (\sigma_\mu)_{d\dot{c}} \bar{\theta}^{\dot{c}} \tag{B.60}$$

Showing all the indices, the identity (A.55) corresponds to

$$\bar{\alpha}_{\dot{a}} \bar{\gamma}^{\dot{a}} \bar{\beta}_{\dot{b}} \bar{\gamma}^{\dot{b}} = -\frac{1}{2} \bar{\gamma}_{\dot{c}} \bar{\gamma}^{\dot{c}} \bar{\beta}_{\dot{d}} \bar{\alpha}^{\dot{d}} \tag{B.61}$$

The left-hand side is exactly the same as Eq. (B.60) if we identify $\bar{\gamma} = \bar{\theta}$ and

$$\bar{\alpha}_{\dot{a}} = \bar{\lambda}_{\dot{a}} \qquad \bar{\beta}_{\dot{b}} = (\partial_\mu \chi^d) (\sigma_\mu)_{d\dot{b}}$$

The identity (B.61) then gives us

$$\bar{\theta} \cdot \bar{\lambda}(\partial_\mu \chi^d)(\sigma_\mu)_{d\dot{c}}\bar{\theta}^{\dot{c}} = -\frac{1}{2}\bar{\theta} \cdot \bar{\theta}(\partial_\mu \chi^d)(\sigma^\mu)_{d\dot{b}}\bar{\lambda}^{\dot{b}}$$

$$= -\frac{1}{2}\bar{\theta} \cdot \bar{\theta}(\partial_\mu \chi)\sigma^\mu\bar{\lambda}$$

which is what we needed to prove.

13.1 One has to be careful with this calculation because the components with upper indices are not independent of the components with lower indices. If one does not keep this in mind, it is tempting to conclude that

$$\bar{D}_{\dot{a}}\bar{D}^{\dot{a}}\bar{\theta} \cdot \bar{\theta} = \frac{\partial}{\partial\bar{\theta}^{\dot{a}}}\frac{\partial}{\partial\bar{\theta}_{\dot{a}}}\bar{\theta}_{\dot{b}}\bar{\theta}^{\dot{b}}$$

will simply give 1. But this is incorrect. The simplest way to do the calculation is to expand $\bar{\theta} \cdot \bar{\theta}$ out in terms of lower components:

$$\frac{\partial}{\partial\bar{\theta}^{\dot{a}}}\frac{\partial}{\partial\bar{\theta}_{\dot{a}}}2\bar{\theta}_{\dot{1}}\bar{\theta}_{\dot{2}} = 2\frac{\partial}{\partial\bar{\theta}^{\dot{1}}}\bar{\theta}_{\dot{2}} - 2\frac{\partial}{\partial\bar{\theta}^{\dot{2}}}\bar{\theta}_{\dot{1}}$$

Now we use $\bar{\theta}_{\dot{2}} = \bar{\theta}^{\dot{1}}$ and $\bar{\theta}_{\dot{1}} = -\bar{\theta}^{\dot{2}}$, we finally get

$$2\frac{\partial}{\partial\bar{\theta}^{\dot{1}}}\bar{\theta}_{\dot{2}} - 2\frac{\partial}{\partial\bar{\theta}^{\dot{2}}}\bar{\theta}_{\dot{1}} = 2\frac{\partial}{\partial\bar{\theta}^{\dot{1}}}\bar{\theta}^{\dot{1}} + 2\frac{\partial}{\partial\bar{\theta}^{\dot{2}}}\bar{\theta}_{\dot{2}} = 4$$

as was to be shown.

14.1 Since we are only interested in reading off the masses, not in the interactions, we will consider only the terms quadratic or bilinear in the fields:

$$V = \frac{1}{2}m^2 R_3^2 + \frac{1}{2}m^2 I_3^2 - g^2 M^2 R_3^2 + g^2 M^2 I_3^2 + \frac{1}{2}m^2 R_1^2 + 2g^2 R_3^2 \langle\phi_2\rangle^2$$

$$+ 2mg\langle\phi_2\rangle R_1 R_3 + \frac{1}{2}m^2 I_1^2 + 2mg\langle\phi_2\rangle I_1 I_3 + 2g^2 I_3^2 \langle\phi_2\rangle^2 \cdots \quad\text{(B.62)}$$

where the dots indicate terms that don't contribute to the mass matrix.

We see that since there are no terms containing R_2 or I_2, these fields are massless. The fields R_1 and R_3 get mixed, and their mass matrix is (in the basis R_1, R_3)

$$\mathbf{M}_{sq} = \begin{pmatrix} m^2 & 2mg\langle\phi_2\rangle \\ 2mg\langle\phi_2\rangle & m^2 - 2g^2M^2 + 4g^2\langle\phi_2\rangle^2 \end{pmatrix}$$

whose eigenvalues are

$$m^2 - g^2M^2 + 2g^2\langle\phi_2\rangle^2 \pm g\sqrt{\left(2g\langle\phi_2\rangle^2 - gM^2\right)^2 + 4m^2\langle\phi_2\rangle^2} \qquad \text{(B.63)}$$

This is not a very pleasant expression. It simplifies greatly if we use the freedom in the choice of $\langle\phi_2\rangle$ to set it equal to zero, in which case the matrix is already diagonal, and we get

$$m_{R_1}^2 = m^2 - g^2M^2 \qquad m_{R_3}^2 = m^2$$

Now consider the terms in I_1 and I_3 in Eq. (B.62). The corresponding squared-mass matrix is (in the basis I_1, I_3)

$$\mathbf{M}_{sq} = \begin{pmatrix} m^2 & 2mg\langle\phi_2\rangle \\ 2mg\langle\phi_2\rangle & m^2 + 2g^2M^2 + 4g^2\langle\phi_2\rangle^2 \end{pmatrix}$$

whose eigenvalues are

$$m^2 + g^2M^2 + 2g^2\langle\phi_2\rangle^2 \pm g\sqrt{\left(2g\langle\phi_2\rangle^2 + gM^2\right)^2 + 4m^2\langle\phi_2\rangle^2} \qquad \text{(B.64)}$$

Setting again $\langle\phi_2\rangle = 0$, we get once more a diagonal matrix and

$$m_{I_1}^2 = m^2 + g^2M^2 \qquad m_{I_3}^2 = m^2$$

14.2 All we need are Eqs. (B.63) and (14.16). If we first take the limit $g \to 0$, then we simply get

$$m_{I_1} = m_{I_3} = m_{R_1} = m_{R_3} = m_{\chi_1} = m_{\chi_3} = m$$

with the other fields being massless, as previously.

On the other hand, if we take the limit $M \to 0$, the masses of the four real scalars split exactly the same way as the fermion masses; i.e., we have

$$m_b = m_f = \sqrt{g^2 \langle \phi_2 \rangle^2 + m^2} \pm g \langle \phi_2 \rangle$$

For each fermion, there is a boson of equal mass, and supersymmetry is unbroken.

14.3 Two of the bosons are massless, whereas the other four have masses squared given by Eqs. (B.63) and (B.64) so that

$$\sum_{\text{scalars}} m^2 = 4m^2 + 8g^2 \langle \phi_2 \rangle^2$$

On the other hand, one fermion is massless, with the other two having masses squared given by Eq. (14.16) so that

$$2 \sum_{\text{spinors}} m^2 = 4m^2 + 8g^2 \langle \phi_2 \rangle^2$$

and we see that the supertrace still vanishes.

APPENDIX C

Solutions to Quizzes

Chapter 2

1. We have

$$\bar{\chi} \cdot \bar{\lambda} = \chi^\dagger i\sigma^2 \lambda^\dagger = -\chi_2^\dagger \lambda_1^\dagger + \chi_1^\dagger \lambda_2^\dagger$$

2. A Weyl spinor has a definite helicity only when it is massless. However, by construction, a Weyl spinor has a definite chirality, no matter what its mass is. So Weyl spinors are *always* states of definite chirality.

3. They transform the same way under rotations but differently under boosts (the boost parameter has an opposite sign).

4. A Dirac mass term is

$$-m\overline{\Psi}\Psi = -m\Psi^\dagger \gamma^0 \Psi = -m(\eta^\dagger \chi + \chi^\dagger \eta)$$

5. The operation of hermitian conjugate on a product of spinor fields is defined to switch their order without introducing a minus sign, i.e.,

$$(\chi_1 \chi_2)^\dagger = \chi_2^\dagger \chi_1^\dagger$$

Chapter 3

1. We first need to move the component with the upper index to the right of the one with the lower index (recall that dotted indices must go diagonally upward), so we get

$$\bar{\eta}^{\dot{a}} \bar{\chi}_{\dot{a}} = -\bar{\chi}_{\dot{a}} \bar{\eta}^{\dot{a}} = -\bar{\chi} \cdot \bar{\eta}$$

This also may be written as $-\bar{\eta} \cdot \bar{\chi}$.

2. $\epsilon_{\dot{1}\dot{1}} = 0$ and $\epsilon_{\dot{1}\dot{2}} = -1$.

3. Recall that we first must get the last indices of the ϵ to match the indices of the σ_μ. We therefore must switch the order of the indices of the first epsilon, which introduces a sign:

$$\epsilon^{ac} \epsilon^{\dot{d}\dot{b}} (\sigma^\mu)_{a\dot{b}} = -\epsilon^{ca} \epsilon^{\dot{d}\dot{b}} (\sigma^\mu)_{a\dot{b}}$$

and now we can contract the indices to get

$$-(\bar{\sigma}^\mu)^{\dot{d}c}$$

Note the order of the indices!

4. Since the common index appears in the same position in both ϵ, this is equal to minus one times the Kronecker delta, i.e.,

$$\epsilon_{\dot{a}\dot{b}} \epsilon^{\dot{a}\dot{c}} = -\delta_{\dot{b}}^{\dot{c}}$$

5. Applying a hermitian conjugation changes a dotted index into an undotted index, and vice versa, without changing the position of the index. So $(\chi^a)^\dagger = \bar{\chi}^{\dot{a}}$, and so on.

6. The matrix $-i\sigma^2$ lowers indices.

7. We can do it in two different ways. We can choose to contract the ϵ together first:

$$\epsilon_{ab} \epsilon^{cb} \chi^a = -\delta_a^c \chi^a = -\chi^c$$

where there is a minus sign in the first step because the index common to the two ϵ is in the same position. Or we can contract one at a time the two ϵ with the spinor:

$$\epsilon_{ab}\epsilon^{cb}\chi^a = -\epsilon^{cb}\epsilon_{ba}\chi^a = -\epsilon^{cb}\chi_b = -\chi^c$$

where the minus sign comes from switching the order of the indices of ϵ_{ab}, which is necessary before contracting with the spinor.

Chapter 4

1. It is the corresponding left-chiral antiparticle state. Of course, this state also will be the left-chiral particle state if we are dealing with a Majorana spinor.

2. The Dirac mass term contains particle and antiparticle Weyl spinors dotted together, i.e.,

$$\chi_p \cdot \chi_{\bar{p}} + \bar{\chi}_p \cdot \bar{\chi}_{\bar{p}}$$

whereas the Majorana mass term contains the same expressions but with all spinors being particle spinors, i.e., $\chi_p \cdot \chi_p + \bar{\chi}_p \cdot \bar{\chi}_p$.

3. There is no difference besides the fact that a Majorana spinor is written in four-component language, whereas a lagrangian of a Weyl fermion is expressed in terms of two-component spinors. Physically, the two are equivalent. This is the convention followed in this book. However, some references define a Weyl spinor to be necessarily massless and use the term *Majorana spinor* to denote the massive case, no matter if it is expressed in two- or four-component language.

4. The answer is yes to both questions. We saw explicitly how a Dirac spinor can be written in terms of left-chiral spinors only in Section 4.6 (i.e., in terms of the particle and antiparticle left-chiral Weyl spinors). It is also possible to use right-chiral spinors only.

5. The mass term is

$$-\frac{m}{2}\left(\chi_p \cdot \chi_p + \bar{\chi}_p \cdot \bar{\chi}_p\right)$$

It is clear that this is real because $(\chi_p \cdot \chi_p)^\dagger = \bar{\chi}_p \cdot \bar{\chi}_p$ and $(\bar{\chi}_p \cdot \bar{\chi}_p)^\dagger = \chi_p \cdot \chi_p$, so taking the hermitian conjugate of the mass term we just wrote gives back the same expression. Note that we absolutely need both terms to make it work.

6. Parity. This is why in parity-invariant theories, such as QED or QCD, Dirac spinors are necessary.

Chapter 5

1. *Rigid supersymmetry* is another name for global supersymmetry, i.e., supersymmetry with a spacetime-independent infinitesimal parameter ζ.

2. It has dimension $-1/2$.

3. Since $[\phi] = 1$, $[\chi] = 3/2$, and $[\zeta] = -1/2$, we need to combine $\zeta\phi$ with something with a dimension equal to one. The only available quantity that is not a field is a derivative ∂_μ. So we have $\delta\chi \simeq \zeta\partial_\mu\chi$. Of course, this does not have the right Lorentz transformation, so it needs to be modified as shown in the text.

4. It is the abbreviation of *supergravity*, which is the theory obtained when supersymmetry is made local (which forces the introduction of a graviton).

5. The right-hand side transforms like a right-chiral spinor, whereas χ is a left-handed spinor.

Chapter 6

1. The supercharges commute with the four-momentum operator but not with the Lorentz generators. This is why the supercharges do not change the four-momentum of a state but change the helicity.

2. It is a field that has no kinetic energy term and is therefore nondynamical. The equation of motion for an auxiliary field is algebraic, and the field therefore can always be eliminated trivially.

3. By counting the number of degrees of freedom off-shell. A Weyl spinor has four degrees of freedom off-shell (two complex components), whereas a complex scalar field has only two of them. Therefore, we needed two more scalar degrees of freedom.

4. The transformation must contain the spinor χ to contract with the ζ. Thus we could try $\delta F \simeq \zeta \cdot \chi$, but this is of dimension 1. To increase the dimension, we cannot use ϕ because we want the transformation to be linear in the fields. We need to use a derivative, so we may try $\delta F \simeq \zeta \cdot \partial_\mu\chi$, but now we need to contract the Lorentz index with some Pauli matrices. We need to use $\bar{\sigma}^\mu$ when we act on a left-chiral spinor in order to have a quantity with well-defined Lorentz transformation properties, so let's try $\delta F \simeq \zeta\bar{\sigma}^\mu\partial_\mu\chi$. This is not quite right because $\bar{\sigma}^\mu\chi$ transforms like a right-chiral spinor. In order

to get a Lorentz scalar for δF, we need to take the hermitian conjugate of ζ. Thus the final answer is $\delta F = \zeta^\dagger \bar{\sigma}^\mu \partial_\mu \chi$.

5. We needed a total of four charges, Q_1, Q_2, $Q_1^\dagger$, and $Q_2^\dagger$. We needed that many because the SUSY parameter ζ is a Weyl spinor, so we needed enough charges to construct Lorentz invariants containing ζ and $\bar{\zeta}$.

Chapter 7

1. It does not change the momentum but increase the helicity by $1/2$.

2. Only two on-shell states, differing in helicity by one-half. If we impose *CPT* invariance in addition, we need four states.

3. It is supersymmetry with more than one set of supercharges (where one set is defined as Q_1, Q_2 and their hermitian conjugate).

4. It is necessarily larger or equal to zero (and strictly larger than zero if SUSY is broken spontaneously).

5. By considering charges that anticommute.

Chapter 8

1. $\mathcal{W} = \frac{1}{2}m^2\phi^2 + \frac{1}{6}y\,\phi^3$.

2. It depends only on the fields Φ_i and not on their hermitian conjugate $\phi_i^\dagger$.

3. $W_i = \frac{\partial \mathcal{W}}{\partial \phi_i}$, $W_{ij} = \frac{\partial^2 \mathcal{W}}{\partial \phi_i \partial \phi_j}$.

4. Since F_i has dimension 2, W_i can be at most of dimension 2 as well (if we consider only renormalizable theories). Therefore, the superpotential is at most of dimension 3 and can contain at most three scalar fields.

5. It is the same as in Question 1 except that we may add a liner term $c\phi$.

Chapter 9

1. A Dirac propagator has only one contribution, from the contraction of Ψ and $\bar{\Psi}$. A Majorana propagator also contains the contractions of two Ψ_M and two $\bar{\Psi}_M$, so there are three distinct contributions to a Majorana propagator.

2. Only two: m and g. The parameter m is the mass of all the particles but also appears as a coupling constant in some interactions.

3. No, logarithmic divergences are present, but these can be cured by wavefunction renormalization.

4. A correction to the mass terms of the scalar fields. For example, we could replace $mA^2/2$ by $mA^2/2 + \delta m\, A^2$.

5. It's an interaction that breaks supersymmetry but does not disturb the cancellation of quadratic divergences.

Chapter 10

1. The photino is invariant under a gauge transformation, like the abelian field strength $F_{\mu\nu}$.

2. A new term must be added to the transformation of the auxiliary field, namely $\sqrt{2}q\phi\bar{\lambda} \cdot \bar{\chi}$. In addition, all the transformations of the field of the vector multiplet must be multplied by a factor of $-1/\sqrt{2}$.

3. It is a term in the lagrangian linear in the auxiliary field D, which is gauge-invariant for an abelian gauge theory.

4.

$$D_\mu = \partial_\mu - \frac{i}{2} g_s A_\mu^a (\lambda^a)^*$$

5. Off-shell, a gauge field has three degrees of freedom. We therefore only need one extra off-shell bosonic degree of freedom to match the four off-shell degrees of freedom of the Weyl spinor. In the case of the chiral multiplet, the boson was a complex scalar field that has only two off-shell degrees of freedom, which is why we needed to incorporate two extra bosonic degrees of freedom off-shell.

Chapter 11

1. Zero.

2. Using the fact that $\theta^1 = \theta_2$, we see that the expression is differentiated and is identically zero. But even if we were only differentiating θ^1 we would get zero.

3. Eight terms. A constant, three terms linear in each Grassmann variable, three bilinear terms (that we may take to be $\theta_1\theta_2$, $\theta_1\theta_3$ and $\theta_2\theta_3$), and finally, the trilinear combination $\theta_1\theta_2\theta_3$.

4. A total of four, counting the hermitian conjugates as independent.

5. It gives $-\partial^a$; see Section 11.6.

Chapter 12

1. There is a complex scalar field ϕ, a left-chiral Weyl spinor χ, and off-shell, a complex auxiliary field F.

2. Because it is a total derivative, which we can ignore.

3. Five. Extracting the D term increases the dimension by 2, and left-chiral superfields have dimension 1.

4. So that the superpotential will be itself a left-chiral superfield. In this way, we know that the F term of the superpotential is a supersymmetric lagrangian.

5. It is the function of left-chiral superfields that contain at least one hermitian conjugate $\Phi_i^\dagger$. In the case of the Wess-Zumino lagrangian, it is simply $\mathcal{K} = \Phi^\dagger \Phi$, the kinetic term of the superfield.

Chapter 13

1. It transforms as a left-chiral Weyl spinor.

2. It is dimensionless (so it may appear as the argument of an exponential).

3. A gauge field A_μ, a left-chiral Weyl spinor λ, and a real auxiliary scalar field D.

4. $N^2 - 1$, which is the number of fields in the adjoint representation.

5. Two. This is so because the electron is a Dirac particle, which is equivalent to two Weyl spinors (which we may take to be the left-chiral electron and positron states).

Chapter 14

1. Because the auxiliary fields D^a are not gauge-invariant, so we may not write a Fayet-Illiopooulos term.

2. It implies that some scalar superpartners of the known fermions would be lighter than these fermions, which is ruled out experimentally.

3. The minimum of the potential $V(\phi_i)$ must be equal to zero when at least one of the scalar fields is nonzero.

4. In F-type SUSY breaking and in D-type SUSY breaking when the sum of the $U(1)$ charges of the particles is zero, $\sum q_i = 0$.

5. It is a massless spin 1/2 state that is necessarily present when SUSY is broken spontaneously.

Chapter 15

1. We need two left-chiral superfields of opposite hypercharge to couple with all the leptons and quarks (which, after SSB of the electroweak gauge symmetry, produces the mass terms). We could not take the hermitian conjugate of the first Higgs superfield because the superpotential must be holomorphic in the superfields.

2. The number of free parameters is the same as in the standard model.

3. The lepton and quark superfields have odd R-parities, whereas the Higgs superfield has even R-parities. The F term of $\mathcal{H}_u \circ \mathcal{H}_u \, \mathcal{L}_1 \circ \mathcal{L}_1$ is of dimension 5, gauge invariant, supersymmetric and invariant under R-parity.

4. $\mathcal{H}_u$ has quantum numbers **1**, **2**, 1, whereas $\mathcal{H}_d$ has **1**, **2**, -1.

5. They are both odd [they transform with a phase $\exp(\pm i\pi) = -1$].

Chapter 16

1. One hundred and five in general and only five in mSUGRA.

2. There are four complex Higgs fields, so eight degrees of freedom. Three become the longitudinal degrees of freedom of the gauge bosons, leaving five observable Higgs fields.

3. The inverse of the $U(1)_Y$ coupling constant decreases with energy in both cases, and the inverse of the strong coupling constant increases. The inverse of the $SU(2)_L$ coupling constant increases in the standard model but decreases in the MSSM.

4. The mass of the h^0 has an *upper bound*. Not only that, but this upper bound is well within reach of the LHC, which means that it will be possible to confirm or rule out the MSSM in the near future.

5. Three: $\delta m_{H_u}^2$, $\delta m_{H_d}^2$, and b.

APPENDIX D

Solutions to Final Exam

1. $[\phi] = 1$, $[\chi] = 3/2$, $[F] = 2$, $[\zeta] = -1/2$.

2. The result is equal to -1 (the minus sign comes from moving θ_1 to the left of $d\theta_2$).

3. We have $\bar\chi \cdot \bar\lambda = \bar\chi_{\dot a}\bar\lambda^{\dot a} = -\bar\lambda^{\dot a}\bar\chi_{\dot a}$. We now need to move the index of $\bar\lambda$ down and move up the index of $\bar\chi$ using the ϵ symbol. There are several ways to proceed. One possibility is to write $-\bar\lambda^{\dot a}\bar\chi_{\dot a}$ as $-\epsilon^{\dot a \dot b}\bar\lambda_{\dot b}\bar\chi_{\dot a}$ and then to go through the following steps:

$$-\epsilon^{\dot a \dot b}\bar\lambda_{\dot b}\bar\chi_{\dot a} = \epsilon^{\dot b \dot a}\bar\lambda_{\dot b}\bar\chi_{\dot a} = \bar\lambda_{\dot b}\bar\chi^{\dot b} = \bar\lambda \cdot \bar\chi$$

4. The supercharges Q_1 and $Q_2^\dagger$ raise the helicity by one half (wihout affecting the four-momentum) while Q_2 and $Q_1^\dagger$ lower the helicity by the same amount.

5. It is given by $\chi^\dagger i\bar\sigma^\mu \partial_\mu \chi$. To prove that it is real, we use the fact that when we apply a hermitian conjugation to the product of two spinors, we switch their order without introducing a minus sign. We therefore obtain

$$(\chi^\dagger i\bar\sigma^\mu \partial_\mu \chi)^\dagger = -(\partial_\mu \chi^\dagger)i\bar\sigma^\mu \chi$$

where we have used the fact that the σ^μ are hermitian and the minus sign comes from the complex conjugation of i. Performing an integration by parts, we recover the original expression which shows that it is indeed real.

6. The auxiliary field allows the supersymmetric algebra to close off-shell for the spinor χ. In this way, the supersymmetric algebra closes for all the fields without the use of any equation of motion.

7. Because the hypercharges of all the particles add up to zero. In such a case, the supertrace vanishes, and this would imply the existence of some scalar superpartners lighter than the known fermions, which is ruled out experimentally.

8. Yes, SUSY is broken spontaneously because the minimum of the potential is not zero.

9. Since ζ is of dimension $-1/2$ and D is of dimension 2, the transformation must necessarily contain the photino λ which is the only field with a half integer dimension. For the dimensions to come out right, there must be a derivative acting on the photino. So we try as a first guess $\delta D \simeq \zeta \partial_\mu \lambda$. But the Lorentz indices don't match. We must introduce a set of Pauli matrices to act on the photino. On a left-chiral spinor, it is the $\bar\sigma^\mu$ that we must apply. Since $\bar\sigma^\mu \lambda$ transforms like a right-chiral spinor, we must combine with it the hermitian conjugate of ζ. Or final answer is therefore

$$\delta D = C\zeta^\dagger \bar\sigma^\mu \partial_\mu \lambda + \text{h.c.}$$

where we needed to add the complex conjugate to make the transformation real since D is a real scalar field.

10. If invariance under SUSY is not imposed, there are ten free parameters: three masses and seven coupling constants. When supersymmetry is imposed, there are only two parameters: one mass and one coupling constant.

11. The constraint is $\bar{D}_{\dot{a}} \Phi = 0$.

12. A Dirac mass term is of the form $-m(\chi_p \cdot \chi_{\bar{p}} + \bar\chi_p \cdot \bar\chi_{\bar{p}})$, whereas a Majorana mass term is $-m(\chi_p \cdot \chi_p + \bar\chi_p \cdot \bar\chi_p)$. This makes sense because a Majorana fermion is its own antiparticle, and therfore, $\chi_{\bar{p}} = \chi_p$.

13. This would be inconsistent with SUSY transformations. If we start with a superfield depending only on x^μ and θ and apply a supersymmetry transformation, the resulting superfield will now depend on $\bar\theta$ as well.

14. First, we use the fact that the matrix $i\sigma^2$ is used to raise indices so that $\bar\chi^{\dot{1}} = \bar\chi_{\dot{2}}$ and $\bar\chi^{\dot{2}} = -\bar\chi_{\dot{1}}$. Then we use the fact that barred and unbarred spinor components are simply related by hermitian conjugation, so we finally have $\bar\chi^{\dot{1}} = (\chi_2)^\dagger$ and $\bar\chi^{\dot{2}} = -(\chi_1)^\dagger$.

15. *Gaugino* is the generic term for the spinor superpartners of gauge fields. Examples of gauginos are the photino, the gluinos, the bino, and the wino.

16. The D term is the coefficient of $\theta \cdot \theta \, \bar{\theta} \cdot \bar{\theta}/4$. Some other references use the convention that the factor of one-fourth is not included in the definition of the D term.

17. Helicity is the projection of the spin along the direction of motion; in other words, it is the eigenvalue of the operator $\vec{S} \cdot \hat{P}$. Chirality is the eigenvalue of γ_5.

18. The lagrangian is

$$\partial_\mu \phi^\dagger \partial^\mu \phi + i \bar{\chi} \bar{\sigma}^\mu \partial_\mu \chi + F^\dagger F$$

19. The LSP is the lightest supersymmetric particle.

20. If R-parity is a valid symmetry, the LSP is stable; i.e., it cannot decay into any other particle.

21. No, because none of the superfields of the MSSM is invariant under the gauge group $SU(3)_c \times SU(2)_L \times U(1)_Y$.

22. The field must have a transformation that contains both ζ (which has a dimension equal to $-1/2$) and one of the other fields, which all have a lower dimension. Schematically, $\delta H \simeq \zeta L$, where we used H to denote the field of highest dimension and L to represent one of the other component fields, which has a lower dimension than H. To make the dimensions match, we need absolutely to include a derivative ∂_μ (as well as some Pauli matrices to take care of the Lorentz indices). Therefore, the component field of highest dimension transforms with a total derivative.

23. To eliminate interactions in the superpotential that violate lepton and baryon number conservation.

24. At the unification scale, we necessarily have $\sin^2 \theta_w = 3/8$, so $\theta_w = \sin^{-1}(\sqrt{3/8})$.

25. Because we need to extract the F term of the superpotential, and extracting the F term increases the dimension by 1 (because the Grassmann variables θ_i have dimension $-1/2$ and extracting the F term results in dropping a factor $\theta \cdot \theta/2$).

26. Because the tree-level relation implies that the mass of the h^0 is smaller than the Z mass, which is ruled out experimentally.

27. Because they have an electric charge or a color charge (except for the sneutrino), and we don't want the color and charge invariances to be broken spontaneously because that would conflict with observations. On the other hand, a

vev for the sneutrino would break R-parity spontaneously, which would cause problems with lepton and baryon number-violating effects.

28. It is $\frac{m}{2}\,\overline{\Psi}_M\,\Psi_M$. Note the factor of one-half!

29. The tree-level relation between the two masses is exactly the same as in the standard model, namely, $m_w = m_z \cos\theta_w$.

30. Masses for the gauginos (the spin 1/2 superpartners of the gauge bosons), mass terms for all the scalar fields (which include the Higgs as well as the sleptons and the squarks), and holomorphic superrenormalizable interactions between the scalar fields.

31. Off-shell, a left-chiral superfield contains two complex scalar fields and a left-chiral Weyl fermion which has two complex components. So there are four bosonic and four fermionic degrees of freedom. On-shell, the auxiliary complex scalar field is eliminated and we are left with only one complex scalar field. On the other hand, the spinor must obey the Weyl equation which reduces in half the number of fermionic degrees of freedom. The end result is that there are two bosonic and two fermionic degrees of freedom on-shell.

32. The general superpotential is

$$\frac{1}{2}m_{ij}\Phi_i\Phi_j + \frac{y_{ijk}}{6}\Phi_i\Phi_j\Phi_k + c_i\Phi_i$$

although the last term is generally not included.

33.

$$\mathcal{L} = \mathcal{W}(\Phi_i)\Big|_F + \text{h.c.} + \sum_i \Phi_i^\dagger e^{2q_i V}\Big|_D + \frac{1}{4}\mathcal{F}_a\mathcal{F}^a\Big|_F$$

34. It transforms as a left-chiral Weyl spinor.

35.

$$\chi^a\lambda_a \qquad \bar{\chi}_{\dot{a}}\bar{\lambda}^{\dot{a}} \qquad \bar{\chi}_{\dot{a}}(\bar{\sigma}^\mu)^{\dot{a}b}\lambda_b$$

36. The particle content is different. Even though the gauge bosons are the same in both theories, the beta functions are different because different particles circulate in the loops.

37. The F term is $2A(x)$ and the D term is $4C(x)$.

38. The potential is given by

$$V = |\mu|^2\big[(H_u^0)^2 + (H_d^0)^2\big] + \frac{g^2 + g'^2}{8}\big[(H_u^0)^2 - (H_d^0)^2\big]^2$$

This potential is clearly minimized when both fields are equal to zero, which implies that the gauge symmetry is not broken. In addition, at the minimum the potential is equal to zero which means that SUSY is not broken either.

39. The general formula is

$$R = (-1)^{3B+L+2s}$$

For the quarks, $s = 1/2$, $B = 1/3$, and $L = 0$ so $R = +1$. For the leptons, $B = 0$, $L = 1$, and $s = 1/2$ so we again get $R = +1$. For the Higgs, $B = L = s = 0$ so they have $R = +1$ as well. The superpartners simply have a spin s differing by $1/2$ in so they all have $R = -1$. On the other hand, the gauge supermultiplets are taken to be invariant under R-parity.

40. The only gauge invariant combination out of two of these superfields is $\mathcal{H}_u \circ \mathcal{H}_d$ (plus the hermitian conjugate, as always). We must extract the F term, which increases the dimension by 1, so the lagrangian is of the form $\mu \mathcal{H}_u \circ \mathcal{H}_d$ with μ having the dimension of a mass. This is of course the μ term of the MSSM. With three fields, our only gauge invariant combination is $\mathcal{H}_d \circ \mathcal{H}_d \mathcal{E}_i$ but this is identically zero since for any column vector A we have $A^T i \tau_2 A = 0$. As for gauge invariant terms containing four superfields, we have two possibilities, $\mathcal{H}_u \circ \mathcal{H}_d \, \mathcal{H}_u \circ \mathcal{H}_d$ and $\mathcal{H}_u \circ \mathcal{H}_u \, \mathcal{H}_d \circ \mathcal{H}_d$ but the second term is identically zero. After extracting the F term, this will be of dimension 5. So our lagrangian is

$$\left(\mu \mathcal{H}_u \circ \mathcal{H}_d + \frac{1}{M} \mathcal{H}_u \circ \mathcal{H}_d \, \mathcal{H}_u \circ \mathcal{H}_d \right)\Big|_F + \text{h.c.}$$

41. A Fayet-Iliopoulos term is a term linear in a left-chiral superfield. Its contribution to a supersymmetric lagrangian is simply $m^2 \Phi|_F + \text{h.c.}$ Such a term is only allowed if the superfield is a gauge invariant. Such a term is absent in the MSSM because no superfield is invariant under the full gauge group.

42. Moving the θ_2 through the first derivative, we get

$$-\frac{\partial}{\partial \theta^1} \left(\theta_2 \frac{\partial}{\partial \theta_1} \theta_1 \right) = -\frac{\partial}{\partial \theta^1} \theta_2$$

Now we must recall that $\theta_2 = \theta^1$ so the final result is -1.

43. The chirality operator is simply γ_5 whereas the helicity operator is $\vec{S} \cdot \hat{p}$.

44. Since the supercharges change a boson into a fermion or vice versa, the supercharges anticommutes with $(-1)^{N_f}$. Therefore, we may write

$$(-1)^{N_f} Q_a Q_b^\dagger + (-1)^{N_f} Q_b^\dagger Q_a = -Q_a (-1)^{N_f} Q_b^\dagger + (-1)^{N_f} Q_b^\dagger Q_a$$

Taking the trace of this and using the invariance of a trace under cyclic permutation, we immediately get that the two terms cancel out.

45. Because the loop correction to the square of the Higgs mass are power-law divergent and an extreme fine-tuning of the bare mass must be invoked to keep the physical mass small. In contrast, the masses of the other fundamental particles receive logarithmic corrections and therefore do not require delicate cancellations.

46. With Dirac spinors, we need only consider the contraction of $\bar{\Psi}$ with Ψ. With Majorana spinors, we need to consider in addition the contractions of $\Psi_M \Psi_M$ and of $\bar{\Psi}_M \bar{\Psi}_M$.

47. Using the fact that the hermitian conjugate of a product of spinors components switches their order without introducing any minus sign, we have

$$(\chi^a \lambda_a)^\dagger = (\lambda_a)^\dagger (\chi^a)^\dagger$$

Recall that taking a hermitian conjugate changes undotted indices into dotted indices without changing their position and adds a bar, so that we get

$$\bar{\lambda}_{\dot{a}} \bar{\chi}^{\dot{a}}$$

which is by definition $\bar{\lambda} \cdot \bar{\chi}$. This is also equal to $\bar{\chi} \cdot \bar{\lambda}$ but that requires a bit more work to prove.

48. The gauginos are the spin 1/2 superpartners of the gauge bosons. Like the gauge bosons, they transform in the adjoint representation of the gauge group.

49. It is simply

$$\sum_{\text{scalars}} m^2 - 2 \sum_{\text{fermions}} m^2$$

where the sum is over the physical (on-shell) states. When SUSY is not broken, each Weyl spinor is paired with a complex scalar field with the same mass. There are thus two real scalar fields degenerate in mass with each Weyl fermion, which leads to a vanishing supertrace.

50. The superpotential must be a holomorphic function of Φ (it cannot contain the hermitian conjugate of the superfield). Since Φ is charged, there is no possible gauge invariant superpotential.

INDEX